DOORS TO HIDDEN WORLDS

Edition Angewandte – Book Series of
the University of Applied Arts Vienna
Edited by Gerald Bast, Rector

edition:'ʌngewʌndtə
Universität für angewandte Kunst Wien
University of Applied Arts Vienna

Doors to Hidden Worlds

The Power of Visualization in Science, Media, and Art

Edited by Alfred Vendl and Martina R. Fröschl

DE GRUYTER

Foreword

Gerald Bast
President (Rector) of the University
of Applied Arts Vienna

From the blood-brain barrier to genetic scissors, from plankton to viruses, from COVID-19 to the consequences of polluting our oceans with noise and microplastics, from Vienna to London, from Los Angeles to Singapore — the Science Visualization Lab at the University of Applied Arts Vienna sets new standards in content and aesthetics.

In 2000, Alfred Vendl, Emmy-Award-winning documentary filmmaker and Professor of Technical Chemistry, began setting up science visualization as a competence field at the University of Applied Arts Vienna in a working group at his Faculty of Technical Chemistry. Then, in 2016, the Science Visualization Lab was founded as its own organizational unit under his leadership, with support from Martina R. Fröschl.

The work done by the Science Visualization Lab has drawn attention and gained recognition around the world due to its scientific precision and aesthetic power. The Science Visualization Lab makes processes visible that can only be perceived by applying the methods of the visualization techniques perfected there. These visualizations have thus provided researchers from different branches of science and scholarship with a foundation and inspiration for developing further scientific and scholarly theories. Moreover, its thematic topicality means that the work done by the Science Visualization Lab will have a lasting impact, even outside of expert circles, in important scientific and sociopolitical discourses — from environmental policy to medicine and genetic engineering. Alfred Vendl and his team know how powerful images can be, which is precisely why they are aware of their scientific and creative responsibility.

With written chapters, photos, and video material that can be viewed using QR codes, this volume provides excellent insights into the world of science visualization, conveys knowledge, incites astonishment, inspires curiosity, and fascinates, thereby advancing competences when it comes to communicating the important issues of the future.

Introduction: Doors to Hidden Worlds

Alfred Vendl, Martina R. Fröschl

We are thrilled to be taking you, dear reader, on a journey to familiar yet unknown worlds. In this book, doors will be opened in order to gain alternative perspectives and therefore new ways of looking at the world that surrounds us. We have made this possible for you with a range of multidisciplinary chapters by our collaborators and guest authors. Most of these texts are personal narratives and observations accompanied by a generous number of images. The chapters have been deliberately written in different styles and shaped by the different approaches taken by the media representatives, artists, and scientists we invited to contribute. The concept of visualization has been interpreted very broadly in this volume because we do not want to limit the meaning of science visualization the way it has been in many other publications on the topic. The authors' own styles have been consciously preserved because we want to show their diversity, but we are all united by one thing—we want to expand perception and therefore humanity's stores of knowledge for anybody who is interested. Expanding perception but also the mind does not take place here by way of substances or in practices like brain machines or meditation, which were widely promoted in the 1960s. The core argument of this book is that this expansion does more for humanity's knowledge when it is facilitated by technological means, for only then is it possible to generate truly new knowledge, which is not the case when barriers are dissolved by chemical substances that only bring to the surface what was already there in the unconscious.

We have a similar goal but different means to the ones that William Blake (1757–1827) described in his texts. The nature mystic and painter wrote openly about honoring the human mind with a visionary view that was often interpreted by his contemporaries as insanity, but which for him was an expansion of understanding, an expansion of the mind. One of his most famous statements—which author Aldous Huxley, who experimented with mescaline, named one of his books after, as The Doors did their band—was, »If the doors of perception were cleansed, everything would appear to man as it is, infinite.« They were all making reference to the »purification« of the doors of knowledge with the use of mind-bending substances that Blake probably intended.

Blake, who was extremely skeptical of materialism, is unlikely to have envisaged that knowledge would one day be expanded by means of technological innovations. And yet, it is precisely technological developments that, now and in the future, will make it possible to experience the world more and more boundlessly and unrestrictedly by opening the doors of perception up wider and making the hidden, real world behind them visible. It is precisely »making visible,« the visualization of often encrypted data, that will open up new, hitherto hidden, but certainly real worlds to humankind.

Doors to Hidden Worlds is intended to provide a clear overview of the added value of visualizing realities that are not normally directly visible in the fields of science, media, and art. William Blake's proposition thus gains a new meaning: »If the doors of perception were opened even further, everything would appear to man as it is, boundless and unrestricted.«

Whereas new technologies in science are making hidden realities directly comprehensible and visible to scientists by means of visualization, the media are using this knowledge to make them understandable and visible to laypeople with the help of customized visualizations. Art takes up all this new knowledge, conveys it in fitting visualizations, and thereby expands consciousness—bringing us back to William Blake, who once said that his own everyday perceptions were always accompanied by and overlapped with visions.

In this book, renowned representatives from science, the media, and art provide an overview of excellent examples of important knowledge-generating visualizations from their respective fields with examples of images that have been expanded once more by augmented reality technology in a number of videos.

Contents

Doors to Reality

Alfred Vendl

The visible can explain the world's secrets. We need to open the doors to the invisible, open up unknown worlds, and make them understandable for everyone. The keys to this are science, art, and media. Each step we take further into the unknown gives us an idea of what the world really is—infinite.

The Most Important Source of Knowledge

Humans have been trying to push the limits of their perception since time immemorial. Their impressions and experiences leave behind feelings of incompleteness and thus the need and urge to learn what might be »hidden behind« them. Alongside our other senses of perception, sight delivers around eighty percent of the information about what we experience, which we then process in our brains. The dominance of the visual is therefore the basis of knowledge acquisition.[1] The need to search out the unknown outside of the reality we experience is an important motivation for a range of activities that humans perform to achieve this goal. The diversity of attempts and paths taken to reach this unexplored reality transcends all cultural, ideological, methodological, and scientific frontiers: »If the doors of perception were cleansed, everything would appear to man as it is, infinite.« This statement by mystic William Blake (1757–1827) has been repeatedly interpreted as a recommendation to take mind-altering substances in order to transcend the limits of perception in our existence. We are only able to recognize a small part of reality with the limited possibilities provided by knowledge. These substances do not open the doors to a new real world but manipulate an individual experience in a distorted world, which is made up of different memories of experiences the user has had.

New opportunities to access knowledge from unknown realities are developing in parallel with new inventions and technologies in the field of visualization, allowing us to experience unknown possibilities visually. This applies to the simple magnifying glass just as it does to the new James Webb space telescope. It is about areas of previously unknown realities that we are gradually pushing into with the help of technological innovations, making them visually accessible and therefore hopefully influencing our consciousness in the long term.

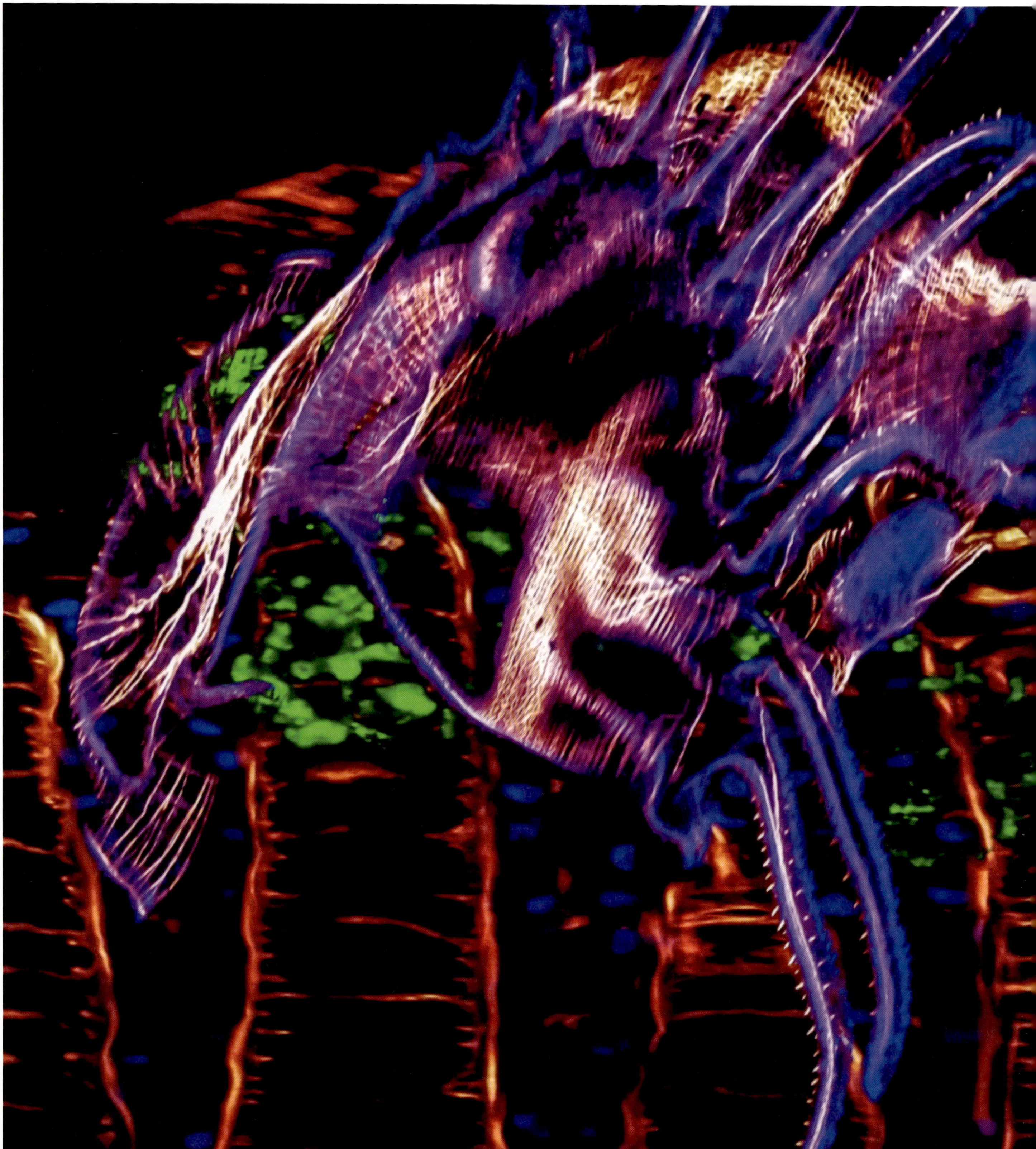

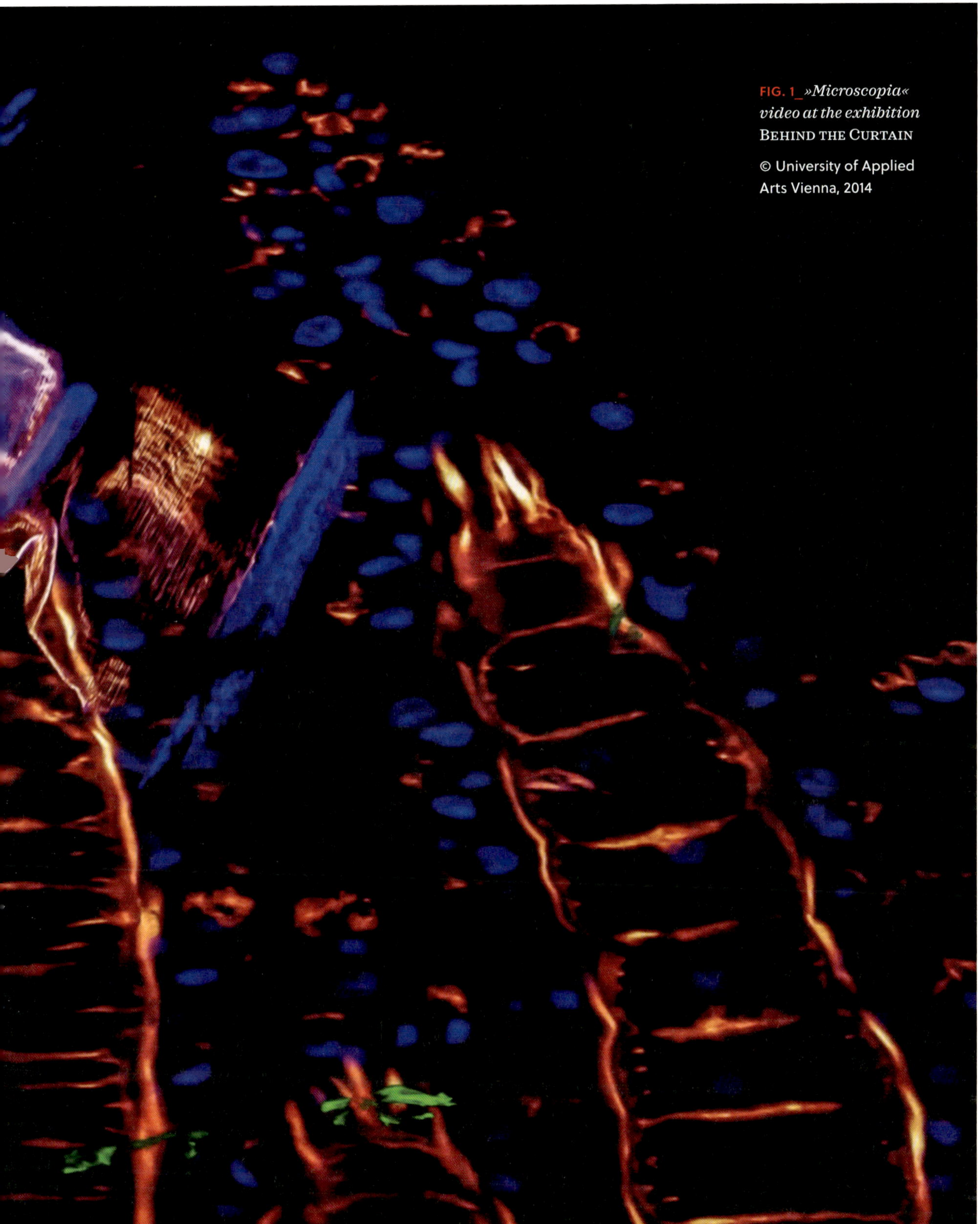

FIG. 1_ *»Microscopia« video at the exhibition* BEHIND THE CURTAIN

VIDEO_ *Excerpt from »Microscopia« video.*

Music by Oliver Vendl

The beautiful preceding images show a seemingly insignificant world, beyond what is visible, that cannot be experienced by the naked eye but that has been made accessible using the latest technological methods. It provides an almost psychedelic — drug-free — experience_FIG. 1.

Brought up and Shaped Visually

Even as a child, I was gripped by a love and passion for film. My father, a businessman, spent all of his free time making short 8mm films about the events that took place in our family. He also discussed his experiences intensively with the like-minded members of his film club, to which he took me even as a very small child. My upbringing was therefore very visual. Quite early on, as a small child, I was able to marvel at myself on the screen and, in the short films about our family outings, I was constantly discovering new things on the screen that I had not noticed during the outing itself. This expanded my visual consciousness of what I had already experienced early on.

I gained a lasting impression of the visualization technology available back then when I was ten years old during a summer vacation in Kals in Tyrol, where a number of Austrian *Heimatfilme* (literally: homeland films) were being filmed — following the example of American Westerns but with protagonists like peasants, foresters, and poachers instead of farmers, sheriffs, and cowboys. The camera was usually fixed; it only moved tentatively in pan shots filmed from the tripod or in slow tracking shots. It was like filming theater, that is, without the events getting any real support from the dramaturgically motivated use of moving cameras.

By the age of twelve, I had collected enough technological equipment to shoot my own first short film with three classmates — with all of the technological shortcomings of the time, the camera still fixed; it was a small, filmed theater. At middle school, I volunteered to work on a film production in possession of a movie cameraman. From him I learned the craft of camerawork from the ground up.

Visualizing Science

When I began studying technical chemistry at the Technical University of Vienna, my spatial perception, which had now been trained by my cinematic work, would turn out to be a great asset. The natural sciences in particular have flourished with the help of visualization techniques — whether following chemical reaction sequences or viewing the atomic structures or microstructures of materials. Visualization gained a special significance for me later on when I was working as a scientist in the field of materials research.

A three-dimensional, visual representation of the arrangement of atoms in the crystal structure of a substance can often open the first door to understanding material properties, as the example of graphite shows when compared with diamond. Both substances are made from carbon. While graphite is arranged in lattices of atoms that slide over each other smoothly, the atoms in a diamond are stacked together tightly in the three spatial directions. The ability of the carbon lattices in graphite to slide over each other easily is what makes it suitable as pencil lead, as it can be worn away layer by layer. The tight, three-dimensional interpene-

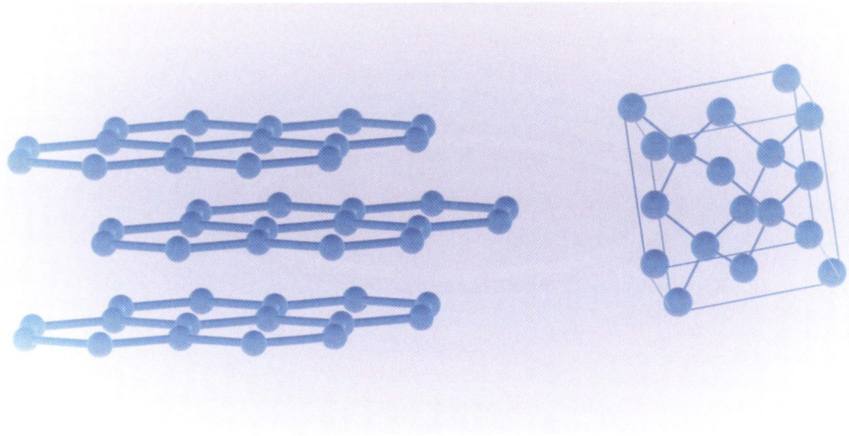

FIG. 2_*Crystal structures of graphite and diamond*

tration of the carbon atoms in a diamond, on the other hand, is what makes it so extremely hard_FIG. 2.

Opening the doors to the microworld has not only provided new insights into the natural and technical sciences — the cultural disciplines have also been fundamentally enriched by these door-openings. For example, it only became possible to gain in-depth insights into the complex painting techniques of a great master like Raphael by taking a microscopic look at micro cross sections of the layers of paint in his painting *Madonna del Prato*_FIG. 3, while the different microstructures in the important bronze sculpture *Youth of Magdalensberg* allowed a metals expert to draw precise conclusions about the casting technique used but also about the work done to the sculpture afterward — depending on whether there was a primary cast structure or a recrystallized secondary structure_FIG. 4.

The Electron Microscope

The microscope is generally the preferred technical means of accessing the microcosm and opening ever-new doors to ever-tinier realities. For materials scientists researching the state of material surfaces and how they change, the scanning electron microscope (SEM)[4] is the device of choice. It allows details to be imaged on a scale of micrometers. In the late 1960s, a SEM was installed at a research institute near Vienna. It was operated by a friendly technician. As a student at the Technical University, I managed to get my first glimpse into the microworld of materials — the ultimate tensile strength of which I was researching at the time — by paying an »entrance fee« of a carton of Austria C cigarettes, which I brought along for the technician each time. It was only by imaging fracture surfaces at a micro level that I was able to successfully draw conclusions about a material's ultimate tensile strength and therefore to bring about the necessary changes in the manufacturing of that material. During a later research stay at Imperial College in London, where I was studying the ultimate tensile strength of brittle materials, I had a SEM at my disposal that allowed me to study the opportunities that this technology provided at length. My love for the visual study of the microworld kept hold of me in my cinematic work as well, and I continued trying to open the doors to the microworld

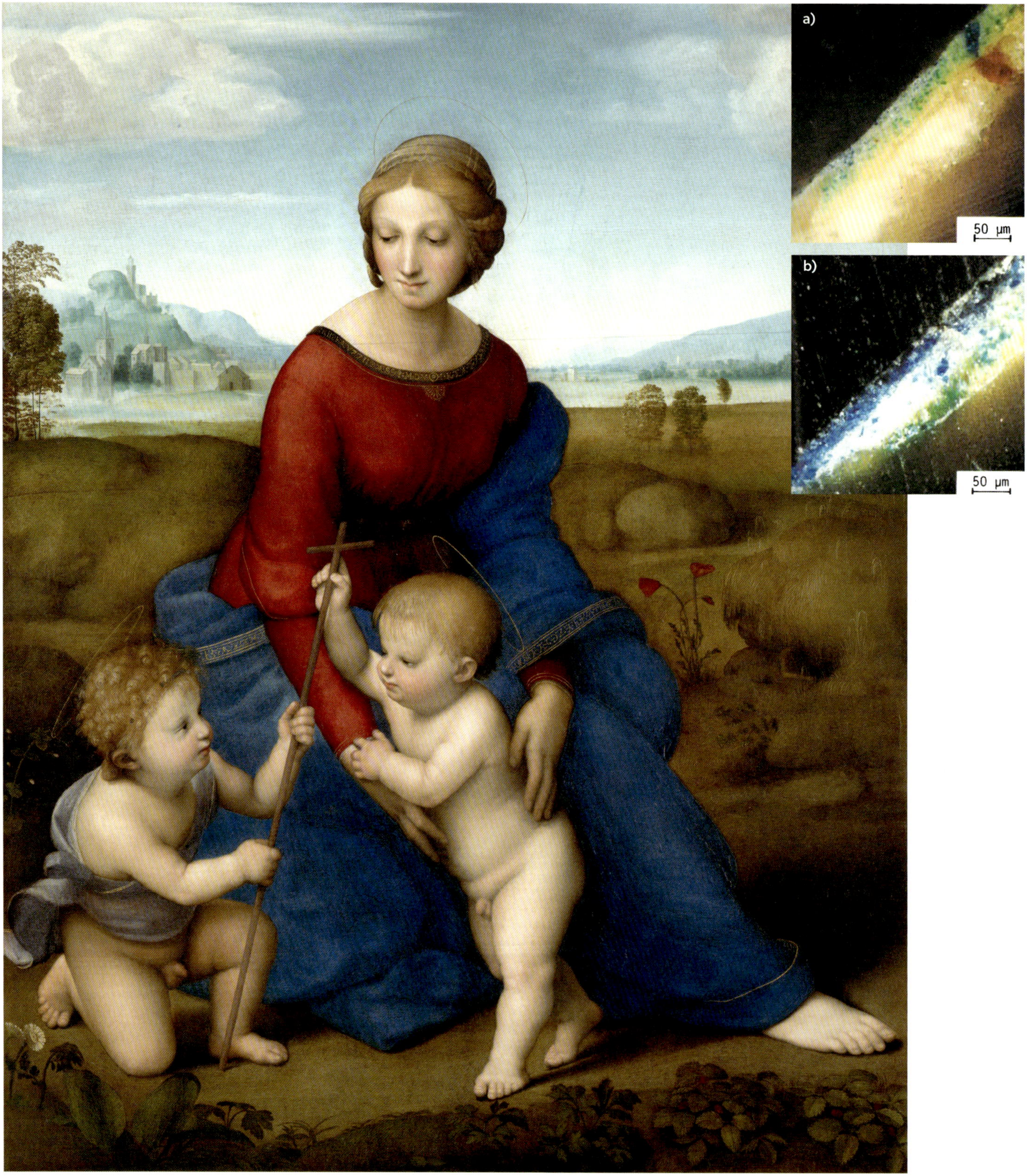
a)
50 µm
b)
50 µm

FIG. 3_Madonna in the Meadow, *painting by Raffaello Santi*

Cross sections of paint layers, sample a) taken from the blue sky, which shows ultramarine toward the painting surface; sample b) taken from the blue robe of the Madonna, which shows ultramarine in the outer paint layer, the most expensive pigment of the time.[2]

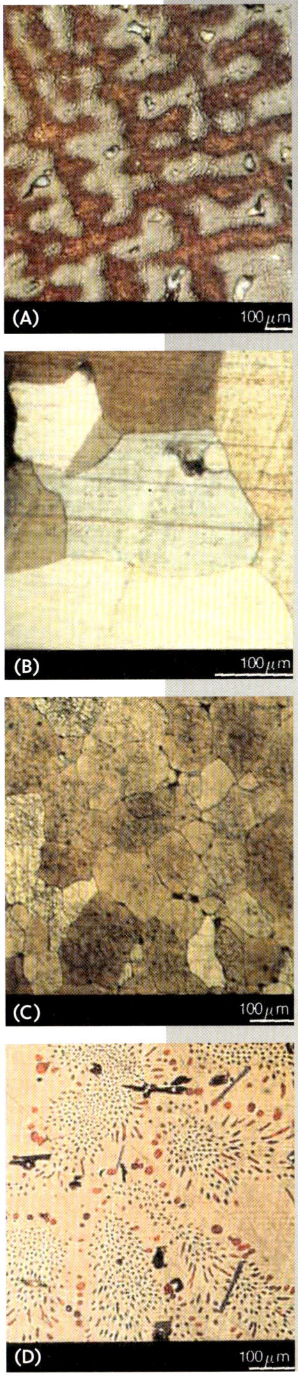

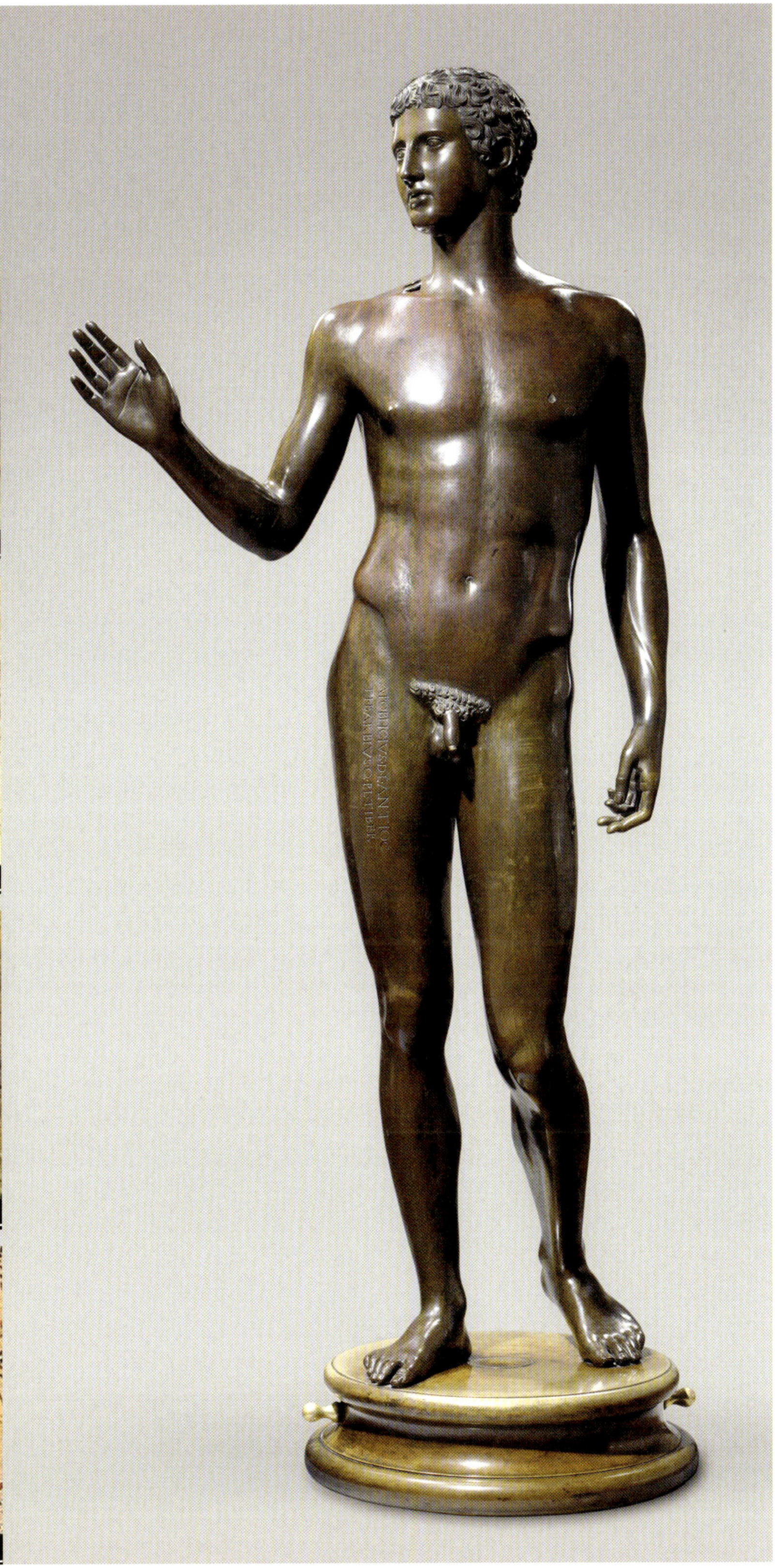

FIG. 4_Youth of Magdalensberg, *bronze sculpture*

Microstructures: different structures depending on casting conditions.
(A) Dendritic structure
(B) Homogenous alpha (Cu, Sn, Ni) solution
(C) Polyhedral structure
(D) Effect of oxygen content: formation of red Cu_2O[3]

for viewers of my television films. I kept working in film while conducting research in the field of ultimate tensile strength at the Technical University of Vienna. A special, super slow-motion camera, which mainly consisted of an extremely quickly rotating optical prism, allowed users to achieve image sequences of 10,000 images per second, which was just enough to record brittle aluminum oxide breaking in a single image. But the SEM remained the central device for all of my future research activities. Because samples had to be examined in a high vacuum, it was very difficult to study materials that do not remain stable in such a vacuum, such as biological materials containing liquids. These materials had to be made vacuum-stable and electroconductive, which could only be achieved by vaporizing them with gold or carbon, which killed off any life.

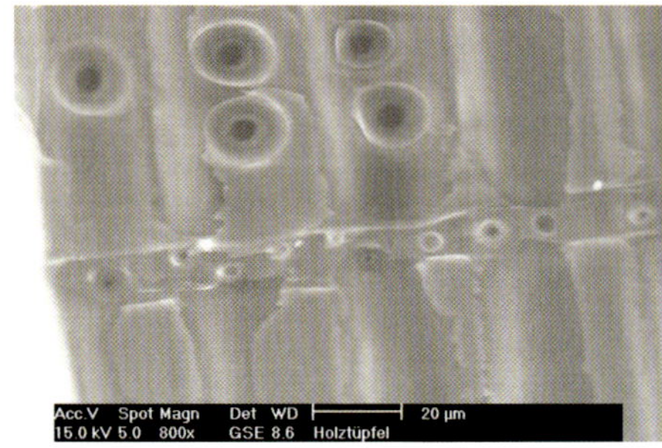

FIG. 5_*Bordered pits of Norway spruce* (PICEA ABIES), *ESEM picture*

© Oliver Vendl, 2002[7]

While the SEM studies were limited to the high vacuum, the development of the ESEM (environmental scanning electron microscope)[5] brought about a breakthrough. With this device, it finally became possible to make visible the microworlds of living organic substances under almost normal conditions on a scale of micrometers, even nanometers. I was able to obtain an affordable ESEM for our institute by coming to an agreement with the manufacturer that would allow interested scientists from other fields to test the device at our institute for their purposes. This allowed me to meet top researchers from the field of biology. For the first time, it became possible to examine sensitive plants under near-normal pressure conditions_FIG. 5.[6]

For my later cinematic work in nature documentaries, however, my association with biologist and mite expert Manfred Walzl from the University of Vienna was trailblazing. We used the ESEM to take the first electromicroscopic pictures of living mites by applying an additional elaborate method. The mites had to be incapacitated before the recording, which we were able to achieve with the help of a simple refrigerator. Then, we scanned the cooled organisms in the ESEM and compiled only the clean images, those without any scanlines at the end of the scan, into a film. The acceleration that this produced ultimately compensated for the movements that had been reduced to slow motion and reproduced the mite's normal speed of movement_FIG. 6.

FIG. 6_*Mite ESEM picture*

© University of Applied Arts Vienna, 2001

VIDEO_*ESEM video,*
LIMITS OF PERCEPTION, *film*

© ORF, 2001

In materials science, there are innumerable examples that emphasize the significance of visualization, even in fields outside the mainstream of technical research. For example, in order to study the formation of the green patina on copper roofs dynamically from the outset, I had to embark upon a journey into the microstructure of the copper surface on a scale of micrometers. A marked spot on the copper surface was photographed by the ESEM for more than two hundred days and artificially weathered at night using corrosive gas in the usual composition and at the usual concentration found outside—ultimately producing more than two hundred individual images. They were then compiled into a film of about twenty seconds_FIG. 7.

Another quantum leap for door openings to the nanoworld on a scale of millionths of a millimeter was still to come: the development of the atomic force microscope. In 1986, the Nobel Prize for Physics was awarded to Gerd Binning (among others)[8] for the development of the scanning tunneling microscope (STM). This device can be used to visualize electroconductive materials on a scale of nanometers.[9] But Binnig also developed the atomic force microscope, which uses a needle to scan

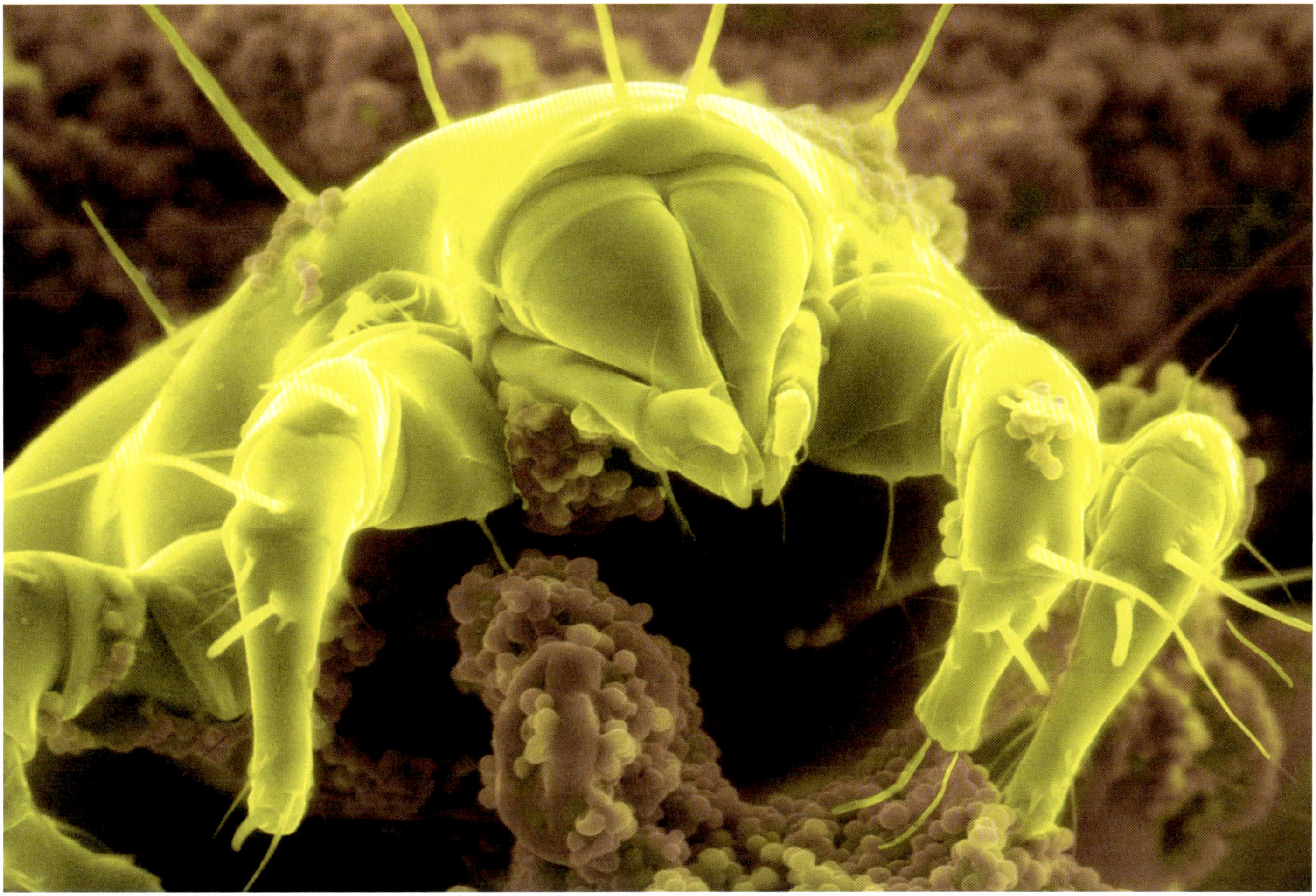

FIG. 7_*Otto Wagner Church am Steinhof*

Image courtesy of Rudolf Klingohr, tvandmore.net TV- und InternetproduktionsgmbH, Vienna, 2018

VIDEOS_*Formation of copper patina, video produced by combining ESEM pictures with 3D animation technique,* LIMITS OF PERCEPTION, *film*

ESEM by Rudolf Erlach

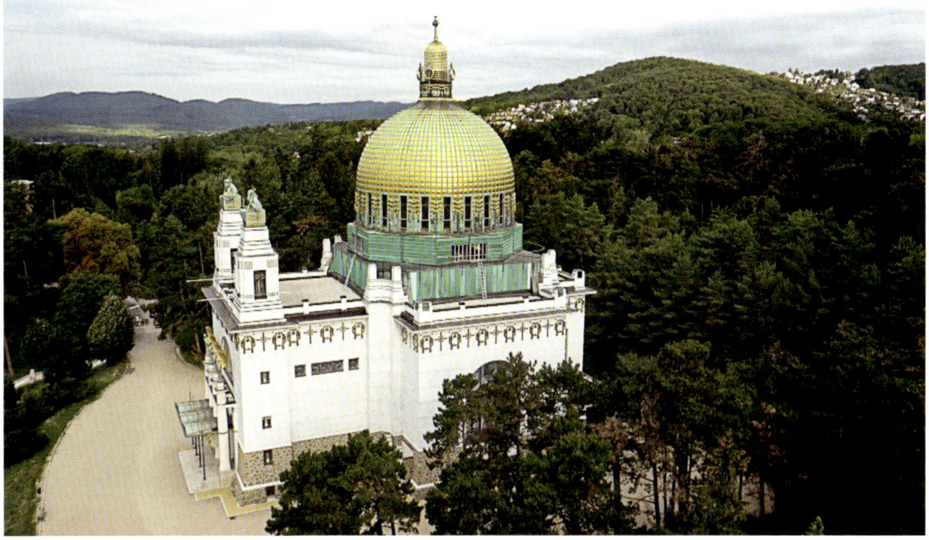

surfaces and to image nonconductive surfaces on a scale of nanometers. In corrosion research, this device is ideal for identifying the first signs of corrosion and its further spread on a surface. A colleague of Gerd Binning constructed a special atomic force microscope for our institute that, unlike later models, still permitted interactive manipulation. We combined this device with an artificial weathering apparatus developed in collaboration with Chalmers University in Göteborg, which emitted corrosive gases similar to the normal corrosive gas pollution found in our environment at a level of ppb (parts per billion). Using this method, we were able to image different corrosion processes on copper crystal surfaces growing in various directions. Using the atomic force microscope, we were also able to image red blood cells_FIG. 8.

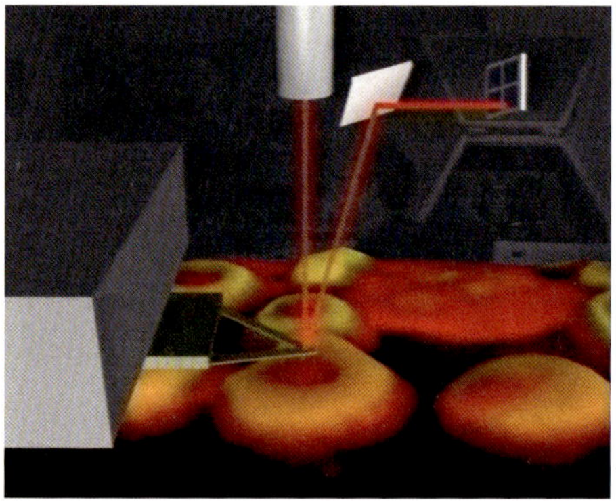

FIG. 8_*Atomic force microscope, 3D representation from a still frame, AFM video of red blood cells*

VIDEO_*Red blood cells, 3D representation from an atomic force microscope video*

Foray into Experimental Film

Alongside my academic research work, I dedicated every free minute I had to producing films, which were listed in the category of »experimental film—genre film« at film festivals. The films in this category were symbolically arranged storylines presented in image sequences that had been constructed in a specific visual way. This category was shaped by the interplay of black-and-white and color film, the alternating use of telephoto views cutting to wide-angle shots, and different editing rhythms, which gave the films additional visual dimensions. With short films produced like this, I was able to win a number of first prizes at the Cannes short film festival in the 1960s, which even received special mention in local newspapers like *Nice-Matin*.[10] During this time, I was especially impressed by a short German film with the title *V for Vietnam*, which had been metrically arranged, with the letters in each of the alphabet sequences that were shown understood as units of time. In the first alphabet sequence, the letters appeared briefly and changed quickly; it was only once the film reached the letter »V« that it lingered for a while, showing footage of the war in Vietnam. The more frequently the alphabet sequence appeared, the longer the naked letters spent on screen and the longer the war footage at the letter »V,« until the footage ultimately disappeared completely, and the letter »V« took its place. The filmmaker thus showed how the most horrific war atrocities could become something that we can get used to and ultimately lose any presence in the global consciousness. This perfectly visualized message still impresses me today. Back then, my personal examination of the Vietnam War was shaped by a very visually interpreted symbolism—which can be explained by my age—that was more apparent and simplistic in my film *Tattoo*.

In the late 1960s, I was heavily influenced by the visual language of the film *A Man and a Woman* by Claude Lelouche. This was a high-quality *film d'auteur*, meaning that one person was responsible for all of the major production tasks, including cinematography and camerawork. This led to the emergence of extremely personal films that were completely in tune with my own approach. I tried to realize the partially symbolic, partially abstract visual language used there in my own film *Kalomel*.[11]

Expanded Cinema

It was at this time that the Austrian art film was developing, in which the »Expanded Cinema«[12] movement played an important role. This movement was shaped by filmmakers like Peter Weibel and Valie EXPORT. It is in this context that I remember an interesting visualization experience. At the Maraisiade of Young Austrian Film in 1968, I went to the premiere of Valie EXPORT and Peter Weibel's »Tap and Touch Cinema«[13] up close. On stage, Valie EXPORT had mounted a structure from her chest with a curtain hanging off it, through which the viewer was permitted to touch the artist for a limited period of time. One of the »tap and touchers« shamelessly went over his allotted time, which did not please Peter Weibel at all. In spite of a high level of tolerance, the intruder was immediately removed. These kinds of »expansions of cinema« in the direction of social problems, in which the original medium of film no longer played a role, provided very meaningful and important impulses for thinking about the state of society back then.

Film and Painting

It was at this time that I got to know a medical student named Peter Patzak who was successfully experimenting with painting. He wanted to commission a film about himself as a painter and was prepared to pay for the film material if I was prepared to produce the film without any further costs. Everybody involved was to work for the love of the cause. As a passionate lover of the medium of film without any primary commercial goals, I was immediately taken by his suggestion. Together we tinkered with the structure of the film, and I assumed responsibility for the camerawork, editing, and postproduction. He wanted the camerawork to be chiefly characterized by the various scenes being filmed in unusual visual resolutions — which ultimately worked out well_FIG. 9.

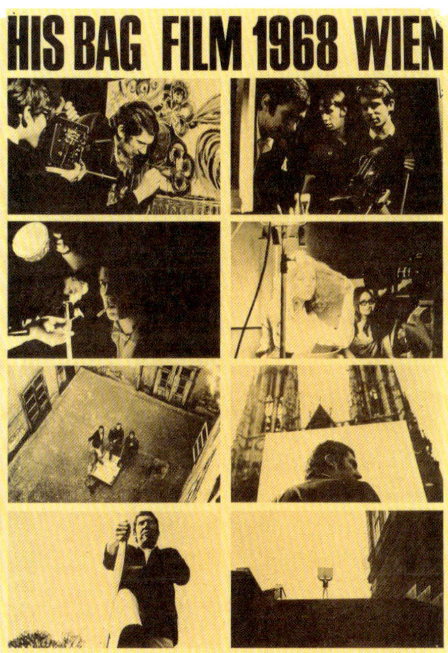

FIG. 9_*Film poster for* HIS BAG

© Peter Patzak and Alfred Vendl, Vienna, 1968

The film was shot using extremely mobile handheld cameras alternating with conventional camerawork, depending on the emotional content of the sequence. What was helpful here was the use of certain optical lenses, like a super wide-angle lens that was still unknown at that time in Vienna, the »fish-eye« lens, which a friend from a group of artists lent to us and for which I constructed the necessary optical camera frame from cardboard.

Filming took over a year, which we spent almost entirely together. The many, not always frictionless cinematic discussions may have cost me a year of study in chemistry, but they were an important chapter in the development of my future films. Ultimately, I was able to fully assert my visual concept, and we created a short, thirteen-minute-long film that was purely visual, without any commentary or use of language. The finished film was titled *His Bag* at the suggestion of Peter's girlfriend and later wife, an American. Although the film won a number of festival prizes, it also served as the basis for Peter Patzak's later cinematic work as an internationally recognized film director. However, back then I never would have thought that anybody would still remember this thirteen-minute-long mini-film several decades later. When obituaries were being written to commemorate Peter Patzak's significance in film and television after his death in 2021, an Italian film critic actually did make mention of the short film *His Bag*[14]:

Film and painting coexist harmoniously in His Bag. Peter Patzak steps before the camera and becomes the protagonist. Everything that he sees becomes valuable, important material. […]

The Birth of an Artwork

[…] of particular mention in this context is the short film His Bag from 1967, which is characterized not just by its constant search for a new cinematic language but also by a topic that was particularly close to the director's heart […]. And so, this short, fine film immediately reveals itself to be a declaration of love to art. A declaration of love to art in all its forms. Being and becoming […]. In His Bag, no superfluous dialogues or image captions are necessary. The images speak for themselves and all together follow a precise thread […]. And His Bag shows us just that: the constant search for beauty, for perfection, the longing for creation. This [is a] short, powerful but also especially lively and colorful film […].

This film would fire the starting gun for Peter Patzak's later career in the movies. For me, the film was above all about sounding out unusual and previously unused or seldom used visualizations of scenic sequences, which clearly succeeded.

A Friend and Mentor

Back then, it was customary to show your films at special film club evenings every now and then. At one of these filmmakers' evenings, where I was presenting my latest films, I got to know Kurt J. Mrkwicka. With his company MR Filmproduktion, he would later become one of the most influential European cinema and television producers, and he was my first mentor. A fraternal friendship connects us to this day. As a former European springboard diving champion, he was a Viennese institution and above all an enthusiastic filmmaker. It was his dive training that brought him to film. Because there were no trainers left in Austria at his level, he got hold of an 8mm movie camera and had his dives filmed during training. He sent the results to America to a top trainer who then made suggestions for improvement based on the film recordings.

So, by the end of his sporting career in the early 1960s, Mrkwicka had developed a general interest in the medium of film. Among other things, he produced a multi-award-winning film about everyday life in Asia, which hardly anybody in Europe knew about back then, with the title *Der gelbe Bruder* (The Yellow Brother). His successful sporting career meant that he had great connections at the Austrian Federal Sports Organization, which led to him being commissioned to shoot a wide range of sports documentaries. One initial focus was producing ski training films for elite state-certified ski instructors. »Ski pope« Stefan Kruckenhauser headed up this project back then and was himself an excellent photographer and cameraman. He had invented *wedeln*, a skiing technique that took the world by storm. Kurt was primarily impressed by my visual language, which I had now further advanced, and asked me to make sports documentaries with him, which I immediately and enthusiastically agreed to do.

Visualization in Sports Documentaries

Back then, sports documentaries were a good platform for using visualization techniques that had not or had hardly been used before. Because it was primarily about presenting sequences of motion, the use of slow motion played an influential role, giving the film sequences an additional dimension.

At that time, we were also developing our own ski documentary style. The prerequisite for this was becoming a master at mechanically operating the focus ring and the zoom lever at the same time using one hand. This made it possible to shoot full-format, frontal facial studies during slaloms without any loss of focus while retaining the same image detail, which, in combination with the use of slow motion, gave the sequence an especially emotional, dramaturgically reinforced depth.

When I got my first job as a television cameraman, which Kurt organized for me, I was tasked with producing special, close-up shots of a world record high jump attempt, which, in combination with slow-motion studies, was supposed to convey an additional dimension of the event to the viewer. The television people were happy with the shots I delivered, and a few years later, I was commissioned with my first sixty-minute, black-and-white, prime-time documentary about fencing—to mark the occasion of the World Fencing Championships taking place in Vienna at the time. I managed to convince the editor in charge to produce the documentary using only images, the original sound, and dramaturgically fitting music, but without any explanatory commentary. He agreed, but was deeply concerned. My dare paid off—both the viewers and the media critics were taken with it.

The following year, Kurt and I were commissioned to produce a prime-time documentary on the topic of horses, the first one in color. Once again, I managed to convince the television people to produce the film using image, original sound, and dramaturgically deployed music only, without any explanatory commentary, thereby allowing the visual to dominate in order to intensify the emotional level for the viewer. The film was a huge hit and was praised by critics as one of the best productions of the year.

That was, I believe, the last time that a documentary without any explanatory commentary was broadcast during prime-time television programming. The progressive overloading of the viewer with informational content that had to be explained increasingly pushed the visual right back as the leading film component. The »cinematic« television documentary increasingly became an at best visually supplemented, written treatise.

The Underwater Marine World

One important expansion to our visual dimensions was achieved with an underwater series about the *Juwelen des siebenten Kontinents* (Jewels of the Seventh Continent) that were to be found in the different marine worlds. A German producer commissioned this series for television. Austria had already established a good reputation in the visual processing of underwater ocean documentaries due to Hans Hass[15] and his colleagues. Kurt had already worked for Hans Hass, so it was clear that we would get the commission. Filming under water posed new challenges_**FIGS. 10, 11**.

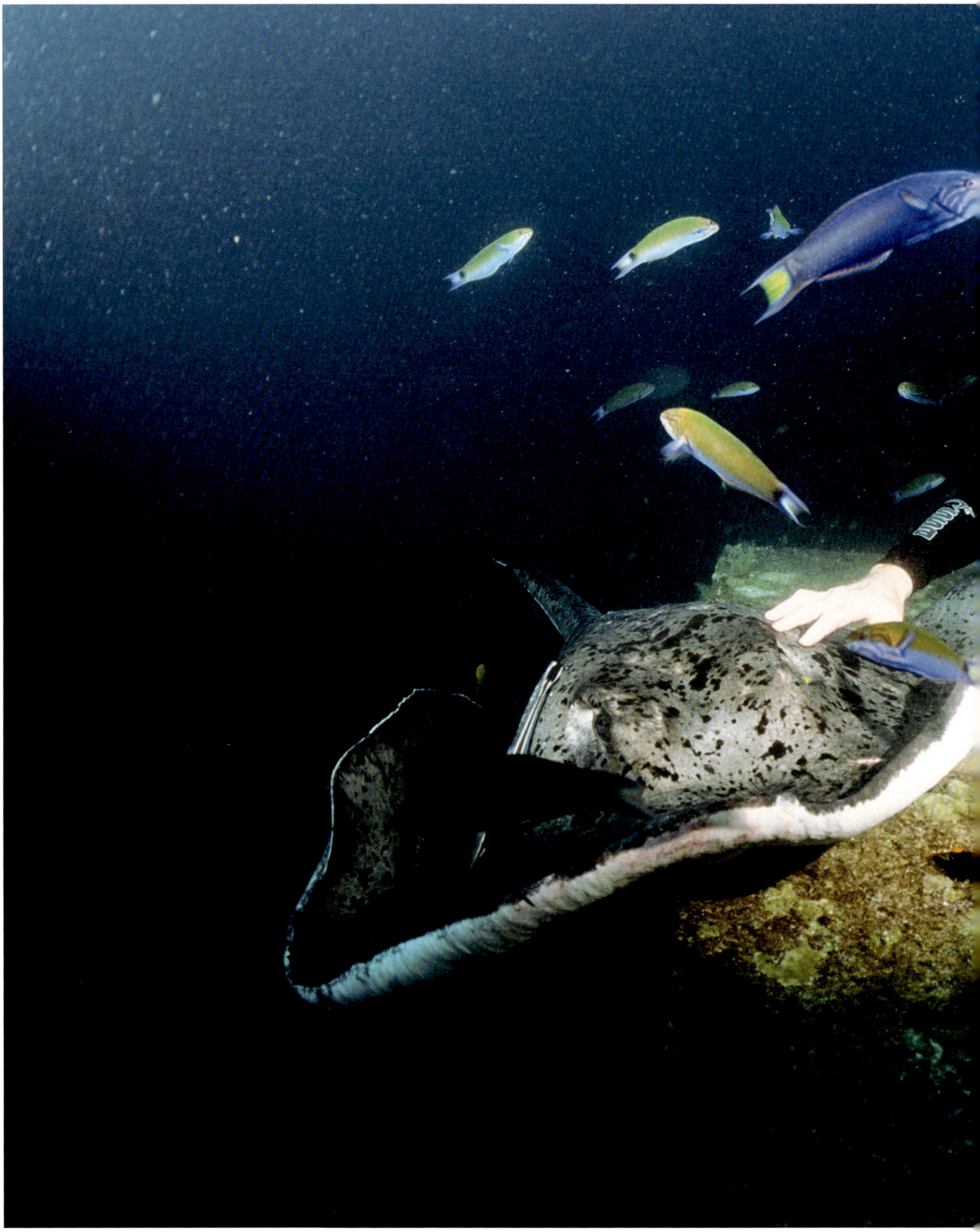

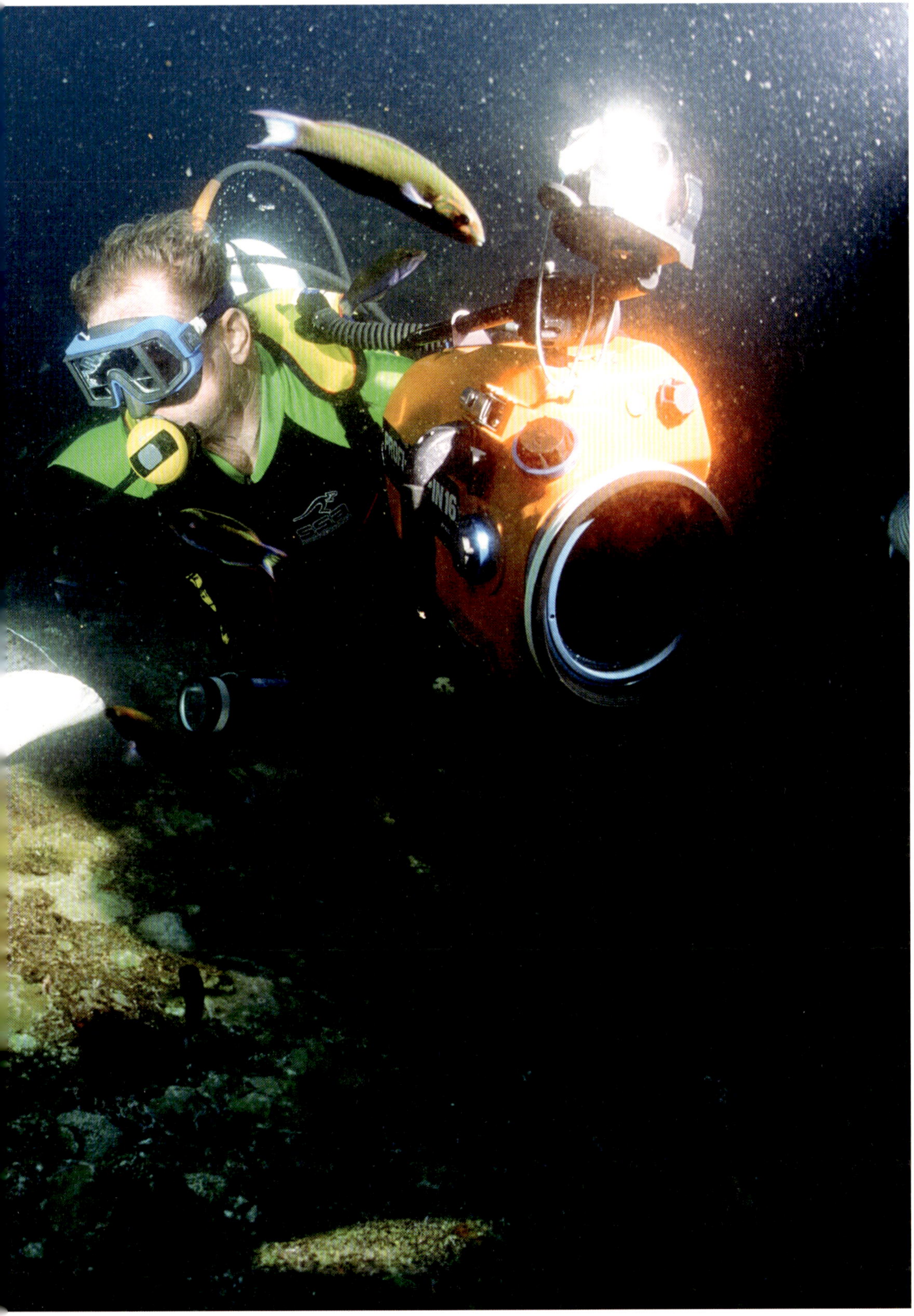

FIG. 10_*Camerawork underwater*

Photo courtesy of Kurt J. Mrkwicka, 1992

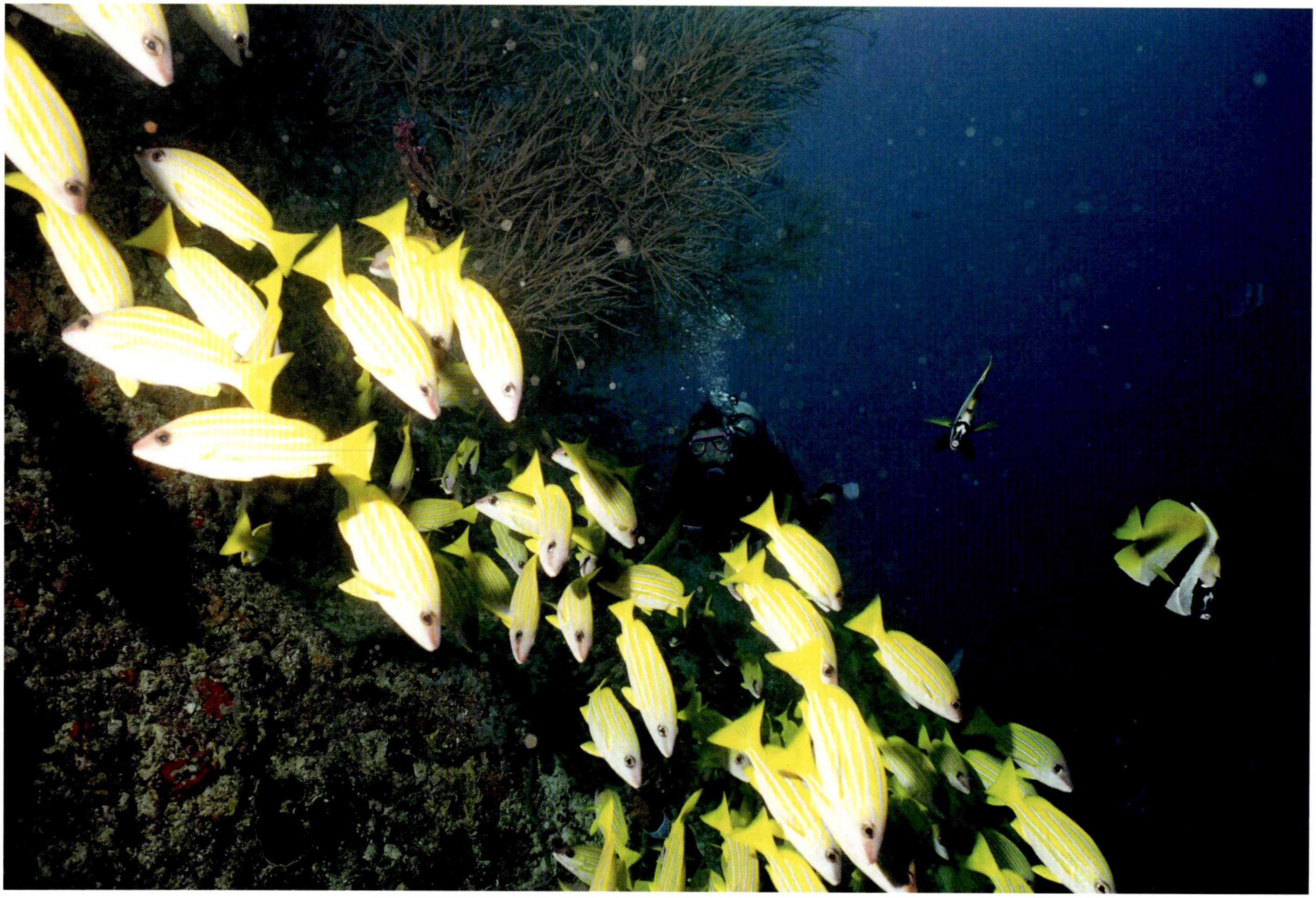

New underwater cameras were being constructed on the basis of the latest findings, making it possible to spend longer periods of time under water, as people back then had been working with thirty-meter reels that only allowed about two and a half minutes of work to be filmed. We extended that to ten minutes. However, submerging and surfacing with each new film cassette load was extremely strenuous after such a short period of time and also dangerous during dives to lower depths. Moreover, we only learned whether our cinematic attempts had been successful weeks after we had returned home.

FIG. 11_*Underwater wonderland*

Photo courtesy of Kurt J. Mrkwicka, 1992

Equipped in this way, we later accompanied a single-handed sailor, Wolfgang Hausner,[16] for another television production filmed in the Solomon Sea near New Guinea. He had previously been the first person to circumnavigate the world in a catamaran without any special technical aids_FIGS. 12, 13, 14.

We spent several weeks in parts of the ocean that had not yet been charted either above or below water. The film *Taboo* was a huge success — above all due to the way that it opened up some of the bizarre splendor of the underwater marine world. It also enriched my visual imagination enormously and would keep hold of me for decades to come.

My work for the television moved from the fields of sports, culture, and music to scientific documentaries, although visual input would remain rather limited for some time due to conventional viewer behaviors.

Universum

In the late 1980s, Austria's ORF went in a new programming direction for weekly prime-time nature documentaries with *Universum* (Universe). To begin with, it primarily broadcast relevant BBC productions. MR Filmproduktion was then commissioned to produce the first Austrian show for *Universum*. It was to be a five-part series about the Mediterranean with the title *Die Gärten des Poseidon* (The Gardens of Poseidon), filmed according to the specifications of internationally renowned ocean biologist Rupert Riedl.[17] I was appointed as artistic director and chief cameraman.

Nature documentaries are better suited to lavish visuals than any other film category — here, once more, with the film's submersion in the fascinating, alien underwater world, but also in lengthy portrayals of beautiful landscapes that speak for themselves and do not require any commentary. They give filmmakers the opportunity to present the visual lavishly and for the viewer to understand the offerings emotionally as well.

It was during this production that I met Walter Köhler, who set up the relevant editorial office at the ORF and then headed it up as well. I would form a close friendship and spend years collaborating with him. *Universum* became a massive hit that continues uninterrupted to this day. The films produced for this series soon

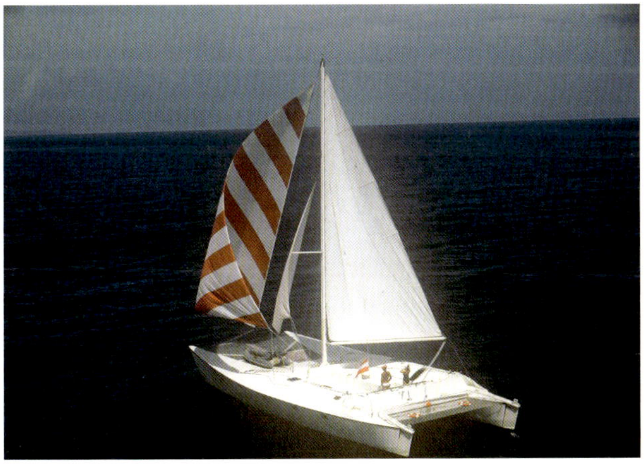

FIG. 12_*The catamaran* TABOO III

Photo courtesy of Kurt J. Mrkwicka, 1984

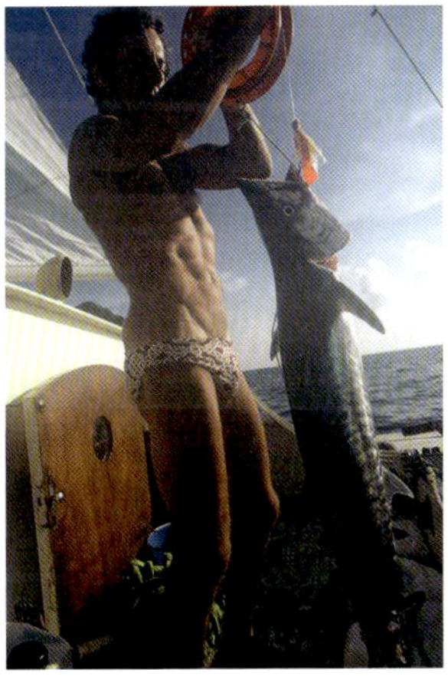

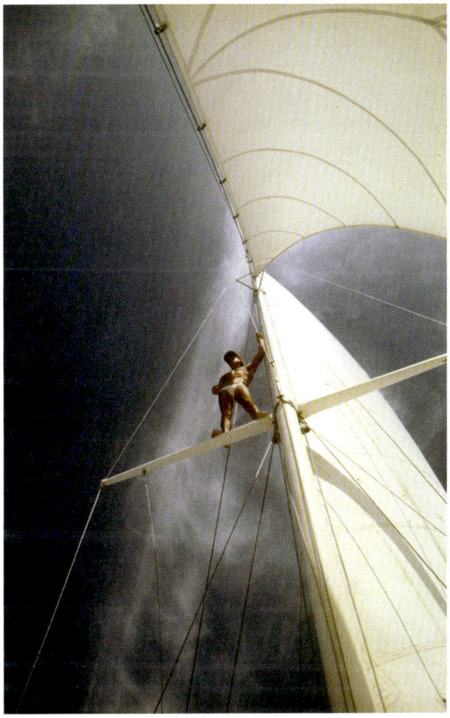

FIGS. 13, 14_*Wolfgang Hausner on board* TABOO III,

Photo courtesy of Kurt J. Mrkwicka, 1984

conquered the relevant TV stations worldwide, which ultimately led to more and more international coproductions.

While filming for *Universum*, I met Steve Nicholls, a biologist with a PhD who had worked as a series producer for the BBC for some time before going out on his own. We began producing nature documentaries together, in which we wanted to portray natural events first and foremost as a basis for generating a better understanding of physical, chemical, and biological processes. By presenting animals and plants functioning in their natural context, we were trying to help viewers to understand the given natural laws. With this goal in mind, we created documentaries like the internationally produced *Limits of Perception*,[18] *Time Limits*,[19] and *Limits of Light*.[20]

Expanding Visualization with Videos and Digital Technology

The possibilities of enriching visualization even further by purely cinematic means were soon exhausted. The replacement of film with video and the development of digital visualization technologies brought the necessary expansions. Now, digital engineers like Reinhold Fragner came on board. He had built up a relevant company and wanted to help us to expand the existing limits of visualization. Together with biologists Stephan Handschuh from the University of Veterinary Medicine Vienna, Thomas Schwaha from the University of Vienna, and physicist Rudolf Erlach, a master in sounding out the limits of electron microscopy, I founded the Science Visualization Lab, in which I also included Reinhold Fragner. Our goal was to combine microscopic techniques with 3D animation in order to throw open new doors to expanded visual perception. We made broad use of these recently developed techniques in the internationally produced *Universum* documentary series that Steve Nicholls and I created, *Nature Tech*_FIG. 15. This series won a 2008 Emmy Award from the American National Academy of Television Arts and Sciences for its innovations.

The basis of nature documentaries is, by their very nature, presenting unadulterated portrayals of processes that occur in the natural world. Digital »expansions« should only be used when the requirements for direct visual presentation are not given. This is the case, for example, when it comes to portraying long extinct animal species like dinosaurs. Here, digital technology provides wonderful opportunities to throw open the doors to ages long gone and to thus bring what has passed back to life in a visually accurate and authentic way on the basis of scientifically obtained data. This also makes it possible to visually compare the past with the present day_FIGS. 16, 17.

Another important task of digital visualization is presenting anything that is too big or too small to be viewed directly_FIGS. 18, 19. In order to visually understand the microworld, the application of 3D animation techniques to digital models based on scientific data provides opportunities to open the doors to what would otherwise not be seen.

For example, we authentically portrayed an attack on *E. coli* bacteria by T-viruses_FIG. 20 and authentically visualized a journey through a wood termite _FIG. 21.

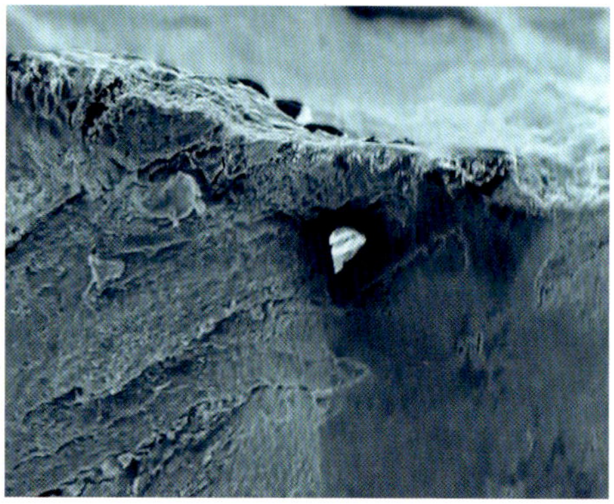

FIG. 15_*Shell SEM picture*

SEM by Rudolf Erlach
© University of Applied Arts Vienna, 2006

VIDEO_*A break in a shell, 3D animation, produced by combining SEM pictures with a 3D animation technique,* NATURE TECH, *film*

© ORF, 2006

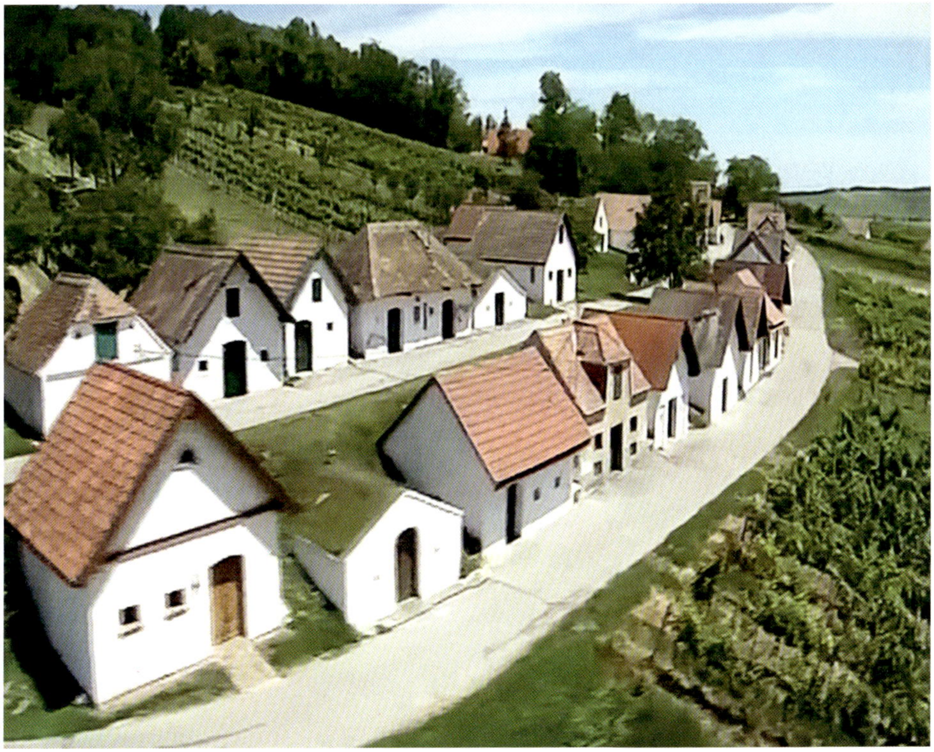

FIG. 16_ *Mammoths at Weinviertel, Lower Austria, 3D representation from a still frame,* WEINVIERTEL — WEITES LAND, *film*

VIDEO_ *Augmented reality representation of mammoths in a contemporary environment,* WEINVIERTEL — WEITES LAND, *film*

FIG. 17_ *Skeletons of sea cow archetypes, stored at Museum Eggenburg, Lower Austria, 3D representation from a still frame,* WEINVIERTEL — WEITES LAND, *film*

VIDEO_ *Prehistoric tsunami catastrophe that killed the sea cow population in the area surrounding the present-day village of Eggenburg in Lower Austria, 3D animation,* WEINVIERTEL — WEITES LAND, *film*

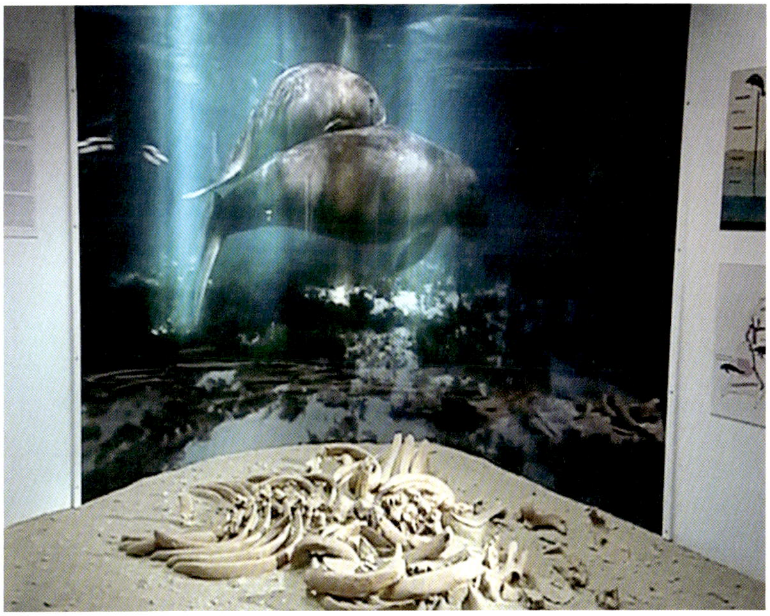

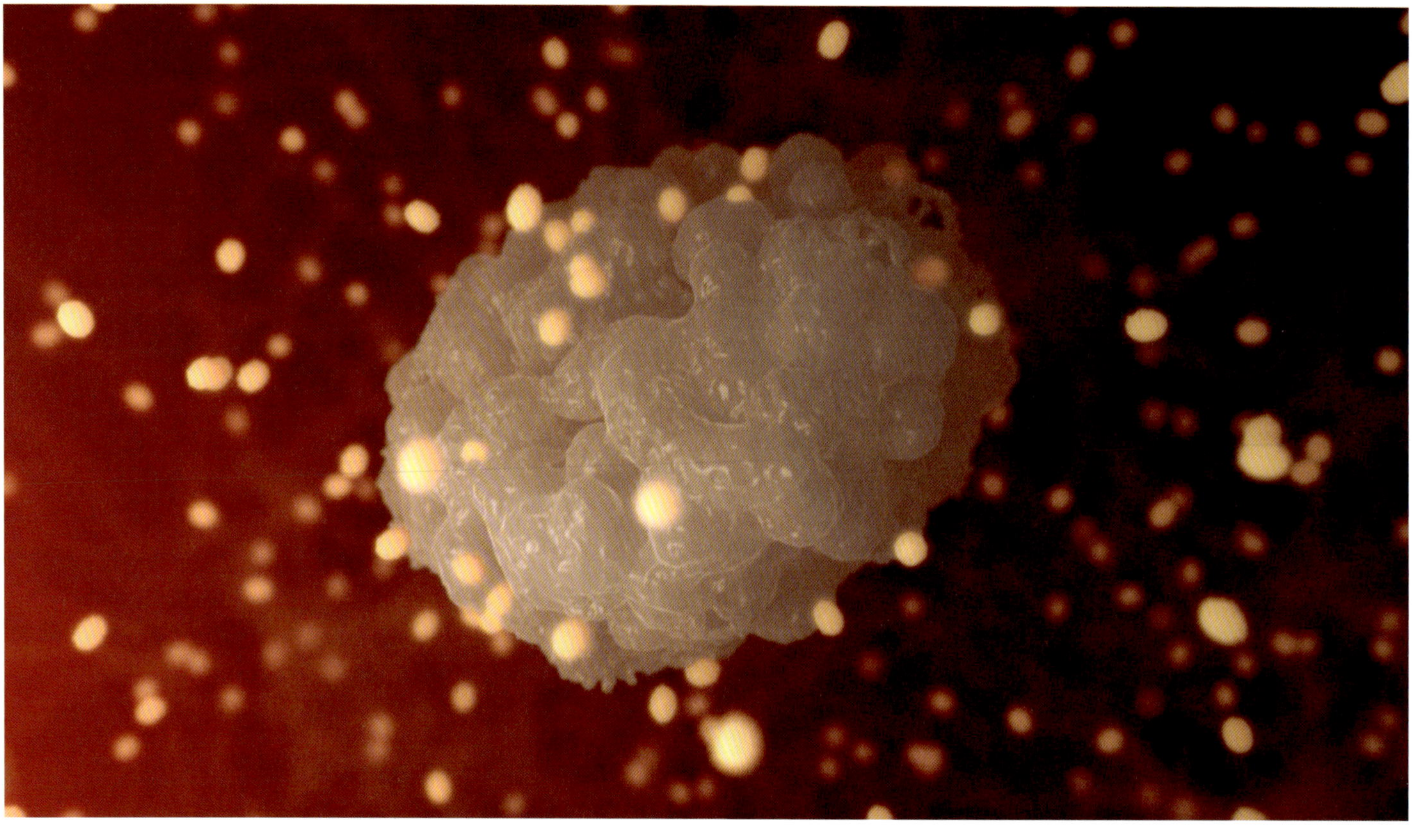

FIG. 18_*Killer cell, 3D representation from a still frame,* PLANET YOU, *film*

VIDEO_*3D animation showing the destruction of a tumor cell by killer cells,* PLANET YOU, *film*

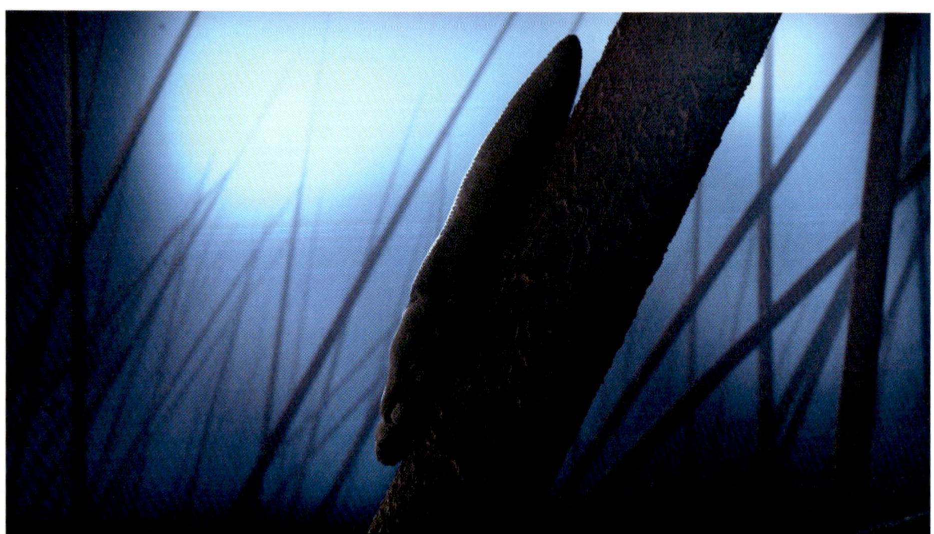

FIG. 19_*Face mite, 3D representation from a still frame,* PLANET YOU, *film*

VIDEO_*3D animation of face mites,* PLANET YOU, *film*

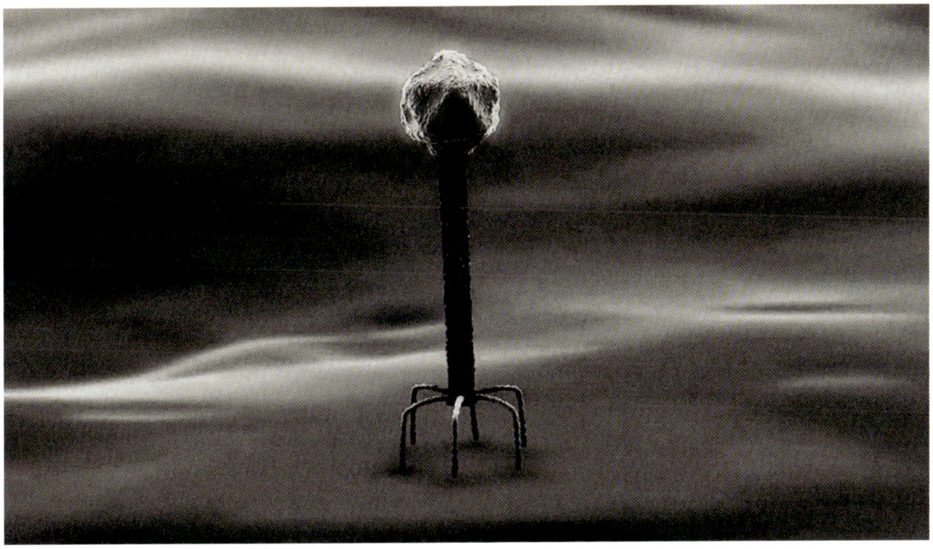

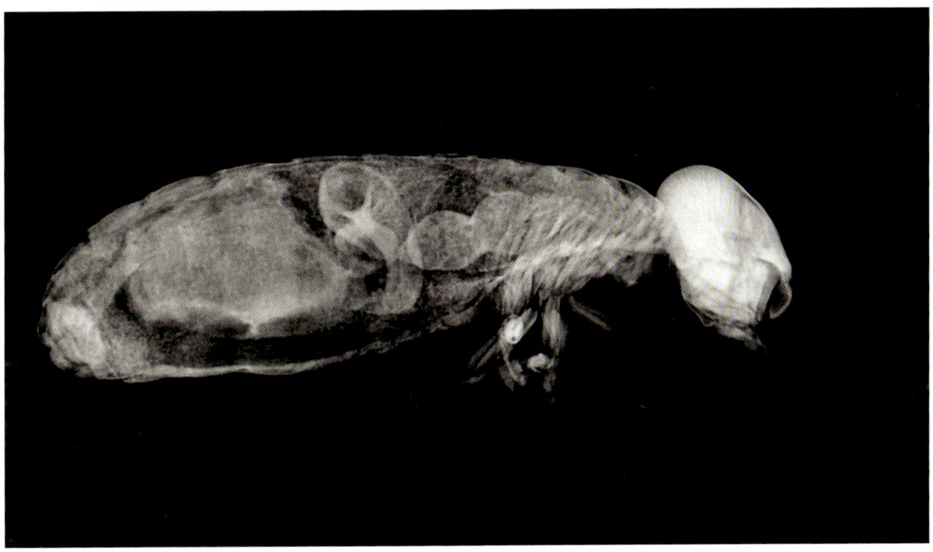

FIG. 20_*T4 phage, 3D model based on ESEM pictures*

© University of Applied Arts Vienna, 2003

VIDEO_*T4 phages attack* E. COLI *bacteria,* ATTACK, *3D animation based on electron microscopic data, video*

Computer animation by Reinhold Fragner
© University of Applied Arts Vienna, 2011

FIG. 21_*Wood termite, 3D representation based on micro-CT Data*

VIDEO_IMPOSSIBLE JOURNEY THROUGH A WOOD TERMITE, *3D animation based on micro-CT data, video*

Computer animation by Reinhold Fragner
© University of Applied Arts Vienna, 2011

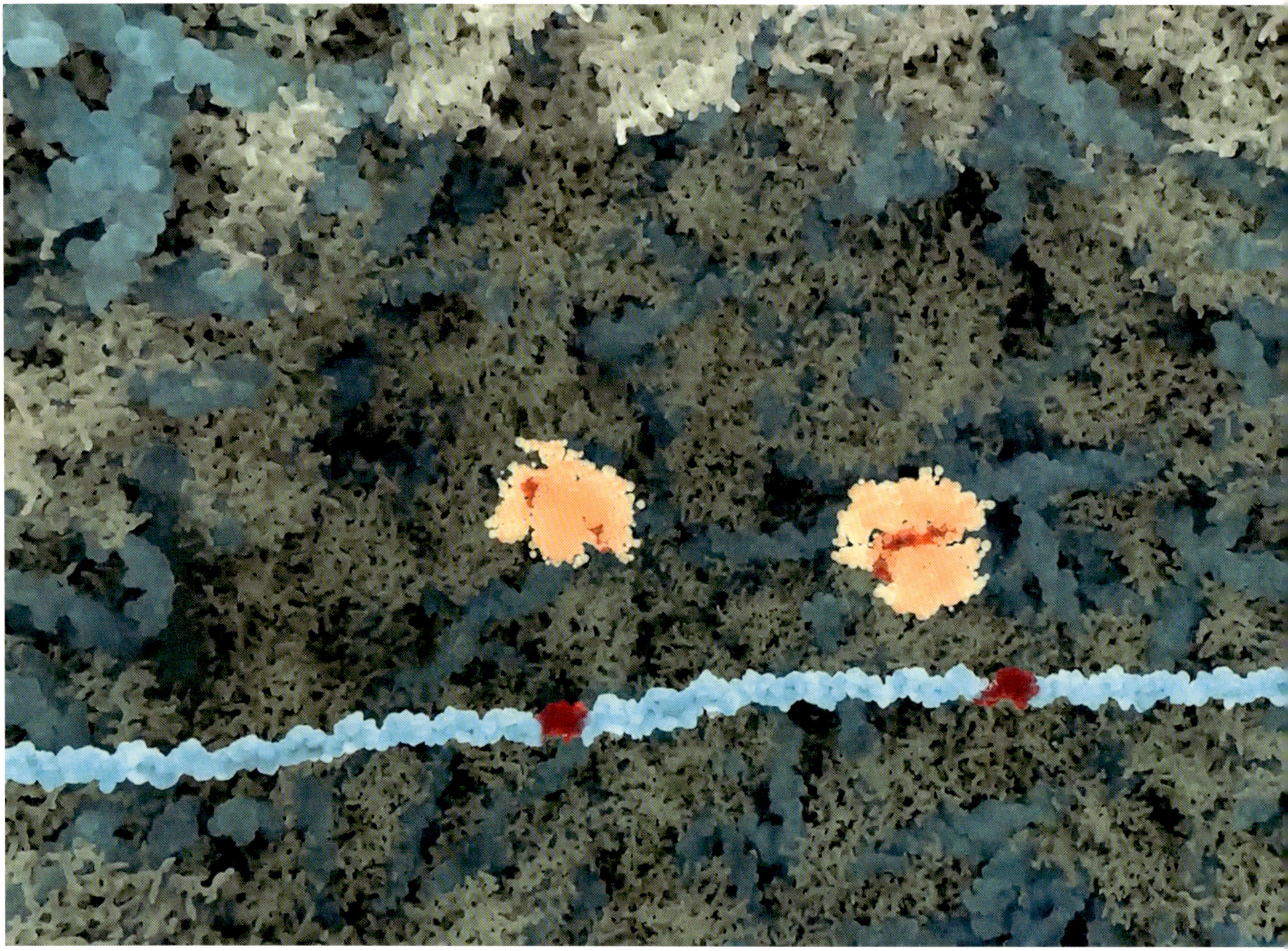

FIG. 22_*Enzymes in a cell nucleus, 3D representation based on electron microscopic data*

© Science Visualization Lab, University of Applied Arts Vienna, 2017

VIDEO_*Enzymatic cutting of a gene,* CRISPR/CAS9-NHEJ: ACTION IN THE NUCLEUS, *3D animation based on electron microscopic data, video*

© Science Visualization Lab, University of Applied Arts Vienna, 2017

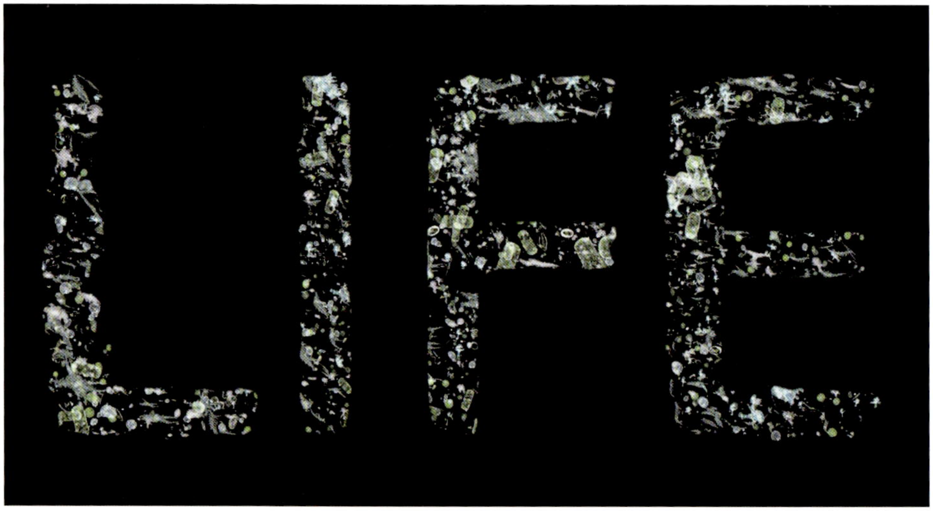

Micro Computed Tomography (micro-CT)

The development of micro computed tomography[21] finally closed the gap to achieving the fully authentic presentation of microorganisms. While electron microscopy still involves computing a three-dimensional image from a number of different angles of examination, micro computed tomography provides three-dimensional data that can be used to create authentic digital models.

The Science Visualization Lab

Our goal when founding the Science Visualization Lab at the University of Applied Arts Vienna in 2016 was to advance the visualization of the microworld and to prepare it for potential artistic treatment. We hired Martina R. Fröschl as a digital expert, who has since managed the digital production of our stories, for example, a portrayal of the CRISPR-Cas9 genetic scissors_FIG. 22; the animation *First Greed*, which shows the first attack to take place among single-cell organisms—a paramecium recognizes an amoeba as its first prey and devours it; *LIFE*, the story of the bizarre world of marine plankton based on authentic models_FIG. 23; and a visualization of an authentically modeled coronavirus penetrating a cell, with two enzymes in the cell helping it_FIG. 24.

FIG. 23_*Still from the video* LIFE, *3D representation based on micro-CT data*

© Science Visualization Lab, University of Applied Arts Vienna, 2018

VIDEO_*The extensive life of sea plankton: an amoeba identifies a paramecium as suitable prey, 3D animation based on micro-CT data,* LIFE, *video*

© Science Visualization Lab, University of Applied Arts Vienna, 2019

FIG. 24_*Coronavirus, 3D representation based on electron microscopic data*

© Science Visualization Lab, University of Applied Arts Vienna, 2021

VIDEO_*A coronavirus penetrates a human cell with the help of human cell enzymes, 3D animation based on electron microscopic data, video*

© Science Visualization Lab, University of Applied Arts Vienna, 2021

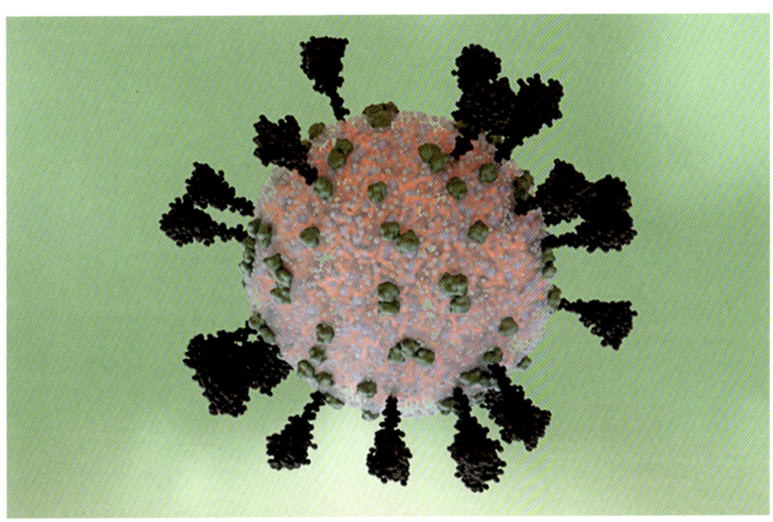

Plankton and *Noise Aquarium*

When I was commissioned by a film producer from Berlin to create authentic microorganism models for a multi-award-winning American director, the world of plankton opened itself up to me. I first created three models based on microscopic and micro-CT data—a bacterium, an amoeba, and a paramecium. In the planned film about the origins of life, the paramecium would identify the amoeba for the first time as potential prey.

At the same time, Victoria Vesna, head of the Art|Sci Center at the University of California in Los Angeles, who I had known for some time, was planning an investigation into the problem of underwater noise in the ocean, looking at the example of fish. I told her about our existing plankton models, and we agreed to shift her plans to the world of marine plankton. Our group at the Science Visualization Lab produced further authentic models of marine plankton, and the *Noise Aquarium* project was launched. Initially produced on the basis of videos, we created an interactive installation together with Victoria Vesna, which would enjoy great success all over the world in years that followed, everywhere from Australia to Asia, from Europe to America and North Africa, although one particularly impressive showing for me was in the Deep Space 8K room of the 2018 Ars Electronica_FIG. 25. A representation of this installation even made it onto the cover of *Leonardo* art magazine.[22]

During a commission for Curiosity Stream, an American production company, about digital micro 3D animation, we faced special difficulties. We were supposed to correctly visualize the development of a butterfly embryo in its impenetrable egg, which is only two millimeters long. For this purpose, we divided its development in the egg, which lasts six days, into three stages, stopping the development of the embryos in the different eggs after one, four, and six days and scanning the data of the various states of embryo development using a micro-CT. We then used that data to construct development models, which were transformed into a development flow with the help of 3D animation technology_FIG. 26.

In the future, we will continue using new technologies to open the doors of perception, making new realities visible and gaining new knowledge, thereby reminding people of what the world really is—infinite.

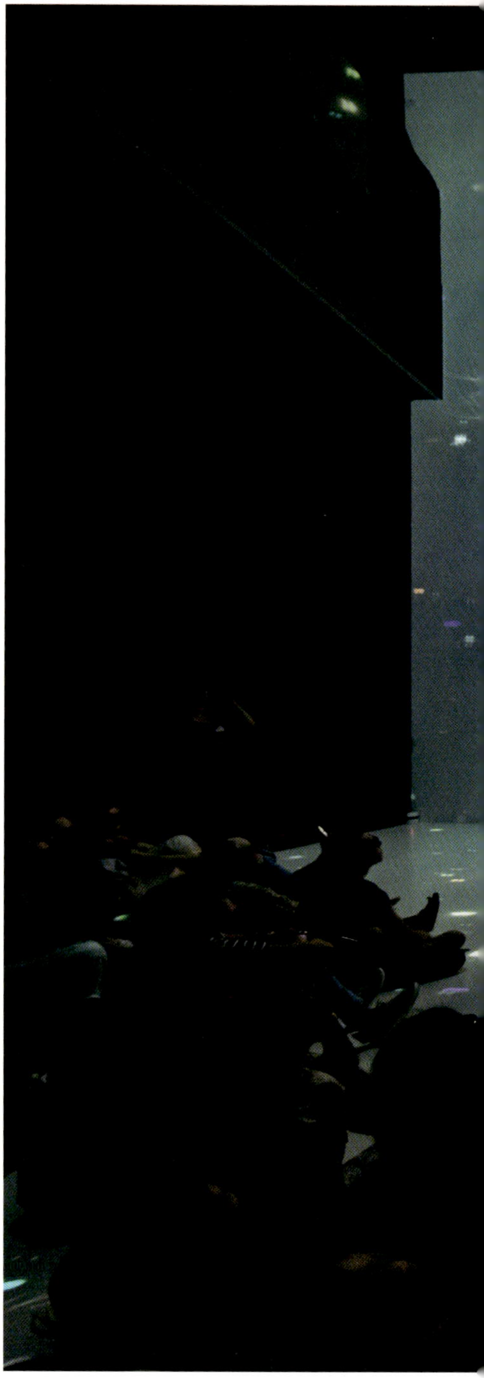

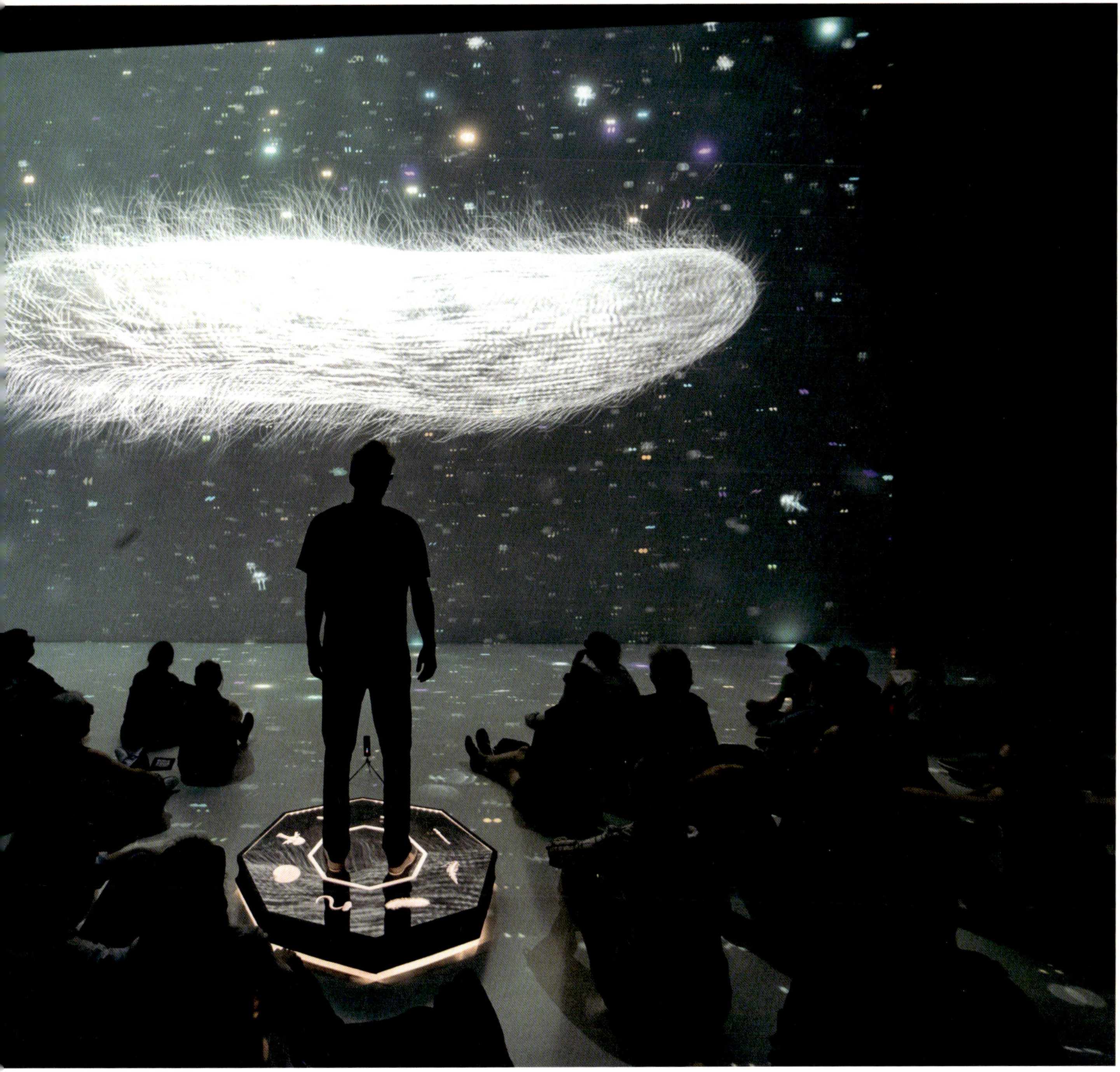

FIG. 25_ *Representation from the interactive 3D animation* NOISE AQUARIUM*, shown in the Deep Space 8k at the Ars Electronica, 2018*

Image courtesy of Glenn Bristol, 2018

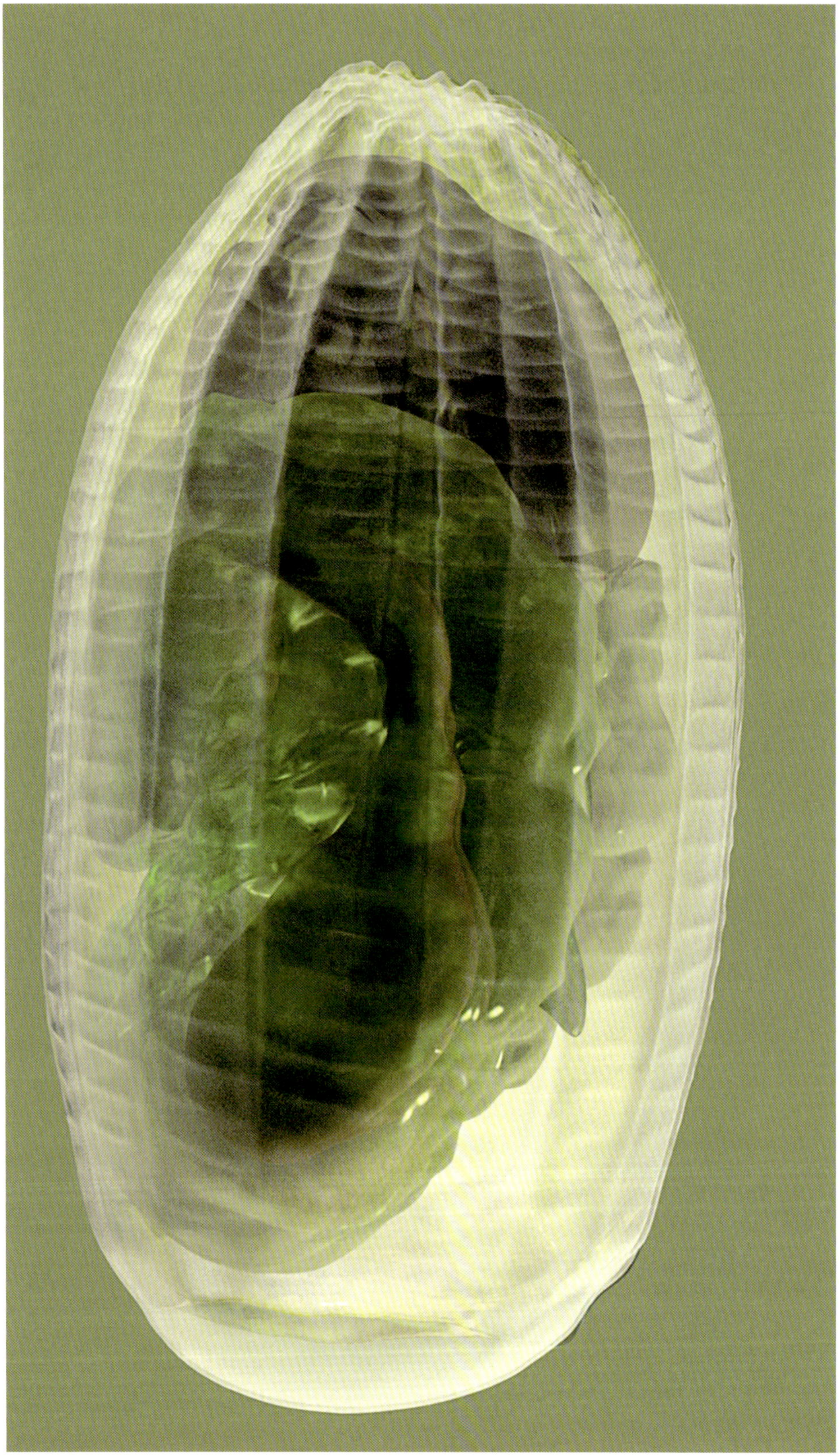

FIG. 26_PIERIS *egg, 3D representation based on micro-CT data*

© Science Visualization Lab, University of Applied Arts Vienna, 2022

VIDEO_*Development of a* PIERIS *embryo inside the opaque egg, 3D animation based on micro-CT data, video*

© Science Visualization Lab, University of Applied Arts Vienna, 2022

1 The sense of sight provides us with around eighty percent of all of the information about our environment, which we process in the brain (cf. Andrea Wengel, »Sehen,« *planet wissen*, December 10, 2020, https://www.planet-wissen.de/natur/sinne/sehen/index.html).

2 Alfred Vendl, Bernhard Pichler, Manfred Grasserbauer, and Annemarie Nikiforov, »Mikroanalytische Malschichtuntersuchungen,« *Wiener Berichte über Naturwissenschaft in der Kunst* 1 (1984): 88, plate VIII2.

3 Bernhard Pichler and Rudolf Erlach, »Metallographische Befundungenam Bronzeguss,« *Wiener Berichte über Naturwissenschaft in der Kunst* 4/5 (1987/88): 308, color plate XII.

4 SEM (scanning electron microscopy) makes it possible to maintain a continuous three-dimensional depth of field with magnification by factors of tens of thousands and up to a resolution of micrometers. An electron beam generated by a cathode is first accelerated to the anode with the help of high voltage and is then bundled by magnetic lenses and focused on the sample. The electron beam systematically scans the sample; secondary electrons from the sample and electrons backscattered from the sample impinge on specific detectors whose signals are used to generate the image. The electron beam hitting the sample excites a pear-shaped region in the sample. Due to the beam electrons interacting with the atoms of the sample, various secondary radiations are generated in addition to heat that can be used for the imaging and elemental analysis of the sample. The absorbed electrons are diverted to prevent the sample from being charged. Therefore, non-conductive samples must first be made conductive and are therefore coated in carbon or gold. The observation must take place in a sample chamber in high vacuum.

5 The environmental scanning electron microscope—ESEM—can maintain a low pressure in its sample chamber of up to twenty Torr, making it possible to observe non-conductive organic materials and even microorganisms such as mites under conditions that are just about tolerable for them. As in the high-vacuum SEM, in the ESEM, an electron beam generated by a cathode is also accelerated to an anode with the help of high voltage and is then focused by electromagnetic lenses on the sample, which is scanned by the electron beam. Due to the heightened pressure and the resulting presence of gas molecules in the sample chamber, the image is generated with the help of a special detector. The electron beam hitting the sample generates secondary electrons that are sucked in the direction of the detector by voltage applied between the sample and the detector. They collide with gas molecules and ionize them by releasing electrons from the gas molecules. The additionally generated electrons snowball and ionize further gas molecules. This amplifies the original secondary electron signal. The positive ion cores created from the gas molecules are propelled toward the sample, where they neutralize its negative charge. This is why, unlike in high-vacuum observations, the sample does not have to be conductive.

6 Oliver Vendl, *Investigations on Permeability Enhancement of Spruce Wood by Influence of Thermophilic Fungi*, diploma thesis, University of Vienna, 2002.

7 Oliver Vendl, *Investigations on Permeability Enhancement*.

8 »Gerd Binnig: Biographical,« *The Nobel Prize*, accessed March 14, 2023, https://www.nobelprize.org/prizes/physics/1986/binnig/biographical.

9 Atomic force microscope—AFM—images are generated through a special device referred to as a cantilever, which has a tip at its end that is directed at the sample. Ideally, this tip comprises a single atom. On the top of the cantilever there is a mirror that guides a laser beam over another mirror to a detector. This allows the detector to follow each movement the needle makes. When the tip moves over the surface of the sample, it is deflected by the atomic forces of the surface atoms. These deflections correspond to the shape of the surface. With the help of the mirrored laser beam, the detector translates the surface deflections into a corresponding image of the surface. Systematically scanning the surface thus generates a detailed overall image. Using the AFM, it is possible to make atoms visible on a scale of fractions of a micrometer.

10 »Voici le palmarès coupe du jeune réalisateur (offerte par M. le secrétaire d'Etat à la Jeunesse et aux Sports): ›Pictures Of Lily,' d'Alfred Vendl, Autriche« (»Cannes: Au XXIIe Festival International du Film—une séance qui bat tous les records de diversité,« *Nice-Matin*, September 1969).

11 »LE GENRE DE ›KALOMEL.‹ Voyage dans les genres, voyage dans les lieux, voyage dans le temps: cette soirée fut un peu tout cela, mais la véritable imagination créatrice n'y faisait pas défaut. Voyage dans le futur... Le film du Monégasque Luis Moiné, qui remporta le Prix de la ville de Cannes l'an dernier, était moins intéressant à ce titre que ›Kalomel,‹ œuvre d'un Autrichien de 22 ans, Alfred Vendl« (»Cannes: Au XXIe Festival International du Film—D'un style à l'autre la qualité technique triomphe et l'imagination fait quelques pas en avant,« *Nice-Matin*, September 6, 1968).

12 Peter Weibel, »Narrated Theory: Multiple Projection and Multiple Narration,« in *New Screen Media: Cinema/Art/Narrative*, ed. Martin Rieser and Andrea Zapp (London: BFI Publishing, 2002), 42–53.

13 Valie EXPORT, »Tapp und Tastkino 1968,« *Valie EXPORT*, accessed March 14, 2023, https://www.valieexport.at/jart/prj3/valie_export_web/main.jart?rel=de&reserve-mode=active&content-id=1526555820281&tt_news_id=1956.

14 Marina Pavido, »His Bag,« *Cinema Austriaco*, August 8, 2021, https://cinema-austriaco.org/de/2021/08/08/his-bag.

15 »Hans Hass,« *Biologie Seite*, accessed March 14, 2023, https://www.biologie-seite.de/Biologie/Hans_Hass.

16 »Wolfgang Hausner,« *Die Homepage von Bobby Schenk*, accessed March 14, 2023, https://www.bobbyschenk.de/n003/kroko.html.

17 »Biografie,« *Rupert Riedl: Ein Leben für die Forschung*, accessed March 14, 2023, https://rupertriedl.org/biografie.

18 Steve Nicholls and Alfred Vendl, *Limits of Perception*, 2001, https://www.amazon.com/Limits-Perception-Steve-Nicholls/dp/B00H8UZLQ8.

19 Steve Nicholls and Alfred Vendl, *Time Limits*, 2008, https://www.primevideo.com/detail/Time-Limits/0TIVDDR31GPR682ECN-NI7W5EFP.

20 Alfred Vendl and Steve Nicholls, *Limits of Light*, 2011, https://www.amazon.com/Limits-Light-Alfred-Vendl/dp/B00V9U9I1W

21 X-ray microtomography—micro-CT for short—uses X-rays to create cross sections of objects. These cross sections can be used to construct virtual three-dimensional models of something without destroying the original object. The pixel size of the cross sections can be measured on a scale of micrometers.

22 *Leonardo* 52, no. 4 (2019), https://direct.mit.edu/leon/issue/52/4.

The Door of Science Visualization

Martina R. Fröschl

The expansion of human perception through computer-animated scientific visualization in combination with social, interactive, personal, and local experiences increases interest in complex topics and makes content easier to remember. Visualizations based on scientific data and expert input gain value when compared with other creations.

The Science Visualization Lab of the University of Applied Arts Vienna in Austria — referred to in the following as the Science Visualization Lab — brings multiple disciplines together. The lab is a meeting place with the goal of using the language and tools of computer animation to make important human issues tangible and understandable. These computer animations include the added value of three-dimensional scientific datasets used for these *computer-animated scientific visualizations*. In order to achieve new insights into natural phenomena, Alfred Vendl experimented with pioneering visual effects techniques in his transdisciplinary documentary films. As early as in the 1990s, he was commissioning visual effects and animations to illustrate scientific content for wide audiences in productions for educational institutions and for television. The Science Visualization Lab evolved out of these ambitions.

My work for the Science Visualization Lab began with stories depicting the microscopic realm in computer animations. The starting point for my work at the lab was my doctoral thesis, *Computer-Animated Scientific Visualizations of Tomographic Scanned Microscopic Organic Entities*[1] and the locomotion of mites — more specifically, two different mite species, one of which had already been featured in the award-winning short animation *The Incredible Water Bear*.[2] Immediately after finishing with the mites, I started working with scans of the organisms *Amoeba*, *Paramecium*, and *Cylindrospermum*, and, later, on plankton species for the projects *First Greed*, *LIFE*, and *Noise Aquarium*. Another topic was the molecules of life: I visualized various macromolecules in the projects *CRISPR/Cas9-NHEJ: Action in the Nucleus*, *Blood-Brain Barrier*, and *Virus Dice*.

The objective of this book is to examine different possibilities for augmenting human perception using various means of visualization. We invited collaboration partners to contribute, thereby bringing together a distinguished group of international authors. We asked contributors to share their personal stories and perspectives on opening doors to expanded human perception through augmentation by technical means, and we highly appreciate their different views, which have allowed us to curate an overview of what *science visualization* could mean in particular, personal approaches.

In this chapter, I will give insights into my perspective of how to open visual doors to the *micro-* and *nanoworlds* that surround us by means of *computer-animated scientific visualization*. I will then discuss the major projects I have been working on since 2016 by examining images and the project-specific characteristic of opening doors to hidden worlds in order to extend our realm. I have divided the descriptions of the projects by topic into two groups of scale: the projects of the nanoworld and those of the microworld, although the presentations of these scales often overlap slightly, in particular due to introductory shots, which often offer an overview in order to provide a sense of orientation and localization. Zooms that are impossible to film with a real camera and therefore have to be simulated with a computer graphics camera are a tool often used to visually communicate where the story of the computer animation that will follow is set. After describing past and ongoing projects, I will venture a conceptual outlook on the future of the Science Visualization Lab.

Science Visualization

Modern scientific research is becoming more and more difficult for the general public to understand as activities are being carried out on scales and in time ranges that we cannot perceive with the naked eye. Visualizing the multitude of problems that humankind faces today is often the key to understanding and awareness. The Science Visualization Lab focuses on making invisible scientific phenomena visible in order to enable visual thinking[3] in the arts and the sciences, education, and communication.

In my doctoral thesis,[4] I defined the term *computer-animated scientific visualization*. The main objective of this definition was to clarify that, unlike *scientific visualization* — a subfield of *data visualization* and a field related to *information visualization* — what we at the Science Visualization Lab refer to as *science visualization* takes a distinct approach. We use scientific datasets in our visualizations but, at the same time, we want to emphasize that there is more creative work involved than in typical scientific visualization workflows, as we are constantly experimenting with new forms of visual style, transformation, and presentation. The latest innovations in science as well as in software and hardware development are important for digital visualization, but intensive research into art and design perspectives is of equal importance.

I think that, in a world in which the sciences are often presented as something unquestionable, even though people are increasingly losing trust in them, it is important to make findings tangible for each individual and thus to address the physicality, locality, and subjectivity of the topics being presented. Robert Root-Bernstein, Todd Siler, Adam Brown, and Kenneth Snelson give a definition that resonates with me in their »ArtScience Manifesto«:

[...] 1. Everything can be understood through art but that understanding is incomplete. [...] 2. Everything can be understood through science but that understanding is incomplete. [...] 6. ArtScience is not embodied in its products so much as it is expressed through the convergence of artistic and scientific processes and skills. [...] 17. The objective of ArtScience is to inspire open-mindedness, curiosity, creativity, imagination, critical thinking and problem solving through innovation and collaboration![5]

The Science Visualization Lab sees itself as kind of a hub for theory and practice. Combining emotional, sensory, and subjective approaches with objective, rational, and analytical approaches is intended to facilitate public understanding. Integrating re-humanization, physicality, and subjectivity is a major challenge for members of our Western culture in societies that have not been taught to do so. People rarely take the time to really »get involved« with the issues being researched or presented. At the same time, visualization is always polysemic — with multiple meanings and ways to interpret it. A wide variety of influences such as the culture in which the recipient grew up, but also the presentation environment, and so on, lead to the visualization being read in different ways to the one intended. This is a challenging part of creating visualizations, and the problem of misinterpretation can never be ruled out completely.

While showing and explaining the *Noise Aquarium* installation in Taos, New Mexico, a question posed by a Native American was a key experience concerning the cultural differences of interpretation. The concept of the art installation includes some elements that allow visitors to immerse themselves in the world of plankton and to experience noise and plastic pollution through their presence in a soundscape and by interacting with computer-animated scientific visualizations. But this one visitor did not seem to be as interested in the actual content of the installation and instead asked questions about the symbols on the platform that visitors could use to control the interactive installation with their bodyweight. He was fascinated by the shiny symbols representing seven species of plankton. In this conversation, I became particularly aware of how people subjectively perceive visual elements.

It is essential, for various professional reasons, to collaborate with others interdisciplinarily. An ideal collaborative team might consist of artists and scientists who are of equal importance for the project. There has to be trust and respect. Computer-animated scientific visualization might help to inform an interdisciplinary process or communication with colleagues. Generally speaking, the visualization of whichever science can be an artwork itself or part of an artistic concept.

The definition of *cinematic visualization* given by Donna Cox and her successor Kalina Borkiewicz at NASA's Advanced Scientific Visualization Lab takes a similar line:

Cinematic visualization is a combination of many disciplines: science, domain science (e.g., astrophysics, biology, climate, geospatial...), mathematics, art, art and design, filmmaking, technology, data science, computer science, computer graphics, computer hardware, humanities, communication, education, psychology. Though we are not domain scientists, our skill lies in being able to understand the sciences and communicate to broad audiences in a visual way, using advanced computing techniques.[6]

Most contemporary contexts describe scientific visualization as computer-generated images or animations that visualize scientific data. These visualizations can be used for discovery, understanding, communication, and teaching.[7]

The visualized data may come from various scientific disciplines or research fields. Ideally, knowledge is created by visualizing data, and the process encourages thinking that is not separated into perception and cognition. In the case of visualization, reality is illuminated, imitated, and interpreted. A language of symbols is created for entities for which there might not even be words yet. Radical innovations embody the pre-verbal stages of new concepts. Expanded visualizations make new perspectives possible. Visual experiments enable advanced visualizations that lead to creative, expanded mindsets.[8]

When creating scientific visualizations, information is transformed and made available for alternative perceptions, and visual doors to worlds that are not otherwise perceivable are opened. Science visualization can be the foundation of art installations or immersive experiences that engage audiences in alternative realities and creative thinking, but these presentations and formats are heterogeneous in their genre-specificity. In an age with great expectations on scholars to think and work interdisciplinary, a lack of clarity about what exactly science visualization is might actually be a feature and not a bug.

Combining »art« and »science« became one of the most fashionable artistic movements in the early decades of the new millennium, and this aesthetic is unsubtle and still evolving alongside various technical achievements. Four main modes of combining the arts and the sciences can be found:

First, art as a communicator of science, second, science tools as a means of art production, and third, science as art—for example, when a scientific image is admired and exhibited in an art gallery.

Fourth, there is a strand where the arts and science are in fluid interchange, and the disciplines are honored for their similarities as well as their essential differences. The arts and sciences are similar in that they are expressions of what it is to be human in this world. Both are driven by curiosity, discovery, the aspiration for knowledge of the world or oneself. Usually, artists and scientists express themselves in different ways: the arts through the body and mind, often driven by the exploration of the ego, contradictions and the sheer messiness of life; science through equations, directed, collaborative research and experimentation that works in a progressive, linear fashion.[9]

Science visualization nests somewhere in between these four modes. All modes are possible, but as you will see in the following two sections, mixing the four modes is a typical approach taken in the projects carried out at the Science Visualization Lab.

The starting point for a collaboration can be anything. At some point, computer-generated visualizations for the projects are created in the lab. The approach of bringing the arts and the sciences together is practiced in many institutions around the world, and these places are increasing in number, which is a good sign, as holistic thinking and the convergence of disciplines is urgently required in an age in which humanity is facing many problems. Pioneers in this area were, e.g., the MIT's Media Lab, which has mainly approached the arts and sciences through technical advances; La Laboratoire in Paris; SymbioticA in Perth, with its wet-lab focus; and Harvard University's Initiative for Innovative Computing, which takes a more medical view of art-science.[10] The Science Visualization Lab focusses on networking and knowledge transfer with such impactful institutions and has organized public lectures with renowned personalities in the field since its inception. For instance, guests in the 2010 lecture series included Oron Catts, Ken Perlin, Bahman Kalantari, and Carl Djerassi. Other important guest lectures have been held by, for example, Victoria Vesna, Virgil Widrich, and Janet Iwasa. Since 2020, the lab has been cohosting the PIXELvienna conference, which invites outstanding leaders from the wider computer graphics field to attend each year.

At the Science Visualization Lab, we apply scientific data in innovative ways by conducting experimental interdisciplinary research. Our specialty lies in our intensive use of 3D information imaging techniques, such as computed tomography (CT), magnetic resonance (MR), and confocal laser scanning microscopy (CLSM). Detailed examinations of datasets generated using these scientific imaging methods result in a variety of variants and styles for visualizing the data and stories underlying the scanned entities.

Science Visualization Lab Projects since 2016

In the following sections, I will describe the various projects that have been launched since I started working at the Science Visualization Lab in 2016. So far, there have been two major categories in which the projects of the Science Visualization Lab have been roughly organized in terms of their scale. They all aspire to open up fascinating microscopic visuals for diverse audiences.

The Nanoscale

For nanoworld projects, we apply data from the ever-growing Worldwide Protein Data Bank,[11] where international researchers publish their research. So far, we have run three projects on this scale, and great collaborations have opened up possibilities to carry out projects in the nanoworld for us. On the one hand, we have worked with scientists from the Vienna BioCenter, the University of British Columbia, Vancouver, and Johannes Kepler University, Linz—above all with Renée Schroeder, Josef Penninger, Ivona Kozieratzki, Stefan Mereiter, Peter Hinterdorfer, Krzysztof Chylinski, and Thomas Marlovits. On the other hand, we have collaborated with the company Nanographics, which was founded in the computer graphics department of the Technical University of Vienna by Ivan Viola, Tobias Klein, and Peter Mindek. These collaborations have made it possible to work with biochemical datasets while receiving advice from experts.

CRISPR/CAS9-NHEJ: ACTION IN THE NUCLEUS_Genetic engineering is a technology that is providing great opportunities but also generating many fears. At the end of the twentieth century, and especially with the invention of the CRISPR/Cas9 method around 2011 by Jennifer Doudna and Emmanuelle Charpentier—who together

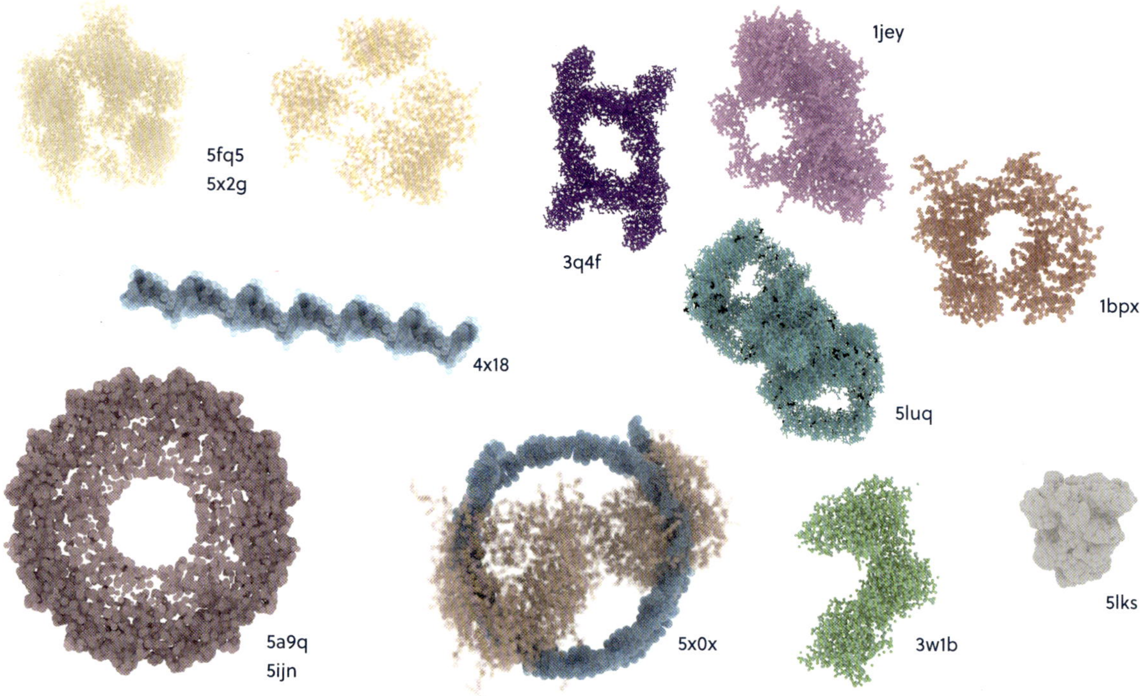

FIG. 1_*Various molecular datasets from the Worldwide Protein Data Bank used in the project* CRISPR/CAS9-NHEJ: ACTION IN THE NUCLEUS

won the 2020 Nobel Prize in Chemistry for their discovery of the game-changing gene-editing technique[12] — new opportunities in the organic natural sciences have become possible. CRISPR/Cas9, often referred to as »gene scissors,« sparked a revolution in multiple research fields.

The first project that we carried out in the nanoworld was *CRISPR/Cas9-NHEJ: Action in the Nucleus*. It was commissioned by Rector Gerald Bast for the exhibition *AESTHETICS OF CHANGE: 150 Years of the University of Applied Arts Vienna* and was presented in the Digital Art department's *Future Room Installation* under the direction of Ruth Schnell in collaboration with Martin Kusch and Peter Weibel. The installation was shown repeatedly in the years that followed, for example, at the 2020 Ars Electronica. The video alone was additionally shown at venues such as the EPICenter Sydney, the SIGGRAPH 2018, and the Kendall Planetarium in Oregon. During most showings of *CRISPR/Cas9-NHEJ: Action in the Nucleus*, we have chosen to augment the video presentation with information linked to the various datasets and their respective papers. We created a legend, which can be seen in_FIG. 1.

In explanation and feedback sessions, the abstract essence of the gene manipulation technique was translated into *CRISPR/Cas9-NHEJ: Action in the Nucleus*. Renée Schroeder helped to give a general understanding of the building blocks of life and has written several transdisciplinary books on the topic. Later feedback on the models and animations was generally given by Krzysztof Chylinski, who has co-authored papers about the discovery of the Nobel-Prize-winning technique.[13]

The animation *CRISPR/Cas9-NHEJ: Action in the Nucleus* was created for different presentation formats, namely stereoscopic video, 2D animation, and full-dome projection. First, the animation shows a schematic cell_FIG. 2, before it goes

FIG. 2_*A schematic cell with a vivid color scheme provides context at the beginning of the animation* CRISPR/CAS9-NHEJ: ACTION IN THE NUCLEUS.

VIDEO_CRISPR/CAS9-NHEJ: ACTION IN THE NUCLEUS

through a cell nuclear pore into the interior of the cell nucleus of an animal cell. In this cell nucleus, the biochemical datasets of different macromolecules have been animated to show the process of gene manipulation using the Cas9-NHEJ method. At the end of the animation, the nucleases, polymerases, and ligases of the cell's own repair system accomplish their work by rejoining the altered DNA strand. This self-repair mechanism takes place about 10^4 to 10^5 times per cell per day.[14] Such facts really do make me admire our human bodies and the wonder of our existence.

The protein parts are designed like jelly clouds with floating locomotion behavior in order to present the organized chaos of life in a way that is both visible and tangible. A pinch of noise and flicker indicates the density of the visible atoms and their movements. In the sound of the CRISPR visualization, text being recited by computer voices adds an additional level to the narration of the animation. Philosophical phrases and questions are intended to make recipients think more deeply about this topic. In addition, an ambient sound composition using pink noise additionally suggests the complexity of life through the dimension of sound. Noise has been deliberately chosen at every level in order to indicate the connection between chaos, organization, and life, and to communicate this omnipresent intricacy.[15]

When we started working on the animation, it was not yet clear how the Cas9 molecules actually approach the DNA strand. It was encouraging that, while we were completing our project, scientists began recording videos that showed movements that looked very noisy, just like ours. The animation, like almost all visualizations by the Science Visualization Lab, was intended for a general audience. Noise

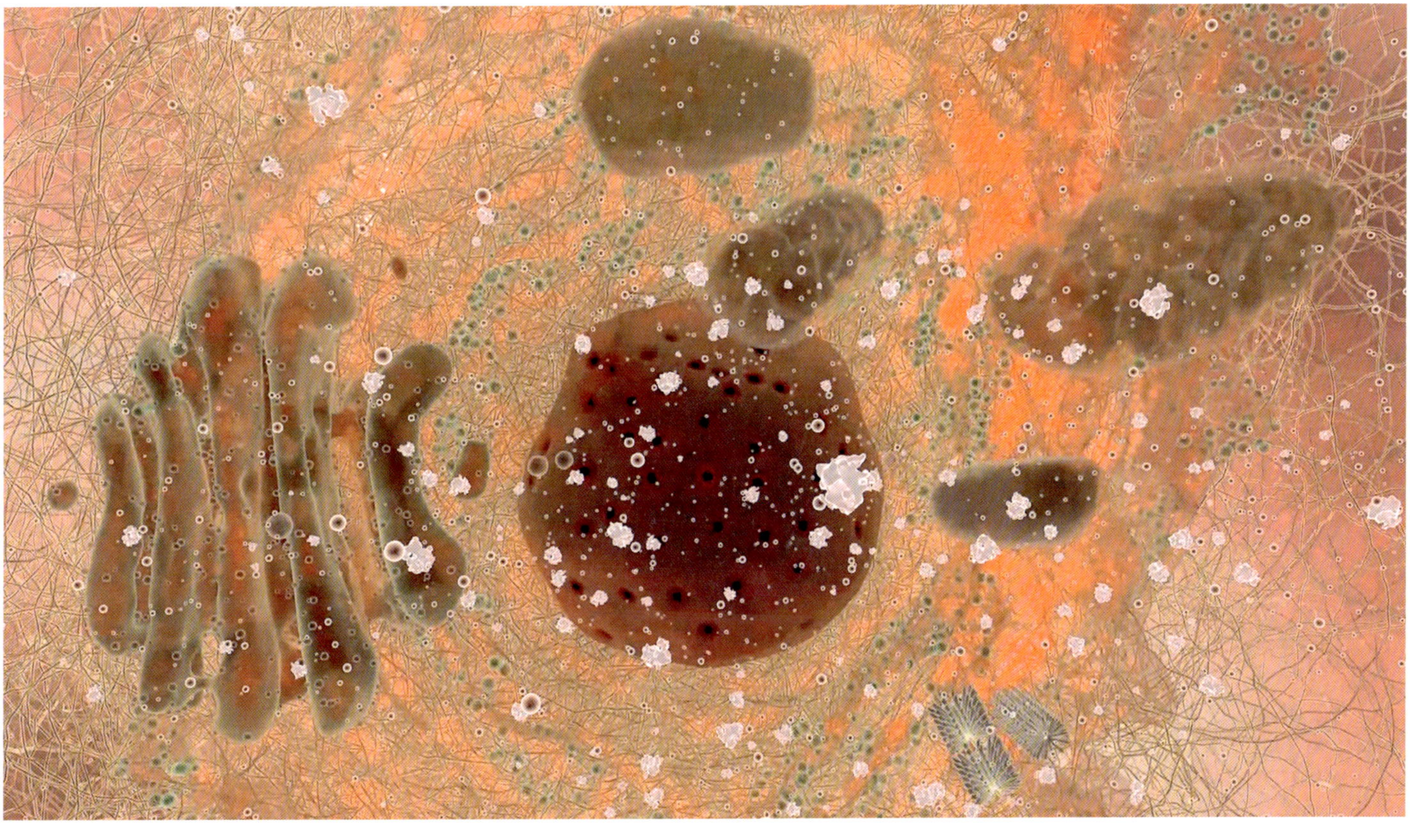

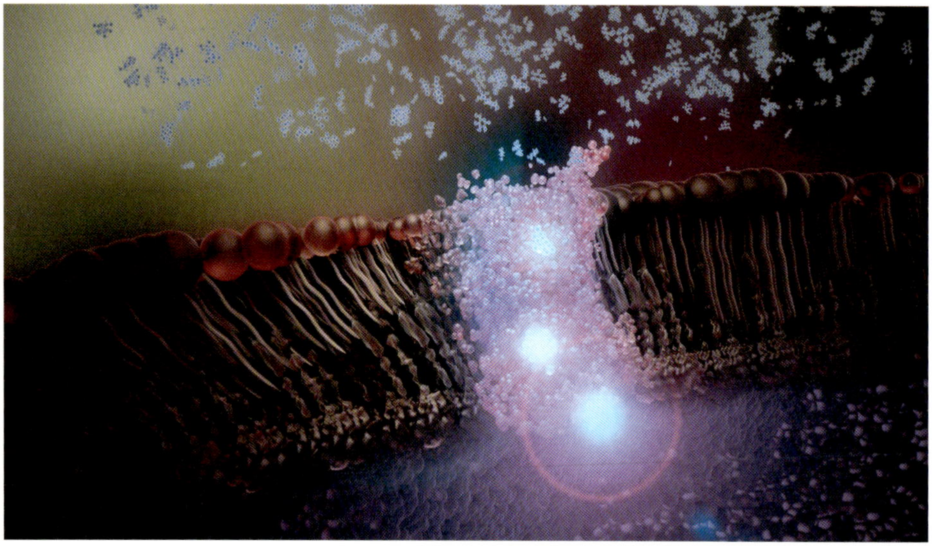

FIG. 3_ *The blood-brain barrier taking up glucose molecules*

VIDEO_ BLOOD-BRAIN BARRIER—ENERGY FOR THE BRAIN

was therefore used to visually hint at the dynamics of chemical reactions in a simplified way.

BLOOD-BRAIN BARRIER—ENERGY FOR THE BRAIN_ This medical information video shows the silhouette of a person and then zooms through the bloodstream into the brain and the blood-brain barrier. The video opens with the following text by director Alfred Vendl:

> *The brain works like a computing center for humans, without interruption and without breaks. The condition for this is a continuous supply of energy. The brain needs a single substance to provide energy—glucose, a component of sugar. Glucose is transported by blood vessels through the blood-brain barrier into the brain. Over the years, the gates of the blood-brain barrier narrow and influence a continuous supply of glucose to the brain. The reduced supply of necessary energy to the brain decreases the power of this »computing center,« and the »illuminating light« of thoughts loses luminosity.*

The brain is supplied with glucose directly via the GLUT1 transport protein.[16] This computer animation was created in cooperation with an Austrian team researching the sugar dilemma.[17] They are examining the functionality of the blood-brain barrier with regard to »cerebral metabolic syndrome,« which they suspect to be one of the triggers of dementia.[18] The animation metaphorically shows how »the light of thought« depends on the molecule glucose_FIG. 3.

At the beginning, the animation shows the regions of the blood-brain barrier, after which it zooms in on the capillary vessels in the brain and then jumps further into the nanoworld of the GLUT1 transporter. The visualization *Blood-Brain Barrier—Energy for the Brain* was presented at the international VIZBI conference[19] in 2021, which was about visualizing biological data, where it was positively received. The project was also part of the project »Movement, Mapped,« which was published by *SciArt Magazine*.[20]

VIRUS DICE_The third science visualization on the nanoscale was the *Virus Dice* project. The idea was born of the second pandemic outbreak of SARS-CoV. Most people were distracted and insecure because of the lockdowns and the general state of society at the time. At the Science Visualization Lab, we decided that we wanted to approach this suddenly omnipresent topic from our own perspective. This was the reason for launching the project *Virus Dice*, which aims to convey knowledge about the virus in a visual, immersive, and interactive way.

Josef Penninger and his team had already been researching the SARS-CoV-1 outbreak, and they immediately resumed their research. Researchers who have devoted their lives to investigating the smallest building blocks of life, like Ivona Kozieratzki, Stefan Mereiter, and Peter Hinterdorfer, helped us to provide an informed view of the parameters that trigger or prevent these specific virus infections. The project's animations introduce the infection pathway and potential medicine components, ACE2 and lectins. Both molecules are produced by the human body; the medication therefore works by flooding the body with these proteins, which can block the spikes of the SARS-CoV-2 virus and thus prevent the virus from entering human cells, subsequently stopping the virus from reproducing. The project aims to show protein processes on a molecular level and the underlying factors of uncertainty and probability. Moreover, we created an animation in collaboration with Nanographics[21] that incorporated one of the first full models of the SARS-CoV-2 virus, published in the paper »Modeling in the Time of COVID-19.«[22] There were several iterations, and we used the latest simulation of the wiggling of the famous spike proteins when we created our animation.

Visual identity was an influential and powerful tool in the global communication of virus-related information. I participated in the AG Animation's remote writing sessions, where we discussed the visual appearance and identity of the coronavirus. The various observations made in the collaborative text dealt with science communication, entertainment, and personal coping strategies. Certain depictions of the virus have become visual tropes and are thus now part of the collective visual memory.[23]

The very first virus animation that the Science Visualization Lab screened was the *Virus Dice* teaser at the 2020 PIXELvienna Conference on Computer Graphics and Animation,[24] as part of the online exhibition titled *Corona and Climate Change*. We presented a teaser of the project's main visual theme, which consists of dice and the SARS-CoV-2 virus. After finalizing the first video, we had an opportunity to approach the topic experimentally. We created an art installation where visitors could play a physical dice game_FIG. 4. The dice game drew attention to the risk-reducing effects of vaccination even before that form of immunization had started growing in significance as a political issue. The game also warned of possible mutations. Chance and the luck of the dice were the driving forces in the game. This kind of coincidence also plays a role in our everyday struggles for a healthy body.

With the dice element in the project *Virus Dice*, I wanted to focus on chance and to raise awareness about the fact that thinking in black and white makes little sense. I dedicated the project to my grandmother, who had died at the beginning of the pandemic — leaving behind a large family who was never able to properly say goodbye due to the restrictions. My grandmother, who was an early organic farmer,

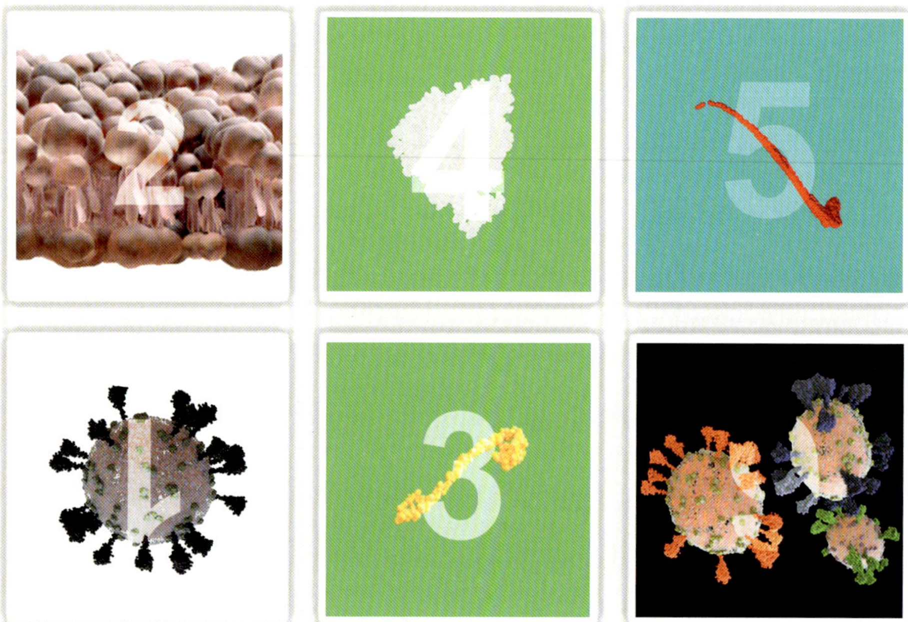

FIG. 4_ *The* Virus Dice *symbols on a dice printout. The goal of the game: there is one surviving cell; if the cell dies, you lose. Membrane is the starting symbol; place it in the middle of the playing area to start a new game. Roll the dice. The symbols that you roll are attached to the starting membrane symbol. Newly rolled symbols have to use free spots on membrane dice. The dice have to be arranged in vertical or horizontal rows (slots)—slots can only be orthogonal. Symbol elements can be stacked in groups of three; if you achieve this, you get a new membrane piece.*

1. *Virus: attaches to an empty membrane or all other parts if there is no free membrane available. Can move between the membrane and the attached mutation.*
2. *Membrane: starting dice symbol; it attaches to other membrane symbols, and the cell grows. The largest surviving cell wins.*
3. *Lectin medication: attaches to the membrane and the virus, and protects the cell from the virus and mutations. Moves between the virus and the membrane if all membrane slots are blocked.*
4. *ACE2 medication: attaches to the membrane and the virus, protects the cell from the virus and mutations. Moves between the virus and the membrane if all membrane slots are blocked.*
5. *Vaccination: attaches to everything or immediately removes a »virus« or »mutation.« Every vaccination erases either one mutation or virus. Vaccination prevents the cell in the game from dying if attached anywhere—for example: if you have rolled the symbols »virus« then »vaccination,« followed by »mutation,« the game is not over as the »vaccination« annihilates the »mutation«—both dice are subsequently removed from the playing area.*
6. *Mutations: end the game if a virus is attached to an unprotected membrane symbol and the membrane is not protected by either medication or vaccination.*

often used a common saying: »Health is the most important thing.« We should try to reduce health risks, but uncertainty is omnipresent in our everyday lives. The parameters in »the game of life« can change at any moment because the influences on and in our body are still too many to calculate, and they vary with ever-changing life situations. The parameters in the dice game change depending on the players and the situation of the performative act of playing. Large cells might survive, or the game might end straight away.

During the *Virus Dice* installation performance at Schmiede Hallein 2021, one woman was able to tell me when she was going to have a bad throw before she took her turn, but I, as the game master in the performance, couldn't tell if she was cheating somehow. Her friends told me she could see the future, but I think she must have been a trained dice player. We can only increase our probability of staying healthy.

For the computer-animated scientific visualization video *Virus Dice*, musician and sound artist Michael Wedenig created a sound mix from his own recordings and samples of the *Climate Sample Pack*, a collaboration between DJs for Climate Action and Greenpeace. The pack contains a selection of diverse samples

VIDEO_
Virus Dice

and loops crafted from the environmental organization's sound library. In the context of the aforementioned *Corona and Climate Change* exhibition, it was important to me to use parts of this sample pack, which contains sounds from the Antarctic, from the rainforests of Papua New Guinea and the Amazon, from Arctic sea ice, and other places where you can hear sounds caused by climate change.

A documentary video about the *Virus Dice* project was exhibited in fall 2021 in the schauraum of the Digital Arts department in Vienna's *MuseumsQuartier*. Virus phage renderings and visualizations of the virus datasets were part of the TV report *Welt der Seuchen* (World of Plagues), broadcast in May 2021 by the Austrian Broadcasting Corporation station ORF 1.

The knowledge-forming effect of computer-generated images can have an intense effect because the recipient not only looks at something but also experiences something at the same time. The data source, the creation of the video sequence, simulations, and backgrounds can be made visible in the computer animation, thus enabling an information chain on various levels. Self-reflection and self-referentiality play a very important role in 3D computer animation.[25]

In the context of the Corona virus pandemic, the shape of the virus already hinted at the main topic for most recipients as it had become so established in various media reports. The color coding was adopted from typical scientific publications in biochemistry, as in previous projects dealing with the nano realm.

Getting immersed in and interacting with the aesthetically interesting and unfamiliar nanoworld allowed the viewers to analyze this more than contemporary issue in an innovative way. Playful interaction reduced the barriers to engaging with the highly complex biochemical topics. The aim was to disseminate knowledge about the virus and vaccination in a visual way. Due to its scalability, the transdisciplinary installation can be presented on various occasions.

The Microscale

***FIRST GREED*, *NOISE AQUARIUM*, AND *LIFE*_**The first work I did at the Science Visualization Lab, initially as a freelancer, later as a full-time member of staff, was on plankton organisms. My first assignment was to study *Amoebae* datasets intensely, and *First Greed* evolved from this. *First Greed* is a short film that deals with the metaphorically depicted first time that one living being absorbed another in order to gain energy from it.

After the three initial scans of the organisms *Amoeba*, *Paramecium*, and *Cylindrospermum*, other plankton organisms followed. Considerations about creating a 3D model database of scanned microorganisms led to the database being expanded with scans of *Oikopleura*, *Noctiluca*, *Tomopteris*, and *Actinotroch*. We continued our collaboration with some of the authors in this book — biologists Thomas Schwaha at the University of Vienna's Department of Zoology and Stephan Handschuh at the VetCore of the University of Veterinary Medicine Vienna — which had been initiated by mite expert Manfred Walzl for projects involving scans of soil organisms. Tomographic scans from the micro-tomograph or the confocal scanning laser microscope were supplemented with information from light microscopes and scanning electron microscopy, and were subsequently computer-animated. For all the animations and the backgrounds of the plankton creatures, it was important to me to really find a visually pleasing look. I experimented a lot with color schemes,

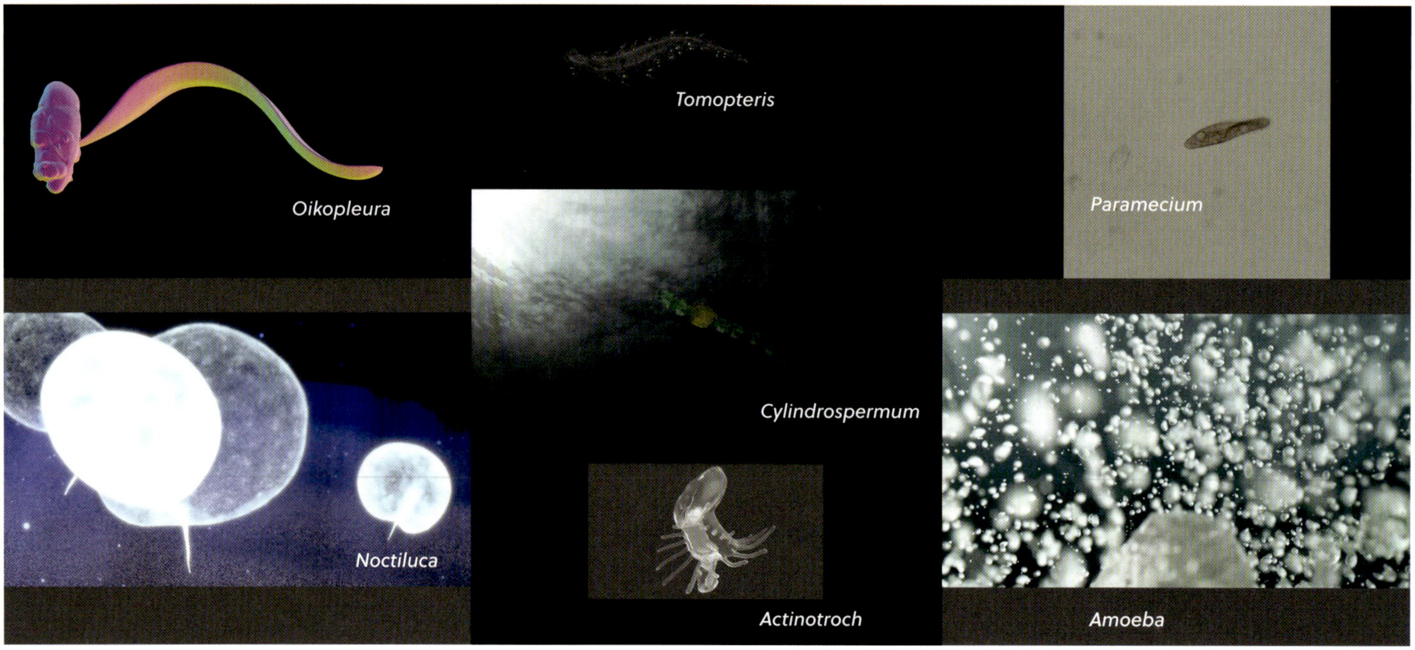

lighting and scenery. In the end, the distinct aquamarine look was frequently used in all the projects that I created, rendered, and composed using the scientific plankton datasets (micro-computed tomography, confocal laser scanning microscopy, scanning electron microscopy, light microscopy).

The Science Visualization Lab commissioned scientific imaging experts to produce intricate detailed datasets in the form of image stacks and their various volumetric representations that fascinate due to their detailed time-capture of the once real, living, and unique subject organism, which were then processed. Various plankton drifters were captured for scientific imaging in salt or fresh water. As most plankton have very thin membranes, it was particularly challenging for the imaging experts to create the scans, and a lot of additional repair work went into visualizing the plankton so that they looked alive again. It took time to prepare, scan, and generate 3D models for animation as deformations had to be corrected that had occurred because the samples were so delicate. You can see the various workflow steps in the augmented reality version of_FIG. 5. These rich and intriguing datasets were processed by experimenting and by means of trial and error, which gradually led to the production of animations that are now being used in three projects. At the Science Visualization Lab, we have developed special methods for creating authentic three-dimensional models of organisms. Individual organism specimens are prepared and then scanned tomographically and stored in image stacks. This data makes it possible to calculate digital geometry and to create animations. The processing procedures and workflows that have been adapted for each project have been documented in various articles,[26] my doctoral thesis,[27] and a book chapter.[28]

After the first projects using three of the datasets had already been completed at the Science Visualization Lab, the project *Noise Aquarium* was launched. I would

FIG. 5_*Visual examples of plankton generation workflow steps*

VIDEO_*Various workflow steps for* Noise Aquarium *animations* like to share a quote from an artistic statement by the project's artistic lead, Victoria Vesna:

> *Noise Aquarium is […] a large interactive virtual display that projects 3-D scans of specimen obtained through unique scientific imaging techniques and immerses the audience in an »aquarium« of diverse plankton projected to appear as large as whales. Noise Aquarium attempts to elicit a visceral response to the vicious circle where there is fracking and all these things related to creating fossil fuels, and then we make plastic from it, use it once, throw it all away to the point where we are just basically killing ourselves—we are drowning in plastic. Through their presence alone, audience participants activate destructive underwater pollution noise such as fracking and sonar, demonstrating how we are implicated by inaction.*[29]

The project deals with a sometimes overlooked threat to living beings: increasing noise pollution in bodies of water around the world and omnipresent plastic pollution. The impact of various noise sources on large aquatic life forms is well known due to the striking examples of stranded whales shown in the media. In contrast, only a few studies deal with the influence of noise pollution on microscopic life forms such as plankton. *Noise Aquarium* shines a spotlight on particularly impressive representatives of the great variety of plankton. In the project's installations, visitors can immerse themselves in fascinating worlds that are not visible to the naked eye. A variety of publications have addressed the artistic qualities of the project.[30] I chose to adapt the look of *First Greed* for *Noise Aquarium*—the color scheme was similar, but since the connection to the destructive influence of humans was to be one of the project's major themes, I made the surface of the ocean aesthetically visible by means of caustic rays of light breaking on the water's surface. At the beginning of *Noise Aquarium*, my computer animations were shown in huge projections at the Queensland University of Technology (QUT), Brisbane, and the Nanyang Technological University (NTU), Singapore. Later on, we got the chance to develop the interactive version, first as a stereoscopic projection with a balance board platform in the Deep Space 8K at the Ars Electronica Center, Linz, and at the Paseo Project Festival in Taos, New Mexico, as an interactive 2D projection. We decided to add plastic particles to the visualizations for the interactive versions. We also designed and created a virtual reality version for later iterations.

The success of the project has been fueled by the strong collaboration between Victoria Vesna, Director of the Art|Sci Center UCLA, and Alfred Vendl, Director of the Science Visualization Lab. *Noise Aquarium* is a true art-science collaboration that can rarely be found in this form. The main project partners brought in further collaboration partners. In addition to the work of biologists and imaging experts Stephan Handschuh and Thomas Schwaha (who write in detail about their perspective on our collaborations in this book), and my computer animation work, Glenn Bristol came on board to program the virtual reality version and to take some photographs of the art installations. Markus Liszt constructed the physical platform and exhibition setups in Europe, while a team of volunteers from the Paseo Project Festival was responsible for the physical US version. They were joined by Paul Geluso, who did the sound editing. Furthermore, biologist Olivia Osborne was conceptually involved at the beginning of the project with discussions and experiments, and Robertina Šebjanič with sound recordings in the first sound versions. The project was supported by our supervisors Ruth Schnell and Gerald Bast, the University of Applied Arts Vienna, UCLA Arts, the Ars Electronica, the Paseo Project Festival, and the California NanoSystems Institute.

Noise Aquarium continues to be shown internationally; you will find recent presentations on the project's website.[31] Some of the highlights of past exhibitions and installations of the project include, to name but a few, showings at SIGGRAPH, the Ars Electronica (an image of the very first interactive showings can be seen in_FIG. 6), the Art|Sci Center UCLA (US), NEXUS Screens NTU (Singapore), QUT Brisbane (Australia), the Paseo Project Festival Taos, New Mexico (US), the Biennale for Change Vienna (Austria), Klang-Moor-Schopfe (Switzerland), aMORE festival, Pula (Croatia), the Centrum Laznia Gdansk (Poland), the Pratt Manhattan Gallery, New York (US), London's Barbican Centre (UK), and Musée de la civilisation, Québec (Canada). The project is part of a traveling group exhibit with the title *Our Time on Earth*, for which it has been newly adapted—the wooden balance board platform and virtual reality have been replaced, and the interaction has been simplified to meet the requirements of the traveling art show. All my computer animations are still being shown, there are large augmented reality displays of all seven plankton creatures, and there is a new information video about scientific backgrounds and the environmental issues.

In parallel with *Noise Aquarium*, the Science Visualization Lab has been continuously developing the *LIFE* project, which reflects the narrative style of director Alfred Vendl. For *LIFE*, new scans of *Brachiolaria*, *Chaetognatha*, *Tigriopus*, and a

FIG. 6_*First showing of the interactive version of* NOISE AQUARIUM *at the 2019 Ars Electronica*

Photo by Glenn Bristol

VIDEO_*This video documents several of the project's international exhibitions.*

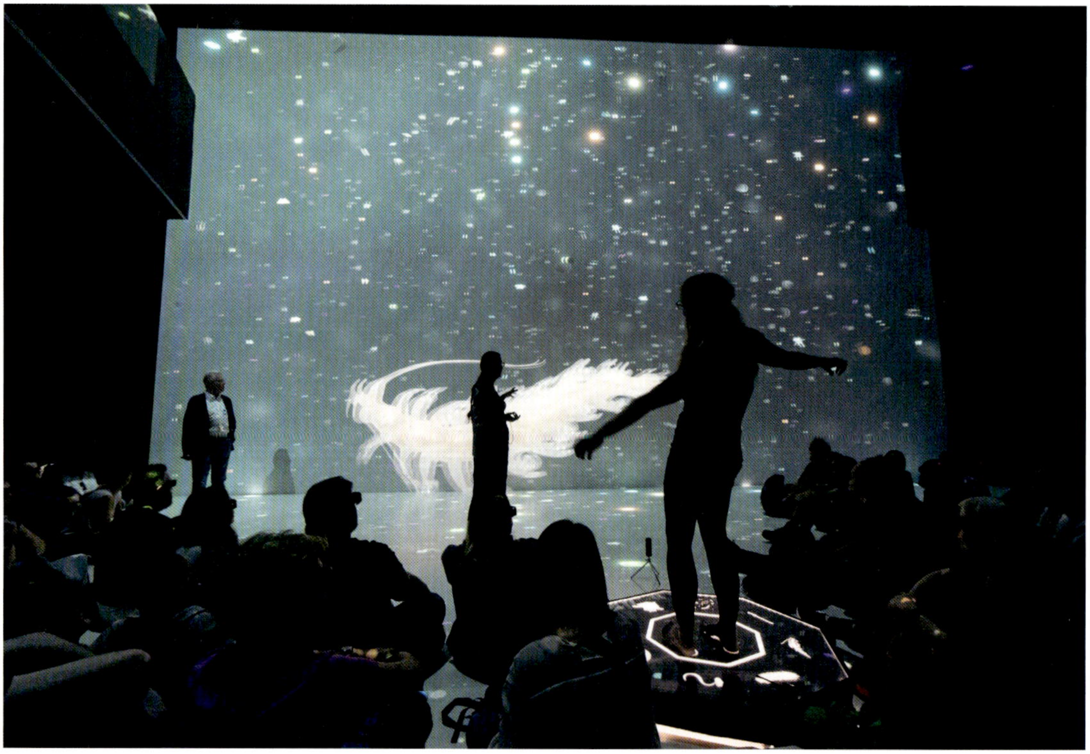

krill specimen were scanned and turned into computer-animated scientific visualizations to better show the diversity of shapes in plankton species. At the same time, the animated short documentary introduces microscopic plants like algae and diatoms, and presents plankton more densely to draw attention to the importance of phytoplankton and to emphasize the density of life in even the smallest drop of water. The voice of narrator Stuart Freeman talks about the urgency of the issues presented in the water bodies of our world. This type of narration, sometimes also referred to as *voice of God*, explains the major topic. This form of imparting knowledge is often criticized for not allowing the recipient any scope to have their own thoughts about the images on screen. At the same time, the dominant speaker's voice often presents an opportunity to achieve knowledge transfer goals, especially when it accompanies the complicated facts portrayed in a visualization, as the recipients are used to this style of television documentary. Experienced television sound designer Hupert Weninger mixed the sound accordingly. *LIFE* was shown in the workshops of the European Cultural Center in Italy in October 2022 during the Venice Biennale in the program of the Center for Didactics of Art and Interdisciplinary Education_FIG. 7. The detailed, tomographically scanned 3D models, completely new presentation elements, and the various important sound designs make both *LIFE* and *Noise Aquarium* impressive experiences that will remain in demand among varying target groups. For the audience and the creative heads of the respective projects, the outcomes might be perceived very differently, but for my part, the work on all the animations was closely connected for both projects.

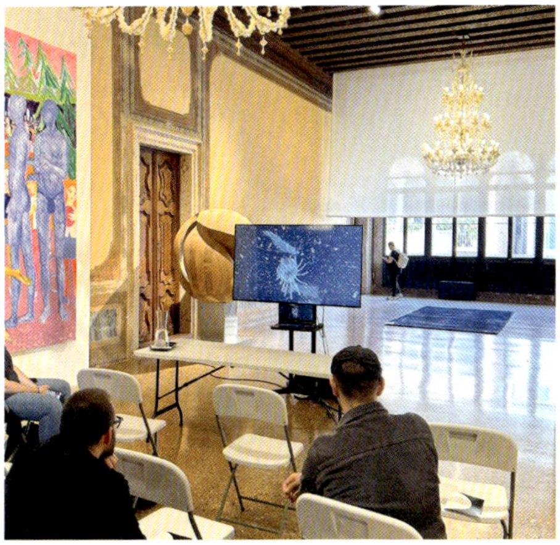

FIG. 7_*Presentation of* LIFE *at the Palazzo Michiel in Venice as part of the European Cultural Center program during the 2022 Venice Biennale*

Photo by Ruth Mateus-Berr

BUTTERFLY}PIERIS{EFFECT_The first initiative for the project *Butterfly}Pieris{Effect* was once again a request from an established director: Steve Nicholls is one of Alfred Vendl's long-standing collaborators and wanted special animations of scanned insects for his three-part documentary *Planet Insect*. After a few post-Brexit problems with specimen samples, the necessary data were obtained as part of our established collaboration with Stephan Handschuh and with the help of his students Christina Kaurin and Valentin Blüml. Three visual effects shots were created at the Science Visualization Lab and can be seen in the first two parts of the documentary series. They include an animation of the gyroscopic organ (haltere) responsible for the fly's fascinating flight skills—this shot was even published in the official trailer. Another shot shows a male moth's antenna, which it uses to track female pheromones over long distances. The third shot was a sequence showing the development of a butterfly as an embryo in the egg. For this purpose, butterfly eggs were scanned at three stages of development and the embryo inside them animated.

After the work on the documentary series had been completed, further scans of a newly hatched caterpillar and additional imaging followed for the interdisciplinary art project *Butterfly}Pieris{Effect*. This project was supported by an artist-in-residency grant from the University of Salzburg's Center for Human-Computer Interaction and nominated for the 2022 Nanyang Technological University Global Digital Art Prize.

The specific species in the project is the cabbage white butterfly (*Pieris brassicae*). Some farmers or gardeners might cry out, »A pest?! As a hero in a project about butterflies and insects?!« Nevertheless, the cabbage white was deliberately chosen because, on the one hand, it occurs frequently and everyone is familiar with it, but, on the other hand, it only becomes a real problem in man-made monocultures, and every insect is of potential importance in functioning ecosystems.[32]

Butterflies and especially caterpillars are important food sources for all sorts of insectivores such as birds. At the same time, most butterflies are highly specialized and depend on a certain habitat, which is why they are bioindicators in the monitoring of ecosystems. Bioindicators are organisms that are very sensitive to changes in their habitat and can therefore serve as a guide to environmental quality. These animals are being reduced in number and are under threat from construction, the intensification of agriculture, the abandonment of meadows or their conversion to monocultures, a decline in grazing, the clearing of landscapes, and the use of poisonous sprays and fertilizers.[33]

The butterfly effect was deliberately chosen because, although it is branded as a pest, this butterfly performs important functions, and we can never be sure that killing a single insect or destroying its habitat—even when it multiplies a lot and seems to be abundant in number—might not trigger major chain reactions without foreseeable outcomes. The shapes and diversity that exist in the insect kingdom are overwhelming. However, this biodiversity is endangered, and this, in turn, will very likely have serious consequences for humans.

We humans need our living environment to remain healthy because we are a part of that environment. We should be more aware of the feeling, living world and the part we play in it. We cannot completely negate our own mortality or the cycles of life of which we are a part, and we should keep ourselves edible and recyclable, i.e., we need to remain a valuable part of our ecosystem. This does not mean that I want to idealize our surroundings in anyway; rather, this project is intended to promote a more realistic worldview. Andreas Weber describes the role of humans on Earth as follows:

> *Nature is not ideal but fruitful. And fruitful, that is: it doesn't exclude any players, but turns everyone's contributions into a great common good. [...] It is becoming increasingly difficult for nature to submit to the two great currents of thought that have shaped our Occidental history: the romantic and the technocratic. This has an impact on the way we think about ourselves. Just as nature is not a great, nurturing mother, but not a dangerous vale of tears either that we must declare war against in order to technologically humanize it, human beings could also prove to be ambiguous by nature: they might not be lawless egoists or deeply egalitarian community architects. [...] Homo sapiens is the species that is rebuilding the planet. It is an ecosystem engineer, much like the beaver, which is responsible for entire wetlands with the biodiversity to match. [...] The exploitation of the land has always begun in order to supply the machinery of power. It is not the distorted world that has produced the distorted image, but rather the—violently—disrupted view that has disrupted the fabric of life. [...] Man refuses to die. More precisely: he refuses to be edible. That is his ecological peculiarity—and that is his ecological bestiality.*[34]

The biodiversity that ensures genetic diversity as well habitat diversity is an essential prerequisite for human life. Over fifty-five percent of the land in Austria is sealed and therefore useless for agriculture or biology. The habitats of flora and fauna are thus being increasingly restricted, reinforced by a persistent trend toward more intensive land use and monoculture, while flower meadows are being transformed into grassy deserts. Butterflies, as an enormously species-rich group of insects, play an important role in ecosystems. They inspire people with their breathtaking transformations from egg to caterpillar to pupa to adult animal.[35]

Not least due to my open criticism of and doubts about current lifestyles in industrialized nations, the topic of this project was of particular concern to me. The importance of insect habitats, especially in urban and suburban areas, cannot be overemphasized. For years, I have stumbled across articles that promote a more ecologically friendly approach in our immediate surroundings. For instance, there should be a campaign that teaches everybody that shortly clipped, manicured lawns are an ecological catastrophe. Manicured lawns support the reproduction of »pest« species, while a reduction in mowing frequency in green urban spaces benefits insect biodiversity.[36]

There are many native plant species that help insect populations to survive and that need almost no care or watering. With increasing droughts and the inability to irrigate using precious water, all we see are yellow-brown areas instead of the variety of forms provided by grass and herb blossom. Anyone who knows me privately knows that I support and pursue community-based agriculture, permaculture, composting, forest gardens, energy-saving measures, and the reduction of consumption as building blocks for a future worth living for all people. The project *Butterfly}Pieris{Effect* focuses on promoting this endeavor, once more opening the intriguing door to details of the *microworld*.

Butterflies are quite popular, but flies and moths often only enter our consciousness when we perceive them as disturbing. The diversity of wild insects is decisive for the overall diversity in the ecosystem. One example of this is the very modern and hyped use of honey beehives in cities. At first glance, it seems to be beneficial that more insects and useful honey bees are settling down in the urban space. But on closer inspection, these hives can actually cause important solitary bees and other pollinators in urban areas to starve because the bees can fly further and empty all pollen sources.[37]

Species-rich systems are self-regulating, while monocultures in combination with the threats being posed by climate change are factors that endanger insects. Insect biodiversity benefits from mitigating climate change, preserving natural habitats, and reducing the intensity of agriculture.[38]

The more than necessary offerings of hardy local plants serve as food sources as well as passageways for animals that cannot fly far enough. It is precisely these rest stops or corridors that enable insects to survive.

The art-science project *Butterfly}Pieris{Effect* aims to increase human awareness of the importance of other creatures in our ecosystem. Most people have a very human-centric worldview; however, it should not be forgotten that human beings cannot survive without the many other living beings, no matter how small or seemingly insignificant they may be. Using the example of the butterfly *Pieris brassicae*, the interdisciplinary project team aims to show that even creatures that are classified as »pests« are important in Earth's ecosystem.

This project's computer-animated scientific visualization videos can be shown as an art installation. In the installation, computer animations offer opposing perspectives. At one end of the spectrum of perception, the development of a

Pieris sp. caterpillar is enlarged and projected to emphasize the importance of insects and to show the details seen by humans. At the other end, there is the perspective of the caterpillar: visitors are invited to assume the caterpillar's perspective and feelings in an immersive VR video. Gradually, the caterpillar develops into a butterfly and flies away.

The human visitor is allowed to embody both the *Pieris brassicae* caterpillar and butterfly as well as the human watching the butterfly egg develop. All of the insect's developmental stages are computer-animated scientific visualizations, which means they consist of data that is scanned in cooperation with biologists and imaging experts, and then developed into animations using special workflows at the Science Visualization Lab. The art-sci installation tries to question the dichotomy of »our« (human) gaze and »their« (insect) point of view. The scientific visualizations reveal the importance of every species and present an opportunity to imagine what it is like to see through the eyes of an insect. With Donna Haraway in »Situated Knowledges: The Science Question in Feminism and the Privilege of Partial Perspective,«[39] we might say that taking a partial perspective can lead to a critical inquiry into our human position in Earth's ecosystem and the systems that are currently dominant.

FIG. 8_BUTTERFLY}PIERIS{EFFECT *art installation*

Photo by Martina R. Fröschl
© Science Visualization Lab,
University of Applied Arts Vienna, 2022

VIDEO_ *A sequence showing the development of a butterfly as an embryo — butterfly eggs were scanned at three stages.*

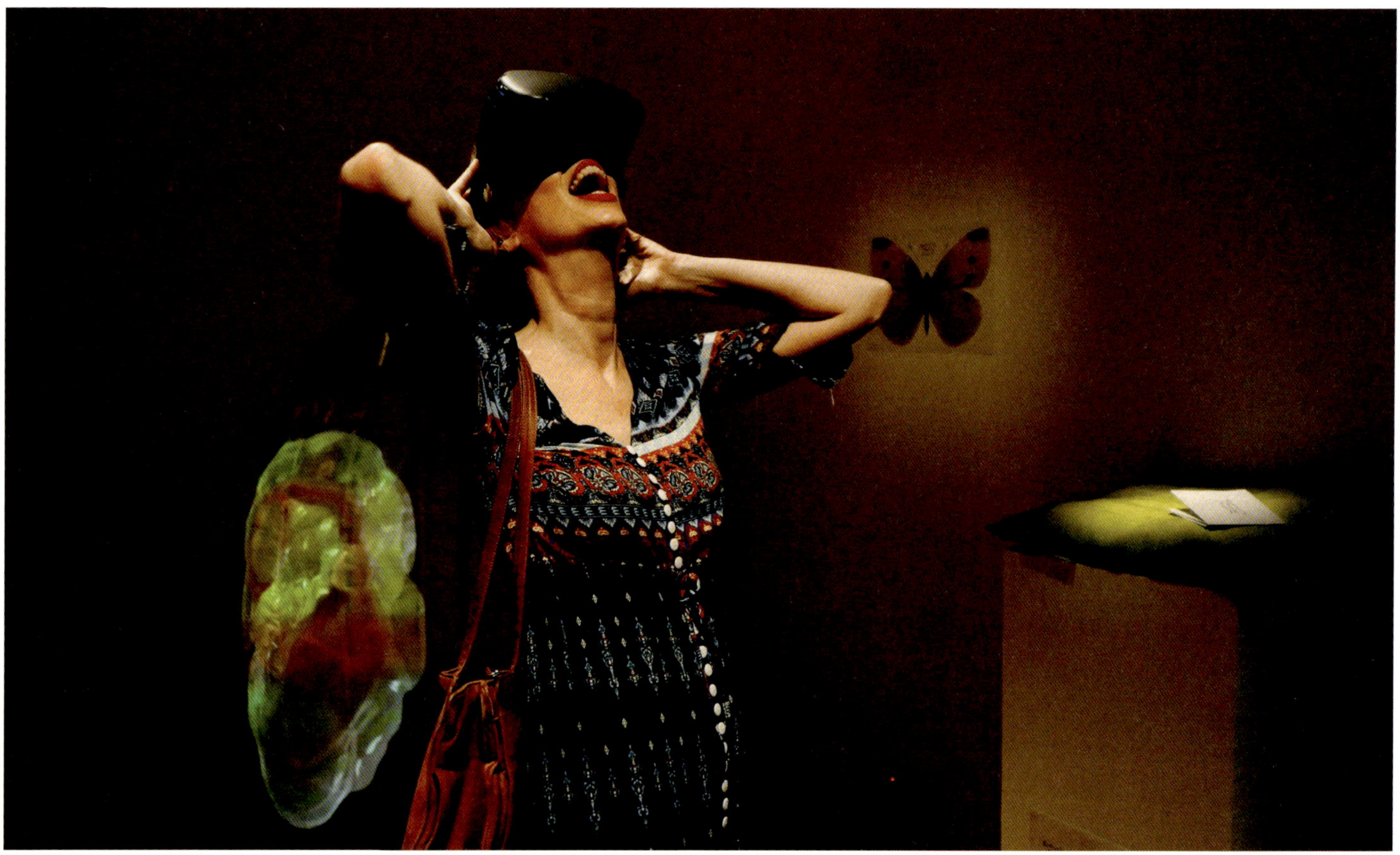

The art installation *Butterfly}Pieris{Effect* is an Austrian-British collaboration between Alfred Vendl and myself at the Science Visualization Lab, internationally renowned documentary film director Steve Nicholls, Thomas Schwaha of the Zoology Department at the University of Vienna, Stephan Handschuh from VetCore Imaging at the Veterinary University of Vienna, artist Michael Bachhofer, and butterfly expert Arthur Bürger. An image of the art installation can be seen in_FIG. 8 and the video. Commentaries by the butterfly expert were transformed into an informative soundscape via a gender-neutral artificial intelligence voice. In the video accompanying_FIG. 9, you will see the ultra-high-resolution photo used in *Butterfly}Pieris{Effect* with a slow zoom-in and hear an outtake of the soundscape. The soundscape plays a crucial role in bringing together the human and the insect perspectives presented in the project.

The project *Butterfly}Pieris{Effect* draws attention to the aforementioned importance of butterflies and caterpillars on different levels—because everyone can do something and should be informed accordingly. Cultivating wilder gardens in combination with support for regenerative agriculture can reduce CO_2 emissions and biodiversity loss. In the Science Visualization Lab's efforts to deal with the great challenges facing mankind, this project at the nexus of ecological message and high-tech visualization techniques fits well into the series of projects that have been carried out at the lab.

Future Prospects of the Science Visualization Lab

This chapter about the projects conducted at the Science Visualization Lab since 2016 summarizes one main thing: all of the lab's projects deal with topics that are important for the future of humankind.

Human knowledge is increasing rapidly and steadily; at the same time, the group of people who do not trust scientific results is growing, for example, in Austria.[40]

There is also a tendency toward alternative, easier-to-understand explanatory models, which might serve to mobilize people for questionable purposes or to lure money out of their pockets. There are therefore already warnings that if the number of people who do not trust science increases sharply, it will be a real threat to democratic society as a whole. Given clear threats such as climate change, our survival as a social society depends on a certain trust in scientific content.[41]

However, it is important that we do not assume that science alone will solve all our problems. Its results also have to be translated into action. Visualization can help inspire people who are already interested in science to get more involved in disseminating scientific findings. Children and young people should be educated to think scientifically and to be curious, and at the same time people who are not so familiar with scientific principles and working methods should be given an appealing, low-threshold introduction, so that more people come to feel that they, too, play an important role in the falsifiability of human knowledge. The *feeling of owning knowledge* is a phrase that I like to mention as one of the most important goals of transdisciplinary approaches.

Although technical advances and automation are making the computer animation workflows at the Science Visualization Lab easier each year, so far, we have

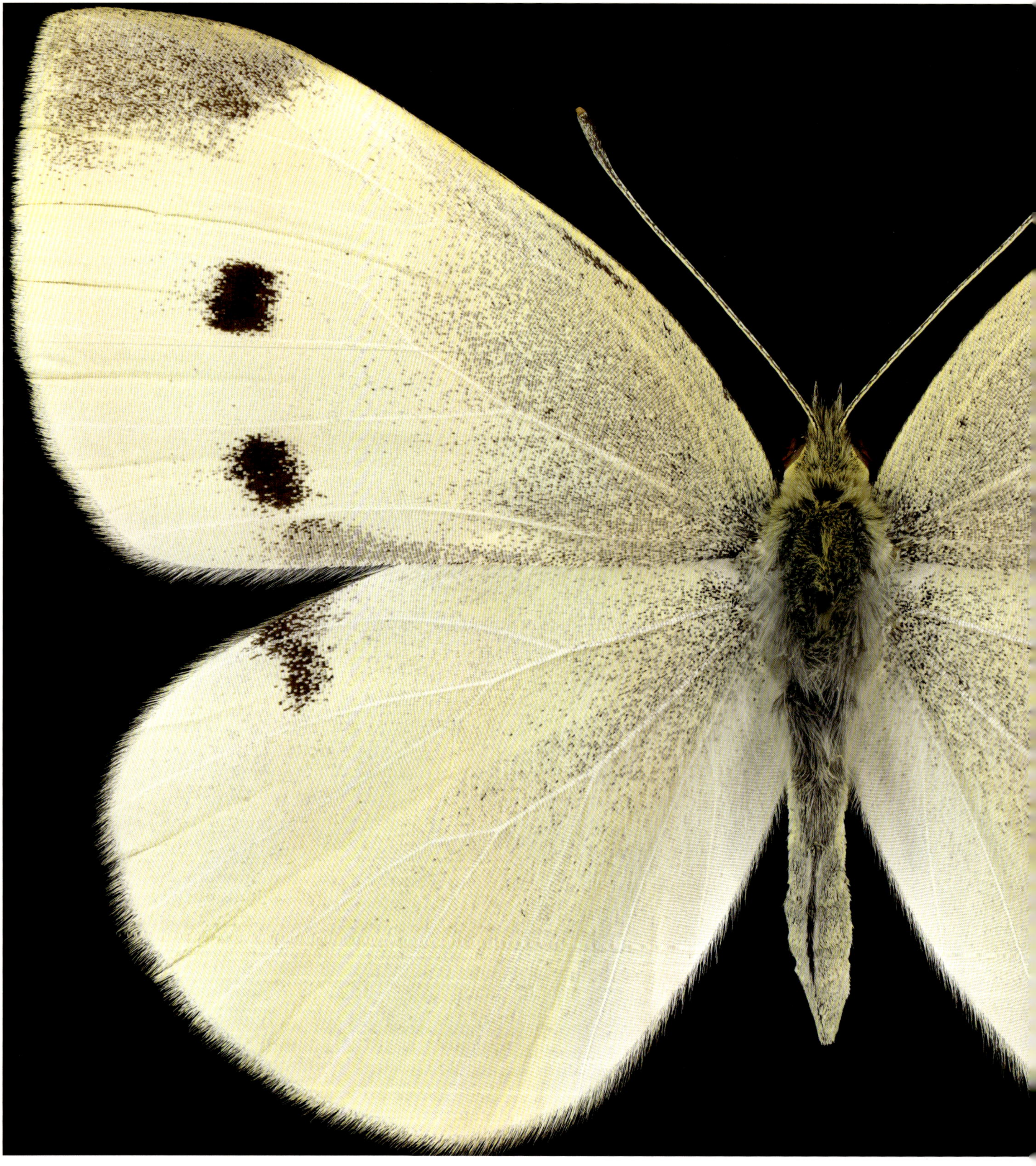

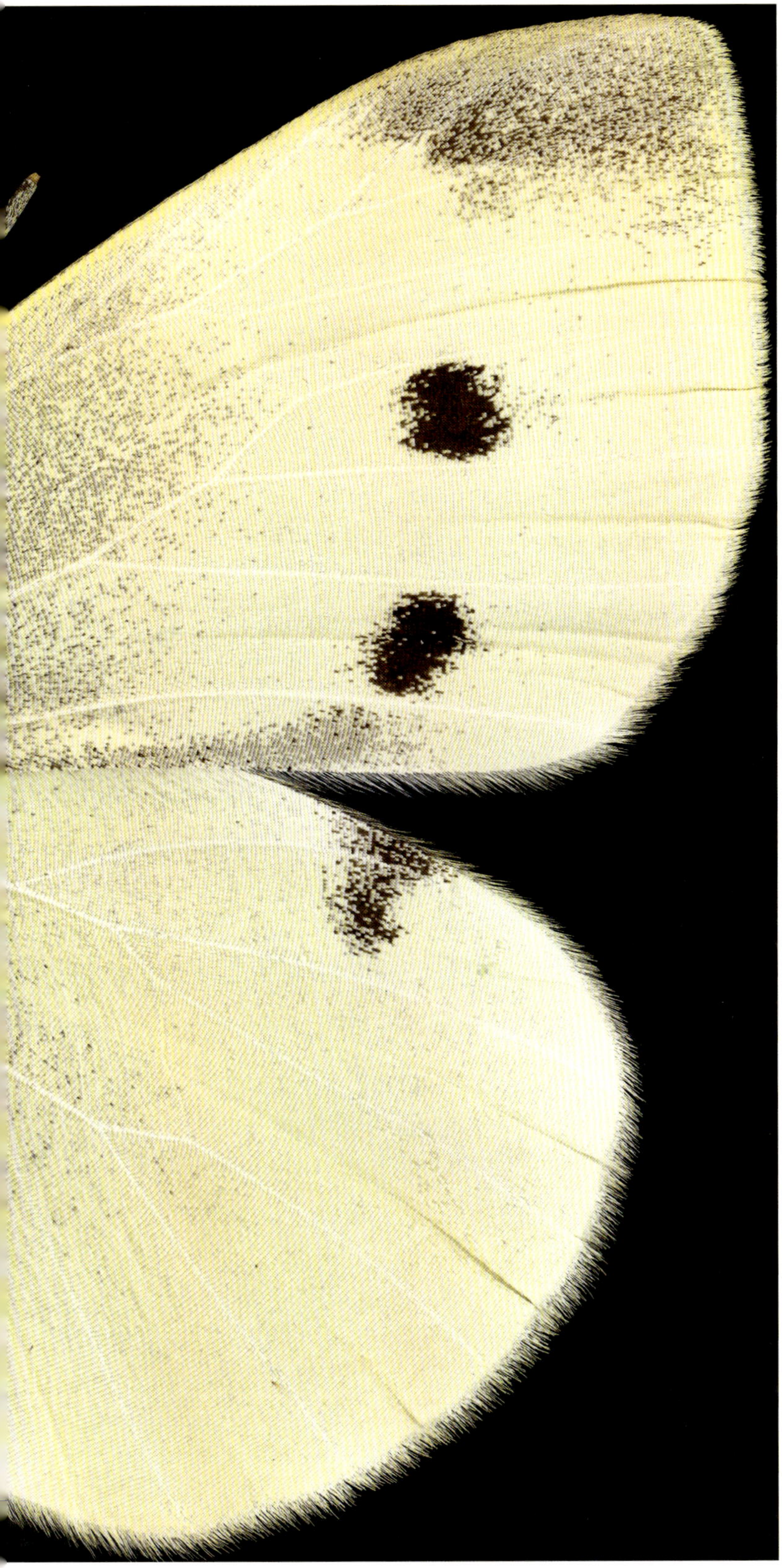

FIG. 9_*Ultra-high-resolution photo used in the project* BUTTERFLY}PIERIS{EFFECT

© Michael Bachhofer, 2022

VIDEO_*In the augmented reality video, you will see a slow zoom-in on the scales of the butterfly and hear an outtake from the soundscape of the art installation. The soundscape plays a crucial role in bringing together the human and the insect perspectives presented in the project.*

© Science Visualization Lab, University of Applied Arts Vienna, 2022 had to edit datasets manually. This manual »cleaning« offers an opportunity to dive deep into the mathematical structural representation of the scanned organic entity. Even though there might not be any need for time-consuming manual examinations of geometry in the near future, it will still be interesting to experience and explore the pure shapes of the scanned objects. There should be enough time and ease to investigate them. Exploring details, approaching them manually, directs the gaze toward the clouds of data and offers very different perspectives every time the viewport of the 3D software is rotated. All the details have to be examined, and the process of moving out of and re-entering the 3D geometry, and of rotating and zooming in on and out of the standard viewport settings or a digital camera perspective, leads to views of the organic form that can be experienced in exceptional ways. I think that there are many opportunities to be had in taking time to explore such datasets, and students in various workshops that I have taught have usually been fascinated by these structures and their inspiring potential.

Just as students and practitioners of the digital arts should take time and learn to see and understand perspective by attending drawing classes, they should also have the time to investigate and document the outcomes of exploring digital datasets. The approaches of the Science Visualization Lab make use of various methods for dealing with the realm of the microscopic and, in the process, for investigating and formulating questions about the microscopic world that surrounds us and all its forms.

When I scroll through the database of scanned organisms and the hard drives with the data from the last few years, and see the many testimonies of attempts and failed attempts, project presentations, lectures, art projects, simulations, technical developments, renderings in different styles, and so on, I get very excited and become aware of how much has happened in the lab in recent years. There is a folder in which I have collected the most visually interesting rendering errors and graphics card glitches. Browsing through these images is a journey through datasets and the projects that have been implemented with them. Our often experimental approach to potential new representations, be they visual, haptic, or interactive, becomes even more visible. Usually these are tests or even »incorrect« renderings from the production process, a single image rendering, or the rendering of an animation sequence. These »happened pieces« are often documents of the workflow and technical processes, but, in my opinion, they are frequently visually interesting. A selection of these happened pieces can be seen in _FIG. 10.

When viewers have insight into the creative development processes involved in computer-animated scientific visualization, it can help them to understand a project. The development processes provide information about the underlying data and thus enable viewers to make mental connections with the scanned entities. In addition, the many creative considerations, test paths, experiments, etc., that lead to the finished production can be presented as examples of the creative process. In digital production, design and creation processes are often less visible than in a final analog artistic output; making them visible can therefore generate more viewer appreciation. In such snapshots of the process of creation, glimpses of the technical medium become just as visible as the thought processes that have ultimately manifested the work.

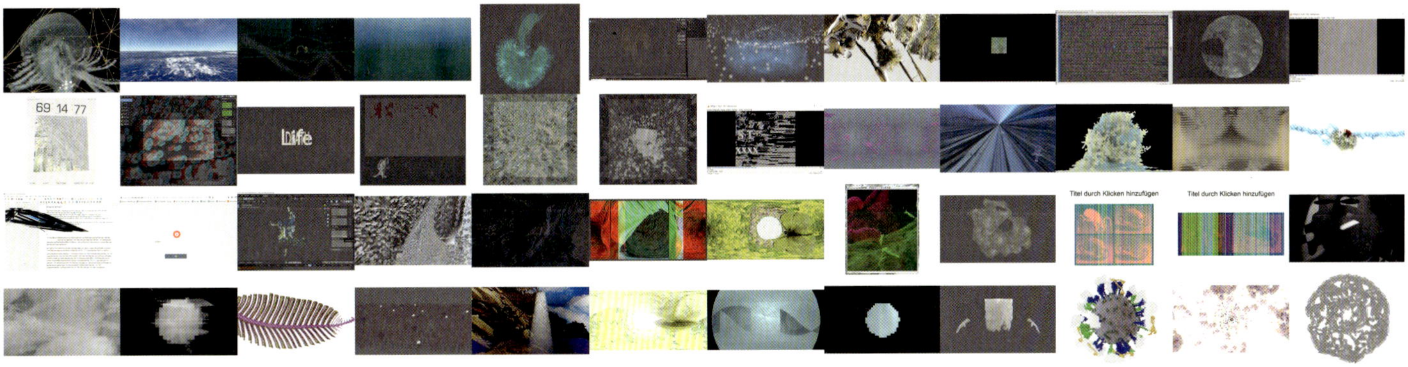

FIG. 10_*Work in progress images: »happened pieces« that show the stages of different computer animation projects from my work at the Science Visualization Lab*

VIDEO_*Animation of work in progress images: »happened pieces«*

Another method of making processes more visible is by means of self-referentiality. I believe that more time for experimentation leads to more creative output, to more innovative solutions and creativity. Still, that does not mean that the output is »better,« however we might measure that, but it is more creative, and the path to output is ideally included in the presentation and documented—that means that we might be able to see the *magic behind* the visual window (or door) of the presentation medium.

The Science Visualization Lab mainly uses scientific datasets that it often commissions exclusively for its projects. A combination of 3D scanning processes—such as micro-CT recording and 2D imaging like light microscopy and SEM—result in a multitude of possibilities and a richness of detail in models that are otherwise rarely created in this way. For example, the volume data of the micro-CT scans can also be used to image the inside of a real living being's organs, and this geometric data can be enriched with movement data from microscopy and/or videos, and textures from electron microscopy images. Such workflows make it possible to work with original scientific data, and this is an added value that is rarely possible, desired, or financed in transdisciplinary media formats such as television documentaries. Nevertheless, once commissioned, these subjectified references in the image are an asset that is often highly valued, and most viewers of our computer-animated scientific visualizations are pleasantly surprised when they hear which datasets are behind our animations. In most of their reactions, it is clear that this input of scientific imaging makes the animations considerably more valuable as sources of information.

The use of visual effects and computer animation in documentary film has been viewed critically; however, this attitude must be reconsidered, at the very least since the emergence of deep fakes that are difficult to detect. There is a tendency to blur the boundaries between the genres of animation film and live-action film as digital postproduction and all kinds of effects enable live-action films to be read as particular cases of animation.[42]

It is precisely because of the arguments made in the last two paragraphs that it is important that the work of the Science Visualization Lab will shift even more toward contemporary discourse in the future: the cultural, embodied, post-digital approach that has manifested itself more and more since I was hired by the lab in

2016 makes data visualization possible that expresses itself in a self-referential, socially critical, systemically critical, feminist way… and continuously finds new forms of expression. It is always a challenge to convey knowledge in an understandable way and, at the same time, to not oversimplify it.

We have started to facilitate experiences with scientific data and are to some extent moving away from unquestioned »fact«-imposing documentary visualizations, because the presentation of scientific output is particularly at risk of taking an objectifying attitude, as Donna Haraway already warned in 1988 in her seminal essay »Situated Knowledges«:

> *The eyes have been used to signify a perverse capacity — honed to perfection in the history of science tied to militarism, capitalism, colonialism, and male supremacy — to distance the knowing subject from everybody and everything in the interests of unfettered power. The instruments of visualization in multinationalist, postmodernist culture have compounded these meanings of disembodiment.*[43]

To venture an outlook: applied experiments entailing consequences like the ones described in Haraway's theory might flow even more into our future work. This has already been attempted to some extent in the project presentations *Noise Aquarium*, *CRISPR/Cas9-NHEJ: Action in the Nucleus*, *Virus Dice*, and *Butterfy}Pieris{Effect*. However, basic steps for localizing content and opening up the possibility for visitors to personally understand the visualizations should be emphasized even more. Thinking globally through local personal experiences can lead to more empathy and understanding. Objectivity, knowledge derived from subjectivity, experiences, and locations ought to play primary roles.

Future expansions to science visualization could incorporate even more hints at the complexity and the ambiguity of data, and could visually experiment with them. Topics should be approached and presented from different perspectives. This will require true interdisciplinary work with biologists and artists, and could restore confidence in science for people living in the area and reinforce the verifiability and comprehensibility of what is being shown.

Many disciplines and institutions deal with the expansion of human states of perception, which means that topics and possibilities for collaboration are probably almost unlimited. The invitation to work with us at the Science Visualization Lab is open to all disciplines in the arts and sciences — I can only repeat the call from my dissertation in the context of this book!

We cannot predict the future, but there are visionary ideas, often formulated long before their time. I hope that our present and future projects will influence the common good of humanity in a positive way. Consciously resisting unnecessary classifications and playing with expectations is a freedom that should be preserved to stimulate and nurture the opening of hidden visual doors in the sphere of our knowledge and environment.

1 Martina R. Fröschl, *Computer-Animated Scientific Visualizations of Tomographic Scanned Microscopic Organic Entities*, May 2019, https://fedora.phaidra.bibliothek.uni-ak.ac.at/fedora/objects/o:34803/methods/bdef:Content/get.

2 »The Incredible Water Bear,« January 27, 2014, *YouTube*, https://www.youtube.com/watch?v=cp1WwNE6Lms.

3 Rudolf Arnheim, *Visual Thinking* (Berkeley: University of California Press, 1969).

4 Fröschl, *Computer-Animated Scientific Visualizations of Tomographic Scanned Microscopic Organic Entities*.

5 Bob Root-Bernstein, Todd Siler, Adam Brown, and Kenneth Snelson, »ArtScience: Integrative Collaboration to Create a Sustainable Future,« *Leonardo* 44, no. 3 (2011): 192, https://doi.org/10.1162/LEON_e_00161.

6 Kalina Borkiewicz, A. J. Christensen, Ryan Wyatt, and Ernest T. Wright, »Introduction to Cinematic Scientific Visualization,« in *ACM SIGGRAPH 2020 Courses* (2020): 1–267, https://doi.org/10.1145/3388769.3407502.

7 Thomas A. DeFanti, Maxine D. Brown, and Bruce H. McCormick, »Visualization: Expanding Scientific and Engineering Research Opportunities,« *Computer* 22, no. 8 (1989): 12–16, https://doi.org/10.1109/2.35195.

8 Leonard Shlain, *Art & Physics—Parallel Visions in Space, Time & Light* (New York: Quill William Morrow, 1991).

9 Ariane Koek, »CERN: Where Art and Science Collide,« *The Art Newspaper*, October 4, 2011.

10 Fröschl, *Computer-Animated Scientific Visualizations of Tomographic Scanned Microscopic Organic Entities*.

11 *Worldwide Protein Data Bank*, accessed December 6, 2022, https://www.wwpdb.org.

12 Heidi Ledford and Ewen Callaway, »Pioneers of CRISPR Gene Editing Win Chemistry Nobel,« *Nature* 586, no. 7829 (2020): 346–7, https://www.nature.com/articles/d41586-020-02765-9.

13 Martin Jinek, Krzysztof Chylinski, Ines Fonfara, Michael Hauer, Jennifer A. Doudna, and Emmanuelle Charpentier, »A Programmable Dual-RNA-guided DNA Endonuclease in Adaptive Bacterial Immunity,« *Science* 337, no. 6096 (2012): 816–21, https://doi.org/10.1126/science.1225829.

14 Nimrat Chatterjee and Graham C. Walker, »Mechanisms of DNA Damage, Repair, and Mutagenesis,« *Environmental and Molecular Mutagenesis* 58, no. 5 (2017): 235–63.

15 Martina R. Fröschl and Alfred Vendl, »Crispr/Cas9-NHEJ: Action in the Nucleus,« in *ACM SIGGRAPH 2018 Posters* (2018): 1–2, https://doi.org/10.1145/3230744.3230747.

16 »4PYP,« *RCSB Protein Data Bank (RCSB PDB)*, accessed October 5, 2022, https://www.rcsb.org/structure/4PYP.

17 Christian R. Noe, Marion Noe-Letschnig, Patricia Handschuh, Chiara A. Noe, and Rupert Lanzenberger, »The ›Sugar Dilemma,‹« *Pharmazie* 75, no. 10 (2020): 456–62, https://pubmed.ncbi.nlm.nih.gov/33305717.

18 Christian R. Noe, Marion Noe-Letschnig, Patricia Handschuh, Chiara A. Noe, and Rupert Lanzenberger, »Dysfunction of the Blood-Brain Barrier—A Key Step in Neurodegeneration and Dementia,« *Frontiers in Aging Neuroscience* 12 (2020), https://doi.org/10.3389/fnagi.2020.00185.

19 Martina R. Fröschl, »Blood-Brain Barrier,« *Visualizing Biological Data*, accessed May 4, 2022, https://vizbi.org/Posters/2021/vC06.

20 »Movement, Mapped,« *SciArt Initiative*, accessed February 2, 2023, https://www.sciartmagazine.com/movement-mapped.html.

21 https://nanographics.at.

22 Ngan Nguyen, Ondřej Strnad, Tobias Klein, Deng Luo, Ruwayda Alharbi, Peter Wonka, Martina Maritan, et al., »Modeling in the Time of COVID-19: Statistical and Rule-based Mesoscale Models,« *IEEE Transactions on Visualization and Computer Graphics* 27, no. 2 (2021): 722–32, https://doi.org/10.1109/TVCG.2020.3030415.

23 Judith Ellenbürger, Erwin Feyersinger, Martina R. Fröschl, Björn Hochschild, Katrin von Kapherr, Sebastian R. Richter, Maike S. Reinerth, and Janina Wildfeuer, »OBSERVE! An Inanimate Virus (Animated),« in *Das Virus im Netz medialer Diskurse*, ed. Angela Krewani and Peter Zimmermann (Wiesbaden: Springer Vieweg, 2022), 267–83, https://doi.org/10.1007/978-3-658-36312-3_16.

24 https://pixelvienna.com/

25 Ann-Sophie Lehmann, »In der Ratte: Der Körper als immersiver Ort in 3D-Computeranimationsfilmen,« *montage AV: Zeitschrift für Theorie und Geschichte audiovisueller Kommunikation* 17, no. 2 (2008): 121–43, https://doi.org/10.25969/mediarep/303.

26 Martina R. Fröschl, »Computer-Animated Scientific Visualizations in the Immersive Art Installation NOISE AQUARIUM,« in *2020 IEEE Conference on Virtual Reality and 3D User Interfaces Abstracts and Workshops* (VRW) (Atlanta, GA: IEEE, 2020), 183–7, https://doi.org/10.1109/VRW50115.2020.00040; Brittany Ransom, »Proliferating Possibilities: Speculative Futures in Art and Design Art Gallery: Introduction,« *Leonardo* 52, no. 4 (2019): 396–9, https://doi.org/10.1162/leon_a_01782.

27 Fröschl, *Computer-Animated Scientific Visualizations of Tomographic Scanned Microscopic Organic Entities*.

28 Martina R. Fröschl and Alfred Vendl, »Computer-Animated Fluidity for Stiff Datasets and the Visualization of Underwater Noise,« in *Plastic Ocean: Art and Science Responses to Marine Pollution*, ed. Ingeborg Reichle (Berlin: De Gruyter, 2021), 179–96, https://doi.org/10.1515/9783110744774-011.

29 Victoria Vesna, Bruce Campbell, and Francesca Samsel, »Victoria Vesna: Inviting Meaningful Organic Art–Science Collaboration,« *IEEE Computer Graphics and Applications* 39, no. 4 (2019): 8–13, https://doi.org/10.1109/MCG.2019.2916962.

30 See, e.g., Anna Nacher, Victoria Vesna, »Diving Deep into the Blue Planet, Flying High into the Cosmos,« in *Towards a Non-Anthropocentric Ecology: VICTORIA VESNA and Art in the World of Anthropocene*, ed. Ryszard Kluszczynski (Lodz: Lodz University Press, 2020), 172–237; Victoria Vesna, »NOISE AQUARIUM: Iterations, Variations, and Responsive Ecotistical Work,« in *Plastic Ocean: Art and Science Responses to Marine Pollution* (Berlin: De Gruyter, 2021), 157–78, https://doi.org/10.1515/9783110744774-010.

31 http://www.noiseaquarium.com

32 Luule Metspalu, Külli Hiiesaar, and Katrin Jõgar, »Plants Influencing the Behaviour of Large White Butterfly (*Pieris brassicae* L.),« *Agronomy Research* 1, no. 2 (2003): 211–20, https://agronomy.emu.ee/vol012/Metspalu.pdf.

33 Peter Huemer, *»Ausgeflattert«: Der stille Tod der österreichischen Schmetterlinge* (2016), https://bienenbrotbrief.de/wordpress/wp-content/uploads/2016/07/Schmetterlingsreport_2016.pdf.

34 Andreas Weber, »Essbar sein,« trans. Martina R. Fröschl, *Oya: enkeltauglich leben* 51 (2018), https://lesen.oya-online.de/texte/3071-essbar-sein.html.

35 Huemer, *»Ausgeflattert.«*

36 Anja Proske, Sophie Lokatis, and Jens Rolff, »Impact of Mowing Frequency on Arthropod Abundance and Diversity in Urban Habitats: A Meta-analysis,« *Urban Forestry & Urban Greening* 76 (2022), https://doi.org/10.1016/j.ufug.2022.127714.

37 Marcus Fairs, »Putting Beehives in Cities is ›Very Dangerous‹ to Other Pollinators Says Bee Expert Paula Carnell,« *Dezeen*, February 8, 2022, https://www.dezeen.com/2022/02/08/beehives-cities-dangerous-paula-carnell.

38 Charlotte L. Outhwaite, Peter McCann, and Tim Newbold, »Agriculture and Climate Change are Reshaping Insect Biodiversity Worldwide,« *Nature* 605 (2022): 97–102, https://doi.org/10.1038/s41586-022-04644-x.

39 Donna J. Haraway, »Situated Knowledges: The Science Question in Feminism and the Privilege of Partial Perspective.« *Feminist Studies* 14, no. 3 (1988): 575–99. https://doi.org/10.2307/3178066.

40 Marc Seumenicht, »Vertrauensbildende Maßnahmen für die Wissenschaft,« *FWF*, September 1, 2022, https://www.fwf.ac.at/de/news-presse/news/nachricht/nid/20220901.

41 Amanda Montañez, »How Science Visualization Can Help Save the World,« *Scientific American*, December 20, 2016, https://blogs.scientificamerican.com/sa-visual/how-science-visualization-can-help-save-the-world/.

42 Chris Gehman and Steve Reinke, *The Sharpest Point: Animation at the End of Cinema* (Toronto: YYZ Books, 2005).

43 Haraway, »Situated Knowledges,« 581.

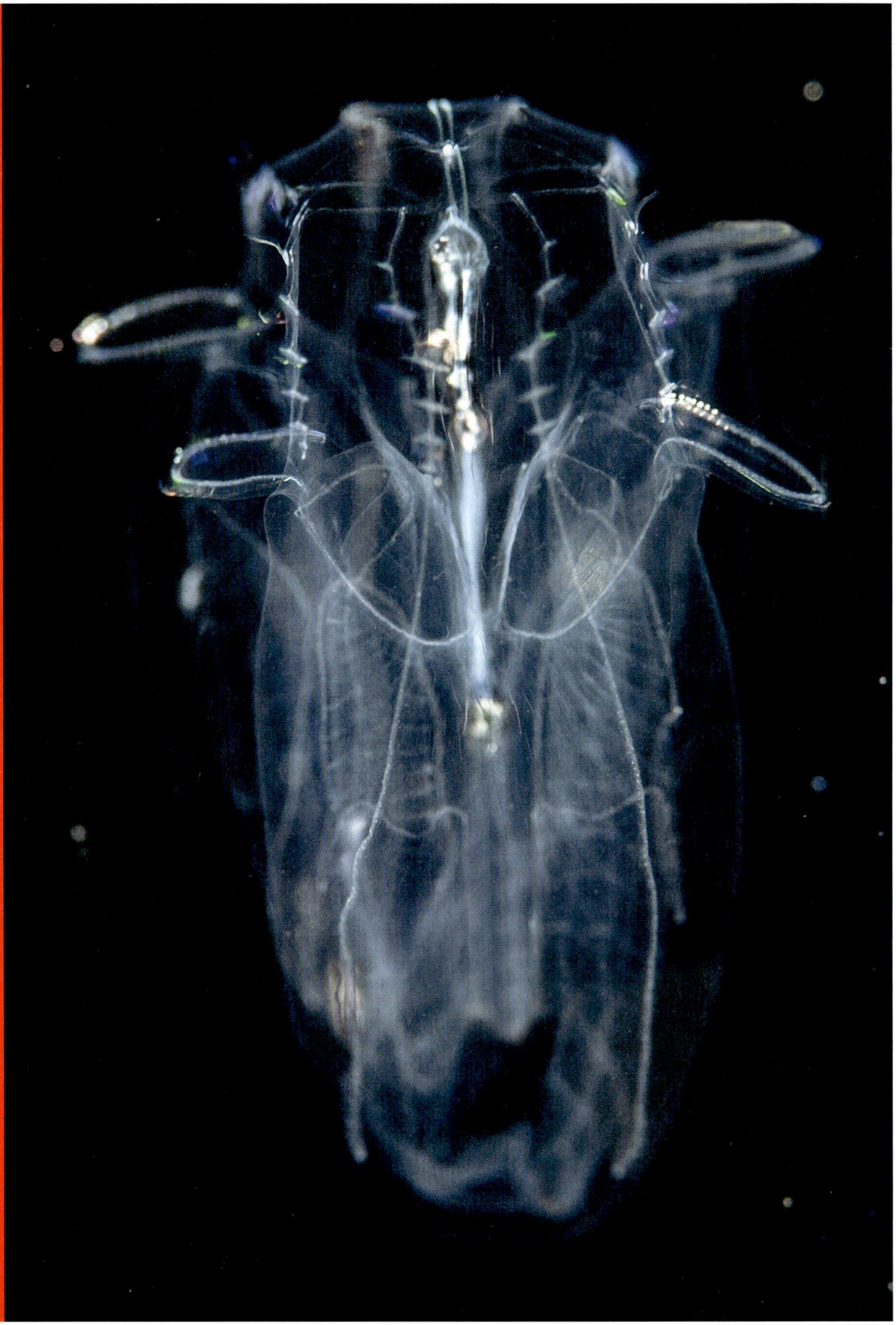

Galaxies from the Depths

Manfred Wakolbinger

Each time I dive into the open sea at night, I think about Jean Cocteau's film *Orpheus*, in which Jean Marais, playing Orpheus, climbs into a mirror to dive into the underworld to rescue Eurydice.

VIDEOS

Sound score by Christian Fennesz

A black water dive takes place at night on the open sea — that is, with no reef nearby and the ocean floor inaccessibly deep. You go on dives like this to see what creatures will come to the surface at night to graze on plankton. A few years ago, we were on a diving boat with friends off the coast of Sulawesi. It was after dinner, in the dead of night; there was no wind, the ocean was quiet, there were no currents, and the sea was 500 meters deep. We decided to see what was going on beneath this smooth, black surface. We sank down to about 15 meters. When you shine the diving lamp into the blackness, little plankton light up. After initial difficulties to float at the depth we wanted, everything calmed down and the lamp's spotlight bored an approx. two-meter-deep hole into the black; otherwise, you would be floating weightlessly like an astronaut in the dark nothingness.

All at once, a form lit up in the cone of light that I had never seen before. This shining thing floated toward me like a galaxy. It was, as I would later find out, an oceanic salpa that I was seeing and photographing for the first time. And thus began my quest for the salpas, or the salpas' quest for me.

For it is not an easy undertaking. It is difficult to plan. At night, the salpas rise to the surface of the ocean from a depth of about 800 meters to harvest plankton.

Salpas belong to the family of Chordata. Alongside the most highly developed organisms, like mammals (including humans), this includes a huge number of simple organisms like the tunicates. As the name Chorda already suggests, these animals have a spine. They also have a brain, comprising a few nerves, which grows back if that part of the salpa is bitten off. Salpas move by jet propulsion. At any rate, the images I took looked to me like the idea of galaxies in photo form. The images were intensely reminiscent of sci-fi. I should probably go back a bit to explain where my affinity for sci-fi comes from.

In October 1957 (I was five years old), an announcement was made on Austrian radio that, at 5 p.m., a man-made Sputnik — in English: fellow traveler, companion

— space probe would be crossing Austria. In my home town of Mitterkirchen, everybody went onto the street at that time and gazed at the sky to view this miracle of technology. Of course, nobody saw it, as the man-made star was much too small. But this experience left a strong impression on me as well as many questions. It was an engine in my thoughts with regard to art and sculpture, the topics of inside and outside, duality, etc.

My associations with the term »galaxy« are infinite distances and expanses. In sci-fi films, you do not just see storylines; these films also open up new spaces for thinking and, if everything goes well, you touch down with a new understanding of yourself. Our own abysses are often more frightening that the most infinite expanses.

This is how I went about making films from photos (over the years, I had created a handsome series). I drew storyboards that showed the viewer's gaze flying through these wide spaces and embarking on journeys in and through galaxies. Damjan Minovski then realized my ideas using my static images in his animated movies, creating films of about five minutes in length. Christian Fennesz made the sound score and gave me four pieces of five minutes in length to use.

Back to Sulawesi

In the depths around Sulawesi, I had my first encounter with salpas. In the high mountains of Sulawesi, which tower out of the sea, live the members of the Torajan tribe. According to their myths, which they still live by today, they come from outer space, and once their time on Earth has come to an end, that is where they will return to.

They build houses in the mountains that are reminiscent of ships, and they have space ships in mind when they build them. Holy men look at the constellations to calculate the date of their journey back to space, which is every thirty to forty years for one family. In the meantime, the bodies of the people who die are preserved, but there is always a doll wearing the deceased's clothing standing in for the person who has died, waiting in a rock chamber to journey to space.

When the time comes, it is time for tabula rasa. The family runs up great debts and buys as many buffalo as possible, which are then slaughtered in honor of the dead. And the lives of those left behind begin again in great debt that has to be worked off until the next trip back. It seems that this is embedded very deep within human nature: total destruction.

You might find out where this journey of the deceased Torajans leads in my films.

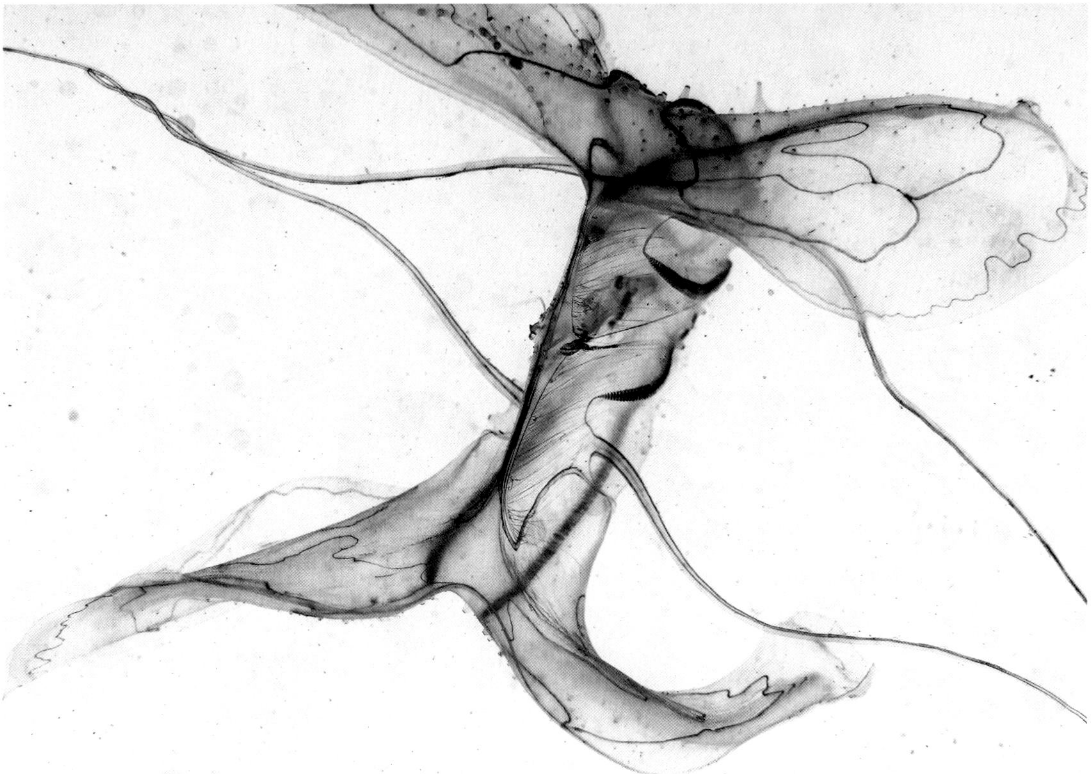

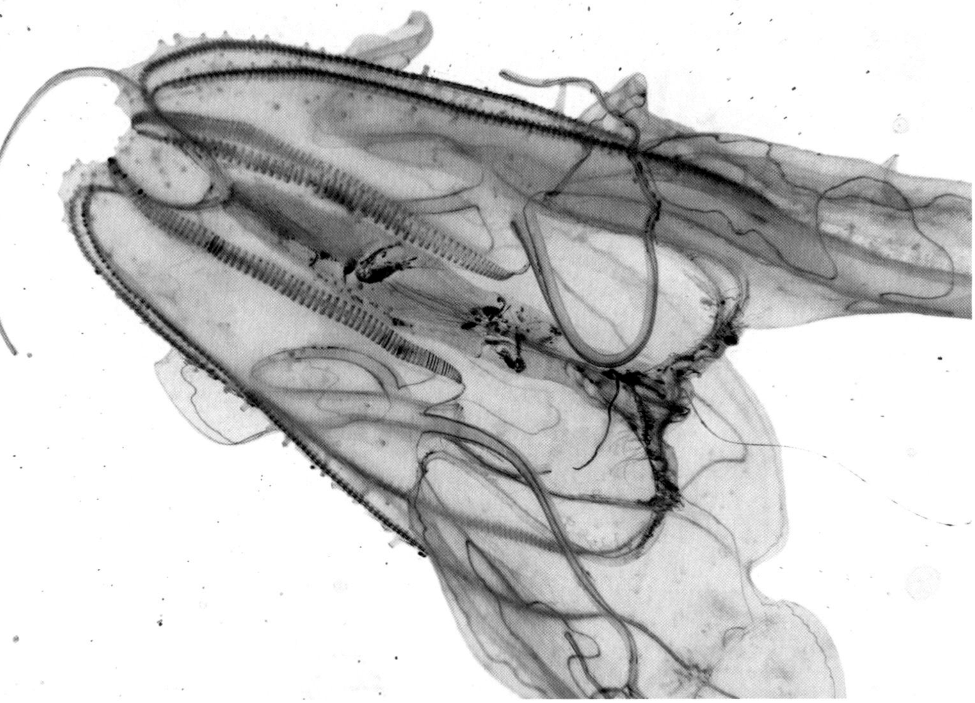

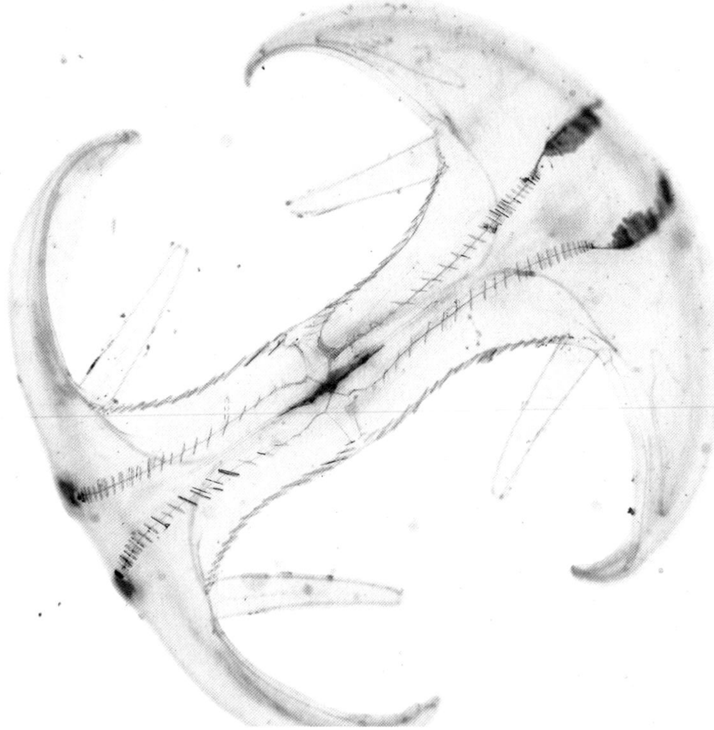

© Manfred Wakolbinger, 2022

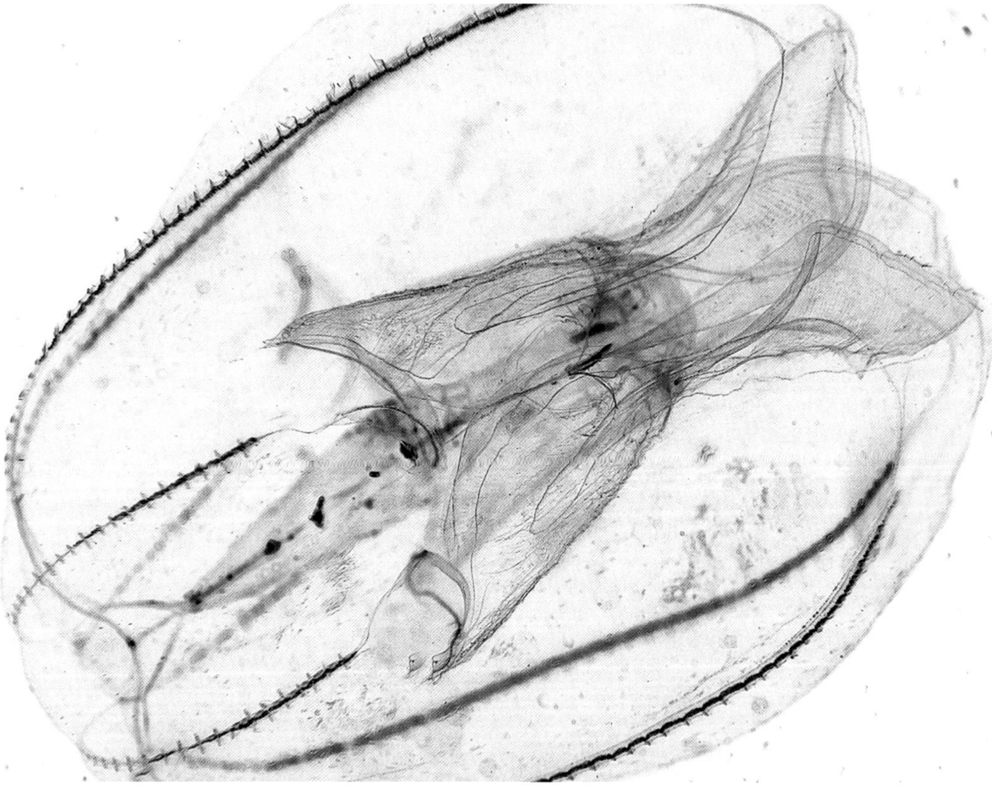

© Manfred Wakolbinger, 2022

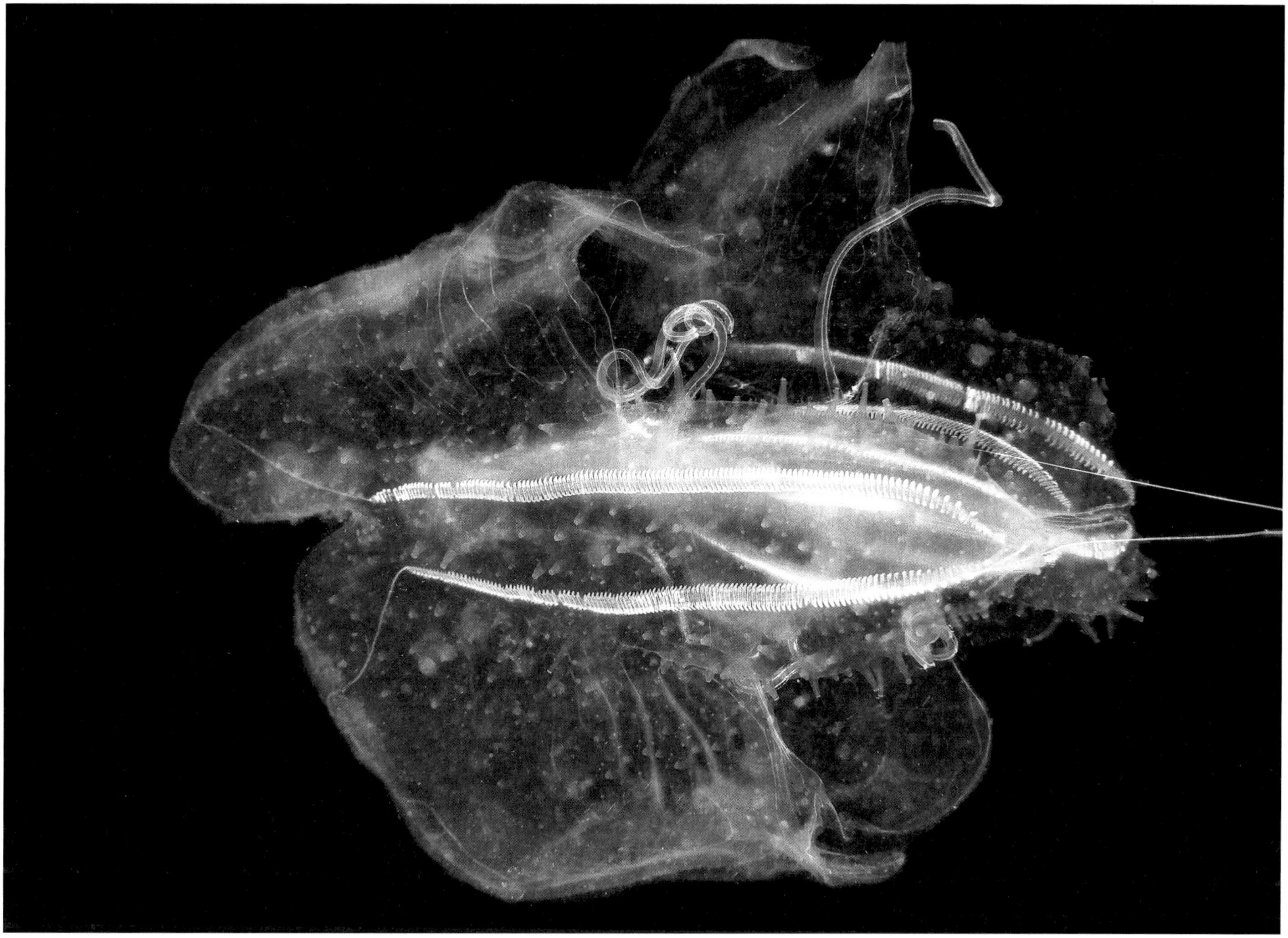

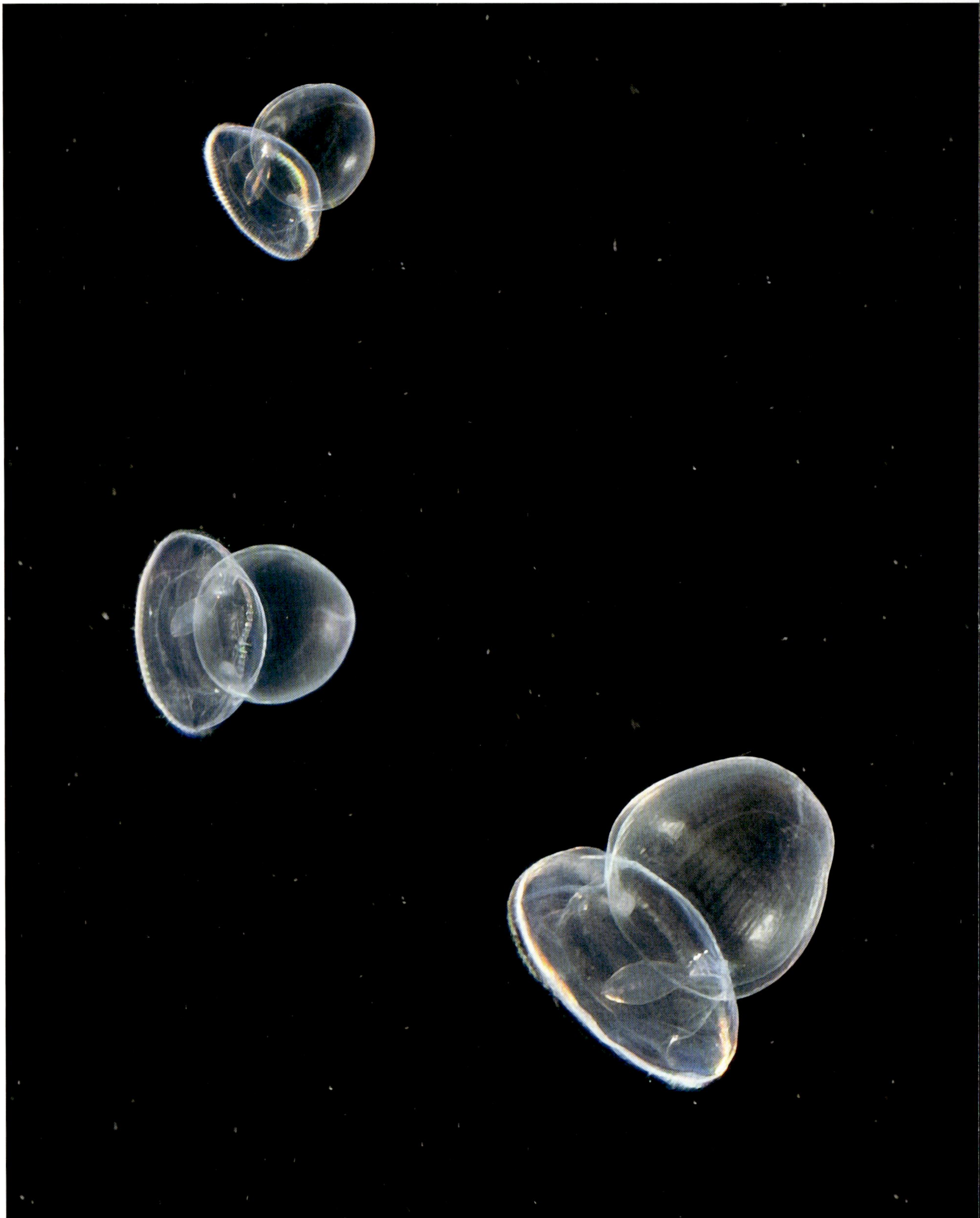

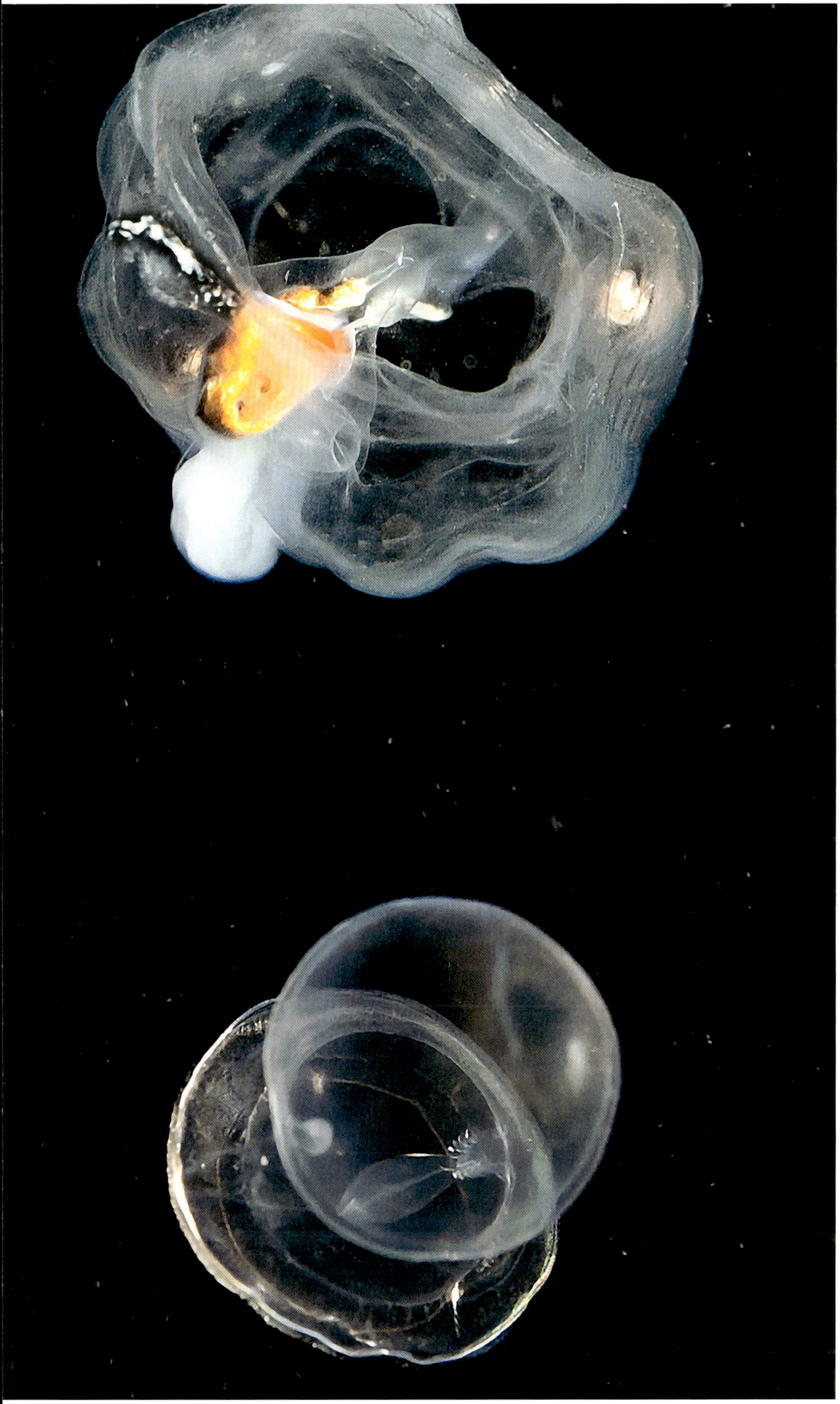

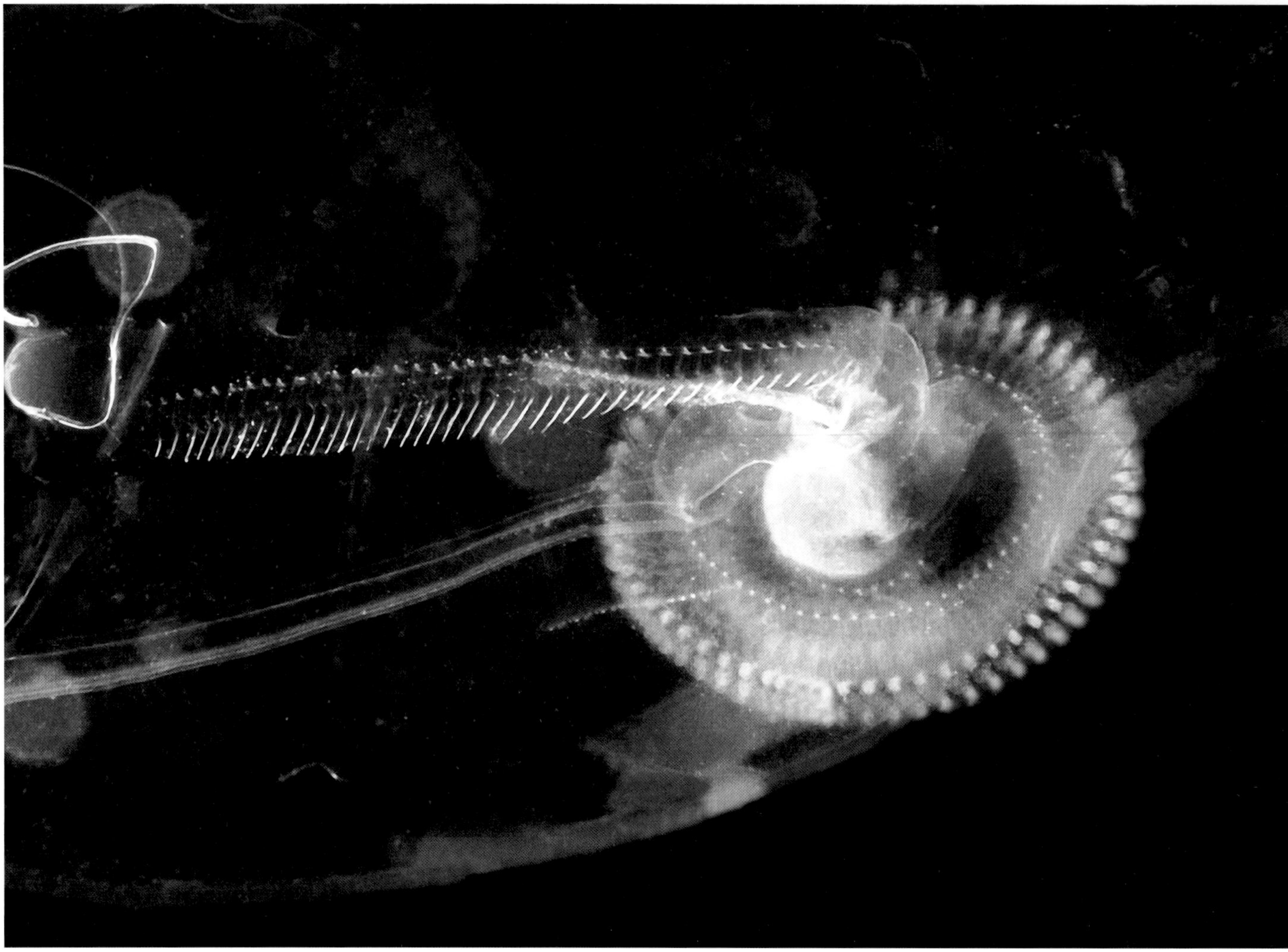

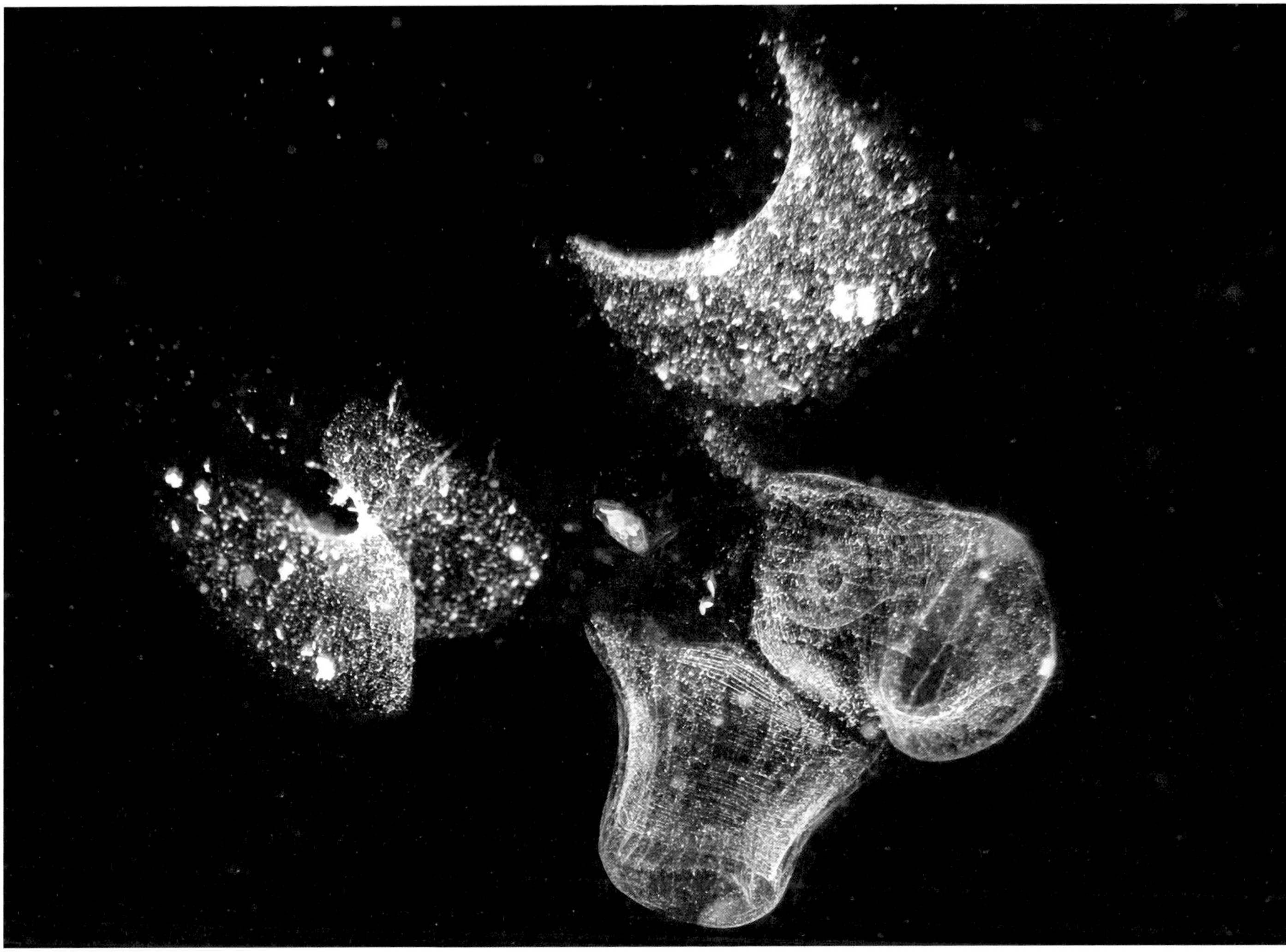

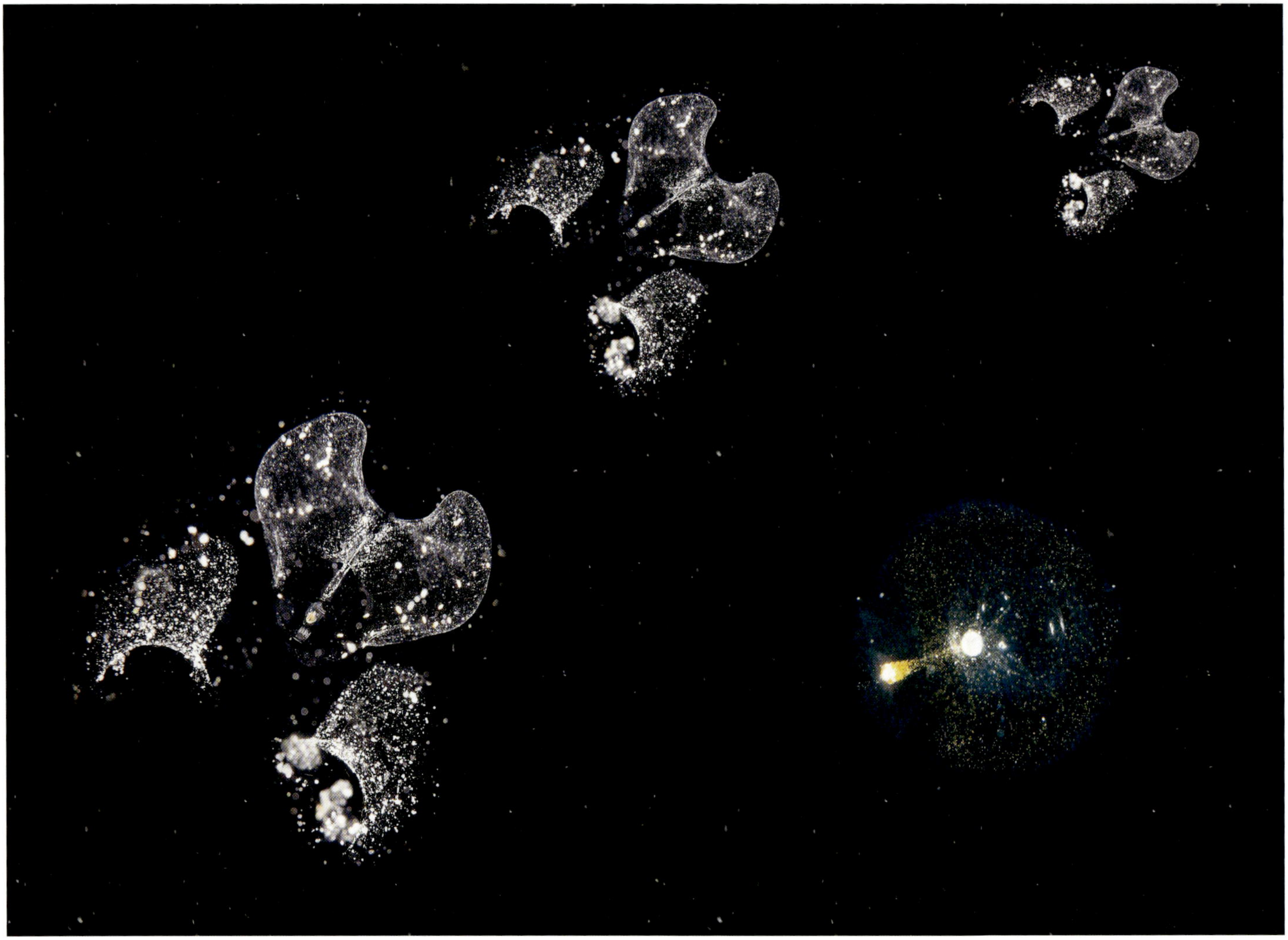

© Manfred Wakolbinger, 2022

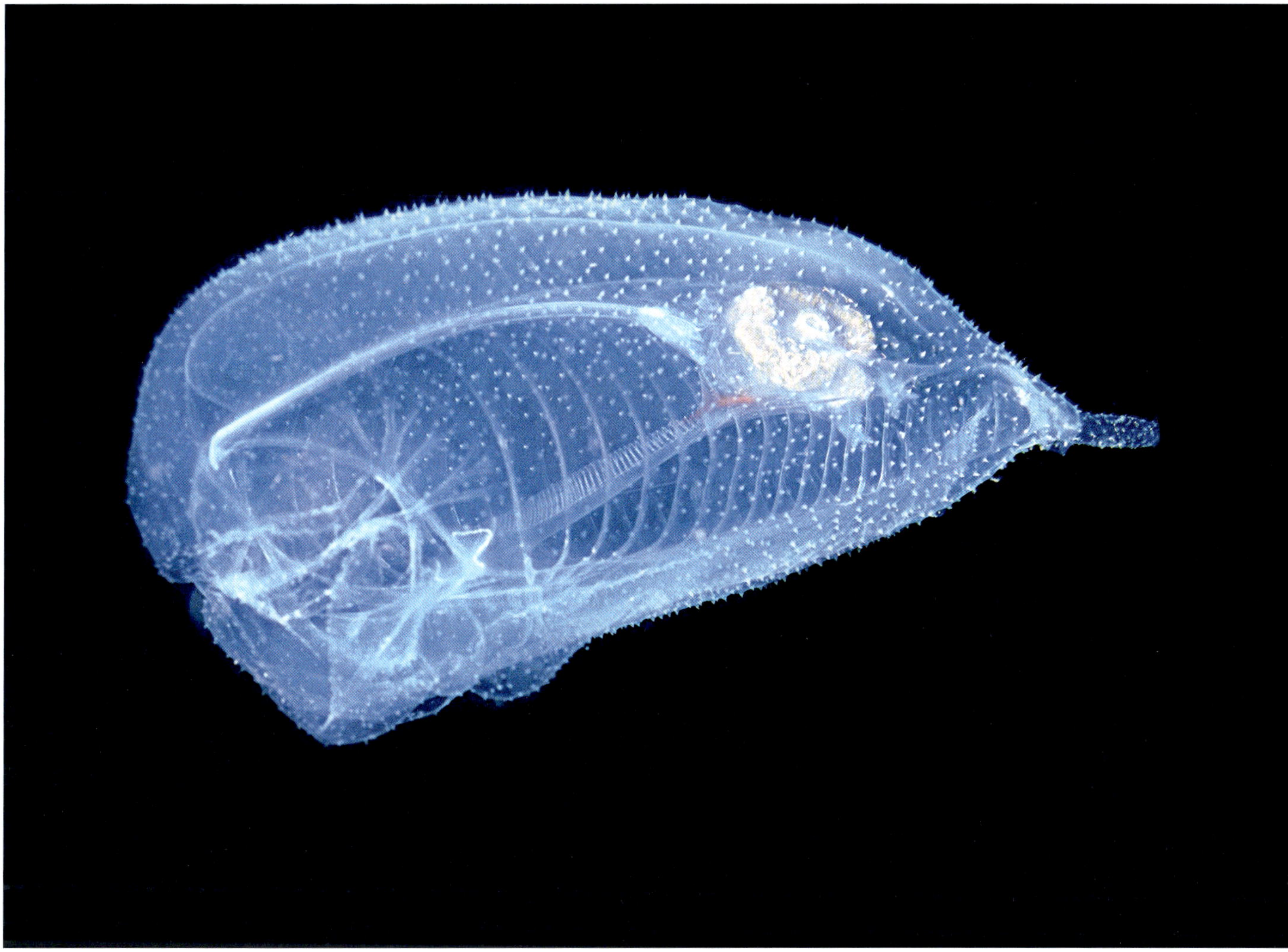

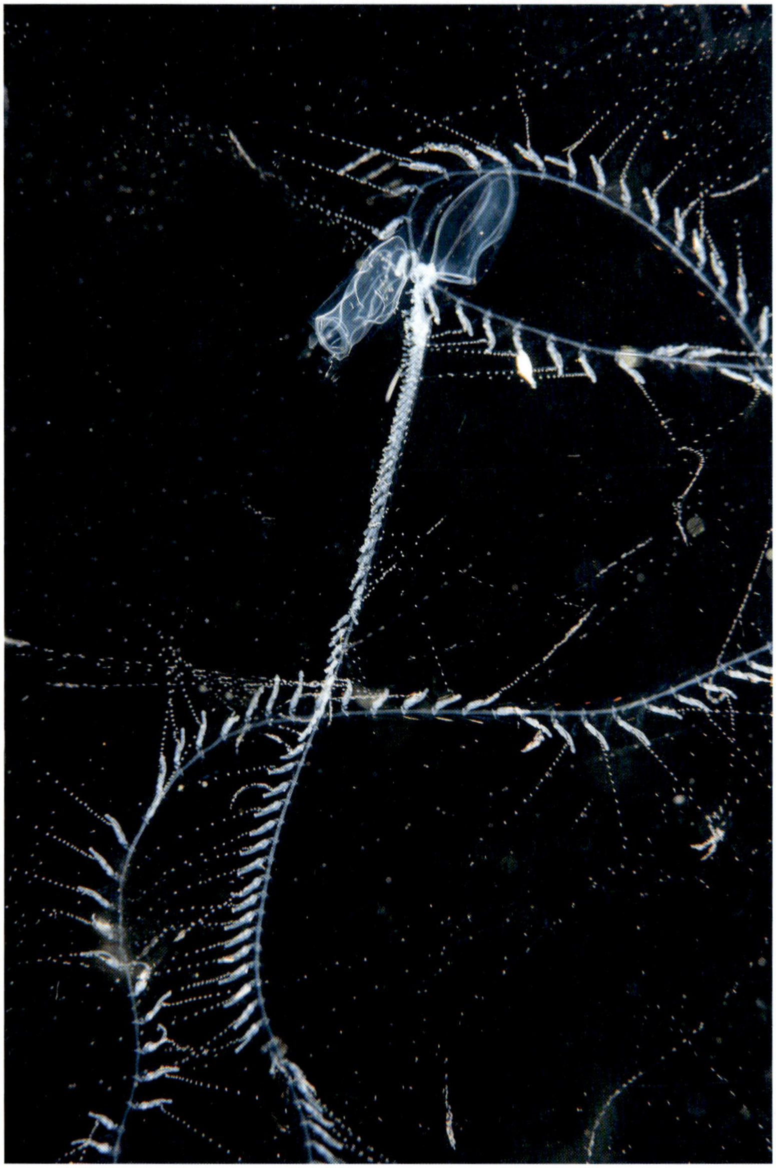

© Manfred
Wakolbinger,
2022

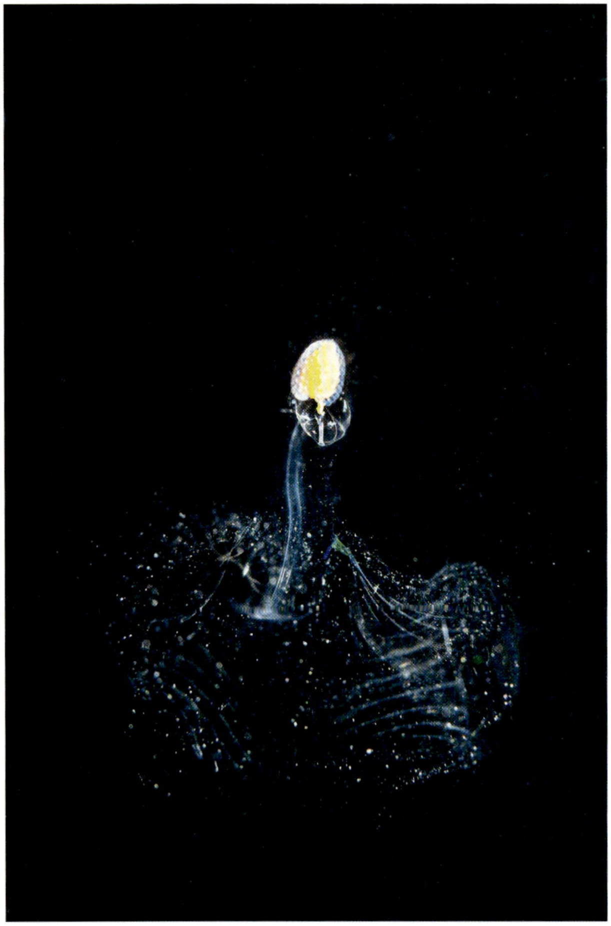

© Manfred
Wakolbinger,
2022

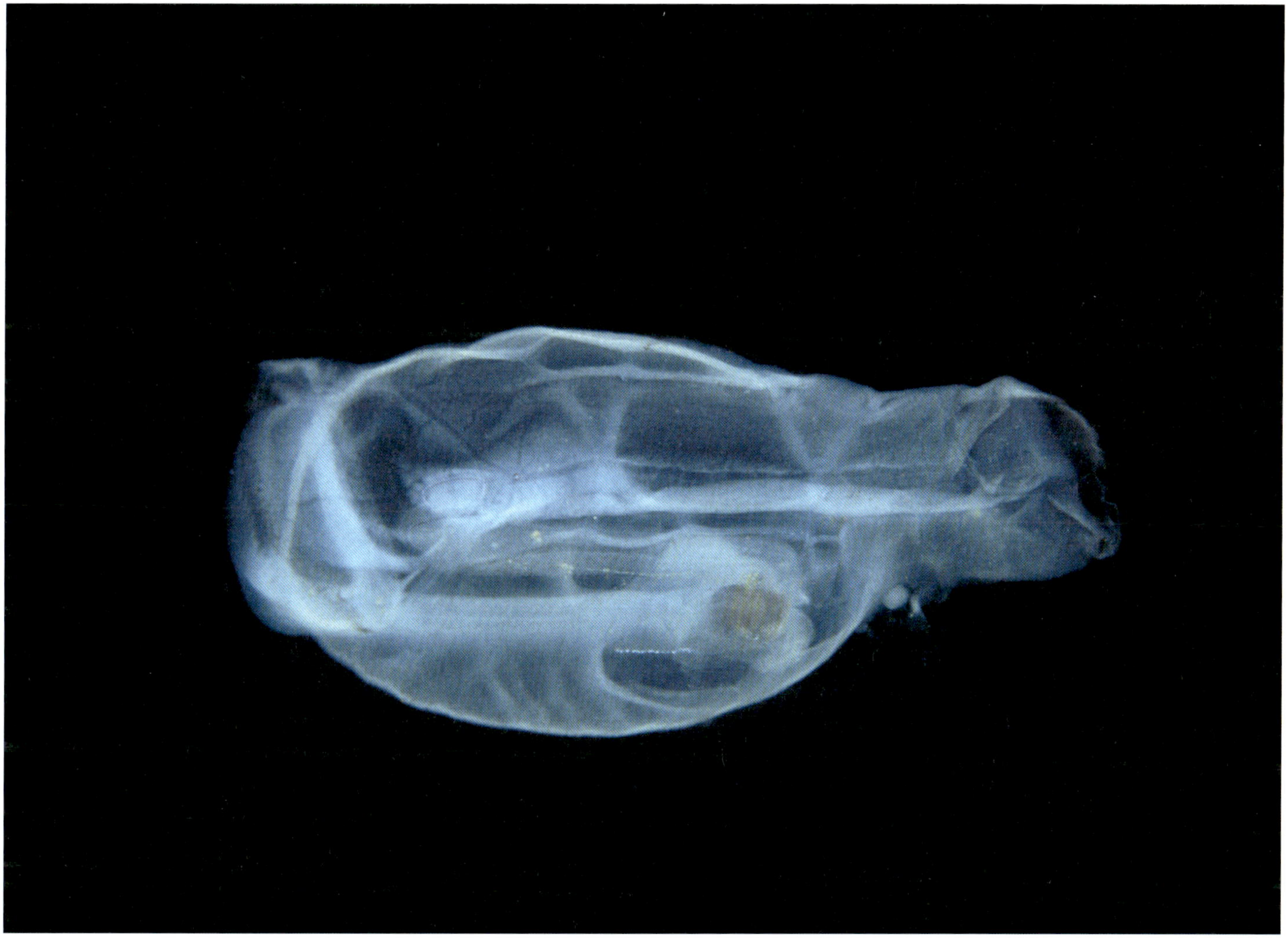

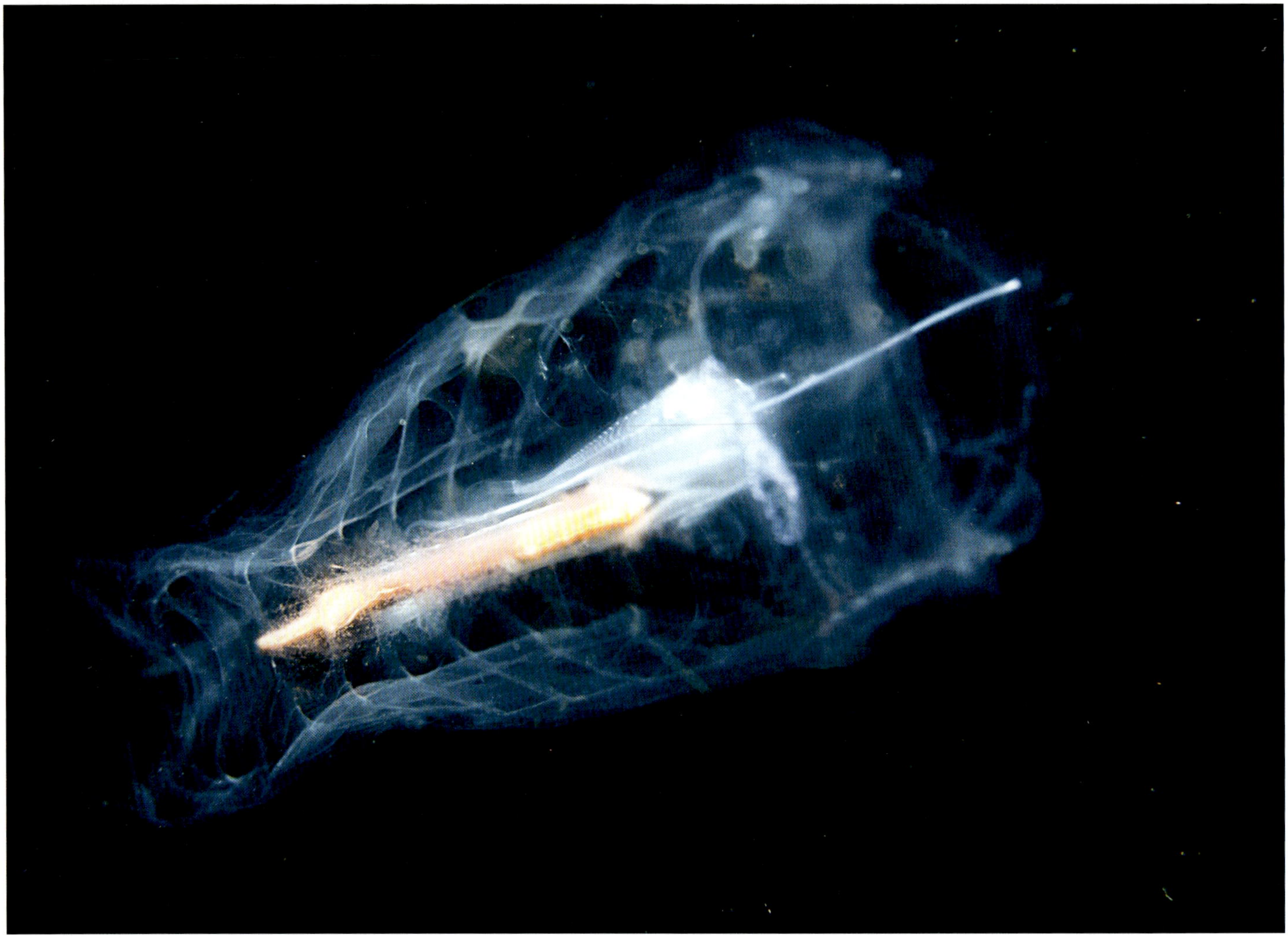

© Manfred Wakolbinger, 2022

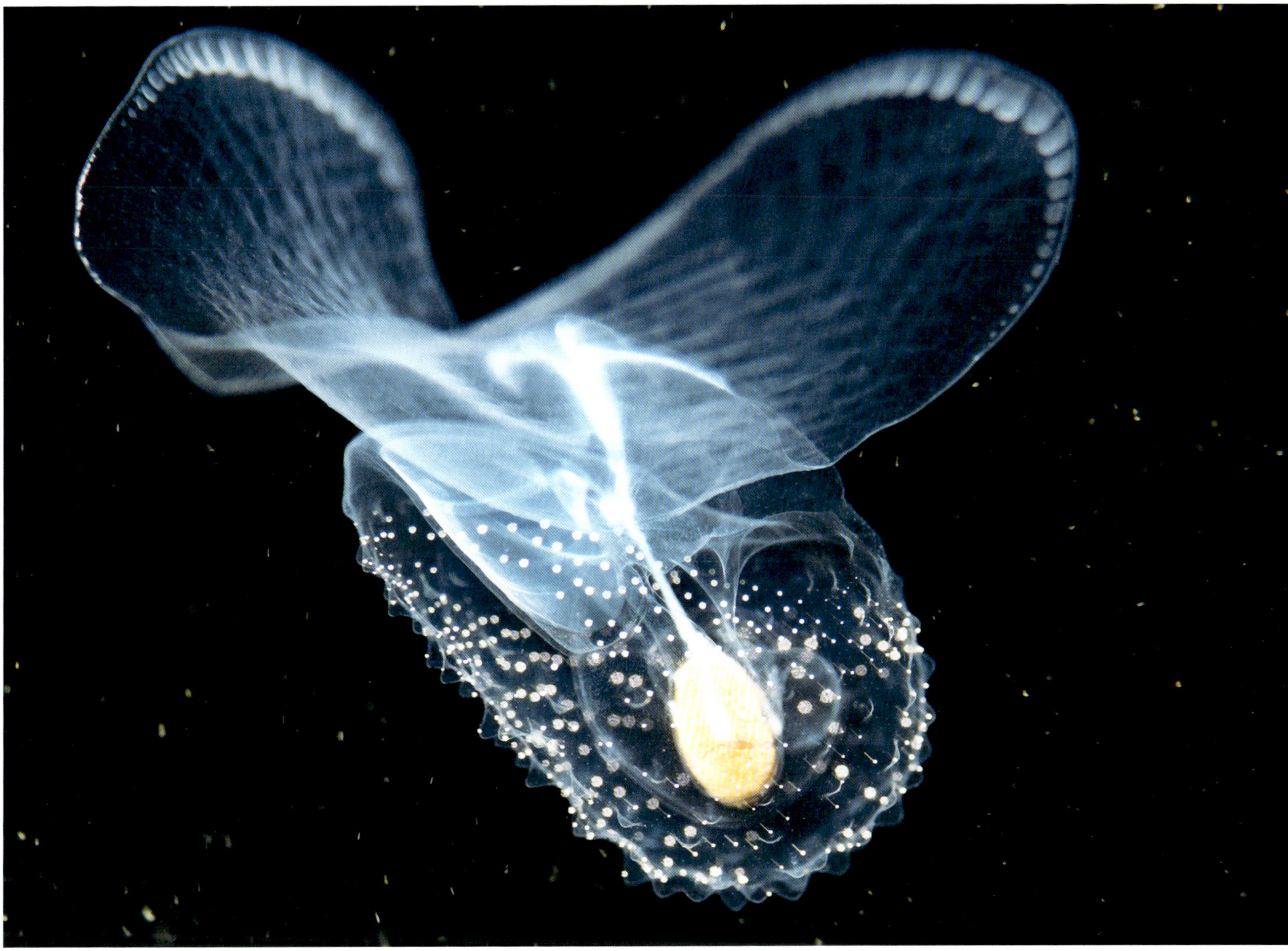

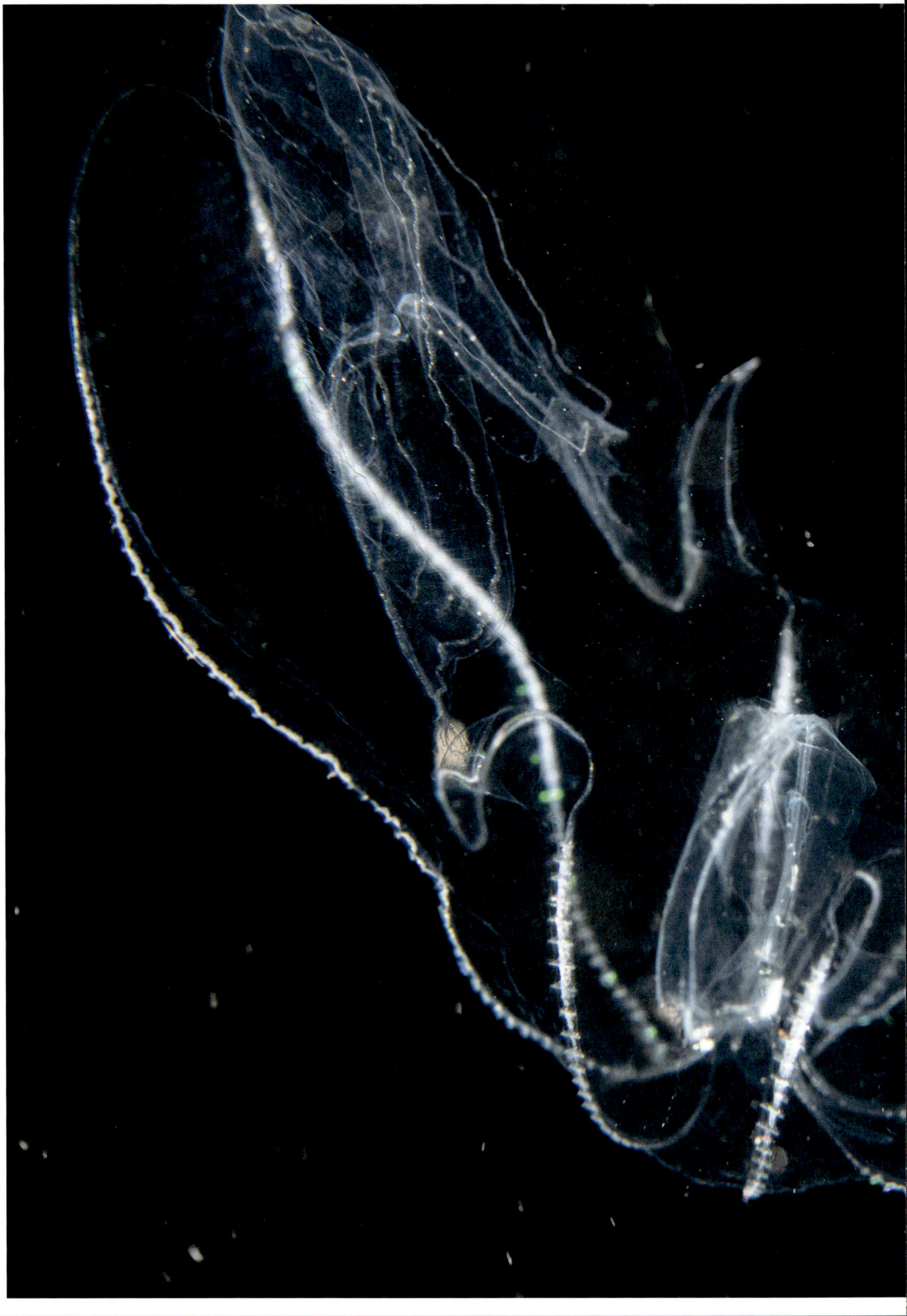

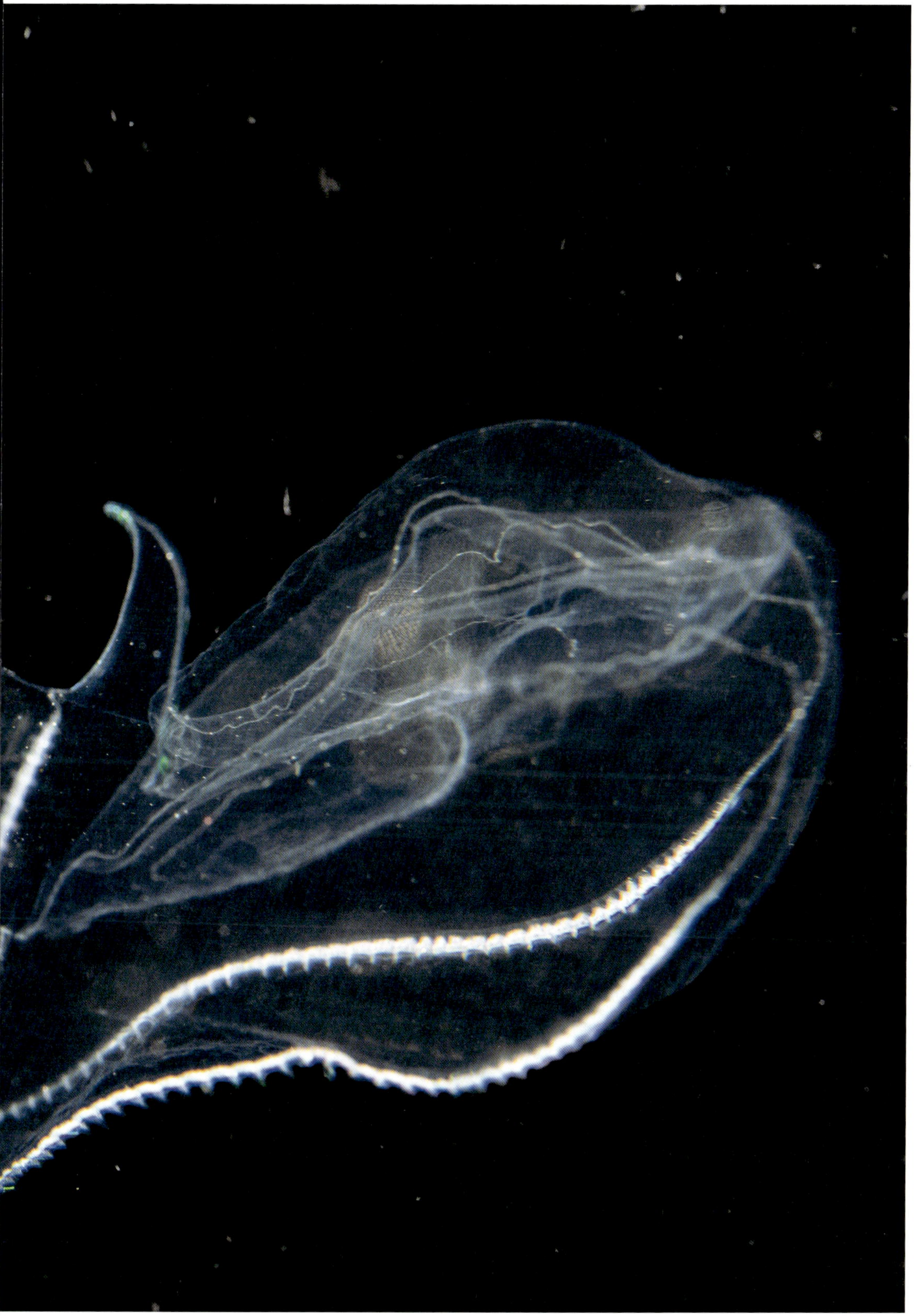

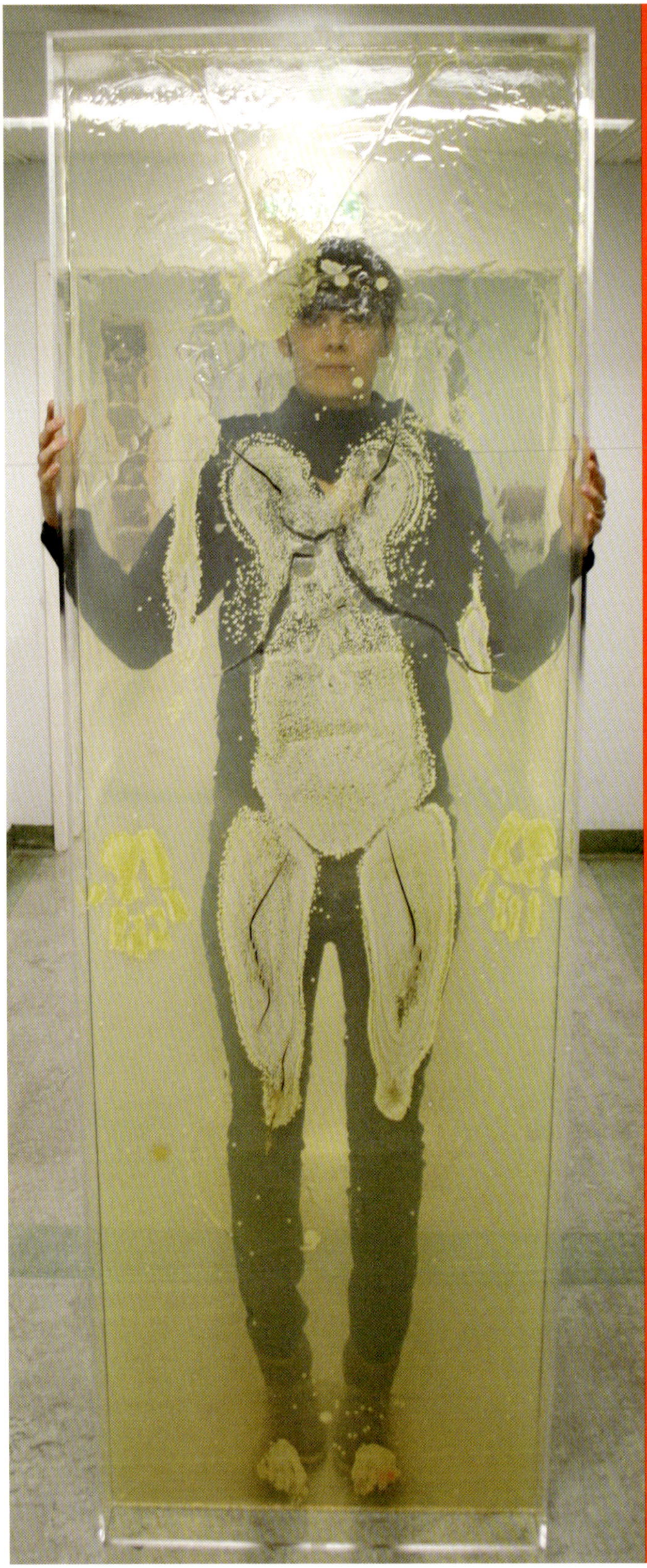

Being in the world as an individual really means being a multi-being community in a vital process of permanent exchange.[1]

FIG. 1_ EXPANDED SELF

Expanded Selves: Searching for Encounters

Sonja Bäumel

We are swimming in biology, that's why we don't feel it.

How to approach the microbial world, the interconnected ecosystem that allows for communication among all existing life forms?

When we touch any surface, or any other organism, our skin picks up many new microbes, leaving others behind. »We are swimming in biology, so we don't see it.«[2] Western science has recently come to understand the ancient Indigenous wisdom that the human body comprises twice as many bacterial cells as human cells, living both in and on our bodies. »Although this fact usually does not really affect our self-recognition directly and is not a threat to our identity, an awareness of it definitely alters the way we think of our bodies, as they no longer can be perceived as sealed vessels, but rather as transspecies environments.«[3] For the most part, our bodies are bacteria, viruses, archaea, eukaryotes, yeast, and parasites. In order to simply exist, we depend on the cooperation of different life forms within, on, and surrounding our bodies. Without them, we couldn't exist. Mitochondria, the energy powerplants of our cells, were created hundreds of millions of years ago from microorganisms. We are symbiotic multi-beings, created from the gigantic, bubbly, lively liquids on planet Earth. We are multitudes of different cells, of different beings, of the same cells, of the same beings, as part of a shared planet. These ideas and forms of knowing human embodiment were formulated decades ago by the American evolutionary theorist, biologist, author, educator, and public speaker Lynn Margulis.

It is clear that this »new« type of awareness might offend our desire for autonomy and environmental independency, as some societies might wish to view their human existence as sterile, not having to worry about biological complexity. However, it has been proven that human bodies, on average, contain around three kilos of bacteria, most of which are found in the intestines. Hence, one could ask the question: Who nurtures whom? Or, who depends on whom? Rooted in such fundamental questions, my work seeks to stimulate the cultural imagination regarding the impact of a deeper understanding of the microbial entanglement we are immersed in and the paradigm shifts that derive from such an awareness. In order to assimilate with biological more-than-human life systems—which is necessary in the face of environmental emergencies/developments—we need to consider and respect microbes and other tiny creatures as equal partners. Microorganisms have successfully adapted to radical changes and transformations in their ecosystems. They are capable of evolving in relation to their needs. We can learn from their expertise in behavioral change, acquired through billions of years of existence.

The launch of the Human Microbiome Project in 2007 by the US National Institute of Health, whose scope was to identify and characterize the microorganisms found in and on the human body, marked a paradigm shift in research practice. This was because bacteria had been studied in petri dishes for hundreds of years, meaning in complete isolation from the complex environments in which they usually reside and thus from the intricate relationships they have with each other. This emergent field of study, called *metagenomics*, aimed not only to observe bacterial behavior in its natural context but also to provide a more complete understanding of diseases and health. The impact of this approach has been immense, for instance, in relation to broader societal issues, such as concepts of privacy, individuality, future desires, and fears.

The ongoing research into the human microbiome has led a number of philosophers, social scientists, historians, and artists to rethink our being-in-the-world, our ontological hierarchies and values, and our conceptual understanding of what individual entities are. Recent work has dealt with the epistemological and ontological shifts that have resulted from engaging with the microcosm. Myra J. Hird,[4] Stefan Helmreich,[5] Monika Bakke,[6] and earlier work by thinkers like Lynn Margulis, Dorion Sagan, and Donna Haraway, have helped to inspire the discussion, since they address questions of trans-being entanglements and the possible ethical, ontological, and aesthetic implications from different angles. What if we imagine the »body as a locus of sensory-aesthetic appreciation and creative self-fashioning« in a »web of transspecies dependencies [...] symbiosis and entanglements«?[7] How can we »meet« on equal footing with bacteria, considering that »[b]acteria's myriad ongoing symbiotic relationships connect living to nonliving matter and sustain the biosphere«[8]? What if we imagine the human body as a locus of messy entangled relations, of »symbionts all the way down,« of the »commonality of life,«[9] or if we think »against categories such as species, sex, [...] as a locus of social and biological categories in motion and in transition«?[10] What is the relation between collective behavior and self-interest, and, most importantly, how can we reconsider the relationship between organism and environment?

Through my artistic research practice, I explore »the living« and, with them, the evolving perception of what bodies are made from. In active collaborative endeavors, I investigate the influence that scientific knowledge has had on the way we have historically perceived and interpreted the human body, and how such previously generated, yet still rather predominant views, affect current (mainly Western) society and the cultural contexts in which we operate. For fourteen years, I have been working with microbial bodies (and I will continue to do so). I have mostly shown them alive, growing, and decaying, which has posed a challenge to the conditions of public institutions and museums as they do not know how to deal with living artworks.

The following text offers a chance to explore a selection of my artistic investigations and the projects that I have developed in recent years by diving into the invisible microbial world and using different media and methods to render the invisible visible, experienceable, and tangible. I will also highlight some of the related challenges that I have encountered along the way.

Rooted in self-experimentation, the work discussed here explores possible ways to encounter invisible and intangible beings while questioning the perception of skin as the ultimate border for any type of body (human and other-than-human alike), embracing risk while directly engaging with living microbial entities, and venturing through landscapes, both macro and micro, to seek clarity as to where the environment starts and where it ends. My work thus renders in-between spaces as vivid entanglements full of life, where the boundaries of trans-beings are dissolved, revealing as yet unexplored forms of intelligence and communication. Finally, discussing the following selected projects will provide me with an opportunity to share some insights into my working process, the challenges of exhibiting living artworks (project #1), and my journey to encounter more-than-human life forms through empathy, touch, gesture (#2), liquids, and movement (#3).

Project #1. *Expanded Self:* Where Does the Environment Start, and Where Does It End?

Expanded Self[11] _FIG. 1 visualizes the invisible surface of human bodies in a unique way. A gigantic petri dish is used as a canvas while the microorganisms living on an individual body function as colors. The focus of this work is skin bacteria. The growing, life-size body print develops and expresses itself in a way that combines art and science.

It was photographed and documented on the seventh day of growth, testifying to the fact that our bodies do not end with our skin; rather, they invisibly extend into space. The resulting artwork consists of a unique mixture of life forms as they are found on a specific human body, in a specific area of Vienna, and on a specific day, seeking to highlight the invisible existing microbial entanglement of not just one but many bodies. In this way, the skin's borders blur, suggesting novel ways of looking at it. This imprint of human micro-landscapes composed of many single entities acting as communities functions as a kind of metaphor, provoking entirely novel perspectives about what it means to be an individual. All of us are so much more than what we think we are. We are only beginning to understand the effect that this astonishing community of different life-forms has on us and how it allows us to cooperate as equals.

In the artistic process, after applying the invisible bacteria colors to the body in high concentrations and by embracing the associated risks, the body is imprinted on agar media, the nutritive substance used to cultivate microorganisms in the lab. A huge petri dish (210 × 80 centimeters) is first filled with agar. After a few days, a living landscape starts growing on the nutritive medium. This allows us to witness skin bacteria transforming from invisible living organisms into visibly active entities thriving on an external support. Something that was part of my body becomes autonomous and clearly observable. Since the very beginning, contemplating a human body outside of its usual context, and the lack of awareness of its »many«

hidden worlds, has been absolutely mind-blowing for me. At the start (2012), my intention was also to get people to reflect on the way in which we tend to relate to microorganisms; considering the time at which this article is being written (after the entire world has lived through a global pandemic), this objective is perhaps more relevant than ever before.

Expanded Self was specially realized for the shooting of the documentary *Wir sind Planeten* (We Are Planets, English OT: You, Planet—An exploration in 3D). It was carried out during an intimate performance at a Viennese university and later exhibited in numerous international museums such as the Frankfurter Kunstverein in Frankfurt, Germany. When exhibiting a living artwork, a process is initiated and the outcome is consequently unknown, as it evolves over the period of the exhibition; in this way, it's as if the exhibition space morphs into a lab, allowing me to present the living performance in a cultural context. This allows visitors to reflect on art as territory for cultural experimentation and on the role of the artist as a researcher. What are the implications of placing living artworks into cultural contexts? This question may seem banal, but it's important to consider that museums as we know them are made to conserve dead things—as dead as possible and for as long as possible. What are therefore the challenges of placing a piece of life, as an artwork, in such a still and transformation-resistant context?

*Expanded Self II*_FIG. 2, a further development that took place a few years after *Expanded Self*, strived to enhance the two-dimensional nutrient agar into a three-dimensional artifact. It clearly required a lot of experimentation with materials as the approach was completely novel. This work is an interesting case study of living artworks exhibited in a cultural context, like in the exhibition *Gare du Nord*_FIG. 3[12] at Waag, Amsterdam. In practical terms: what happened was that the artwork was removed from the show after a few weeks as it appeared to »challenge« the cultural space because it became »too alive«_FIG. 4 and thus started to smell and to attract flies to the exhibition space. While observing what was unraveling through a more conceptual lens, the reasons given for its removal allowed me to imagine what such a space would look like and made me wonder: what could this »too alive« mean for us, both in the experience of the artwork and in regard to the functioning of a museum/exhibition space itself, typically a space to preserve, conserve art, and foster dialogue?

There is no simple answer to these types of questions as living artworks grow, change, and arouse curiosity; they are process-driven, demand care, adapt, and eventually they pass on. A living artwork »does not simply work in between the two fields of art and science, but it kind of redisposes forces drawn from both, the transformative force of its rapture lies in opening up a nexus of power-free relations in both fields. Its freedom is in being this voice in the middle, too much alive to be totally defined either by art or science.«[13]

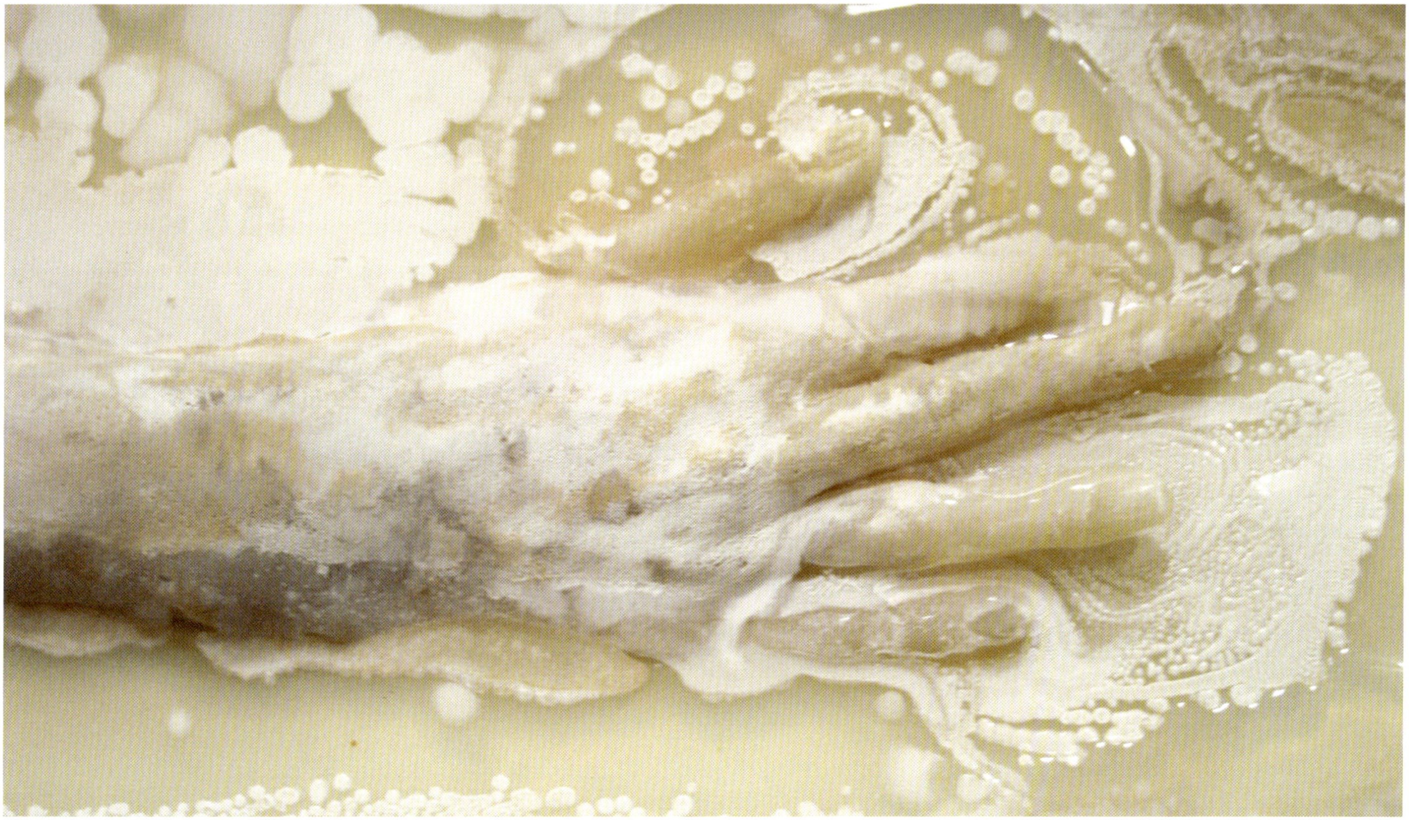

FIG. 2_ Expanded Self II

We are swimming in biology, that's why we can't feel it.

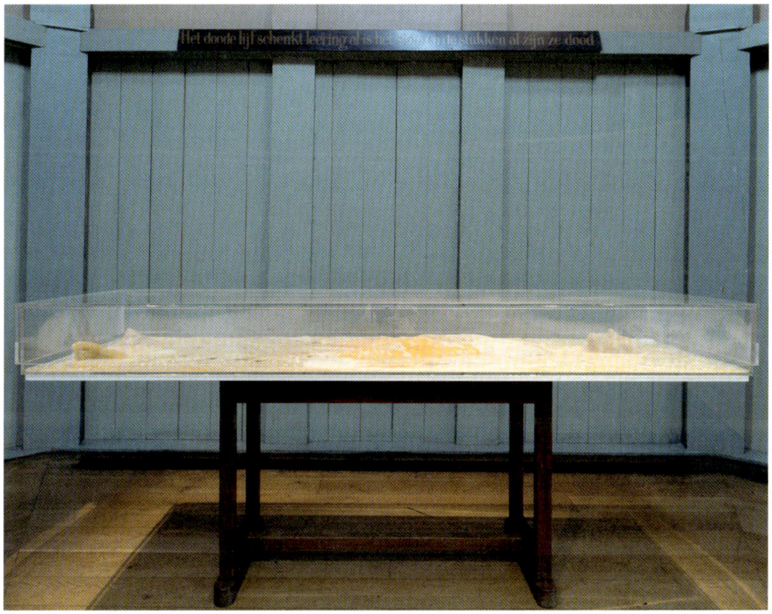

FIG. 3_ Expanded Self II, *Waag Theatrum Anatomicum, Amsterdam*

Photo by Cassander Eeftinck Schattenkerk

GARE DU NORD - PRESS RELEASE

August 5th 2015

A project curated by Chiara Ianeselli in collaboration with Lucas Evers, Waag Society

Participating artists: Sonja Bäumel, Laurent-David Garnier, Nicola Samorì

Exhibition dates: 17 July – 15 August 2015

Theatrum Anatomicum, Amsterdam

Sonja Bäumel, *Expanded self n. 2*, 2015

Agar-Agar, microorganisms of an individual's body on a certain day, acrylic glass cm 75 x 220 x 20

> ***Expanded self n. 2*** explicitly depicts and shows the human marvel: we, humans, are "a semi-continuous spectrum of interactive bacterial nations" (Reg Morrison comments on Margulis/Sagan 2007), as we have ten times more bacterial cells than human cells living both in and on us. Suddenly, we realize that the form we witness is not cast as a unique piece but results from the accumulation of the smallest parts: bacteria that inhabit and rule the human hosts. Through this work, Bäumel creates a space where the potential of bacteria as cooperative partners can be re-imagined and where we can explore the implications for larger processes of cultural significance.

"Larger processes of cultural significance": this is what of the brochure of *Gare du Nord* contained and released in reference to the piece of Sonja Bäumel. This statement gained more significance from the second week of the exhibition period onwards as the artwork has indeed faced a growth that challenged its own structure, materials, its author, its environment, the other pieces in the show and all the different individuals related to it, such as the curator and all the visitors.

Nome omen, names are omens: the **expansion of *Expanded self n. 2*** took place in particular in relation to the anatomical theatre, its air, its humidity, its conditions. The process of cultural significance of the exhibition follows this extremely powerful expansion as now other themes and layers are added to the numerous histories already documented in the space and the show: how can perception still be conditioned? How can an artwork ideally take over its own life and decide for its destiny?

When was exactly the turning point? How can it still be preserved and documented? How did the microorganisms relate to a bigger context? What is now the legacy that the artwork leaves behind? How is its nature being determined?

Expanded self n. 2 planned as a sophisticated gamble, has allowed us to meet *fortuna*.

> Kentridge employs the concept of 'fortuna', which he describes as neither 'a plan, a program, a storyboard; nor sheer chance. "Fortuna" is the general term I use for this range of agencies, something other than cold statistical chance, yet something outside the range of rational control. In other words, we might understand this as a kind of directed "happenstance", the engineering of luck, or a sophisticated gamble, involving both possibility and predetermination. 'Fortuna' alludes to a state of becoming wherein the work of art is endlessly under construction — even when encountered as a finished product by the viewer. There is a sense of discovery, rather than invention, as Kentridge has written." 'Fortuna' also suggests a celebration of eccentricity that supports the works' political engagement: 'This reliance on "fortuna" in the making of images or text, mirrors some of the ways we exist in the world, even outside the realm of images and texts."

Expanded self n. 2 is now resting in the artist studio, whereas some bacteria still float in the space.

FIG. 4_*Press release,* EXPANDED SELF, GARE DU NORD *exhibition, Waag, Amsterdam*

Project #2. *Fifty Percent Human*: Encountering through Touch

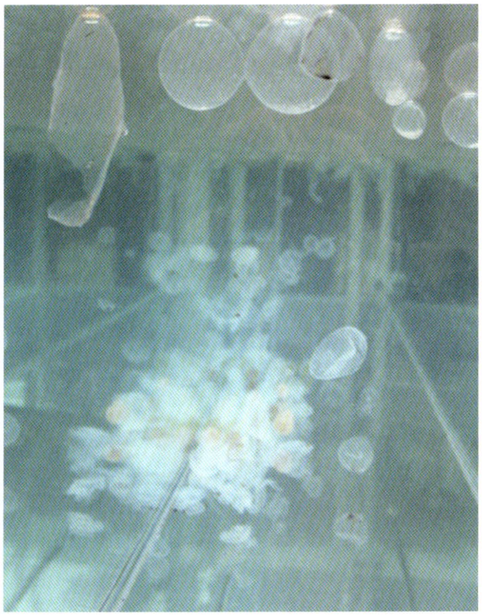

FIG. 5_ Fifty Percent Human

© Sonja Bäumel, 2015–16

»Bio-art explores the boundaries between art and science, generating a new hybrid form which raises questions about the shifting and permeable borders of what it is to be human.«[14] *Fifty Percent Human*_FIG. 5,[15] a collaboration between cultural historian Birgit Nemec, artist Cocky Eek, scientist Hauke Smidt and myself, explores what it means to be human when we realize that we have a close relationship with and dependency on millions of microbes. It therefore opens up a space in which we can encounter otherness through touch.

We humans have recently come to understand that we have twice as many bacterial cells as human cells, living both in and on our bodies. If fifty percent of the cells that constitute our body are not human but microbial, how might we get in touch with our more-than-human cohabitants?

Philosopher Nicole C. Karafyllis has reminded us that tissue is not only able to shift the borders of organisms and species but, when used as a material in art, can also question the limits of what was once thought to be known.[16] In other words, there is a need to recognize microbial life-forms as actors that co-shape our bodies (and thus our physical and mental constitution) — particularly if we aspire to truly experience and thus better understand inter-organismic communication, as it may allow us to better take care of the microcosm and thus, ultimately, to better take care of ourselves.[17]

This project started with the aim of encountering the smallest units of life, the microbial cells (bacteria, archaea, eukaryotes, viruses, yeasts, and parasites) in which we are »constantly swimming.« An empathic encounter could start with questions such as, »Who are you? Who are your companions? What happens when you meet organisms that are not like you?« As Sanne Bloemink, a journalist who was involved in the process, asks: »How can the voices and interests of non-humans be heard and be taken into account?«

During the artistic process, I visited the Laboratories in Wageningen[18] to study microbial beings. We, microbial ecologist Ruth Schmidt and I, measured volatile organic compounds.[19] Furthermore, Hanne Tytgat (Scientist, Microbiology Laboratory) and I used the method of gene sequencing as all organisms harbor a unique genetic barcode in their DNA. There are too many microorganisms to study all of them in the time given; we focused on bacteria, eukaryotes, and viruses (the latter of which we haven't found a single one). We set out to read all the genetic barcodes present in the skin samples to find out which organisms are present via PCR (polymerase chain reaction) and NGS (next-generation sequencing). During this process, the cell's membrane is destroyed, and the organism is reduced to its DNA.

One part of the microbial genome is very particular and is used by scientists to differentiate between species of organisms. By using such tools and visualizing genetic information, we can attempt to provide answers to questions such as, »With whom do you like to surround yourself?« However, following the call made by sociologist Nicholas Christakis, we believe »that many things in the world have properties not present in their parts« and thus cannot be understood simply by taking them apart; one has »to observe interactions of the whole.«[20] Consequently, we decided to not only investigate the genes of microbes but to look at how they exist in the world in the context of their communication, their movement, their gestures, and

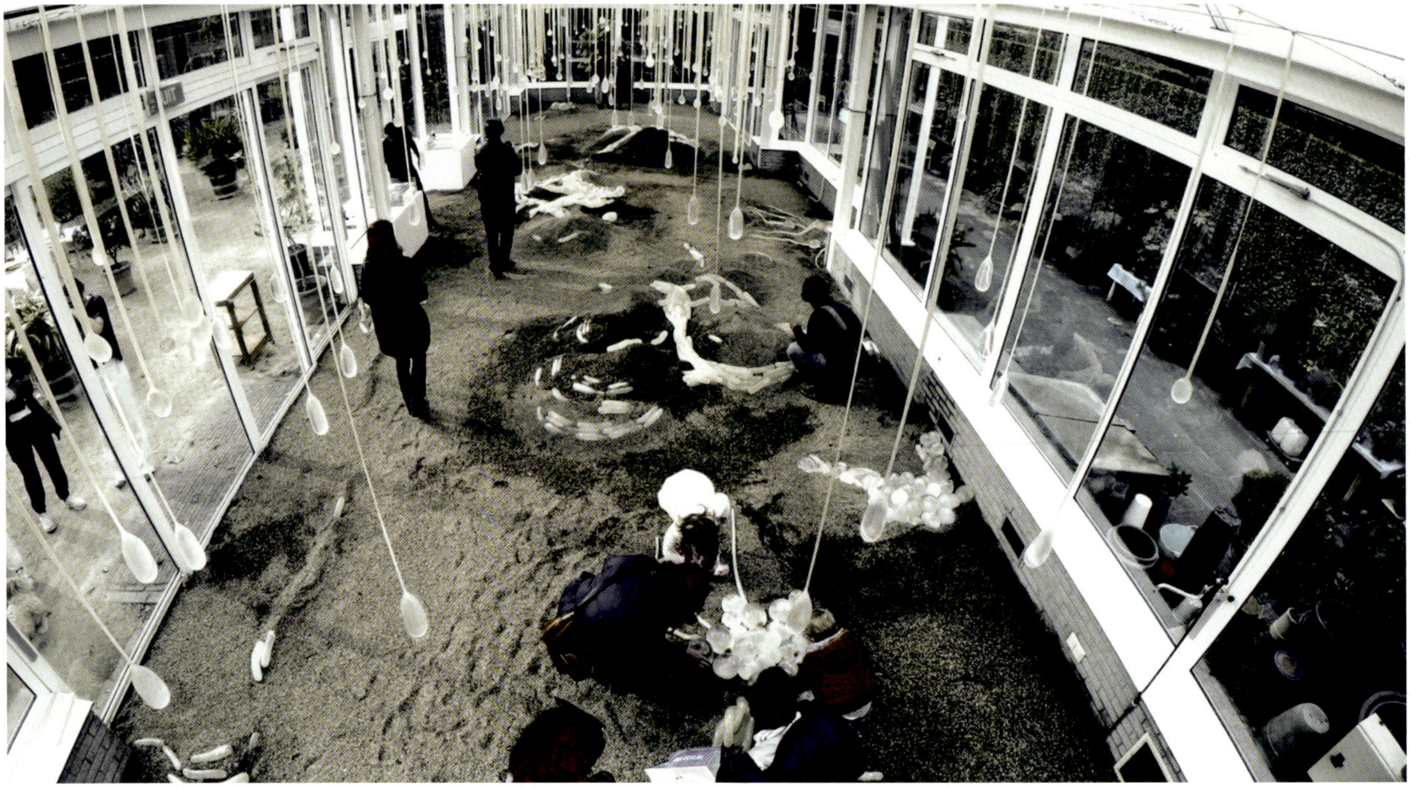

FIG. 6_ *Time-lapse video*
FIFTY PERCENT HUMAN

VIDEO

their surroundings. Questions like those of companionship, strangeness, and otherness have to be approached from a different angle, and those asking need to be open to a more imaginary approach. We agree with media theorist and philosopher Jean Baudrillard when he notes that »the irruption of radical uncertainty in all fields and the end of the comforting universe of determinacy is not at all a negative fate, so long as uncertainty itself becomes the new rule of the game. So long as we do not seek to correct that uncertainty, by injecting new values, new certainties, but have it circulate as the basic rule.«[21]

In the making process, we search for materials and in-between spaces that address uncertainties, ambiguities, and imaginings linked to the microbial paradigm shift on both an aesthetic and an epistemological level. We seek in-between spaces that allow artifacts to be imperfect, to leak, to break up, to dry out, to change forms. In full resonance with Monika Bakke's analysis, we »seek not a sentimental view of always happy intersomatic experiences, but rather a sober view of the reality of messy transspecies entanglements.«[22]

If we later place these artifacts in an installation, they might alllow the visitor to experience and engage with bioscientific discoveries related to the (human) microbiome. Using them as imaginary tools, we sharpen and awaken the explorer's senses in order to raise awareness and encourage the development of private knowledge of the body. To this end, we model the microbial world, its qualities and its interactive relations, translating the in-between spectrum of messy entanglements of microbes that are made from the same substance into tangible artifacts. Different materials and modes are tested to create a sense-driven and experiential setting that invites visitors to explore while it recomposes perceptions of scale, space, and environment. Giuseppe Penone, a key artist from the Arte Povera movement, inspired the process when he said that »the act of touching is an act that helps understanding the reality of things.«[23] Is it possible to sense a language through which we can encounter nonverbally communicating organisms through touch?

As part of *Zone2Source*,[24] held in the Orangerie at the Amstelpark, a hidden gem in one of the greenest parts of Amsterdam, we created a space_FIG. 6 that invites visitors to touch and explore. We crafted a different world, a damp world, filled with enlarged transparent and liquid-membrane-bound microbial cells, collectively swimming, lying down, or floating—an intersecting multispecies landscape to explore. It was an imaginary world intended to break down hierarchies, dimensions, and scales and to get the public to venture into an uncertain, divided, destructed, distorted, never-pure, ever-connecting, swarming, open, living, and empathic in-between space.

The in-between space that we created was intended to encourage visitors to engage in questions such as: How can we activate the second microbial layer on our body, the biggest nonverbal communication platform between humans, as a method for thinking critically and imagining alternative futures, risks, desires, and fears? In this way, *Fifty Percent Human* helps us to rethink, reshape, and deepen our understanding of fundamental aspects of our being-in-the-world. We seek to encourage people to view microbes as a web connecting the health of all living beings and the environment. Alongside encouraging greater diversity in the methods used to imagine alternative futures, this may help us to understand how to improve the microscopic ecosystems we live in as well as the ecosystems that live in and on us.

Project #3. *Microbial Entanglement—In Vitro Breakout*: Encountering through Movement and Liquids

The focus of this artistic research project[25] is non-linguistic microbial communication and collective movement. The project investigates the language of microbes, scientifically defined as »quorum sensing.« This field of investigation was initiated more than thirty years ago, and its true potential is still to be fully unraveled. According to their population density, microbes can alter their behavior using an intercellular molecular signaling process. In this way, quorum sensing effectively allows microbes to be aware of one another's presence, to »count« themselves, and to behave as a multicellular group in high cell numbers.

*Microbial Entanglement—In Vitro Breakout*_FIG. 7 is a site-specific performance that emerged from this research and that was developed for the Frankfurter Kunstverein[26] in collaboration with choreographer and performance artist Doris Uhlich, and performance artist and dancer Andrius Mulokas. In science, *in vitro* refers to experiments conducted in a controlled artificial environment outside a living organism, such as a petri dish. The performance aims to point out the DNA fetishism in science. A living being cannot be reduced to its DNA, and, in order to explore the role it plays in social relations, we need to expand on existing scientific tools artistically and philosophically.

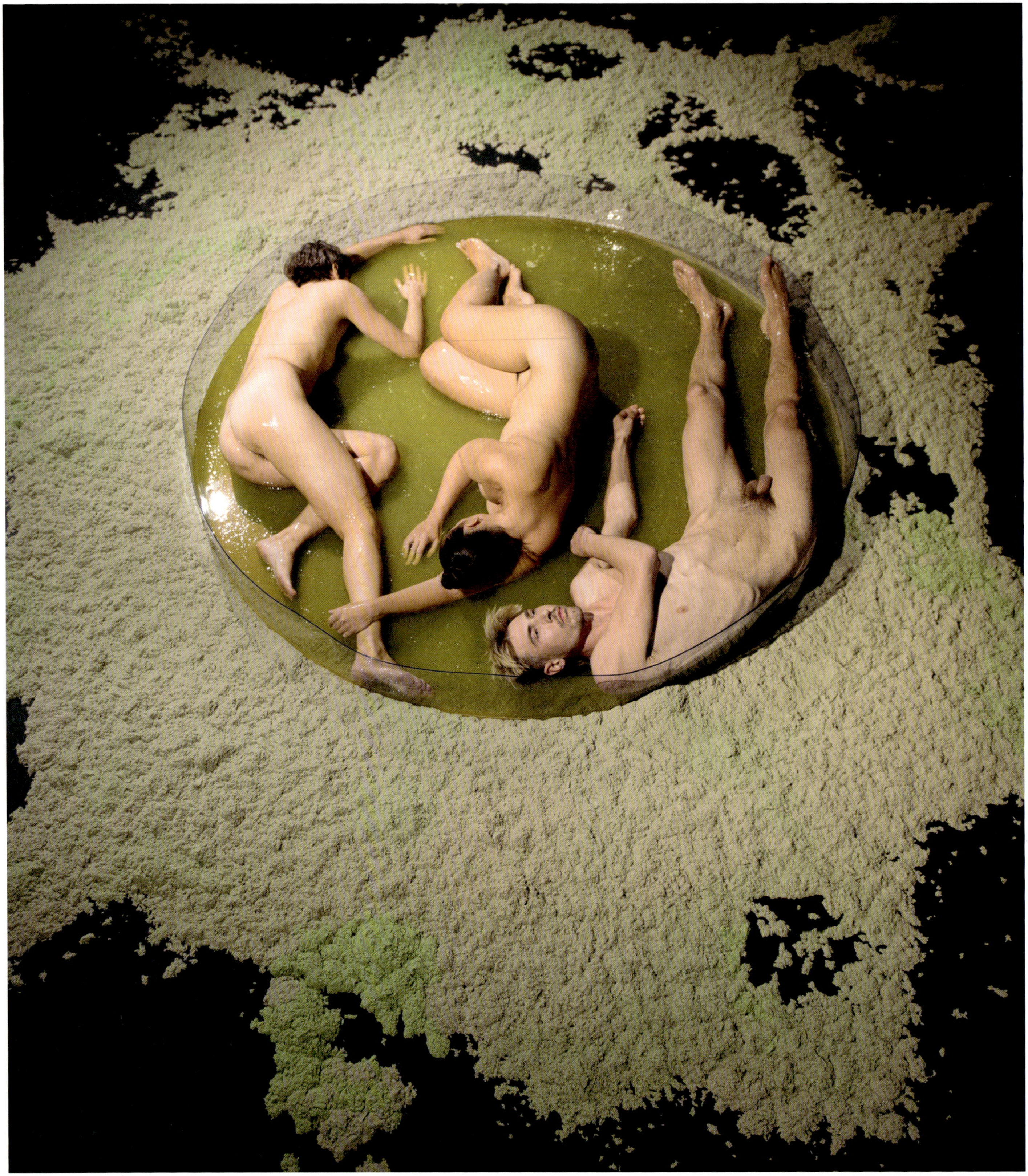

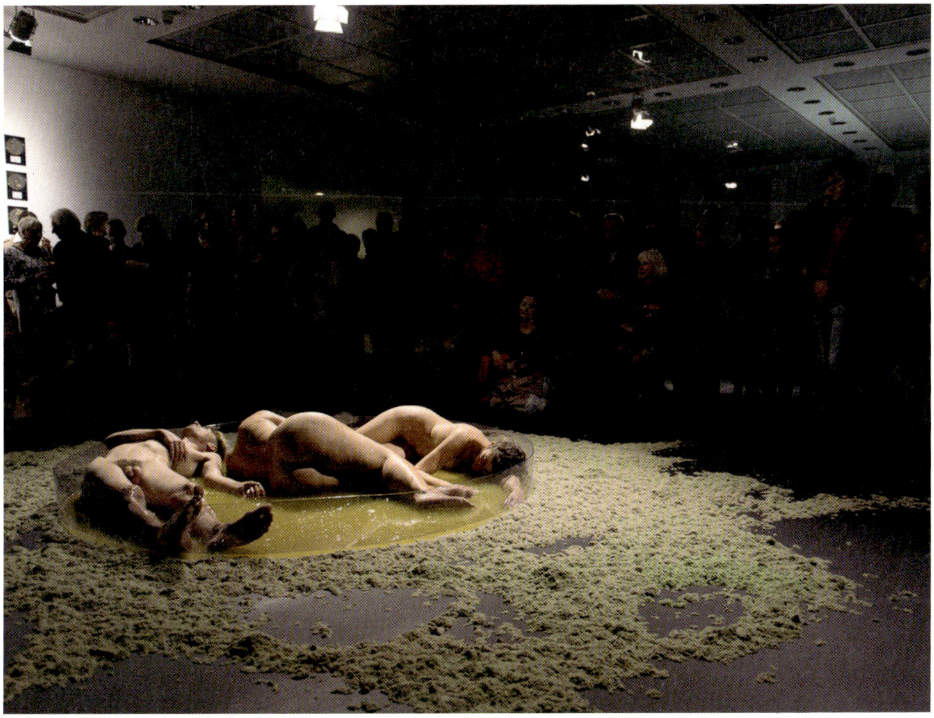

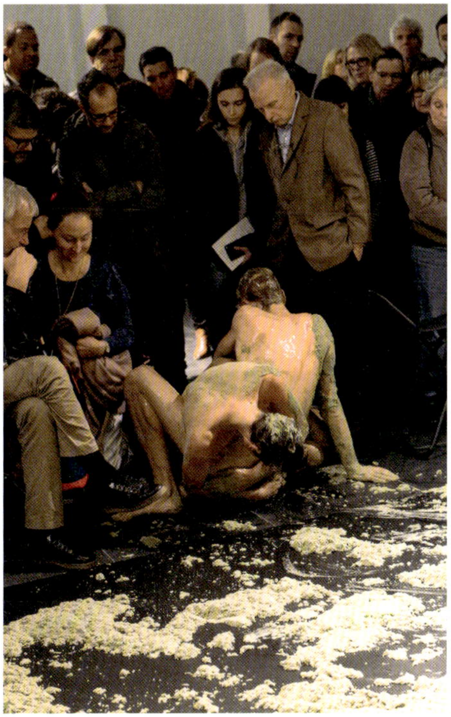

How can an imaginary world facilitate a radically new view on biological rules, hierarchies, interactions, dimensions, and scales?

FIGS. 7, 8, 9_MICROBIAL ENTANGLEMENT — IN VITRO BREAKOUT, *Frankfurter Kunstverein*

Photos by Robert Schittko

The microbes we depend on dissolve trans-being boundaries and exhibit as yet unexplored forms of intelligence and collectivity. With this performative work, as Franziska Nori (Director of the Frankfurter Kunstverein) writes,

> *Bäumel uses principles of biological models of cooperation to metaphorically transform them. Here she negotiates phenomena such as quorum sensing, a field of research that explores the communication systems between microbes. These coordinate activities such as their growth and structure formation as a group via chemical processes. The term quorum comes from the political form of the Roman Senate and describes the process by which a collective must reach a decision through negotiation.*[27]

In the performance, three undressed performers lie in a fabricated petri dish. They break out and move freely and collectively in the exhibition space_FIG. 8, in which materials such as kinetic sand and methyl cellulose have been distributed, with which the bodies contaminate themselves and probe what it means to move in high viscosity like bacteria do. After the performance, the resulting material landscape remains part of the exhibition.

How can performance help to activate a radically new view of biological rules, hierarchies, dimensions, and scales_FIG. 9? Where could this renovated self-understanding conduct us?

Donna Haraway emphasizes the ethical aspect, i.e., that living together also entails risk. It is not a romantic or sentimental notion of being together; it is about taking risks to find the courage to open up to that which is different, that which is not the same as we are: »Allowing uncertainty, confusion and ignorance to be the basis of our actions in the world. The endeavor lies in making ourselves vulnerable in our explorations of our material relationships, with that which is inherently alien, yet so close to us.«[28]

Ongoing Thoughts

How can we pretend to experience a sense of care and responsibility when we are not capable of relating to microorganisms, whose behavior and cooperation are key to our very existence?

The three selected artworks presented above seek to expand our sensory, haptic, physical, visual, and imaginary language to explore our relationship to microbial life-forms and therefore to deepen and broaden our understanding of being in the world. Experiencing bodies as walking biotopes opens up a realm of completely new perspectives and possibilities as part of human culture. Bodily experiences can allow us to broaden our view, to envision further forms of living together. By challenging the concepts of »species« and »individual,« it is possible to increasingly blur the boundaries of what it means to be a »self«—particularly when everything points to the fact that the processual exchange and constant interaction between living beings and the environments they are immersed in determines the essence of all that is, us included.

1 Monika Bakke, »Practicing Aesthetics Among Nonhuman Somas in the Age of Biotech,« in *Practicing Pragmatist Aesthetics: Critical Perspectives on the Arts*, ed. Wojciech Małecki (Amsterdam: Rodopi, 2014), 155.
2 Peter Friedman and Jean-François Brunet, *Death by Design* (documentary), 1995, June 13, 2013, https://www.youtube.com/watch?v=4NytzTLnyKM.
3 Bakke, »Practicing Aesthetics,« 155.
4 Myra Hird, *The Origins of Sociable Life* (New York: Palgrave Macmillan, 2009).
5 Stefan Helmreich, *Sounding the Limits of Life: Essays in the Anthropology of Biology and Beyond* (Princeton: Princeton University Press, 2015).
6 Bakke, »Practicing Aesthetics.«
7 Bakke, »Practicing Aesthetics,« 154–5.
8 Hird, *The Origins of Sociable Life*, 118.
9 Hird, *The Origins of Sociable Life*, 126.
10 Helmreich, *Sounding the Limits of Life*, 62–72.
11 *Expanded Self* was realized in collaboration with bacteriologist Erich Schopf for the documentary *Wir sind Planeten* (2012), directed by Martin Mészáros and Alfred Vendl.
12 »Gare du Nord,« *waag futurelab*, accessed March 15, 2023, https://waag.org/en/project/gare-du-nord/.
13 Federica Colombo, »Being ›Too Much Alive‹—A Conversation with Sonja Bäumel,« *Art & Life Sciences Course* (Leiden University, 2018), 4.
14 Suzanne Anker and Dorothy Nelkin, *The Molecular Gaze: Art in the Genetic Age* (Cold Spring Harbor: Cold Spring Harbor Laboratory Press, 2003); quoted in Frederica Colombo, »Being ›Too Much Alive‹—A Conversation With Sonja Bäumel,« 1.
15 Sonja Bäumel and Birgit Nemec, *Fifty Percent Human*, exhibition brochure, Zone2Source Amstelpark Amsterdam, October 2016, 3–18;
16 Nicole Karafyllis, »Endogenous Design of Biofacts, Tissues and Networks in Bio Art and Life Sciences,« in *sk-interfaces: Exploding Borders—Creating Membranes in Art, Technology and Society*, ed. Jens Hauser (Liverpool: FACT, Liverpool University Press, 2008), 42.
17 Myra Hird, *The Origins of Sociable Life*; quoted in Monika Bakke, »Practicing Aesthetics Among Nonhuman Somas in the Age of Biotech,« 158.
18 For this research we worked in the Laboratory of Microbiology at Wageningen University and the Research Department of Microbial Ecology at the Netherlands Institute of Ecology in Wageningen.
19 It is technically possible to track microbial exchange through chemicals by measuring volatiles.
20 David Brooks, *This Will Make You Smarter: New Scientific Concepts to Improve Your Thinking* (New York: Harper Collins, 2012), Foreword.
21 Jean Baudrillard, *Impossible Exchange* (London: Verso, 2001); quoted in *Process as Paradigm, Art in Development, Flux and Change*, ed. Susanne Jaschko and Lucas Evers, exhibition catalogue, Centro e Arte y Creación Industrial, 2010, 22.
22 Bakke, »Practicing Aesthetics,« 154.
23 Ikon Gallery, »Guiseppe Penone Documentary,« directed by Ikon Gallery, Birmingham, 2011, *YouTube*, February 18, 2011, https://www.youtube.com/watch?v=7Fo-76Gfg3w, 00:05:20.
24 Sonja Bäumel, »Fifty Percent Human,« *zone2source*, accessed March 15, 2023, https://zone2source.net/en/8-16-october-2016-50-human-sonja-baumel/.
25 Part of the project *What Would A Microbe Say?* (2017–20), a collaboration between Sonja Bäumel and Helen Blackwell, Professor of Chemistry at the University of Wisconsin-Madison, US (http://www.sonjabaeumel.at/works/bacteria/what-would-a-microbe-say/), which developed into a site-specific performance called *Microbial Entanglement*.
26 *Microbial Entanglement* was developed for the exhibition opening *Trees of Life—Stories for a Damaged Planet* at the Frankfurter Kunstverein in Germany (2019–20), https://www.fkv.de/en/exhibition/trees-of-life-stories-for-a-damaged-planet/.
27 Franziska Nori, press release, Frankfurter Kunstverein, October 2019.
28 Alice Smits, *Living in the Big Mesh: Fifty Percent Human,* exhibition brochure, October 2016, 19.

Visualization Technologies in Nature Documentaries

Walter Köhler

Forty years of natural history filmmaking—how this genre has revolutionized camera and visualization technologies since the dawn of moving images and continues to do so.

The nature documentary is more visually striking than almost any other film genre and is in fact the ultimate filmmaking discipline. Although it belongs to the documentary genre, the nature documentary is more of a hybrid and has more to do with narrative films than one might think. Of course, whatever is being shown should be—in the best sense—»true«; it should depict reality, document what has actually taken place. But it should also be paired with an interesting storyline, competent protagonists who stage themselves skillfully, editing that perfectly realizes the narrative while putting the camerawork on display, and a film score that emphasizes the scenes without pushing into the foreground.

In no other film genre are there as many challenges to master. Nature documentary-makers do not just have to be excellent camerapeople, both creatively and technically: in order to understand their »actors,« they also have to be biologists. They must be talented improvisers because, outside, everything is constantly changing. In nature documentary-making, scripts and production plans are extremely important, because only those who have a plan know how to react when nothing goes to plan but they still have to deliver a presentable result. And they need to have excellent technical knowledge: there is hardly any other genre with so many different visualization technologies that you have to be familiar with and understand how to use—if not master in terms of camera technology. The nature documentary constantly unleashes its innovative power in order to record what has not been seen or filmed before. And it has managed to transform equipment that was actually developed for major narrative films into something that can be used productively »outside in the field.«

In no other film genre is the camera department as important as it is in the nature documentary. In the nearly forty years that I have spent working in the nature documentary industry, I have always adhered to one principle: if you have excellent camerawork and an interesting script, you will ultimately be able to deliver an excellent nature documentary. If you have a perfect camera and an average script, you can still create a good nature documentary. If camerawork is just something you »like,« even if your script is amazing, you will never be able to produce a good nature documentary!

In those forty years, however, the camera department has continued to advance at a dizzying speed. Everything has changed so fundamentally — image resolution and frame rate, the number of different cameras and shooting formats, and so much more. This is because the transition from analog film to digital recording technology has made many things possible that were previously inconceivable or extremely difficult to realize. For quality reasons, nature documentaries were the last bastion of analog film technology and were still using Super 16, Super 35, and even the 70mm IMAX format when everyone else was using small electronics. On a normal nature documentary shoot — alone in the camouflage tent with a long tele lens — the advantages were not so considerable at the beginning: complicated digital cameras were extremely expensive compared with their robust film pendants, and the convenience of having much more time on the digital cassette than ten minutes per film roll and, later on, of being able to back film up to hard drives was not immediately significant. The cameraperson's skill still ruled the shot, and in many cases, this still holds true today. Technology has made many things more complicated, but it has also made a lot of things easier. However: a cheetah running more or less frontally toward the camera in perfect elegance — this is something that the cameraperson's skill is still primarily responsible for. They have to be able to perfectly plan ahead what is going to happen, to empathize with the cheetah, to anticipate which prey animal it has chosen, and to position their camera tripod in such a way that they can capture the perfect shot — analog or digital. The superiority of digital technology only makes itself felt when the entire scene has to be recorded in slow motion — like when you are on the tail of the world's fastest land predator. For as soon as the goal is to produce a special recording of whatever kind, digital technology is unbeatable.

I can still remember my first slow-motion nature documentary shoot — on film, of course. My task was to record a kingfisher on the hunt in slow motion. Surely that's impossible, the layperson might think, but with a little biological knowledge and the odd utensil, it is in fact possible to completely outwit the kingfisher. What you need is a camouflage tent, a small aquarium, some piscine prey, a camping chair, a tripod, a branch sourced from the surrounding area, and a low cam — back then a film camera with a slow-motion function. Instead of chasing the film through the shutter at twenty-four or twenty-five frames per second, this camera exposes many more images in order to slow down the situation accordingly. This means that the ten minutes of film roll are pulled through the camera faster, so that only a fraction of the recording time is available.

The set is easy to create: the camping chair is placed in the stream, the kingfisher's hunting grounds, and the aquarium is placed on the chair with the fish inside. Next to this is the tripod with the perching branch mounted on it. The camera is pointed at the subject, the cable release placed in the camouflage tent. And then you wait until the resident kingfisher spies the fish. The trick almost always works, but now comes the difficult part: the kingfisher perches on the branch and peers into the water. You press the button on the cable release, the film runs through, but

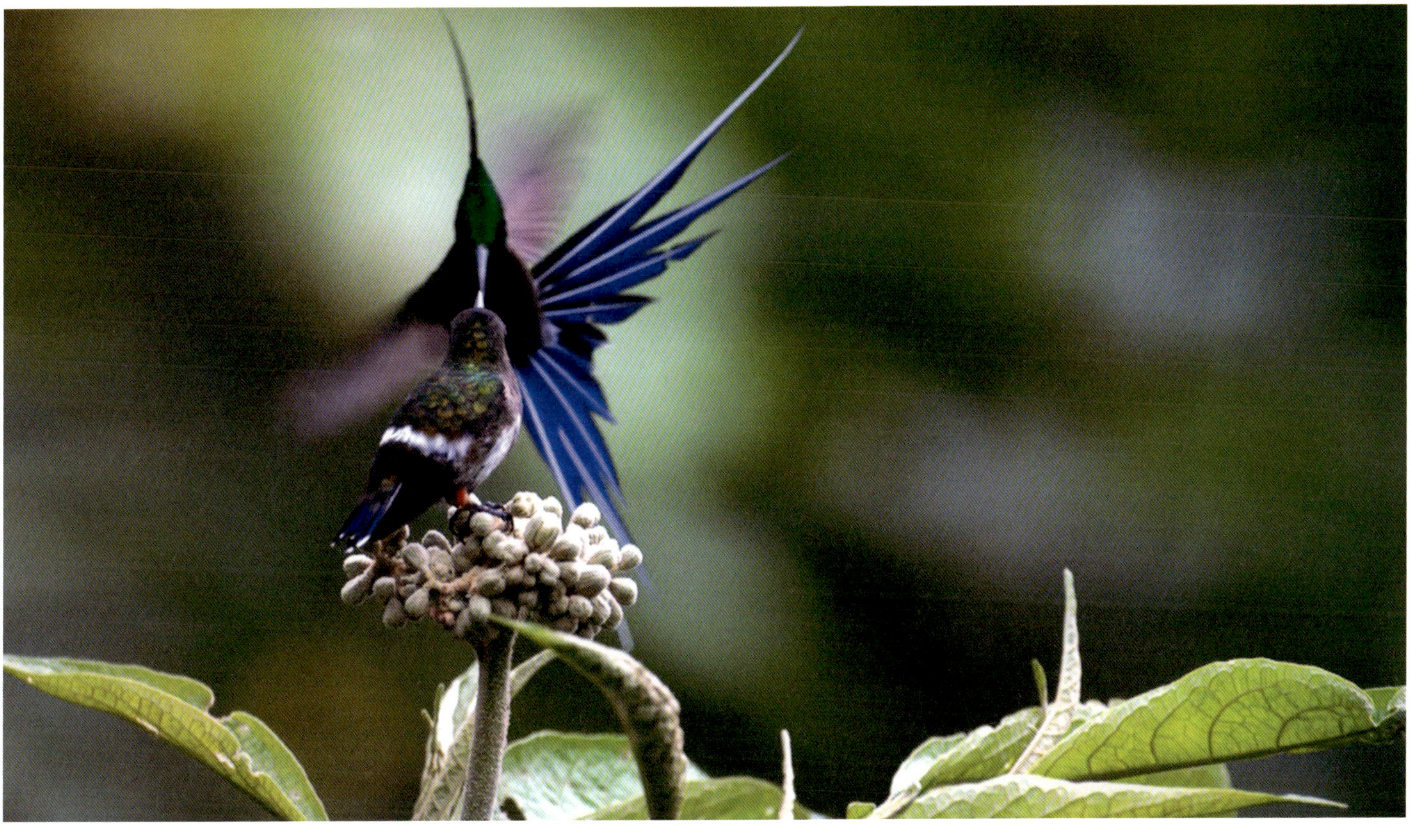

FIG. 1_*The Wire-Crested Thorntail is a hummingbird species where the males are just about ten centimeters long—including their tail feathers. The male has a bizarre way of impressing the female. Its courtship lasts only a matter of seconds and went unnoticed until the Phantom high-speed camera revealed its unusual nature.*

VIDEO

the kingfisher is still sitting on the branch. So, you have to put in some new film… but now the bird swoops in and takes the fish. It took umpteen film rolls, all sorts of little fish, and a lot of time until the shot was in the can—together with my anxiety about having to tell my production manager that we had not been able to keep to the shooting budget.

In order to capture the right moment back then, it took a lot of experience, a feel for timing, and a huge amount of luck. Even bigger portions of these ingredients were required when it came to capturing one of nature's fastest movements on film. That is precisely what Alfred Vendl and I planned in the mid-1980s in our first multipart *Universum* series for the ORF, produced in-house, about the Mediterranean—and the firing of a cnidocyte. To capture this, specialists from the Austrian Institute of Scientific Film had to chase the film through the camera shutter at a speed of ten thousand frames per second. At this speed, the heat of the friction was so intense that the film began to smoke. I can no longer remember how many attempts failed due to the film catching fire, but there must have been a lot of them.

Whether one hundred frames per second or ten thousand—today, in the digital age, shots like these are still exciting, but they are comparatively quickly in the can. The challenge today is of another nature. Modern digital Phantom cameras constantly record eight seconds and then rewrite them if you don't push the release button. The elusive kingfisher would not have had a chance to escape. But there is quite a lot of technical outlay required to use the camera. This was something we

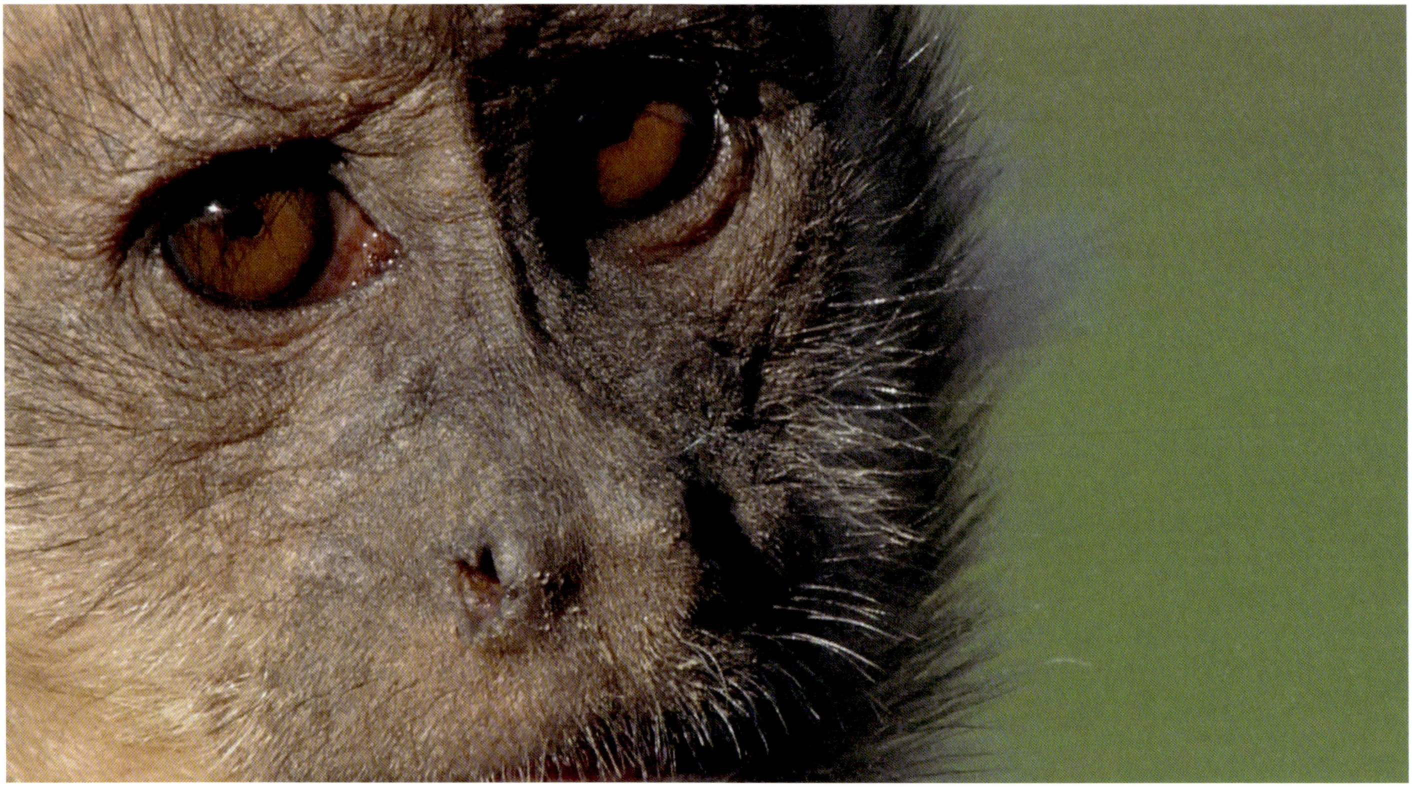

had to learn when I had the idea—soon after Terra Mater Studios had been founded—of shooting an entire film using the Phantoms in more or less slow motion. The protagonists were humming birds, and we wanted to shoot the first full-length nature documentary in complete slow motion about these acrobats of the air_FIG. 1.

What is fantastic about slowing down time is the extreme aesthetic that slow motion creates: each shot seems superelevated, like it is surrounded by an aura of perfection. Moreover, only slow motion can provide insights into animal behavior that have never been observed before_FIG. 2.

We shot thrilling images of previously unknown mating rituals practiced by some humming bird species. The effort was worth it, but it was in fact an enormous amount of work for the director, cameraman, and camera assistant: shooting by day, creating backups and somehow bringing order to the surplus of material by night. After this shoot, we created a new technical assistant position to handle all the material, because otherwise the camera crew would not be able to get any rest during the shoot or to sleep at night.

Nature documentary-makers also like to use the exact opposite of slow motion—the time-lapse. It is fascinating to watch someone squash long periods of time into a few moments by slowing down the recording. Whether it is forcing an entire day, sunrise to sunset, into the space of a minute, observing the orbit of the Milky Way through the night sky, the breaking of cloud formations on the ridge of a mountain, or the back and forth of the tides on a coast—it is often the major natural cycles that the nature documentary can give more space_FIG. 4.

FIG. 2_*In the wild heart of Brazil, capuchin monkeys have learned how to find special food—by using tools. The adults are strong enough to pick up heavy rocks with which they try to crack palm nuts in order to get to the nutrients inside. It may take up to ten attempts before they are successful. The slow-motion shots give an idea of the struggles that these monkeys have to work through to reach their goal.*

VIDEO

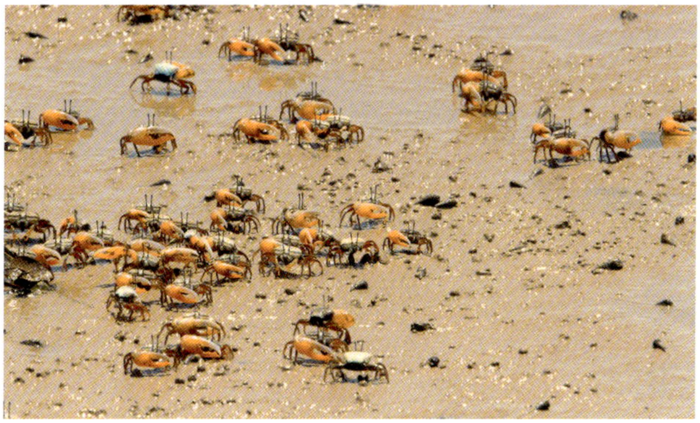

FIG. 3_*Time-lapses help to visualize the continuous ebb and flow of the oceans, thus also revealing the appearances of strange creatures in the world's coastal tidal zones. Here, we see Fiddler crabs that live on many of the countless islands of Southeast Asia that gave the production* ISLANDS IN TIME *its name.*

VIDEO

But the time-lapse gets really exciting when it comes to animal behavior that otherwise takes place far too slowly to be observed. With the time-lapse, we can literally watch the grass grow, we can observe starfish carrying out raids on the bottom of the sea, and turn snails into veritable rockets_FIG. 3.

Time-lapse technology becomes truly fascinating when it is not just static — when the camera moves too. The easiest thing about this is that the camera simply swivels along a pre-planned route. The scene becomes more exciting, gains additional depth. And it gets really exciting when filming species that are more or less motionless for us — when the wind is not involved — as they really are. I will never forget our first major attempt to document the actual nature of a blackberry bush — a fast-growing creature that attacks like a predator and thrashes about in order to conquer new territory. The BBC realized this scene for the great David Attenborough multipart series *The Secret Life of Plants* in the mid-1990s. How it worked: in the studio, the camera was placed on a mini-dolly that was moved about by a stepper motor and synchronized with the predicted growth of the blackberry bush. This works using light, of course, because plants grow toward the light. And as long as the spotlight shows the bush the way, it will grow in that direction, and the camera can easily follow the bush's instincts — until the artificial sun breaks down and its instincts change direction; then you have to rebuild the scene and send somebody to make sure that the spotlight really is shining and that the motor is moving at the right speed. Only then can memorable shots succeed — when a blackberry bush becomes a furious snake!

Digital technology has not been as revolutionary for the time-lapse as it has been for slow motion. Photo cameras of varying quality have always been used for time-lapses. But, of course, now you do not have to change the film, and it is easier to program and steer the scenery. And the result is, of course, much better — free of background noise and, where required, in 4 or 8 or even 16 or 32 K image resolution.

Another technology that has been revolutionized in recent years and that plays a major role in nature documentaries is aerial shots_FIG. 5.

It was only recently that I took a look at our first films from the 1980s. The degree of wobble that we accepted back then in aerial shots is quite shocking from today's perspective. But then the great nature documentary revolution took place — and it started in Austria. Georg Riha's skill when it came to aerial cinematography shocked the international competition. He was one of the first people

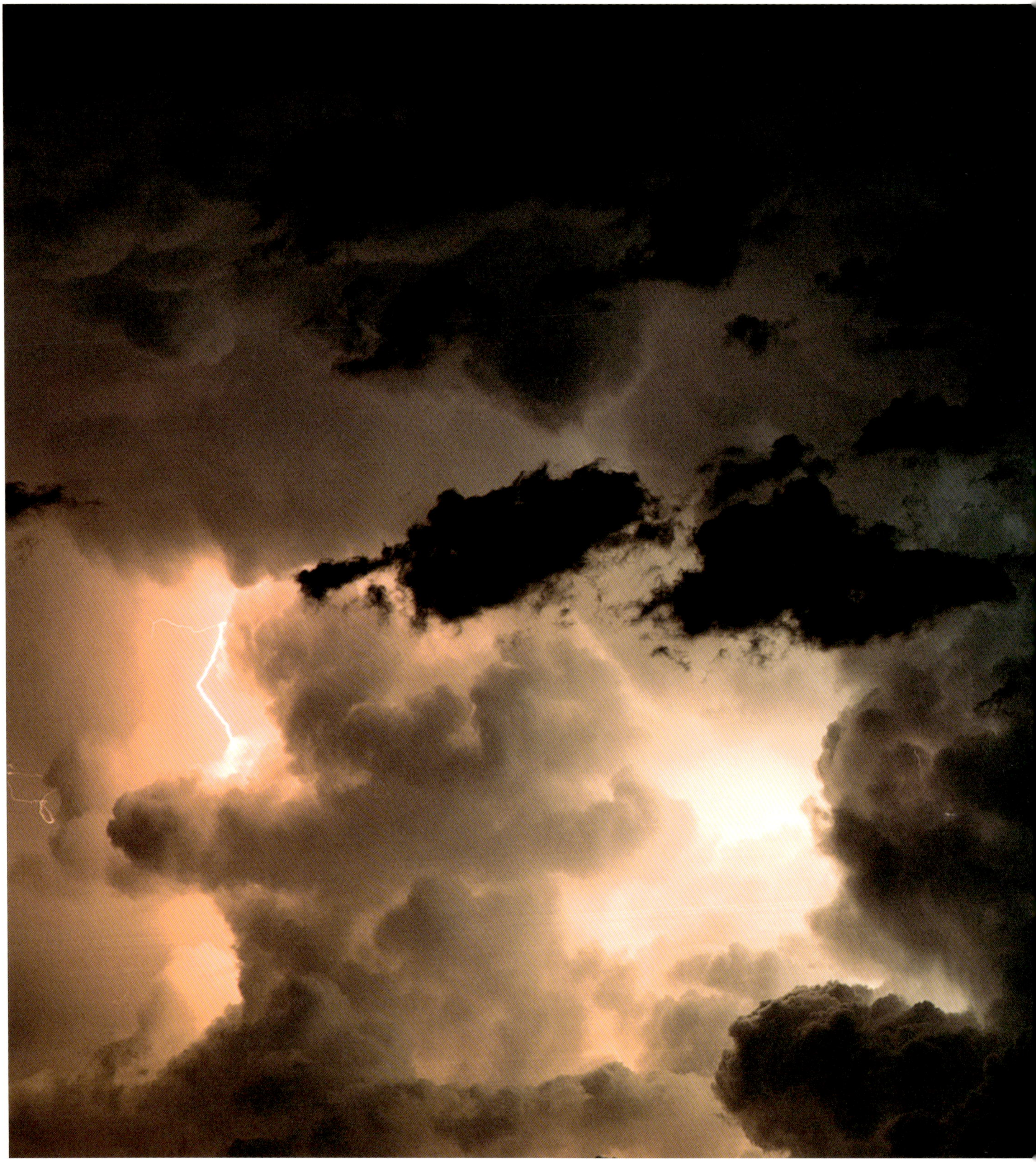

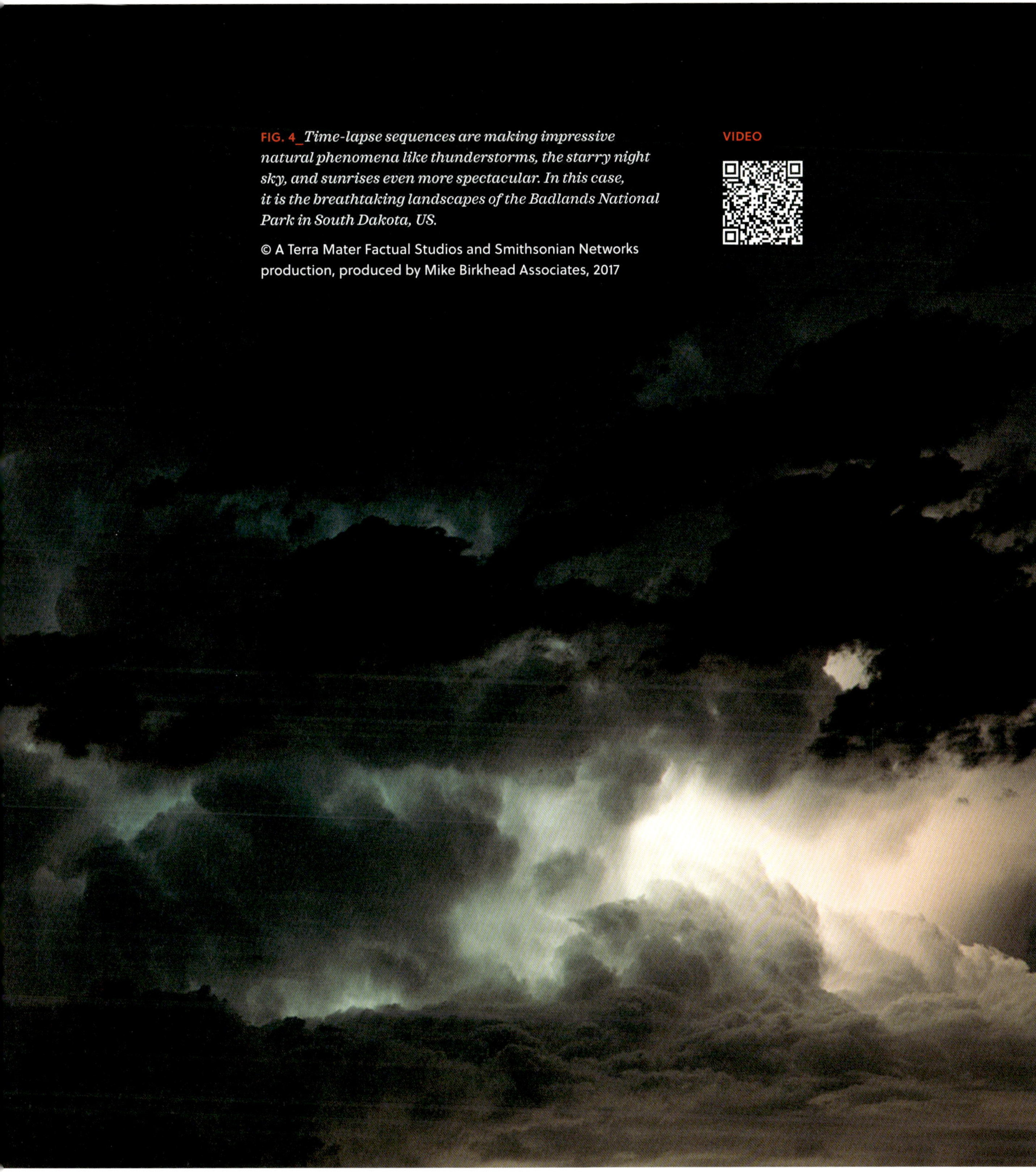

FIG. 4_*Time-lapse sequences are making impressive natural phenomena like thunderstorms, the starry night sky, and sunrises even more spectacular. In this case, it is the breathtaking landscapes of the Badlands National Park in South Dakota, US.*

VIDEO

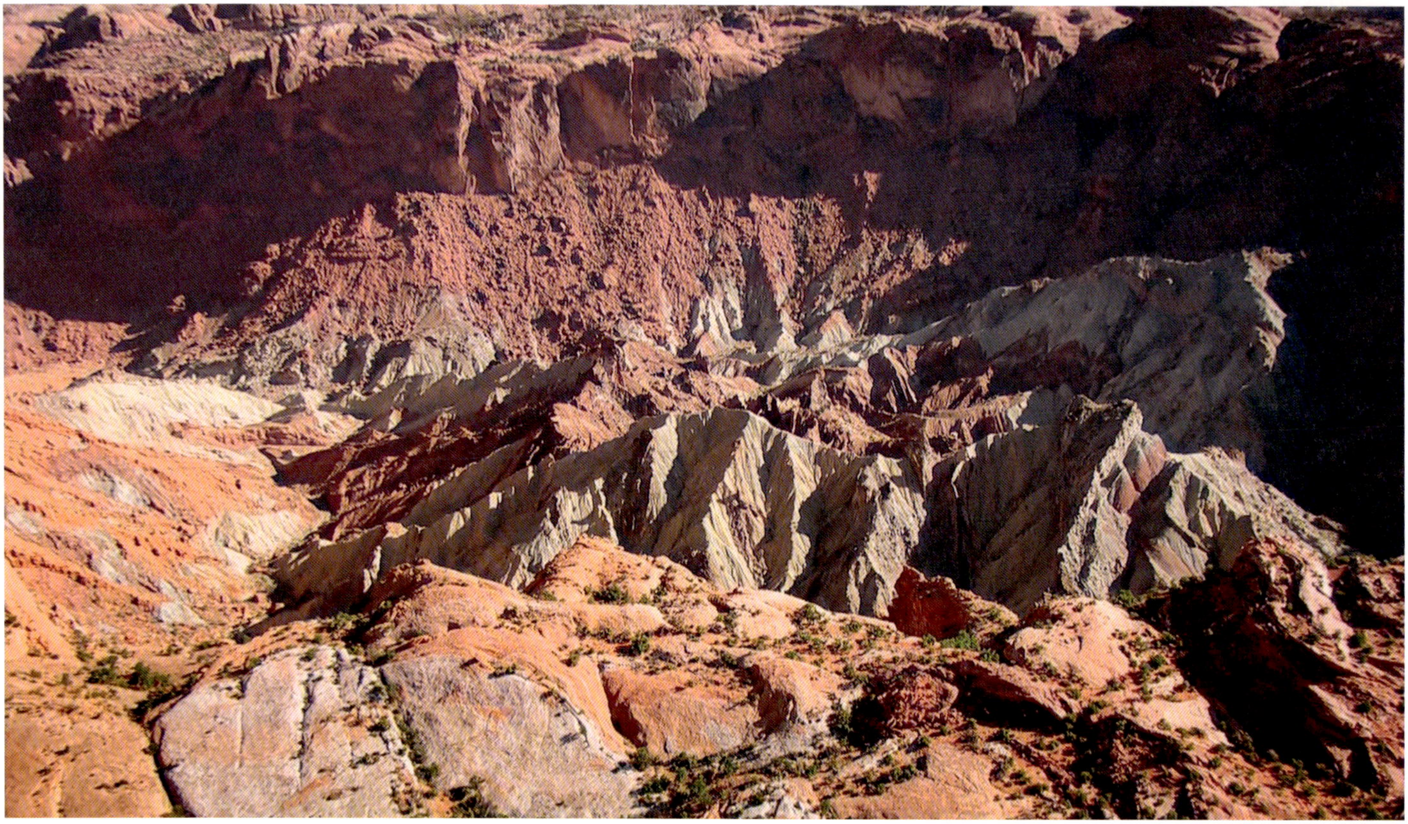

in the world to use the Wescam system from narrative film—a gyro-controlled camera mount that absorbed all of the vibrations of the aircraft and allowed the zoom objective to be controlled fully remotely—in documentary films. However, it was not enough for him to attach the Wescam to the aircraft wing or helicopter—he needed his own projects. His camera blimp became a thing of legend: a small zeppelin with a platform with a Wescam mounted to and hanging down from it, which was attached to three vehicles on the ground by winches. The entire system looked like a giant tripod that allowed for extremely gentle movements, and it was also possible to use it in places where there was a no-fly zone for helicopters. He achieved fame for his recording of Vienna's St. Stephen's Cathedral, with his camera floating just meters about the southern tower's steeple and staying there. Many colleagues back then could not explain how Georg shot these images. The blimp system was spectacular, but susceptible to wind. However, Riha developed another, groundbreaking system. With his Camcat, a kind of cable car for the Wescam, he was able to float above the landscape as far as he was able to span a rope. The shot he took of the Kläfferquelle—one of the exit points for Vienna's mountain spring pipeline—from above as it frothed in spring is still breathtaking today.

But what is really fascinating about aerial shots is that they do not just illustrate the beauty and atmosphere of a landscape but can also capture exciting animal behavior as it unfolds—as long as you can pay for enough helicopter flight hours. It was once again the BBC that first used this technology in a groundbreaking

FIG. 5_*Utah's Canyonlands National Park is a natural wonder—and has a particularly dramatic structure. But it's only from the air that the giant Upheaval Dome is revealed to be an ancient meteorite impact site. A spectacular Cineflex zoom-out shot provides a stunning comparison between the size of a person standing on the rim and the massive rock structure—the distance to the camera helicopter is two kilometers...*

VIDEO

way for the mega series *Planet Earth*. Their Wescams observed African wild dogs on the hunt. For the first time, cameras were able to record from the air how cleverly the pack advanced — with a central pack, single animals coming up the flanks, and others putting themselves, front on, in the path of the fleeing gazelles. This also enabled biologists to see up close why the success rate of African wild dog hunts is so extremely high, at over eighty percent — much higher than for lions or cheetahs. We, too, used the flying cameras in this way, for example, in our IMAX film about the Alaskan Arctic, to film wolves hunting caribous.

But the real revolution in nature documentary-making came just a few years ago — when drones took over from helicopters. Although we still need the giant dragonflies sometimes for special shots, the small drones have long outranked them. They are easy to use, very environmentally friendly, much more cost-efficient, and can therefore be used universally_FIG. 6.

Drones come in all possible shapes and sizes, and there is hardly a film today that does not make use of camera drones. They do not just provide overviews of landscapes from all kinds of altitudes but can also stay in the air long enough to film rare animal behavior without disturbing the protagonists or influencing them at all_FIG. 7.

And drones can — depending on the flight savvy of the pilot — produce images similar to the ones that Georg Riha conjured up using his much more complicated Camcat. We have used them for long-term observations — for example, in our cinema documentary *Sea of Shadows*, for which we filmed the campaign to save the Vaquita, the world's smallest whale. We had two large drones watching the action, and when the battery ran out for one of them, it automatically flew back to our convoy ship, where it was replaced by the second one. And at night — equipped with image enhancers — they allowed us to track poachers and smugglers by camera.

Apropos tracking: Wescams, today, above all the Shotover, are still being used to obtain wobble-free images — but, in nature documentaries, they are not just attached to aircrafts. Mounted on an off-road vehicle, they allow the cameraperson in the vehicle to keep pace with the animals. Flying with hand-reared geese — not a problem, as long as the human goose-mother is sitting beside the cameraperson in a car, boat, or ultralight. Having said that, the bird's eye view is still reserved for drones and aircraft with cameras attached — and animal camera staff as well. The renowned Eagle Cam delivers legendary images from the eagle's perspective. Extremely light but high-resolution cameras have been adapted to fit the bird of prey and show the hunting flight from the eagle's perspective, or they direct the objective at the eagle's face to capture the bird's reactions_FIG. 8.

But nature documentary-makers are also smuggling other robots into animal gatherings to obtain very intimate images. The legendary Alan Root was the first to do this when he mounted a film camera in an empty turtle shell to film the stampede of the wildebeests in the Serengeti up close. Today, all kinds of robots, some of them very well camouflaged, are on the hunt for intimate animal images — from model cars that drive into prides of lions equipped with cameras, to mini jeeps disguised as piles of dung to spy on elephants, not to mention cameras fixed to the backs of sharks or whales. Robot penguins have also been successful: one digital member of the penguin species was actually accepted into an emperor penguin colony — including clear attempts at mating made by the real penguins.

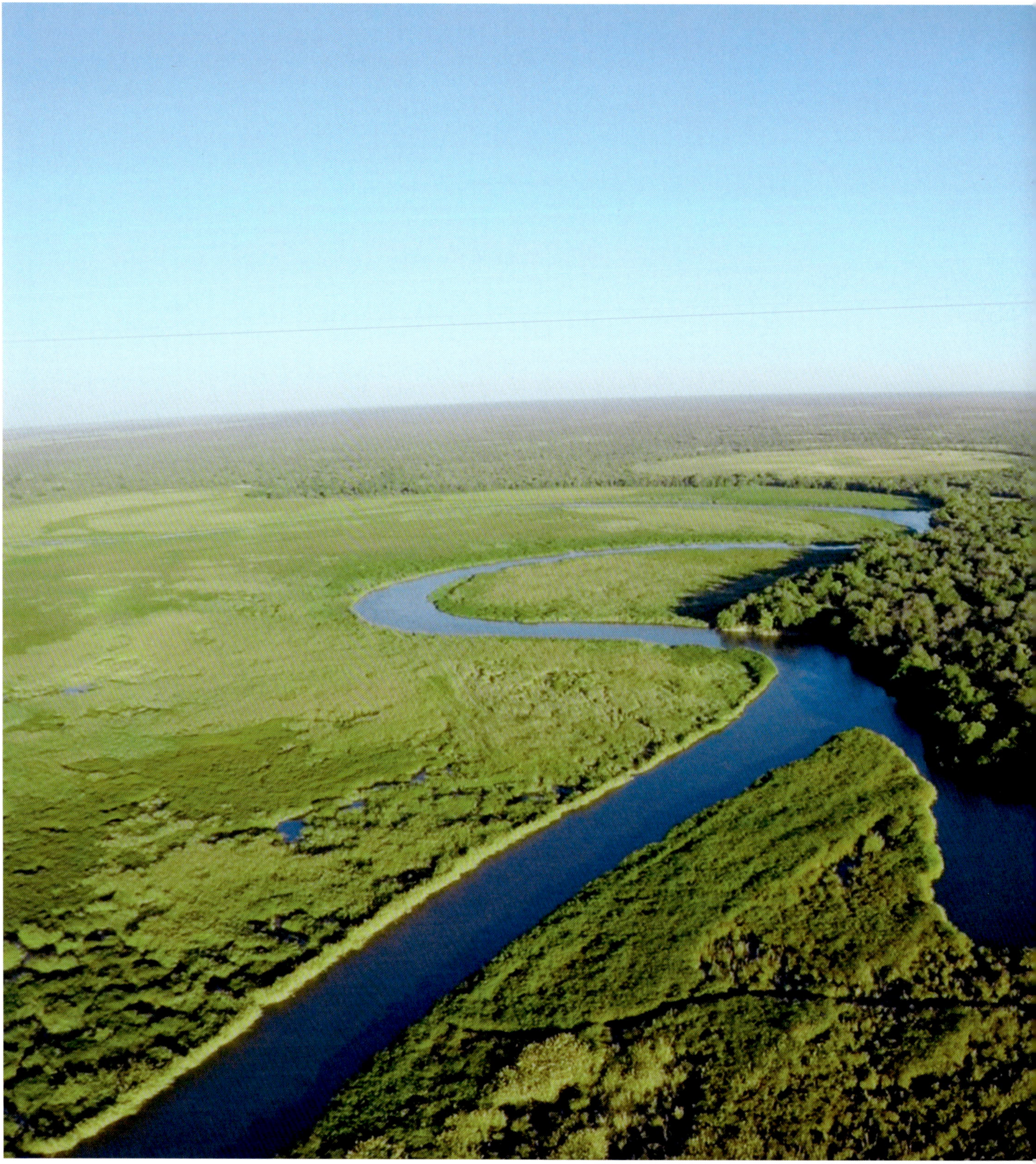

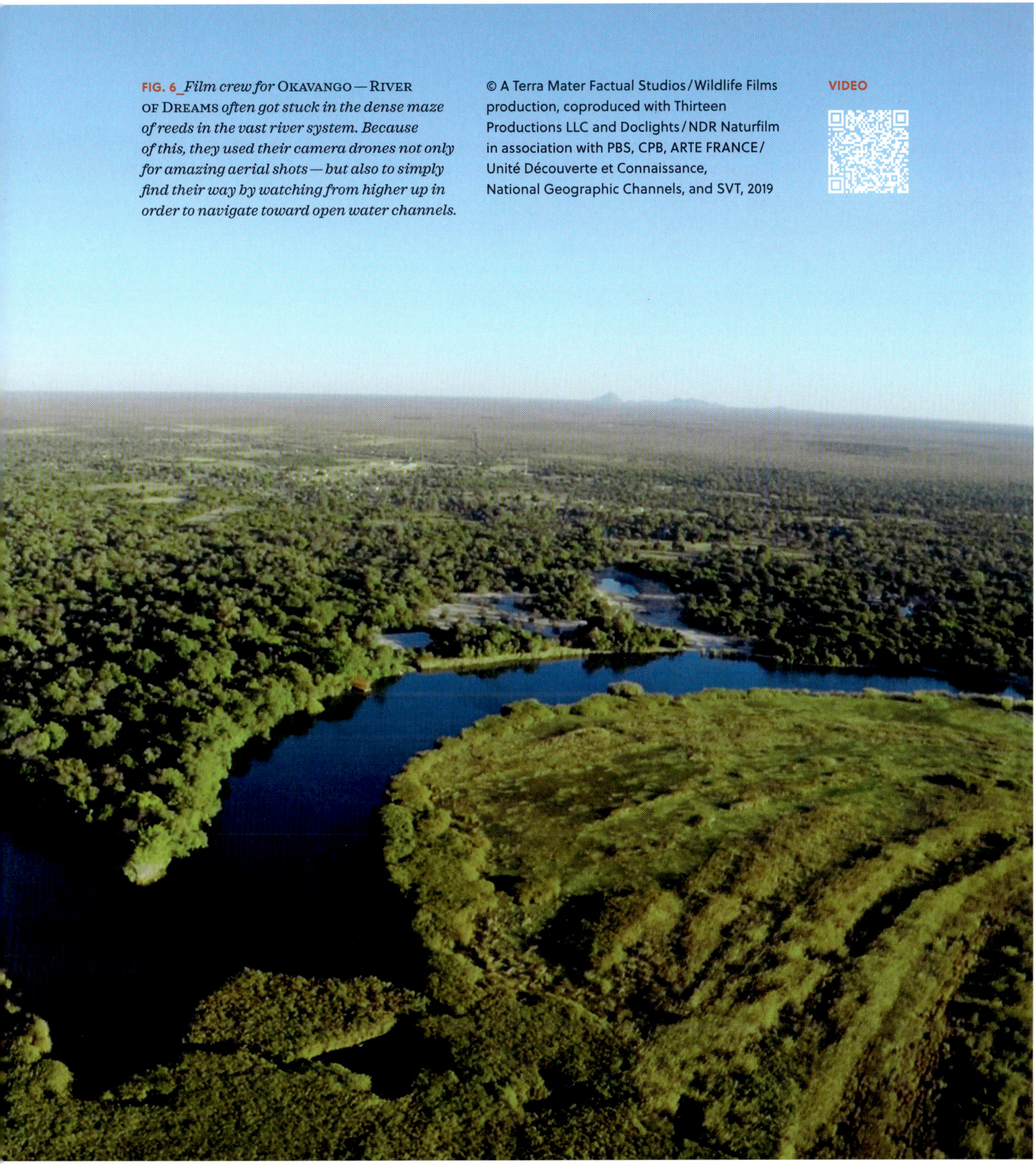

FIG. 6_*Film crew for* OKAVANGO — RIVER OF DREAMS *often got stuck in the dense maze of reeds in the vast river system. Because of this, they used their camera drones not only for amazing aerial shots — but also to simply find their way by watching from higher up in order to navigate toward open water channels.*

VIDEO

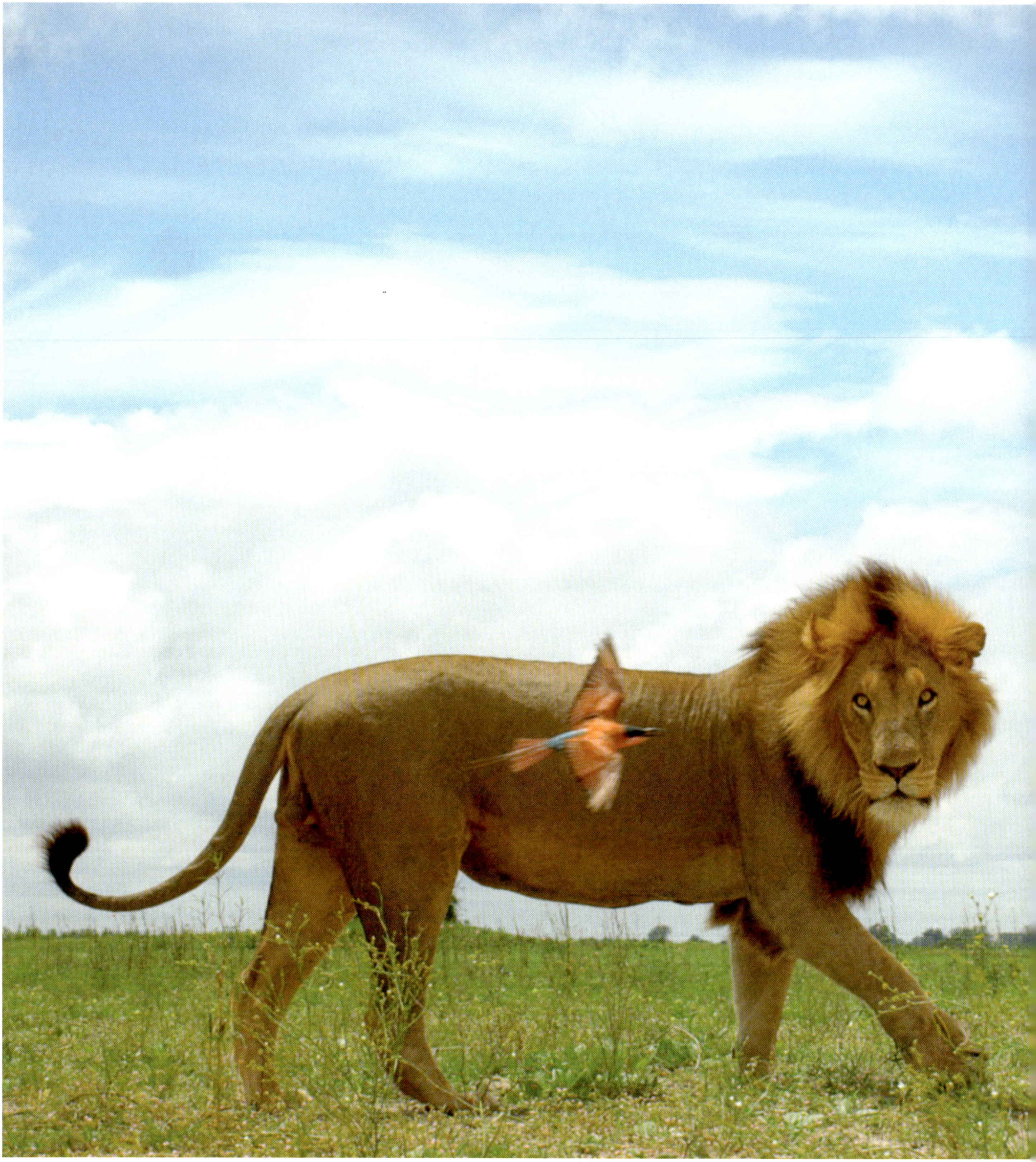

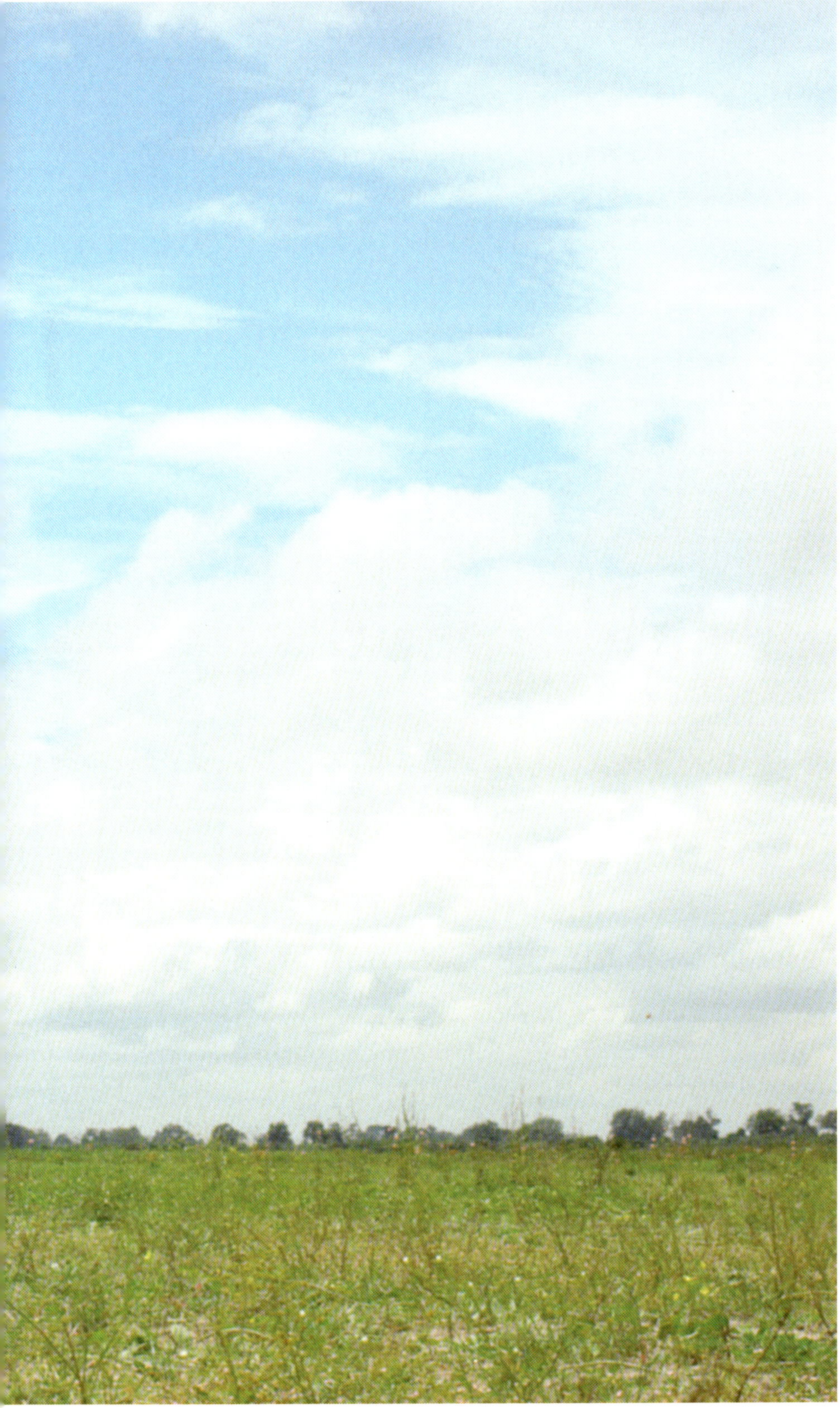

FIG. 7_ *From a car driving on rough terrain, it's almost impossible to get smooth tracking shots of animals on the move — unless the film crew is using special equipment. During the shooting of the four-part series* OKAVANGO — RIVER OF DREAMS, *stabilizing camera rig systems helped to capture amazing slow-motion tracking shots of lions, wild dogs, and leopards.*

VIDEO

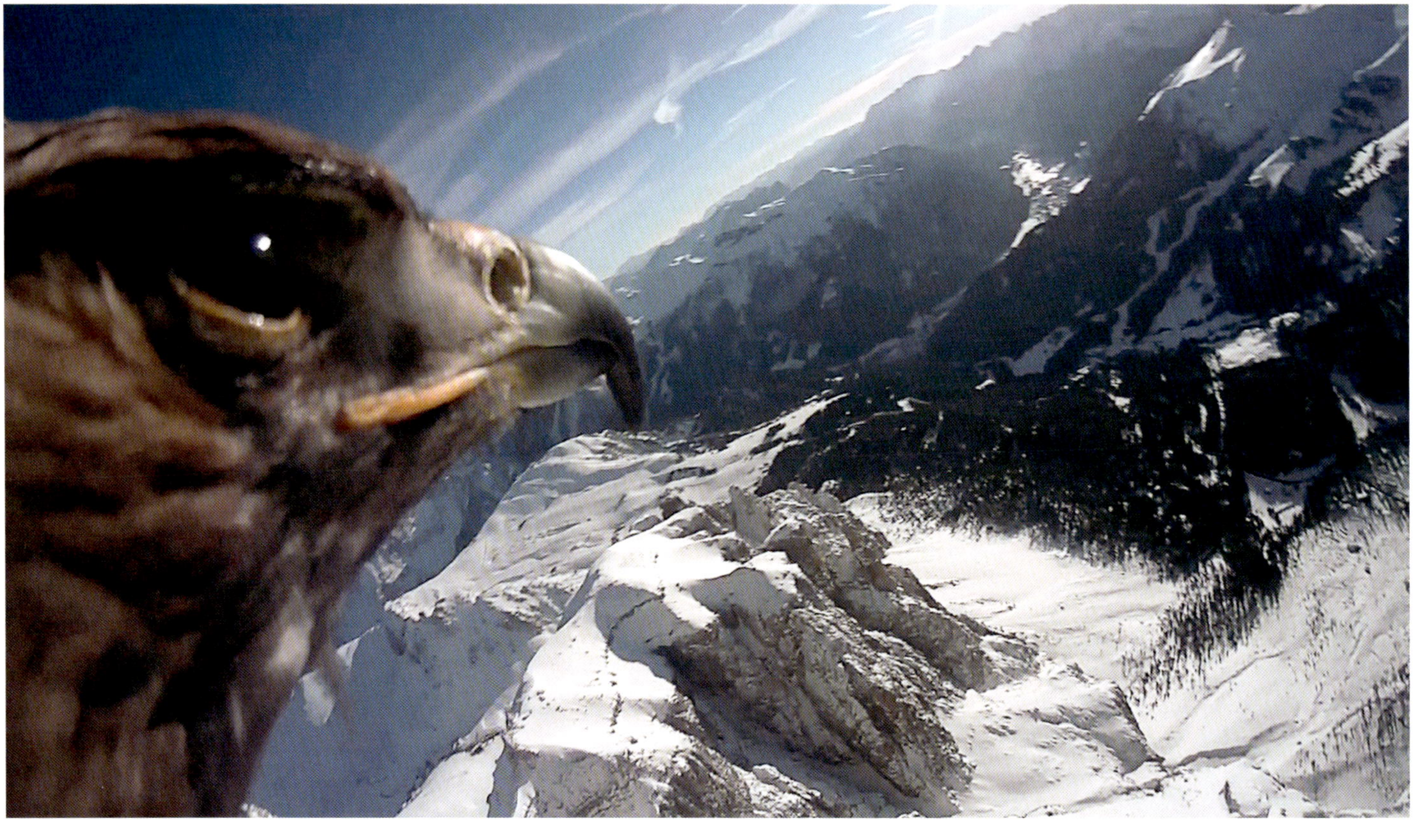

As I said, although these innovative camera systems have given rise to intimate points of view, including some funny moments like the one with the penguin cam, and even though they can give a film a certain kick, they cannot replace the real cameraperson and their craft. This craft faces extreme challenges when the mission is to obtain shots at a time of day when there is a lot going on in the animal kingdom: nighttime. Film is the art of capturing light and creating images from it. In the analog age, people made use of more photosensitive film stock, which, however, had two major disadvantages — a profound loss of color and a massive increase in film grain, the notorious »film noise.« Both problems have more or less endured, although digitally in such a weakened form that it is now possible to create night images that are actually breathtaking in their impact and beauty. New digital cameras are equipped with such photosensitive sensors that they have largely eliminated an old problem from nighttime documentary films — disruptions to or distortions of animal behavior caused by artificial light. The reactions of blinded animals ranged from disturbed to aggressive; I can still remember films from the 1980s that recorded nocturnal duels to the death between lions and hyenas. These films prompted the criticism that it was the strong floodlights used to illuminate the situation triggering this behavior in the first place.

Today's cameras get by for the most part without artificial light; the light of the moon or, during a new moon, even the light of the stars is enough for them. They are called Starlight cameras and can gather up the remaining light on their chip in such

FIG. 8_*It was a challenging task: for* BROTHERS OF THE WIND — *Terra Mater Studios' feature film — the producers wanted to let the audience experience the realm of the golden eagle in the European Alps — but from the* EAGLE'S *point of view. The unique combination of an HD lipstick camera and a trained golden eagle resulted in incredible shots, the like of which hadn't been seen before. Meet the Eagle Cam…*

VIDEO

a way that, depending on the amount of light available, it allows them to produce a perfect image that, in spite of the darkness, still offers nuanced color. It was not until the invention of these cameras that it became possible to observe nocturnal Africa in unique images, to figure out how the hunting behavior of, e.g., lions actually works at night. The darker it gets, the grainier the image becomes here too, although the pixels are much smaller than film grains and the quality can be further enhanced in postproduction. It is only with cameras like these that it became possible to make some of Terra Mater's hit films—above all the production *Life That Glows* with David Attenborough about bioluminescence in the animal and plant kingdoms. The artistry of this new generation of cameras has allowed us to capture this spectacular light phenomenon as never before—glowing mushrooms and glowworms above water, colorfully pulsing jellyfish under water. Most spectacular were the dolphins plowing through a bioluminescent tide_FIG. 9.

But there are places where these Starlight cameras meet their limits—places of enduring darkness: caves. Caves do not just provide an important home to species such as bats; some caves are also chosen by the giants of the animal world, because this is the only place that they can find the minerals necessary for their survival. In Uganda, for example, there is the legendary Kitum Cave, which elephants journey to again and again in order to scratch salt from the cave walls with their tusks in total darkness. The elephants have embarked on this journey for generations in order to consume the vital salt. The path has been well trodden but is still extremely dangerous for these heavy animals. The Terra Mater film team illuminated the cave with infrared spotlights and captured the scene using infrared cameras and fixed camera traps. Infrared light is beyond the limits of our human perception and also that of elephants. On camera, it delivers perfect images from the world of eternal darkness_FIG. 10.

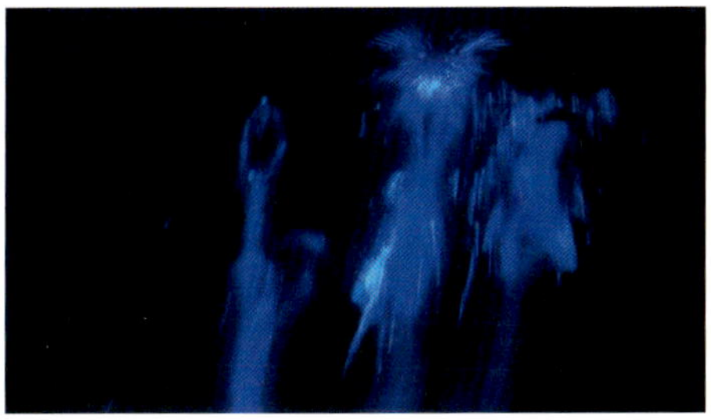

FIG. 9_ *At first glance, filming in the pitch-black night seems impossible. But extremely light-sensitive cameras and a special postproduction process reveals almost otherworldly sequences of »glowing dolphins« escorting the film crew's boat. The blue glow is bioluminescent light, created by tiny microorganisms called dinoflagellates that live in every ocean on Earth.*

VIDEO

From the largest creatures on land to the rather small—because the latter are the El Dorado of macrophotography. And macrophotography also requires a lot of light, because the objects of desire are always very close to the objective. In the analog age, this meant the death of the odd insect subject: the heat of the spotlights, even when they had been cooled down, often spelled the end for the little film stars. It was not until the invention of digital technology—first cassette- and then file-based—that this changed, and it became possible to hold the macro lens's hunger for light at bay. It is in matters of macro in particular that the Austrian nature documentary has a long tradition. Kurt Mündl's legendary *Housefly: An Everyday Monster* was filmed as the ORF series *Universum* was taking off on its international trajectory, and productions like Wolfgang Thaler's *Ants—Nature's Secret Power* and *Bees—A Life for the Queen* became international sensations because they allowed the little crawlers to appear in a completely new light. Thaler, in particular, is a master of macrotechnology. He built a glass city for his ant colony, allowing him to elicit many visual secrets from them. He used his special cameras to penetrate huge natural nests and flew with queen bees in their swarms. Thaler was the prototype of the congenial nature documentary-maker—a gifted cameraman (who also put his craft at the disposal of major narrative films), a beekeeper, and an insect fan, all in one. By paying painstaking attention to detail, he pushed his films to the limits of what was technologically possible at the time. It was only for the title of his ant film that we went back to the analog—we did not use graphics or animation, just ant

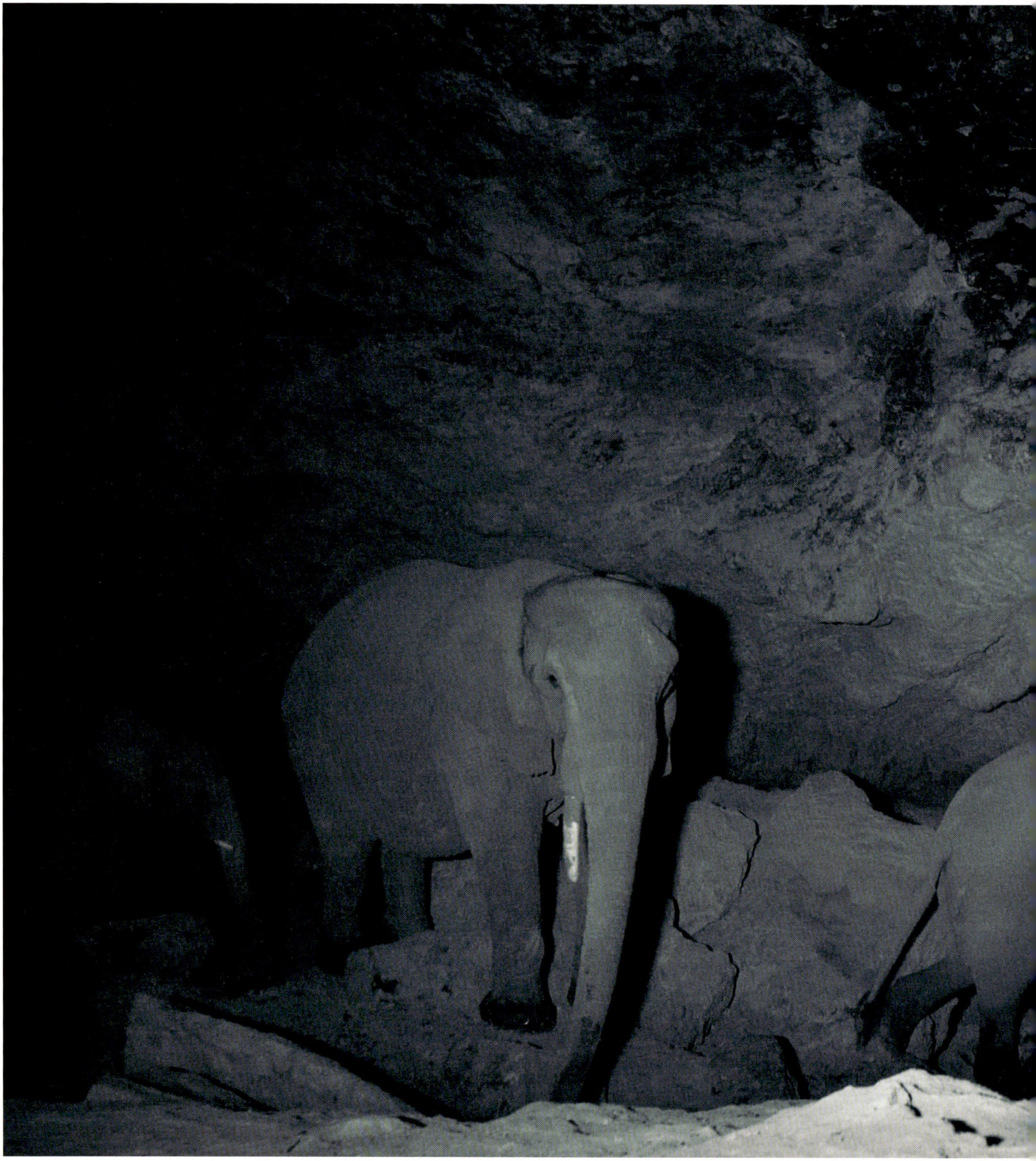

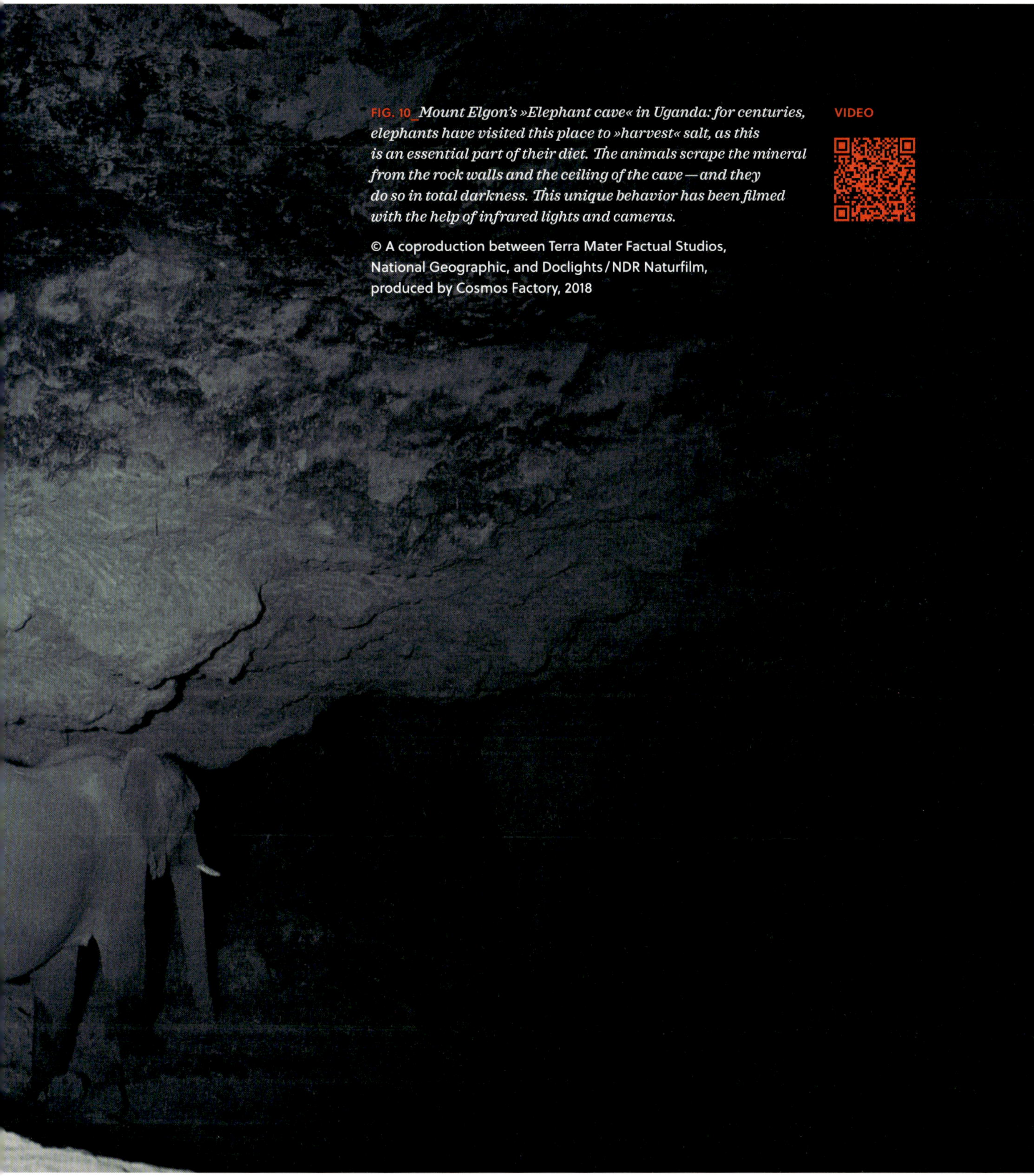

FIG. 10_*Mount Elgon's »Elephant cave« in Uganda: for centuries, elephants have visited this place to »harvest« salt, as this is an essential part of their diet. The animals scrape the mineral from the rock walls and the ceiling of the cave — and they do so in total darkness. This unique behavior has been filmed with the help of infrared lights and cameras.*

VIDEO

aesthetics! With the help of pheromones, we got some leafcutter ants to march over a thin branch, from one glass platform to another. Beforehand we had dipped small pieces of paper in pheromones, each of them with a letter from the title on it, and the ants carried them like leaves past the camera, from A to B. In this way, the ants themselves became part of the title design, the only thing we had to help them with was the grammar.

Thaler made his macro films at the beginning of the high-definition era—it is difficult to imagine what he would have been able to produce today using the latest technology, which has led to the advance of digital technology and the unbelievable miniaturization of the highest-quality objectives. Something similar transpired during our last great ant film, which only came out a couple of years ago. For *David Attenborough's Ant Mountain*, we filmed in the world's largest ant colony in Switzerland. We used the Frankencam, a genius invention by the English cameraman Martin Dohrn, who also revolutionized the Starlight camera _FIG. 11.

The Frankencam is a modular camera system with unbelievable resolution that makes it possible to transform the viewer into the ant by means of an extremely small camera lens and a sophisticated tracking system. The camera is so tiny and mobile that it is able to immerse itself in the column of ants, so to speak, presenting the ant's perspective. With its 4K miniature visuals and high-resolution endoscopes, we used it to follow the ants around in their underground realm and therefore to show animal behavior that it had never been possible to film before—war and peace in the kingdom of ants.

Camera technology and the craft of camerapeople are at the heart of the nature documentary. And yet, in recent decades the genre has been enriched once more, and a new squad of artists has taken the stage. Digital technology has made it possible for us to create our protagonists ourselves and to place them in real natural spaces. I am talking about photorealistic computer animation. I can still remember well the four-part international series about dinosaurs that I directed in the early 1990s. With painstaking attention to detail and at great effort with puppets, animatronics, and stop-trick technology, we brought the world of terrifying lizards, everyday research life, and the history of paleontology to life. Just a few weeks later I sat at the premiere of Steven Spielberg's first *Jurassic Park* in the cinema as the T-Rex trampled over me. I understood that the series that we had only just broadcast was as old as yesterday's newspaper and would certainly not be repeated again. At the turn of the millennium, the BBC followed with *Walking with Dinosaurs*, mixing animation and animatronics and pairing them with real filmed environments—an incredible documentary hit. For the first time, it became possible to take this expensive technology from the cinema and broadcast it into the living room. Later, we teleported this approach into the future with international partners and produced a series with the title *The Future is Wild*. Here, we allowed the animal world to keep evolving and imagined what the Earth and its inhabitants would look like in 5, 50, and 250 million years. Together with an entire parade of top scientists, we came up with the most bizarre creatures and brought them to life in bits and bytes. Back then, some of the effects were still impossibly difficult—depicting fur and feathers photorealistically was extremely expensive but still looked fake. Even getting viewers to believe that these heavy dinosaurs, which weighed tons, were actually walking on the earth was still a little bit difficult sometimes. It felt like the digital mon-

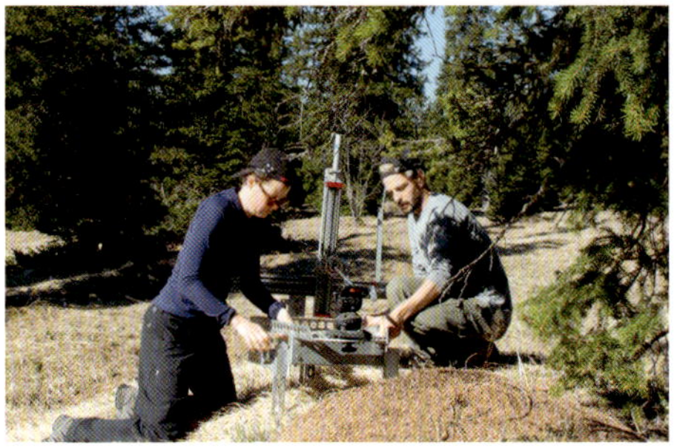

FIG. 11_*Meet Frankencam—a device for positioning tiny cameras and small wide-angle lenses into awkward corners with extreme precision. It enables the film crew to enter the world of ants in a way that has never been achieved before. Frankencam's tiny lens has amazing abilities, and it's remarkably cheap, because lenses like these are made by the millions—for the cameras on your mobile phones...*

VIDEO

sters were floating rather than touching the real desert soil and leaving behind footprints. All of these problems — including the budgets — are now a thing of the past. Disney's real *Lion King* proved that we can now conjure up living animals in their full furry splendor as deep fakes, and the new BBC dinosaur series on AppleTV+ gets the giants to really touch the ground.

While Disney's 3D lions were not lacking in photorealism but somehow in life, the dinosaurs are still lacking in real relevance. They are constantly being squashed into scenes that I am familiar with from earlier films, copying the behavior of other contemporary living creatures. Of course, this happens in consultation with top paleontologists and is just an approximation of how they might have lived, what they might have looked like, and what colors, feathers, or armor they might have had. Nothing is certain, even the most speculative ideas are allowed.

But there is one kind of animation that is based on secure scientific facts, i.e., what is »real« in the actual sense of the word. However, it is not about the largest creatures to have ever walked the Earth but the smallest, which still exist today — from mites to tardigrades, plankton algae and single-cell organisms to viruses. For the Terra Mater 3D production *You, Planet*, we initially tried out and were quickly won over by this expensive technology. I am talking about the work done by the Institute for Science Visualization at the University of Applied Arts Vienna, led by Alfred Vendl, with whom I have had a professional friendship since our first collaboration in the late 1980s. He and his team have managed to scientifically prepare animations of these creatures so precisely that the results can be considered »true« as far as documentary and research are concerned. They use all conceivable visualization methods — from the scanning electron microscope to different medical visualization methods like MRI and CT, in order to ultimately allow viewers to experience a computer-animated, but still real camera trip through, e.g., the intestines of a termite; some of the 100 trillion organisms that inhabit the ecosystem of our bodies, without which we would not be able to survive; a battle between killer cells and viruses in our circulatory system; or the most bizarre plankton species that live in one drop of water. Every detail is, so to speak, »real«; the image is »true« — apart from the colors, which do not actually exist at this level in the macro world. The color is only there to help our reception patterns.

Nature and film documentaries were at the birth of the film advances that took off with Eadweard Muybridge's galloping horse in 1887. Since then, the genre has continued to gift the medium of film and its audience with new innovations. The possibilities of visualization have continued to expand, the art and craft of camera technology have been further improved and refined. This will also continue in the future and will provide us with new, even more spectacular images of the world that surrounds us. I am often asked if there is anything left out there to film today and whether the audience is losing interest. The answer is simple; there are still millions of natural stories that have not been told, millions of species that have not been photographed. The only danger for nature documentaries is the constant species extinctions that have rapidly increased in number in the last forty years that I have spent working on nature documentaries. The only real danger for the nature documentary is that we are slowly losing our wild protagonists. Simply filming animals in the zoo or creating them using computer animation is not an alternative. Conserving biodiversity is the dictate of the hour!

Digital Twins for Civil Infrastructure Applications

Amirali Najafi, Ali Maher

The current civil infrastructure ecosystem is highly decentralized and risk-averse, incentivizing short-term and resource-intensive solutions that accelerate climate change and cannot quickly adapt to the dynamic needs of society, the economy, and the natural environment. Modern technological advancements present an opportunity to reinvent the current built environment from a life-cycle perspective. For example, sensing, automation, and artificial intelligence technologies are helping to encourage efficient behaviors and infrastructure adaption instead of new construction.

Contextualize your home in relation to its surroundings. Whether you reside in a city or the countryside, a number of systems are necessary to support our modern lifestyles. These systems include the large intricate networks of pipes that provide us with safe drinking water, the electric lines that power our homes, and various means of transportation including streets and bus stops, which allow us to commute. Civil infrastructure is a general term that designates a wide array of systems that support modern human life and activities, and civil engineers are the planners, designers, and makers of these systems.

Billions of dollars in investments are needed to design, construct, and operate civil infrastructure systems. The world's leading economies are often referred to as »developed« nations because, in these countries, the infrastructure was funded and built by the mid-twentieth century. The countries in North America, Western Europe, and much of Oceania and Eastern Asia are now regarded as developed nations. Remarkably, many of these nations had to rebuild from scratch following the utter destruction wreaked by World War II. As for countries deemed to be »developing,« the definition is not universally agreed upon. Some of these nations are moving toward progress. Meanwhile, others are retrogressing due to local and global conflicts, neocolonialism, corruption, and the ineptitude of leaders. The point here is that infrastructure projects are expensive and are directly linked to job creation and the growth of modern economies.

Now that we have established a brief definition and history, we will shift our attention to the current state of infrastructure. All around the world, infrastructure is subject to deterioration problems. We have all seen the bumpy road surfaces, corroding steel bridges, and leaking water pipelines. Yes, proper design and construction practices can prolong the life of a system. However, eventual deterioration problems are inevitable. Any material or system repeatedly used by humans and machinery, subject to hot, cold, and wet environments, and exposed to corrosive chemicals will eventually degrade. An average vehicular bridge for example may stand for seventy-five years with constant heavy truck use, will experience repeated freeze-thaw conditions, and be exposed to large varieties of deicing chemicals.

Similar to other international civil engineering organizations, the American Society for Civil Engineers (ASCE) publishes a yearly report card of US infrastructure scores. This report card considers, among other things, key national assets such as aviation, bridges, broadband, dams, drinking water, energy, hazardous waste, inland waterways, ports, rail, roads, schools, and transit infrastructure. The US infrastructure scores a meager C–. If we wish to know the »whys« and »wherefores« behind this score, we must dig through years of leadership inaction, poor planning, and decreased and inconsistent investment. Reactions to these infrastructure problems have mostly been inconsistent and delayed fixes, as illustrated by the example in_FIG. 1. In this image, a highway bridge is supported by a freshly painted green steel girder (also called a »beam«). Note the complete loss of section near the bottom of the girder web. That is due to corrosion. A coating of paint is highly effective at creating a barrier between moisture and steel, and should have been applied sooner. Clearly, this image indicates a period of inaction before new paint was applied.

The current state of civil infrastructure can therefore be safely categorized as poor in many parts of the world. How do authorities then ensure our safety and, for instance, that our bridges will not collapse? The answer is periodic inspections. Government and civil infrastructure agencies dispatch professional inspectors to observe and record the state of various civil systems. Inspection reports are handed to engineering services for detailed evaluation. Engineers report specific recommendations back to government agencies. Finally, infrastructure assets are classified by urgency for remediation and replacement. This is not to say that every infrastructure system is inspected: concealed systems such as underground water pipes may never be inspected. In addition, inspection reports are often lost or forgotten as a result of bureaucratic complications.

Periodic inspections are typically performed every year or two, depending on the infrastructure classification, urgency, and importance. Two years between inspections is a long time for systems that have already badly aged and are rapidly deteriorating. As an example: how good is a security camera system if it only records short videos once or twice a day? Similarly, the reality of our built infrastructure is complex, and many safety-altering events can happen in between inspections. In addition, visual inspection is subject to human limitations and biases. Inspectors may miss a major crack or flaw in an important structural element. An illustrative example of an inspector's observational limits is the Interstate 40 bridge connecting West Memphis to Memphis. For years, a crack was growing in a primary supporting beam of the Interstate 40 bridge, as was later discovered in an amateur videographer's video footage. By May 2021, the small crack had turned into a complete fracture of the supporting beam, leading to a near-collapse event. Despite the seriousness of this structural flaw, no cracking was ever indicated in any prior inspection reports. Whether the failure to observe and record this crack was due to the inspector's negligence or because of detection and access limits remains unknown. It takes a long time before the responsible party is held accountable in such cases. What is clear, however, is that observation-based inspection, just like other

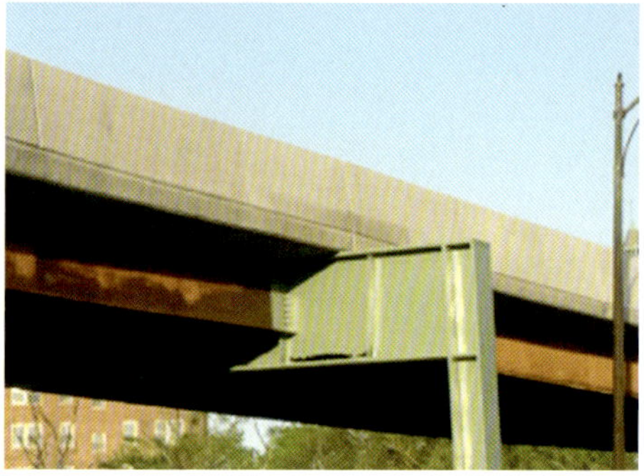

FIG. 1_*Recently painted girder web with corrosion section loss, Boston, MA*

human qualities, is subject to the limits and lapses of judgment. To tackle challenges pertaining to inspections, such as the long intervals between inspector visits and observation limits, different types of sensors have been proposed for monitoring infrastructure systems. Sensors are electronic gadgets that can record information about infrastructure assets. Sensor data tend to be more objective and less subject to bias than human observations. Sensors can also be installed for longer periods and be used for near real-time monitoring of the civil infrastructure. There are generally two categories of the sensor-based monitoring of infrastructure systems: (1) structural health monitoring (SHM) and (2) non-destructive evaluation (NDE). SHM refers to a class of sensor-based procedures that monitor the local and global behavior of various infrastructure assets. Accelerometers, such as those in smartphones, are a commonly deployed SHM sensor used to monitor vibrations. For instance, using accelerometer data, engineers can identify how the vibrational signatures of a bridge evolve over time. NDE refers to a number of procedures for monitoring local and subsurface conditions. For example, ground-penetrating radar (GPR) is an NDE method that sends high frequency radio waves through material depths in order to image subsurface conditions. Reflected GPR waves in the ground can be used to create a map of subsurface objects such as pipes, cables, and cavities.

Sensor-based monitoring procedures are thus a large number of data-driven approaches for assessing the condition of various infrastructure systems. Sensors record waveforms that must then be interpreted in the context of the wider system under consideration. The drawbacks of data-driven methods include their interpretability and whether they are useful in decision making. Although we can monitor how a bridge vibrates, this information does not inherently translate into making decisions about how much longer a bridge can remain operational, for example.

Many municipal and transportation authorities around the world have begun using online web-based dashboards to display sensors and infrastructure monitoring data. However, the information presented on these dashboards tends to be quite simplistic and lacks the sophistication needed to make decisions in the greater context of public policy on civil infrastructure. Data-driven approaches can be augmented using physical and graphical models of the infrastructure system of interest in order to create a greater system level context and interpretability. This is where *digital twins* come in, and they are great for visualizing pertinent information for any infrastructure asset.

To understand how digital twins can make a difference, we should explore the history of industrialization, manufacturing, and construction, and identify the tools that gave nations the technological edge. Before industrialization, craftsmen who had more expertise and experience created products of superior quality, and that's what gave them the edge. During early industrialization, we figured out the laws of physics, including thermodynamics and electromagnetism. We learned to design machines that manipulated these physical laws to assist in creating more sophisticated assembly lines that gave us the technological edge.

Next came the information age with the advent of integrated circuits and the wide availability of computers and the information of things (IOT), which gave nations and companies the technological edge. An increase in complexity, data volume, and automation has been noted in each era. Now we have arrived in the current period, or the big data era. We're surrounded by artificial intelligence, highly advanced science, and lots of data. Therefore, the modeling that we need to perform to keep up with the complexity of modern systems and to maintain our technological edge must be complex too.

Digital twins are exactly that. They are models on steroids, driven by data, tuned, and constrained by physical laws, comprising graphical environments for the visualization of information — which is why this chapter fits into this book. An earlier version of the digital twin concept, though never labeled as such, was introduced by NASA during the Apollo mission era. NASA developed ground-based replicas of various spacecraft modules which were used as simulators for training purposes. When an oxygen tank exploded during the Apollo 13 mission resulting in damages to oxygen supply and propulsion systems, these ground replicas were instrumental in rescuing the three astronauts 200,000 miles from Earth. NASA mission controllers rapidly adapted and modified these simulators to match the conditions of the real spacecraft. For instance, they had to devise fixes to the oxygen supply system using a few items that the astronauts had available to them. The simulator was thus a ground twin to the spacecraft. In a similar sense, digital twins can also be deployed for various scenario simulation purposes in engineered systems in addition to information visualization.

The obvious and rational next question to ask is how and whether the digital twin concept can serve civil infrastructure. To answer this question, we must first consider the stages in the lifecycle of a civil infrastructure asset. These stages include design, construction, operations, and maintenance.

During the design stage, a physical infrastructure asset has not been built yet. However, engineers and designers understand the environment, surroundings, and existing projects around the future construction site. In such instances, a digital twin can be used to virtually verify a preliminary design. By conducting simulation and analysis, the design of the asset can be tested, improved, and verified.

During the construction phase, digital twins can serve to monitor and manage targets such as workers, machinery, and materials. Digital twins can also be useful for monitoring and managing construction progress, quality, and safety. During the operational life of a system and for maintenance, digital twins can be helpful for

FIG. 2_ *Digital twins for operations and maintenance, Verrazano-Narrows Bridge, NY*

VIDEO_ *Verrazzano-Narrows Bridge — Drone footage and information*

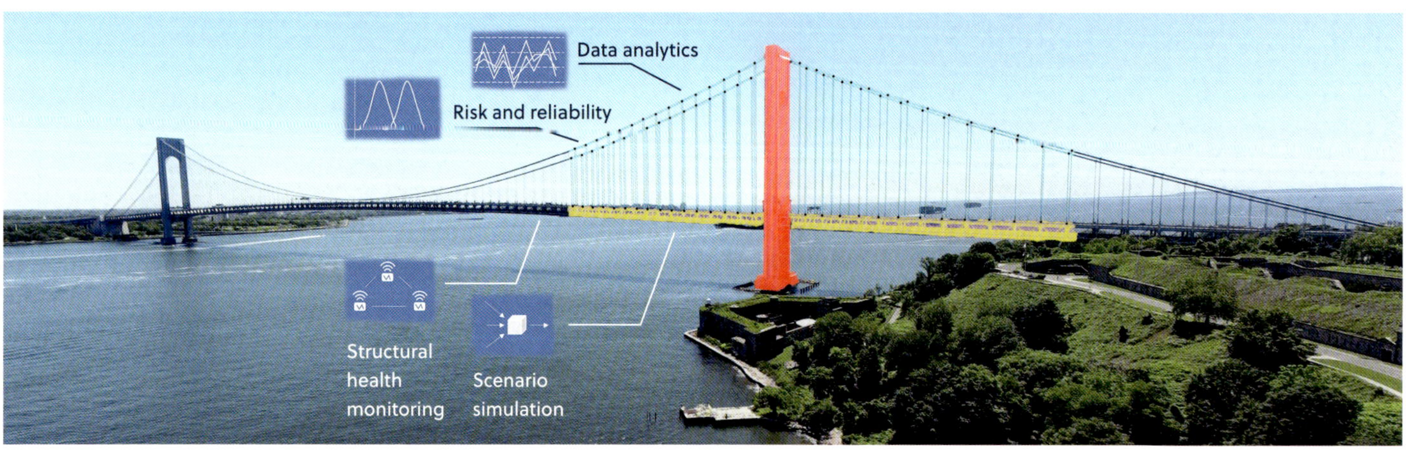

monitoring and detecting sudden or slowly evolving defects. Once the digital twin has been updated to fit the real (physical) twin, scenario simulation and analysis can be performed — to determine »what else can go wrong.« Monitoring and analysis can inform the actions and corrective measures necessary to prolong the service life of a civil infrastructure asset.

Digital twins for operations and maintenance can be used to examine trends, simulate scenarios, assess the likelihood of those scenarios, and to monitor and model existing systems. Tremendous possibilities exist in relation to digital twins and data analytics. A digital twin can be used to process and analyze operational data to help inform maintenance decisions. Findings can be visualized and displayed in the digital environment. Risk and reliability are the interfaces between engineering and mathematics that examine different scenarios from a probabilistic perspective. As an example, we might want to estimate the probability of a structure failing after a particular earthquake. A digital twin's scenario simulation capabilities can be an indispensable tool for such a purpose. These ideas are illustrated in_FIG. 2.

So far, we have talked about the real physical system, which is our infrastructure asset of interest, and the digital twin. We will now talk about some of the other constituents required. The general digital twin framework comprises of five constituents. First is the physical part, which can be a single system like a bridge or a series of complex systems or processes such as those in construction and manufacturing. The second constituent is the digital part, which is typically a graphics-based model of the physical system along with all the physical laws relevant to the problem. The extent to which the digital part mirrors the physical part depends on the application at hand. Next, we have the connections. These are information access points such as sensors, cameras, and other scanning devices that can acquire valuable data from the real world and physical part. A connection is any sensing or data acquisition device installed on the physical part that can inform and update the

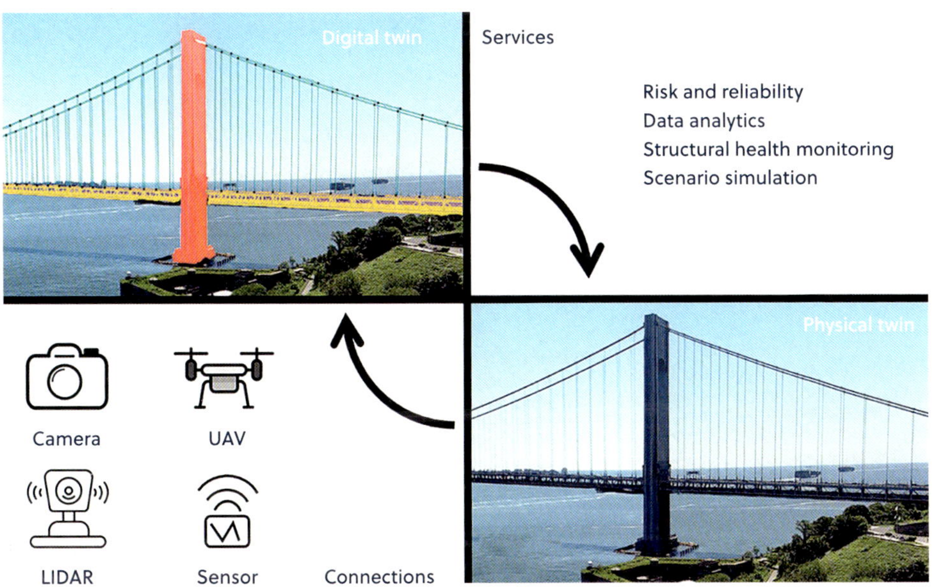

FIG. 3_*Digital twin constituents*

© CAIT, 2022

digital part. The fourth part is services or objectives of the digital part in relation to the physical part. The service can be interpreted as a mathematical optimization problem. In other words, a digital part optimizes something for the physical part through the service. The applications discussed earlier can be considered services. The final constituent is data. How we collect, organize, and rearrange data is where the magic happens. Therefore, data can be from individual connections or fused together to serve an enhanced purpose. These constituents are depicted in relation to one another in_FIG. 3.

Note that the direction of information flow is important in the digital twin framework presented above. Connections strictly direct information from the physical part to the digital twin. And services are strictly directed from the digital part to the physical twin. In many digital twin applications, there may be an additional constituent called the feedback, which is when the digital twin informs and updates the physical part. However, feedback can be considered a service and not an independent constituent or entity.

If you have heard about the building information model (BIM) concept, you may be confused as to how it differs from the digital twin concept. Although both are essential to the digital transformation that has been taking place recently, these concepts are different. The primary difference between a digital twin and a BIM is in the connections and presence of underlying physical laws. Connections are essential in digital twins but may not be necessary in a BIM. A BIM is simply a digital representation of a physical part with instances where processes and functionalities may be considered too.

Without a doubt, creating an accurate digital twin environment with accurate physics-based underpinnings is a challenging task. Such developments depend on the quality of data acquisition and data processing, modeling methods and tools, and the expertise of the engineers.

Data acquisition often involves installing sensors on an infrastructure asset of interest and collecting intended vibrational behavior under operating conditions. Sensors, however, have many limitations that must be understood. For example, acceleration sensors (i.e., accelerometers), depending on the variant, are typically unable to measure slow deformations in a structure. Raw data must be processed before any interpretation or modeling is conducted. The efficiency of data acquisition and processing can influence the physics-based modeling process.

Physics and graphical models are two important components of a digital twin. Physics-based models enable digital twins to have predictive capabilities based on the physical laws of nature, while the graphics-based models are graphical environments that support the visualization of various information pertaining to the physical twin. Models can vary in complexity and accuracy. We will now elaborate on what physics- and graphics-based models mean.

In civil engineering projects on buildings, bridges, roads, tunnels, and railways, understanding the underlying physics and modeling standards is important. Finding commercial modeling software that addresses several of these areas at the same time is challenging as there are no one-size-fits-all solutions. Physics-based models are the most commonly used. And when appropriate physics-based models are unavailable or do not fit well with sensor data, black box models can be used. Black box approaches are a class of model that can have any number of parameters

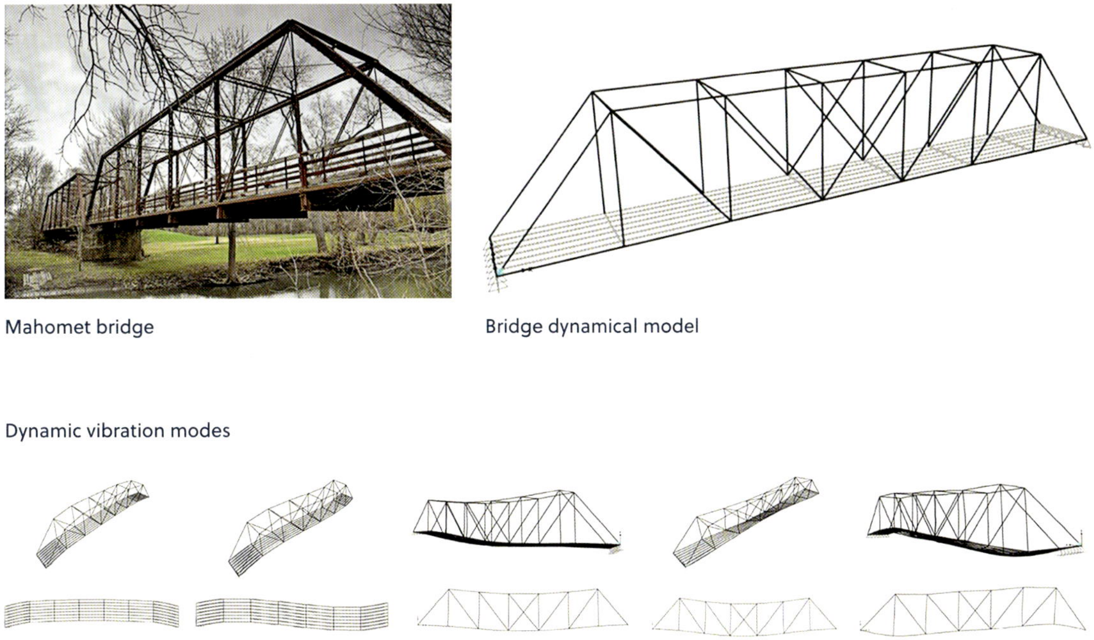

FIG. 4_*Mahomet bridge and its dynamical modes, Mahomet, IL*

and ugly mathematical complexity. Engineers are usually happy to use black box models as long as they produce good modeling predictions.

Many physical laws are mathematically expressed as functions of change over time and/or distance, and are commonly referred to as differential equations. For example, the vibrational behavior of solids can be described via a differential equation called the equation of motion. The finite element method is a common method that allows engineers and mathematicians to approximate solutions to complex differential equations, therefore enabling rapid scenario simulation capabilities for various physical laws.

Physics-based models of existing infrastructure systems can be developed from scratch, just as we would design a system before it is built. Physics-based models can also be developed using the recorded sensor data from an SHM process. The process of modeling using sensor data is typically called *system identification*. Now consider a truss structure like the Mahomet bridge shown in_FIG. 4. A number of accelerometers were installed, and the bridge was vibrated by getting pedestrians to jump. From the recorded accelerometer data, a finite element or dynamical model of the bridge was then developed. Every structure has a series of resonant frequencies and modes at which bridge vibrations are amplified. The mode shapes for each of the first five resonant frequencies of the Mahomet bridge have been illustrated as well.

Once we have a physics-based model, the next step is to perform scenario simulations. Different scenarios can be considered for each civil infrastructure application. For instance, turbulent flow may be considered for a canal during a storm scenario in order to calculate the hydraulic stresses on the surrounding environment. Or building damage can be simulated under different intensities of

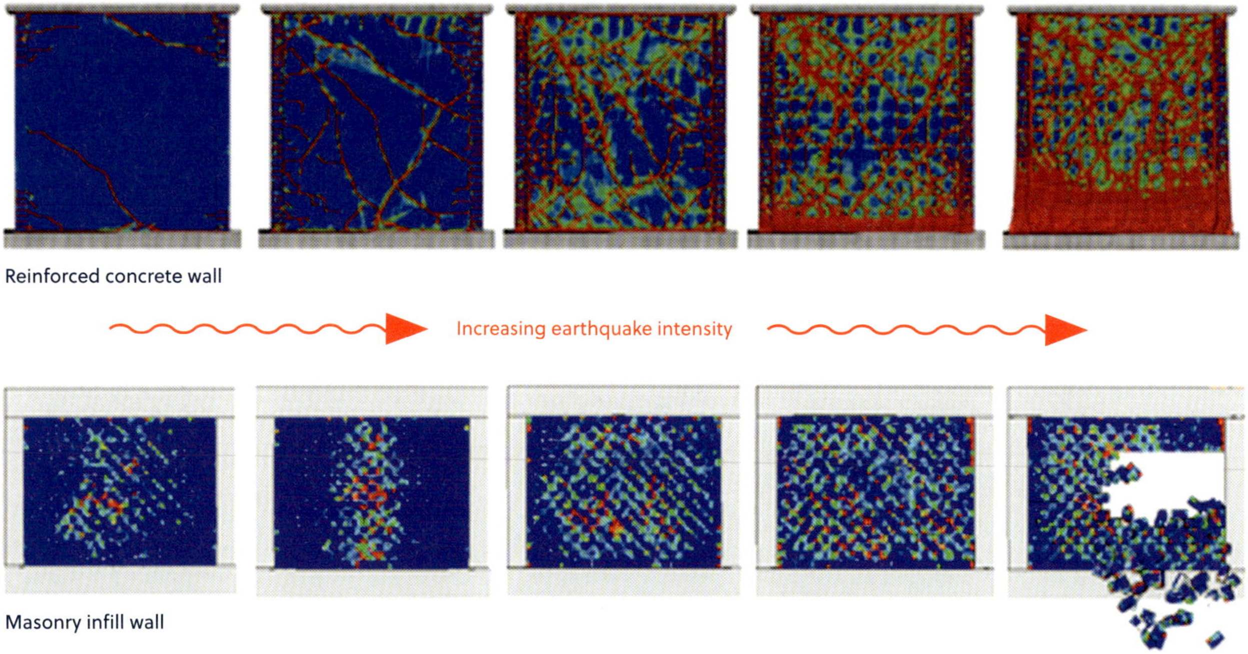

FIG. 5_ *Simulation of two types of walls under increasing levels of earthquake intensity*

earthquake. The outcomes from these simulations can then be used to inform engineering and policy decision-making.

Consider two types of walls: the first kind is a reinforced concrete wall, typically designed to resist earthquake forces in a building. The second kind is a reinforced concrete frame with a more traditional masonry infill. These walls are subjected to increasing intensities of earthquake loading in a physics-based simulation environment to assess how damages appear and propagate. Results from these two walls are presented in_FIG. 5. Blue and red colors represent undamaged and damaged components of the walls, respectively. In the case of the masonry infill wall, the friction between the masonry is also modeled. With increasing levels of earthquake intensity, this friction is overcome, and the infill wall collapses.

A three-dimensional (3D) graphics model may be manually developed using computer generated imagery (CGI). 3D graphics models can also be attained through 3D scanning methods. There are two primary 3D scanning technologies employed in most civil infrastructure applications. These are based on camera and light detection and ranging (LIDAR) devices, both of which can be mounted on stationary and moving platforms such as unmanned aerial vehicles (UAVs). FIG. 6 illustrates a camera-mounted UAV scanning a fire control tower from World War II. Moving platforms provide more flexibility and access; however, they also tend to increase measurement inaccuracies.

Three important components are needed for successful 3D scanning using LIDAR. First, the position of the LIDAR scanner must be known. The exact position of a LIDAR scanner is determined with the help of the global navigation satellite system (GNSS). The onboard GNSS receiver might be connected to dozens of satellites at any moment in time for accurate and precise positioning. A LIDAR system has an onboard laser, which generates a laser beam as the basis for distance mea-

FIG. 6_*UAV with camera mount scanning fire control tower, Staten Island, NY*

© CAIT, 2022

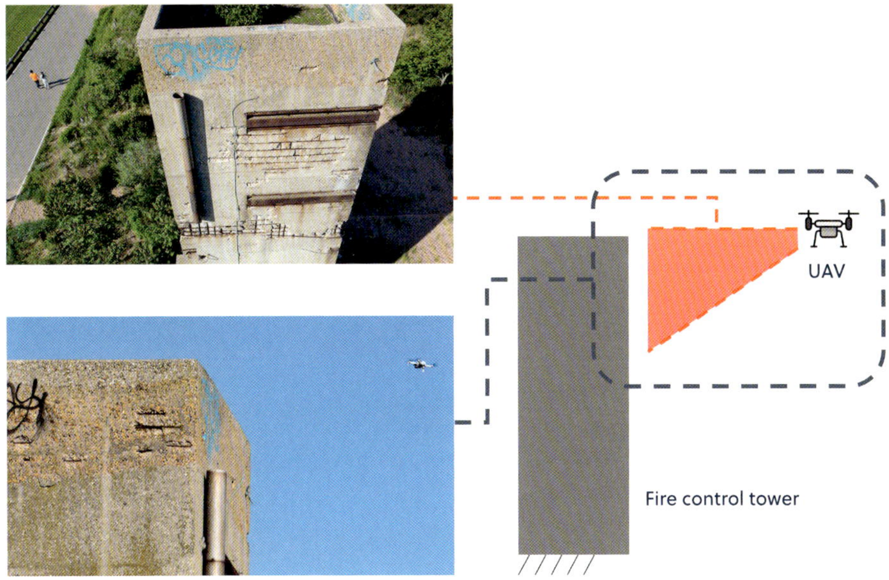

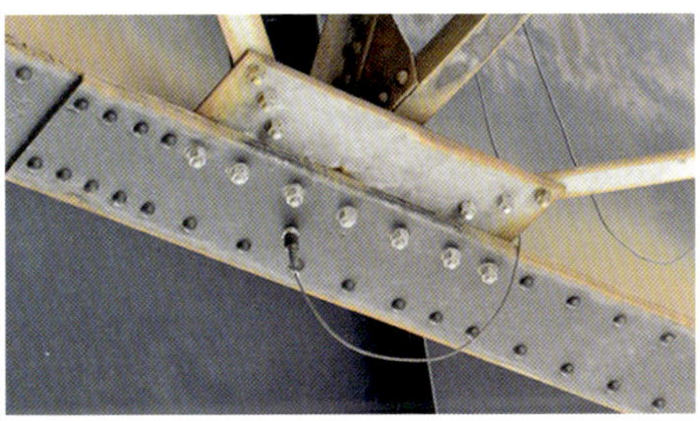

Accelerometer instrumentation

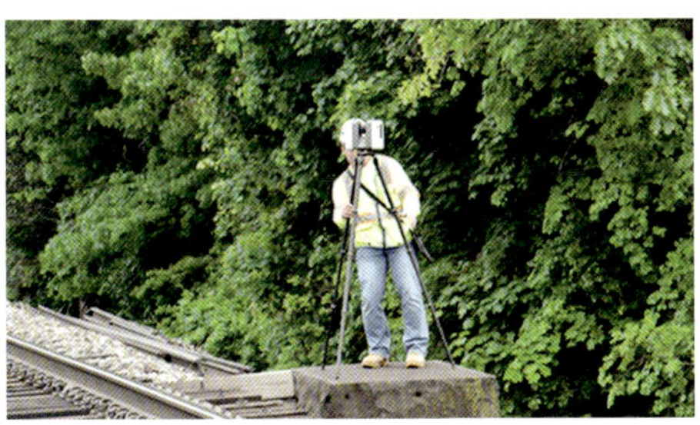

LIDAR survey

FIG. 7_*Sensor- and LIDAR-based SHM*

© CAIT, 2022

surement. The direction in which the laser beam is sent is the next important component. LIDAR systems possess inertial measurement units (IMU) that track the movement of the unit around its three axes every time a laser pulse is emitted.

The final important component is distance. Once the laser beam reflects off of a target, it must return to a receiver inside the LIDAR and be processed by a polychromator. Distances can be calculated the same way that we calculate the distance driven during a road trip based on our speed and time on the road. We know the speed of light, and the time that is needed for a beam of laser to be emitted, reflected, and received by the scanner. The distance to a target is therefore the speed of light multiplied by the total time until the reflection is received, divided by two. In reality, this process is more complicated, of course. The polychromator, for instance, has to disperse light into different frequencies of the light spectrum and analyze the delay at each wavelength. LIDAR scanners emit thousands of pulses every minute to create high-resolution 3D reconstructions. FIG. 7 shows a bridge inspector performing a LIDAR survey along with more traditional accelerometer instrumentation.

Camera-based 3D scanning is a completely different process. Instead of shooting thousands of laser pulses to locate individual points on a surface, cameras photograph entire images, which are projections of the 3D environment, onto a plane. Now, given two views of a scene, what is the relationship between the location of a scene point in one image and its location in the other image? This relationship is what must be understood to go from 2D image geometry to a 3D graphics model.

The first step in creating a camera-based scan is to identify recognizable and matching features between photographs taken of an asset. Features can be corner points, edges, and textures. There are mathematical algorithms that generate these features without human involvement. Next, the feature points between each image must be matched to those in other images. Again, there are algorithms that observe the general shapes, patterns and positions of each feature and match them with large numbers of photographs.

Every point on a photographed object is projected onto the image plane through a straight line, which means that we have two straight project lines when the same point is observed from two different camera positions. Using this information in addition to the translation and rotation between each camera position, we are able to estimate the 3D coordinate of the point of interest. This final step is called triangulation.

These algorithms must work in conjunction with many photographs in a process called photogrammetry to create a large collection of 3D points called a point cloud. There are several pieces of commercial software with proprietary algorithms that automatically manage the required feature matching and triangulation. The UAV-based photogrammetry of the fire control tower discussed earlier is presented in_FIG. 8. This figure shows the feature matching between two photographs and the resulting 3D point cloud. Photogrammetry-generated point clouds are considerably more cost efficient to acquire than LIDAR-based point clouds. However, the LIDAR tends to produce more accurate measurements.

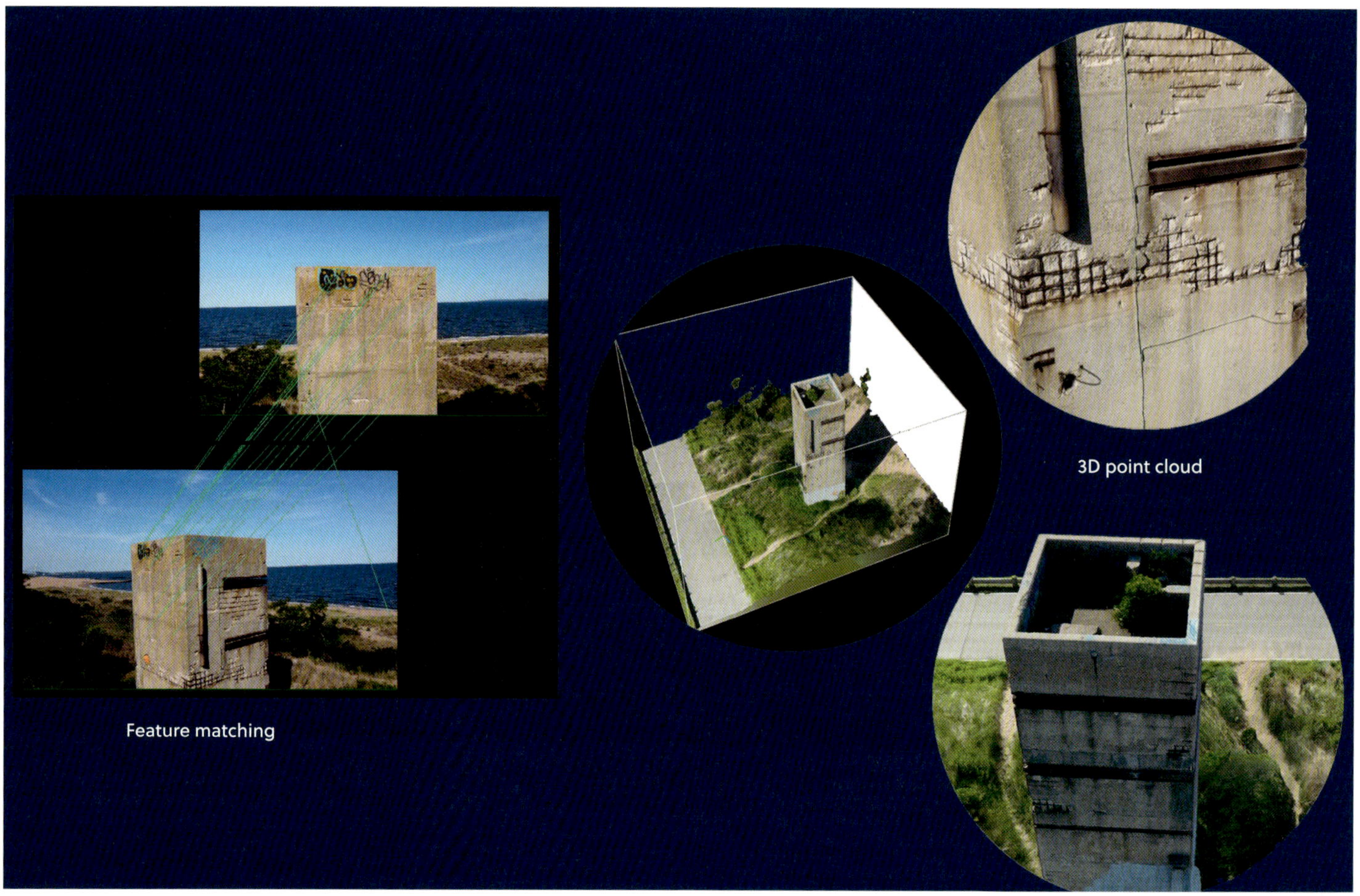

Institutions and government agencies have developed digital twins for their assets using the very tools described here. Some examples of these digital twins include those used by the Minnesota Department of Transportation (DOT) and the Norwegian Public Roads Administration.

The Minnesota DOT uses UAVs to assist in their bridge inspection procedures, which allow for rapid inspection deployment and the efficient 3D scanning and generation of digital twins for Minnesota's bridges. These digital twins are then cast into virtual reality headsets for remote inspection and monitoring purposes. With this method, the Minnesota DOT has been able to see changes over time and obtain holistic views of their bridge assets, which also allows them to view past inspections side by side with current data. The Minnesota DOT has reported cost savings of as much as fifty percent through the use of digital twins.

In Norway, the eighty-year-old Stavå bridge was long a source of stress for the Norwegian Public Roads Administration. As a result, they decided to equip this bridge with SHM sensors and create a cloud-based digital twin, capable of detecting and reporting structural anomalies. Interestingly, in April 2021, an automated notification was received indicating a serious structural deficiency in the Stavå bridge, leading to the rapid field deployment of inspectors to confirm the deficiency. The bridge was then closed, and an interim bridge has been put into operation while the new bridge is being built.

To conclude this chapter, we wish to highlight where the civil infrastructure industry is headed in the next five to ten years in relation to the digital twin domain. First, to realize the full potential of digital twins, we must train the next generation of engineers to develop sophisticated graphics-based models, visualize data through the latest available data science tools, and write AI-based decision-making algorithms. The digital twin approach will promote smarter construction practices. The construction industry is already the top consumer of UAVs and LIDAR scanning tools. This enhanced monitoring will combine with digital twins to create low-cost, visualizable approaches to simulate, calculate, analyze, optimize, and manage construction processes and qualities.

FIG. 8_*UAV-based photogrammetry of fire control tower*

Artificial intelligence will help to automate significant aspects of digital twins. AI will be valuable in analysis and diagnosis, defect detection, and decision-making. Therefore, AI approaches will be needed to minimize human intervention. Blockchain is a method of preserving information in a distributed manner and can be used to distribute the monitoring and assessment of infrastructure systems. This technology will make it easier to access and ensure the security of data and cloud-based digital twins. Rapid connectivity tools such as 5G networks and low-orbit internet satellites will allow for the enhanced coverage and faster transfer of large quantities of data for real-time monitoring and decision-making.

Beyond the Limits of Our Perceptions

Steve Nicholls

Every man takes the limits of his own field of vision for the limits of the world.
Arthur Schopenhauer, Studies in Pessimism

To see a world in a grain of sand
And a heaven in a wildflower,
Hold infinity in the palm of your hand
And eternity in an hour.
William Blake (1757–1827), Auguries of Innocence, circa 1803

William Blake was an English poet and painter with, to say the least, a very idiosyncratic view of the world. During his lifetime, he was widely regarded as insane, although his work is now both admired and appreciated, so much so that one critic recently described him as »far and away the greatest artist Britain has ever produced.« In his poem, *Auguries of Innocence*, he laments the loss of childlike innocence and the wonder in the world that such innocence gives. As we grow to adulthood, the world becomes more familiar, more predictable… more boring—but only if we allow ourselves to be limited by our everyday perceptions.

No matter how jaded we become, the world is still full of surprises and jam-packed with revelation. We just need to see it in the right way. Thankfully, science provides us with the ability to do just that—and the tools to extend our natural senses to reveal these hidden worlds. Throughout my thirty-five-year career in TV science documentary production, I've explored many ways of revealing those aspects of reality that lie beyond the limits of our normal perception—at scales both too large and too small for us to appreciate. In this chapter, I'll examine several pioneering science visualization techniques that have enabled me to bring entirely new worlds to the screen.

I began my career in academia in the late 1970s as an entomologist working on dragonflies and mayflies. My work involved detailed structural studies using both transmission (TEM) and scanning electron microscopes (SEM) that produce images of such astounding magnification that they reveal, in the case of the transmission microscope, the inner structure of individual cells and, in the case of the scanning microscope, mind-boggling, three-dimensional images of surface structures.

The first time I inserted a specimen into a scanning microscope I was hooked—captivated by the wonderous images that slowly drew themselves on the tiny screen. The utterly unexpected architecture and microscopic precision of a mayfly's mouthparts were fascinating from both scientific and artistic viewpoints.

What else could I discover in these hidden worlds?

Any time that the scanning microscope was free from its scientific duties, I used it to examine a huge range of insect specimens, from the iridescent scales of morpho butterflies to the waterproof hairs covering the bodies of pond skaters. Most of these were purely artistic endeavors, although I never admitted that to the technical staff in the microscope suite, who just assumed I was a particularly dedicated researcher. When I joined the BBC's Natural History Unit in the 1980s, the focus was on the more filmable world of larger animals, but the magic of the microworld never left me. After establishing myself as a wildlife producer in the unit, I teamed up with David Spears, a skilled microscopist-turned-documentary-filmmaker, and we hatched a plot to reveal an unexpected microworld to the TV-watching public.

Dave managed to acquire an older model SEM, and I managed to persuade the BBC to fund a one-hour film, *Impossible Journeys,* in which we would use scanning and light microscopy to reveal hidden aspects of our familiar, everyday world. The first problem we faced was that the SEM (at least older models) only worked with dead specimens to produce still images. A SEM works by bouncing an electron beam off the surface of a specimen and slowly scanning the beam across the specimen to build up a complete picture. The specimen first needs to be carefully dried to remove all traces of water and then placed in a vacuum chamber to prevent the electron beam from being scattered by air. To ensure crisp and clean reflections off the surface, our specimens also had to be coated in a layer of gold just a few atoms thick. The prepared specimens looked beautiful, like shining little jewels, but we didn't think that the viewing public were quite ready for frozen images of dried, dead creatures.

Dave came up with an idea, based on techniques that we had been developing for moving time-lapse shots, to create more dynamic images. In essence, this meant taking a still frame then moving the camera by a tiny amount before taking the next still frame. This technique, called *motion control,* should, we felt, be adaptable to work with SEM images. It took a few minutes for the SEM to build up a complete still frame as it scanned the beam over the specimen. We would be able to move the specimen by a tiny amount and capture another still frame. The stage that the specimens were mounted on allowed us to both rotate and track in any direction, and we were also able to adjust the magnification between frames. By combining all these axes, we could, in theory, produce complex, sweeping moves that would zoom in from something resembling normal size to magnifications of hundreds of times. Dave rigged each of the microscope's axes with small stepper motors that could produce tiny but accurate moves. The problem lay in keeping track of what each motor was meant to be doing during a complicated move. Our low-tech solution was to plot the whole move out beforehand on an Excel spreadsheet, with a column for each axis. It was painstaking work and each ten-second shot could take several days to complete.

At the end of all that, we had a moving image—but only in black and white. An electron beam operates on a single frequency and therefore only contains information on the brightness of the reflection. In other words, it produces a grayscale picture, which is fine for scientific research but still doesn't show this hidden world off to the best extent. So, we set about hand-coloring each shot, often frame by frame, another process that could take days. The sheer effort that went into producing each shot for this film was extraordinary and took far longer than many wildlife camerapeople wait for shy creatures to perform in front of their cameras in remote wildernesses. But the results were more than worth it. We flew over herds of dust mites marching across bed sheets like tiny herds of antelopes sweeping over the African plains. We saw the garden in a new light as we followed a delicately sculpted grain of grass pollen released from an anther on its wind-blown journey to the stigma of another flower. We appreciated the hidden beauty of a nettle leaf, with stiff hairs that end in a glass-like bulb, full of the stuff that stings when you brush against the leaf and break the bulb. And we marveled at the precision engineering of a common housefly.

But the specimens were still inert and lifeless. Was there a way to bring this world to life more dynamically? After I left the BBC, I set up my own documentary production company in order to work in more diverse ways and with a wider range of people. And I soon met up in Vienna with Alfred Vendl, a professor at the University of Applied Arts Vienna. Apart from his academic research, involving elaborate and advanced SEM techniques, he was also a long-standing producer of science documentaries, and we discovered that we shared a deep interest in inventing new ways of showing the world to as wide an audience as possible. One of our first collaborations was a film for the ORF (the Austrian national broadcaster) and its long-running *Universum* series.

We opened this film, *Limits of Perception/Grenzen der Wahrnehmung*, with a quote from no less a thinker than Albert Einstein, which succinctly summarizes the ambitions of both that film and this book: »It is entirely possible that behind the perception of our senses, worlds are hidden of which we are unaware.« It is only slightly ironic that Einstein detested the idea of one of the most extraordinary of all hidden worlds—the unpredictable realm of quantum mechanics. This is one world that is truly impossible to illustrate because it goes so far beyond the »reality« that we perceive. Another famously witty physicist, Richard Feynman, is said to have stated, »If you think you understand quantum mechanics, you don't understand quantum mechanics.« This is probably apocryphal, but it nevertheless makes the point. It's a world where matter can pop in or out of existence for no apparent reason, which is why Einstein was so uncomfortable with the idea, summed up in his belief that »God does not play dice.«

Nevertheless, with *Limits of Perception* we planned to push the visualization of hidden worlds as far as science and imaging techniques would allow. The idea behind the film was an old one—a journey through scales of powers of ten, from the subatomic world to the entire known universe, a trip that takes surprisingly few steps. But we hoped to illustrate this journey in new ways. And our starting point was to improve on the SEM images we had used in *Impossible Journeys*.

For part of his research, Alfred employed a newer technique of scanning microscopy using an environmental scanning microscope (ESEM). This needed only a partial vacuum to work and could handle wet specimens. In other words, we were able to generate images of living micro-creatures. With due deference to Einstein, it was a »quantum leap« beyond *Impossible Journeys*, although the images were still black and white, and therefore still needed to be colored. Having said that, in the decade and a half separating *Impossible Journeys* and *Limits of Perception*,

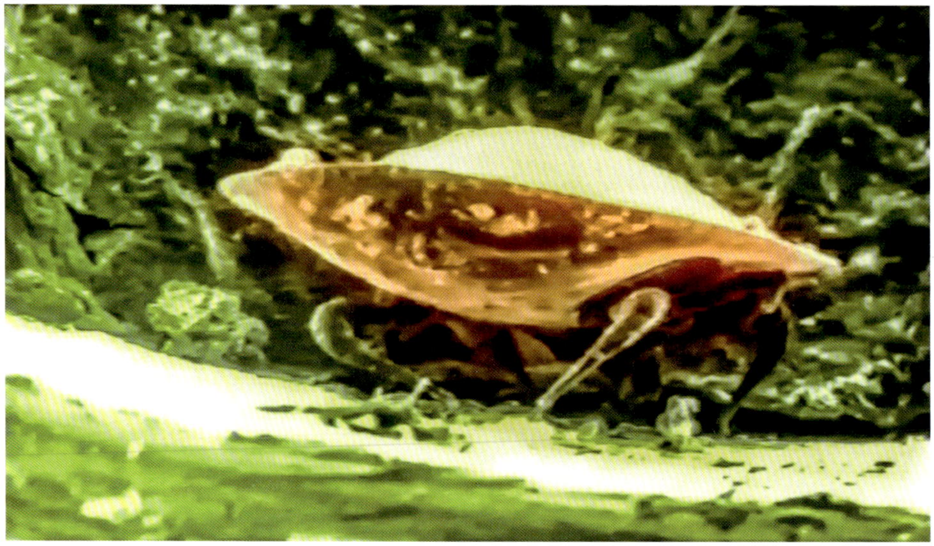

computer graphics had advanced enough to make this job a bit easier. Looking back to the mid-1980s and our experiences making *Impossible Journeys* brings to mind the image of medieval monks, hunched over desks in a scriptorium, hand-copying manuscripts.

Even so, the environment inside an ESEM is far from benign. So, we needed creatures tough enough to survive being bombarded by electrons in a partial vacuum. And they don't come tougher than mites. Mites are alien-looking creatures, mostly too small to discern with the naked eye. Yet they exist in uncountable numbers and in bewildering diversity. It's often said that insects are the most diverse group of animals to have ever evolved (there are currently one million described species), but it's likely that mites are also in the running for that accolade. There may even be more kinds of mites than insects—it's just that, lying beyond the limits of our perceptions, there are fewer people looking for them.

They live in huge numbers just about anywhere you can imagine, from cracks in paving stones to the base of your eyebrows. The Antarctic mite, *Alaskozetes antarcticus*, is probably the most abundant terrestrial invertebrate on that frozen continent. Mites also thrive in the cozy environment of your bed sheets (dust mites, family Pyroglyphidae) and crawl in hordes through your cheese (cheese mites, *Tyrophagus casei*, and others). As gruesome as it sounds, cheese mites are an integral part of the process of making certain cheeses, such as *Milbenkäse* (literally: mite cheese). We had created frozen tableaux of dust mites and cheese mites for *Impossible Journeys*, but the ESEM took us to a whole new world. The mite sequence was set in a tiny sliver of moss growing in a crack on a statue, something most of us would never give a second glance. Yet, seeing bizarre creatures stalking through a towering forest of moss was entrancing. Here was a world teeming with creatures that look like nothing else on Earth and yet exist in countless multitudes right beneath our feet—literally.

At best, mites appear to us as nothing more than tiny specks. But the way the world works at scales that we can actually see owes a lot to its hidden microscopic

FIG. 1_ *ESEM image of a mite in* LIMITS OF PERCEPTION. *Mites are tough enough to survive the extreme conditions inside an environmental scanning electron microscope. Most are no bigger than a speck of dust, but they are as diverse as the more obvious insects—and come in a bewildering variety of forms. This one looks like a flying saucer with legs.*

VIDEO

properties. And the SEM can reveal these surprising aspects of our familiar world. Several years after *Limits of Perception*, Alfred and I were once more working together, this time on a three-part series exploring how nature can inspire technological advances. The science of *biomimetics*, or *bionics*, was growing in popularity and scope when we began work on *Nature Tech/Bionik* at the turn of the millennium. The first program looked at innovative materials based on designs perfected over millions of years by evolution — sophisticated solutions for living in all manner of demanding places. And much of that inspiration came from understanding how natural materials work at microscopic levels.

The Indian lotus, *Nelumbo nucifera*, grows in swamps across tropical Asia, its huge parasol leaves emerging from filth and mud spotlessly clean. This ability to remain immaculate in a world of muck and mire has earned this plant a central place in Hinduism and Buddhism, where it symbolizes purity in a corrupt world. But its trick of shedding dirt of any kind has also attracted the interest of scientists in search of self-cleaning materials. The lotus' secret to clean living turns out to be the microstructure of its leaves. They are covered in tiny, pointed hairs, the tips of which are almost unwettable. This means that water droplets can't spread out over the leaf but remain as spherical droplets perched on the very finely pointed tips of the hairs. Since there is almost no contact between the leaf and the water, there is little friction to hold the drop in place. So, it rolls easily across the leaf and off the edge. As it does so, it picks up any particles of dirt and removes them from the leaf.

We included sequences in the program of the leaf surface in microscopic detail, but we combined this with spectacular shots of a whole living leaf to show how the microscopic structure is responsible for the behavior we can see. Using a high-speed camera shooting at one thousand frames per second, we simulated rain splashing on a living lotus leaf. The results were astonishing. The falling water bounced straight back off the leaf in large, quivering drops then ran over its surface like quicksilver. At the end of the shot, the leaf was perfectly dry — and, of course, clean. The images of water rolling, bouncing, and swirling over the lotus leaf were

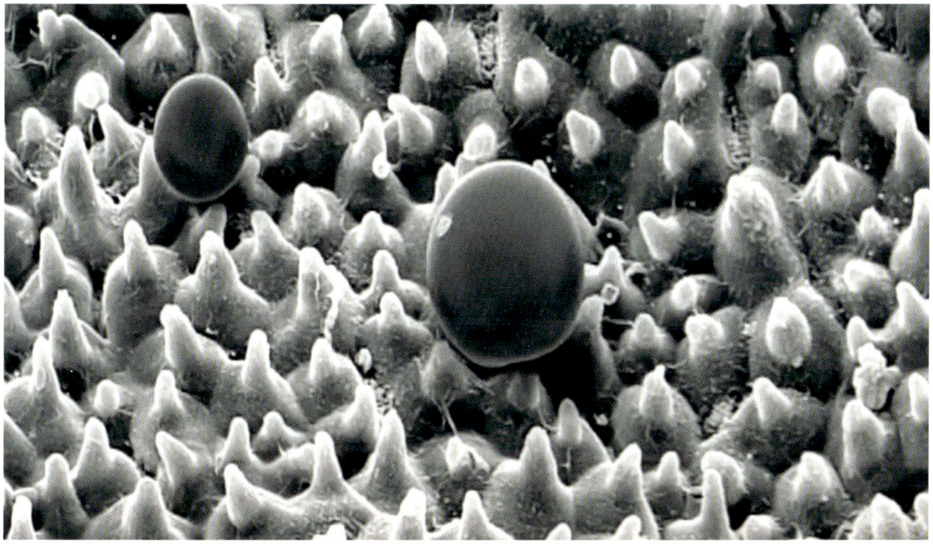

FIG. 2_*An image from an environmental scanning electron microscope (ESEM) reveals that the surface of a lotus leaf is covered in microscopic bumps, each with an unwettable, finely-pointed tip. Because the ESEM allows us to use wet specimens, we can also see a drop of water perched on the very tip of these points, with almost no contact to the leaf.*

VIDEO

so hypnotic that they attracted a lot of public attention. By combining microscopic images with extreme slow motion, we had revealed a remarkable hidden world that also has an important economic aspect. Mimicking the effects of a lotus leaf, scientists have produced a variety of self-cleaning materials, including several kinds of paint and a coating so completely repellent to anything that even honey just rolls off, leaving it perfectly clean, with not a hint of stickiness.

There is so much exciting science in the world, but the challenge for science documentary makers is to find ways of creating striking images to illustrate this to a wide audience. Numbers — equations and algorithms — tell us a lot about how our world works but are unlikely to draw in the crowds. The third program in the *Nature Tech* series explored the way nature uses energy and the lessons we can learn from it. It was a very abstract subject, which meant we had to look for new techniques to reveal these particular hidden worlds. In one case, the scientific data itself came to our rescue.

Termite mounds are extraordinary structures, built of mud, saliva, and feces. The largest, relative to the size of the termites that build them, are more than twice the height of the tallest building we have ever built (the Burj Khalifa in Dubai, just short of 830 meters in height). Yet the termites don't live in their skyscrapers. Their nests are largely underground, at the base of the tower. So, why do they go to all this effort? The tower, it was thought, somehow works to regulate the air inside the nest, full of busy termites producing lots of heat and lots of carbon dioxide. Since termites don't use energy-burning AC units, this bit of termite engineering drew interest from architects as well as entomologists.

Back in the early 2000s, we traveled to Namibia to join up with Scott Turner, a biologist from the State University of New York (SUNY), and Rupert Soar, an engineer, then at Loughborough University in the UK, who were trying to find out how termite mounds worked. Scott had inserted probes to measure temperature and gas concentrations inside the nest and found them to be remarkably constant over the course of each day — and very equitable despite the searing African midday sun. They had previously dissected the mounds and the nests beneath using a large digger, something only an engineer would consider a precision tool. This revealed that the tall mound was filled with a branching network of tunnels, and it was clear that understanding the exact arrangement of these tunnels was key to understanding how the mounds worked. That wasn't going to be easy using their relatively crude excavations. A more ingenious plan was needed.

The next idea was to inject the mounds and nests with gypsum that had been made up into a mixture runny enough to penetrate all the passageways in the nest but that would still set solid after a few hours. Now it was time to have fun with a high-pressure hose, washing away the mixture of soil, saliva, and feces that the termites use as a construction material to reveal a sculpture of exquisite beauty. The tracery of tunnels running through the tower remains behind as a cast of white gypsum — and a termite work of art. In fact, these *endocasts* are so impressive they have been transported for exhibition around quite a few South African museums.

As impressive as they are, it's still hard to use these solid sculptures to analyze how the internal structure provides air-conditioning, so one final step involved building a heavy scaffold frame around a gypsum-filled mound and suspending what amounted to a giant bacon slicer above the mound. Now, Rupert and Scott were able to slice off a few millimeters from the top of the mound and photograph the cut surface, a pattern of brown mud and white gypsum. Slice off a few more millimeters and photograph it again to reveal a slightly different marbling of brown and white. They kept going for a few days and nights until they'd sliced all the way through the mound. Then, using software similar to that used for turning a series of two-dimensional MRI scans into a three-dimensional model, all these photographs were combined into a 3D computer model which could be examined in detail from any position and any angle. It was a way of seeing inside the mound as if it was made of glass.

The mound consists of a large chimney rising all the way up the center of the mound and connecting to a branching network of smaller tunnels, some of which run just below the surface of the mound. One early idea about how the mound worked was that millions of busy termites were generating a lot of hot air in their nest at the base of the tower. Hot air rises, and in this case, it does so through the central chimney. As it rises, it cools, and eventually this cooler air sinks back down again through the network of smaller tunnels, creating a circulation of cooler air through the nest.

Architect Mick Pierce was so impressed by this energy-efficient ventilation system that he designed a building to work on similar principles. The result was the Eastgate Centre in Harare, Zimbabwe, the world's first termite-inspired building. Air warmed by activity during the day rises up through a chimney-like structure to draw in cooler air from below in the evenings. It proved a remarkable success. Termites saved the developers around three and a half million dollars during construction since they didn't have to buy and install expensive air-conditioning. This passive cooling system also keeps running costs low, which translates into lower rents for office and retail space. An all-round win—except this isn't how termite nests work.

Later studies revealed that termite mounds are even more sophisticated than the scientists first thought. As computers grew in power and produced ever more detailed models based on the termite scan data, and as Scott and Rupert made ever more precise measurements, it became clear that air circulation in the termite mound was solar-powered. The sun heats up the air in the tunnels on the mound facing the sun, which causes it to rise up the mound. As it passes into the network of tunnels on the shady side, it cools and sinks back down to the bottom of the mound. As the sun moves across the sky, it heats different aspects of the mound, so the pattern of air circulation changes throughout the day. At night, when the outside air and the mound are cooler, the hot air produced by the termites in their underground nest does rise up the central chimney and drives an entirely different circulation pattern to the one found during the day. So, air circulates around the inside of the mound both day and night.

For *Nature Tech* we used the raw data gathered by Scott and Rupert to generate our own computer model. Back then, our results, though state of the art, were still quite crude, like a piece of three-dimensional abstract art. But Rupert kindly allowed us access to this hard-won data again in 2021 for a new series on insects, *Planet Insect*. Now, we were able to build a really sophisticated model that looked exactly like the termite mounds we had been filming in South Africa, but with the added bonus that we could make this mound transparent to see the tunnel network.

We could even add the effect of circulating air to create an image that was as artistically appealing as the gypsum endocasts, but which showed how these incredible structures worked.

The surface of the mound is not a solid layer of mud. It is pockmarked with myriad tiny channels, carefully built by the termites. The air rising from the nest has high levels of carbon dioxide from a million or so respiring termites, but as the air moves along the tunnels just below the mound surface, carbon dioxide diffuses out through the pores. In addition, oxygen can diffuse in, creating fresh air that circulates back to the base of the mound where the main nest lies. The mound, it turns out, is not so much an air-conditioning unit as a lung, but that hasn't stopped the Eastgate Centre from being an energy-efficient building, even if it was based on an incomplete understanding of how termite mounds actually worked.

Advances in computer graphic imagery (CGI) in the two decades separating *Nature Tech* from *Planet Insect* have also allowed us to make great strides in how we portray the microworld, illustrated by the radically different ways in which we were able to visualize the interior of the same termite mound. For *Planet Insect*, I wanted to illustrate the micro-architecture of insects. Inspired by my misspent time in SEM suites back in the late 1970s, I knew that showing these unseen aspects of insects would be a startling revelation for most people. But these images do more than just allow us to marvel at the micro-engineering of insects. They can reveal why insects are so extraordinarily successful.

I've already mentioned that insects are the most diverse group of organisms ever to have evolved, or at least one of the most diverse groups. Nematode worms and mites may give them a close run, but it's hard to grasp what that really means. Not all insect groups are equally successful. There are only a few thousand kinds of mantises or stick insects, for example, but five groups of insects have evolved numbers of species that are off the scale. Beetles (Coleoptera) are often said to be the most successful insect group, and there certainly are a lot of beetles out there. One in every four animals on our planet is a beetle of some sort.

Butterflies and moths (Lepidoptera) are also enormously successful. One in every ten animals on Earth is a butterfly or moth. These two hyper-diverse groups are joined by flies (Diptera), true bugs (Hemiptera), and ants, bees, and wasps (Hymenoptera). And it's not just numbers of species—insects also exist in unimaginable numbers of individuals. It is, of course, impossible to get accurate figures for insect populations, but that hasn't stopped some scientists from producing estimates based on well-informed guesses. I've seen figures of ten quintillion individual insects alive at any one time—a number that quite literally defies imagination. Their success is down to many factors, some of which can be revealed by examining them in microscopic detail. For the series, I picked just one example, the ubiquitous green bottle fly, *Lucilia sericata*.

Green bottles belong to the highly successful order of Diptera, which some scientists think may be even more diverse than the beetles. A few years back, thirty entirely new species of flies were discovered within the city of Los Angeles in just a single year. It's not that Los Angeles is infested with flies—a city of flies and not a city of angels—rather, it is an indication that there are an awful lot of flies everywhere that we have yet to discover. One reason for this amazing success can be found in a neat evolutionary trick.

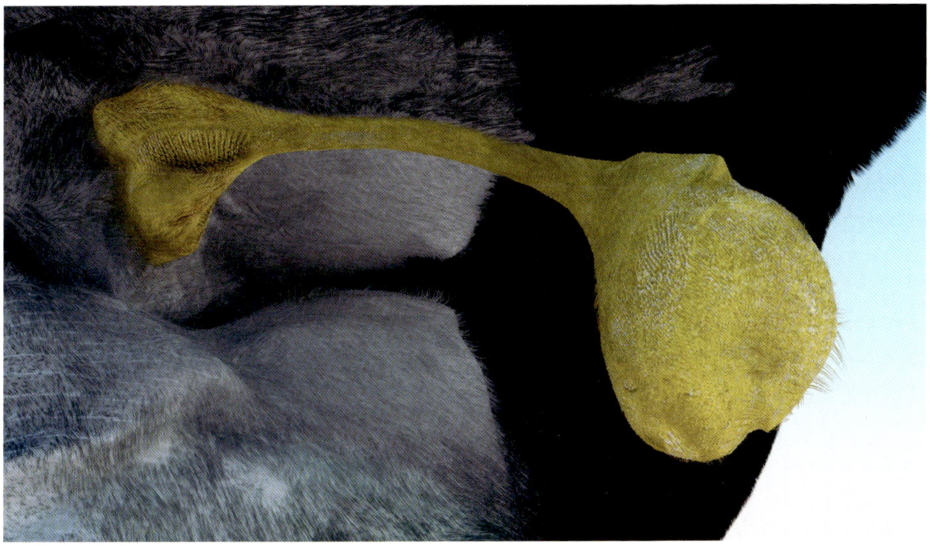

FIG. 3_*One reason for the evolutionary success of flies is that they have converted their hind wings into halteres — structures that work a bit like gyroscopes. Using micro-CT scans to gather three-dimensional data on these structures and a scanning microscope to image the fine details of the surface, we can see the intricate fields of sensors at the base of each haltere, which measure deflections in the stalk and send detailed information on flight attitude to the fly's brain.*

Insects normally have two pairs of wings, but flies have lost their hind wings, at least as structures for flight. So, they fly on just one pair of wings — the forewings. The hind wings are reduced to tiny knobs called halteres that work like gyroscopic stabilizers. Each haltere consists of a long stalk with a heavy bulb on the end. The whole structure beats up and down in flight, but the weight of the bulb at the tip means that the haltere always tries to beat in the same direction. As the fly buzzes around, it resists any changes in direction, much as a spinning gyroscope resists any attempt to tilt it off its axis.

But it's at the very base of the stalk that the magic happens. When the fly twists and turns in flight, the stalk of the haltere is forced to twist at its base to compensate as the halteres continue to beat in their original direction. Seen in the SEM, the base of the haltere is covered in an elaborate network of microstructures — beautiful to look at but essential for the fly in flight. They are tiny strain gauges that measure the forces produced by the twisting. This information is then fed to the fly's brain, giving it a constant readout of how its body is moving in three dimensions.

This is one reason that flies can fly so well, and the most aerobatic flies, such as the robber flies that pursue their prey in flight, have the most elaborate sets of sensors. But even the familiar green bottle is a »top gun« of the insect world. It can fly upside down and can just as accurately land on the ceiling as on the floor. To show how these tiny structures worked, we needed to do more than just see the green bottle motionless in the SEM. Luckily, Alfred Vendl had built a team to continue developing new techniques for visualizing microstructures. Working with Stephan Handschuh in the Imaging Unit of the VetCore facility at the University of Veterinary Medicine Vienna and Martina R. Fröschl at the Science Visualization Lab, we can now combine data from the SEM with animated CGI models.

The fly is first scanned using a micro-CT scanner that produces an accurate 3D model of the specimen and also records details of internal structures that are impossible to see in the SEM. The specimen can then be inserted into the SEM and the surface details recorded. The team have worked out techniques to combine data from both of these sources to produce a computer-generated image that can be animated in a lifelike way. Importantly, these are not fictional images generated in a computer — all too familiar in science fiction films, for example — rather, they are real images of the microworld made visible in a lifelike way thanks to modern computing power. By filming flies with a high-speed camera, we can see how the halteres move and replicate that in our model. The final result allows us to zoom in on a flying fly, right up to the base of an individual haltere, to see the elaborate network of sensors as the haltere beats up and down — an impossible feat without these extraordinary techniques.

At this halfway point in the chapter, it's time to switch from the very small to the very large, much as we did in the film *Limits of Perception*. Our perceptual world occupies a tiny range of scales perched in the middle of reality. Just as the microworld lies beyond the range of our natural senses, much of the world around us works at scales far too large for us to perceive, although scientific ingenuity has again provided us with tools that allow us to extend our perceptions.

We call our home Planet *Earth*, but that simply reflects our limited view of the world and our biases as medium-sized terrestrial vertebrates. A far more appropriate name would be Planet Ocean since four-fifths of the planet is ocean. The average depth of the ocean is over three and a half kilometers, although the deepest abysses lie nearly eleven kilometers below the surface. By far the largest living space on the planet is hidden from our view, a world so hostile and alien to us that we've barely begun to explore it. Submersibles may have penetrated the deepest parts of the ocean and explored the strange landscapes of deep-sea vents and underwater volcanoes, but we've only glimpsed a vanishingly small fraction of Planet Ocean. We've dragged nets through the ocean to sample the life down there, but even this is barely scratching the surface. It's like trying to understand the vast diversity of a rainforest by dangling a butterfly net from a balloon tethered several kilometers above the canopy — at night.

Water hides the most dramatic landscapes on our planet, so the most obvious way to reveal them would be to remove the water. This was how a discussion with graphics designers Paul Bond and Andy Davis-Coward began back in 2006. Since doing this for real is impossible, that left the ever-evolving sophistication of animation computers. But, just as with our portrayals of the microworld, we wanted to create images based on reality — as accurate as those created by micro-CT scans and scanning microscopy. If we could work out a way of doing that, we could quite literally drain the ocean.

I pitched the idea to National Geographic, who instantly liked the unique concept — and so *Drain the Ocean* was born. Now, all we had to do was to work out how to visualize ocean landscapes drained of water. First, this meant taking a trip to the Scripps Institute of Oceanography at La Jolla in Southern California to meet David Sandwell. Along with colleagues, he had been working on mapping the ocean floor at a global scale using a variety of techniques. He also knew where to find the best margaritas in San Diego so was perfect for the project. He agreed to help by both providing data and appearing in the film to help put these ocean landscapes into some kind of context.

It might seem counterintuitive, but the ocean surface isn't flat. It has its own hills and valleys. Sometimes this is caused by ocean currents. For example, in the

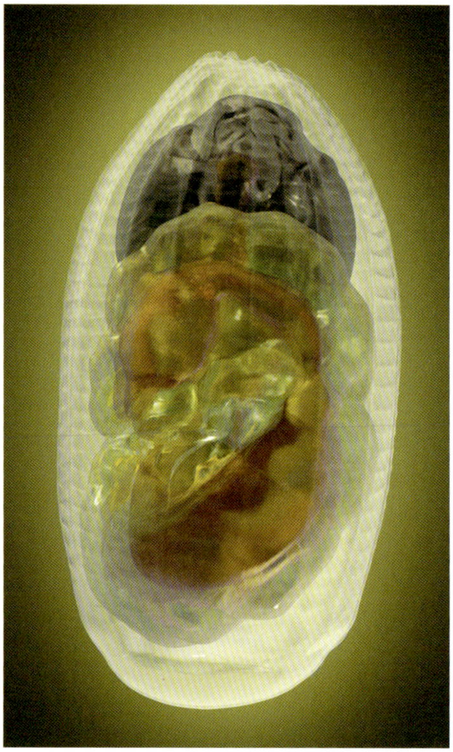

FIG. 4_*Combining scanning microscopy and micro-CT scans allows us to visualize both the internal and surface structures of even the tiniest objects in unprecedented detail. An image of a large white butterfly* (PIERIS BRASSICAE) *egg shows both the intricate sculpting on the outside, which serves to protect the growing embryo, and a tiny caterpillar, almost ready to hatch but still folded up within the egg.*

North Atlantic, water swirls around the whole basin in a circular gyre with the Sargasso Sea at its center. The currents cause water to pile up in the center, so the surface of the Sargasso Sea is considerably higher than the surrounding ocean. But some of the variation in sea surface height is down to the underlying topography of the sea floor.

Since the seventeenth century, when Newton observed an apple falling to the ground, we've all known that Earth exerts a gravitational pull. All objects have their own gravitational field, but the magnitude of that tug depends on the size of the object. The moon is much smaller than the Earth and therefore exerts only about one-sixth of the pull that we experience on our planet. But even on Earth, gravity varies from place to place, depending on the mass of rock beneath your feet. And gravity affects water just as much as it does apples. We see the effects of gravity on the oceans every day in tidal cycles, as the gravitational attraction of the sun and moon combined pulls water toward them. The effect of local variations in Earth's gravity is much smaller than the flowing and ebbing of tides, yet it is still measurable. Water is drawn to large masses such as underwater mountain chains, which therefore raise the sea level above them, making the ocean surface a mirror of the landscapes that lie beneath.

These variations in sea surface height are tiny but can be measured by satellites to give a broad picture of the ocean landscapes on a global scale. This low-resolution map forms part of the Smith and Sandwell dataset, a starting point for our attempts to drain the ocean. The dataset consists of a large series of x, y, and z coordinates that plot the ocean floor in three dimensions. Scientists use a geospatial software package called *Fledermaus* to turn this mountain of coordinates into representations of mountains in a three-dimensional image, where different depths or heights are represented by different colors. The result is a striking image that can be viewed from any angle, but not the realistic rendition of ocean landscapes that we were looking for.

For those, we had to turn to another software package called *Terragen*, which is widely used to render landscapes for visual effects in movies or in games. The software comes with its own lighting effects and surface textures, so, in theory, it would allow us to create a realistic sense of what four-fifths of the planet would look like if it were drained of water. Unfortunately, it didn't prove to be that simple. *Terragen* was designed to create the kind of landscapes we see every day — spectacular mountain ranges or great plains. The problem was that the landscapes beneath the waves are on another scale altogether. The software package simply couldn't handle the vastness of the ocean terrain without modification. After some experimentation and tweaking of the software's code, our graphics company was able to input information from the Smith and Sandwell dataset and generate realistic renditions of ocean floor topography. With this technology, the world looked very different — and very surprising.

Drain the water from around the Bahama Islands and you are left with a very different landscape from the flat, low-lying islands fringed with palm-lined beaches. The Bahama Islands familiar from holiday brochures are only the very tip of a huge underwater limestone plateau called the Great Bahama Bank. This massive structure is built from the skeletons of uncountable numbers of ocean creatures that have accumulated over many millions of years. Drained of water, the Bahamas

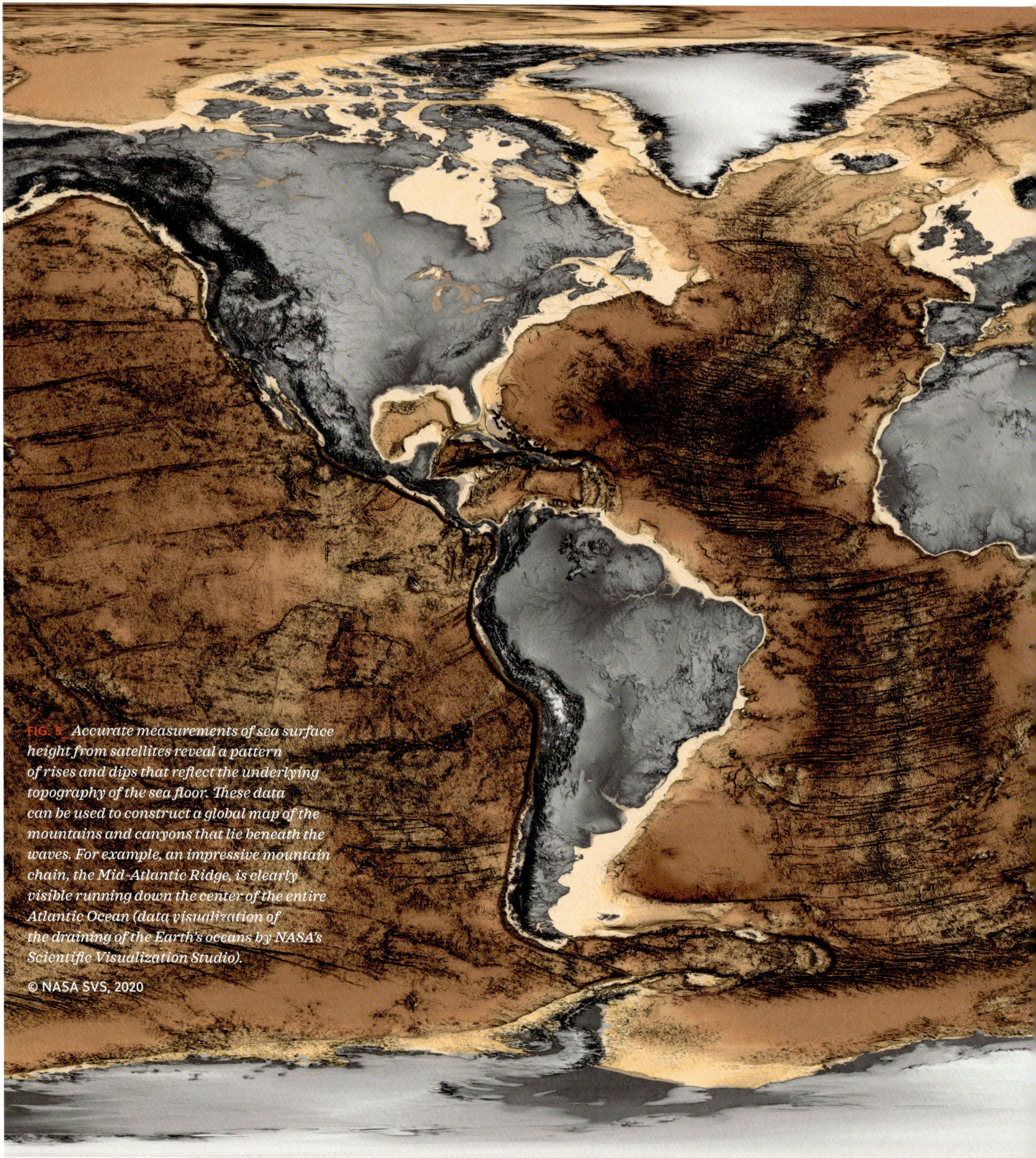

FIG. 5_ *Accurate measurements of sea surface height from satellites reveal a pattern of rises and dips that reflect the underlying topography of the sea floor. These data can be used to construct a global map of the mountains and canyons that lie beneath the waves. For example, an impressive mountain chain, the Mid-Atlantic Ridge, is clearly visible running down the center of the entire Atlantic Ocean (data visualization of the draining of the Earth's oceans by NASA's Scientific Visualization Studio).*

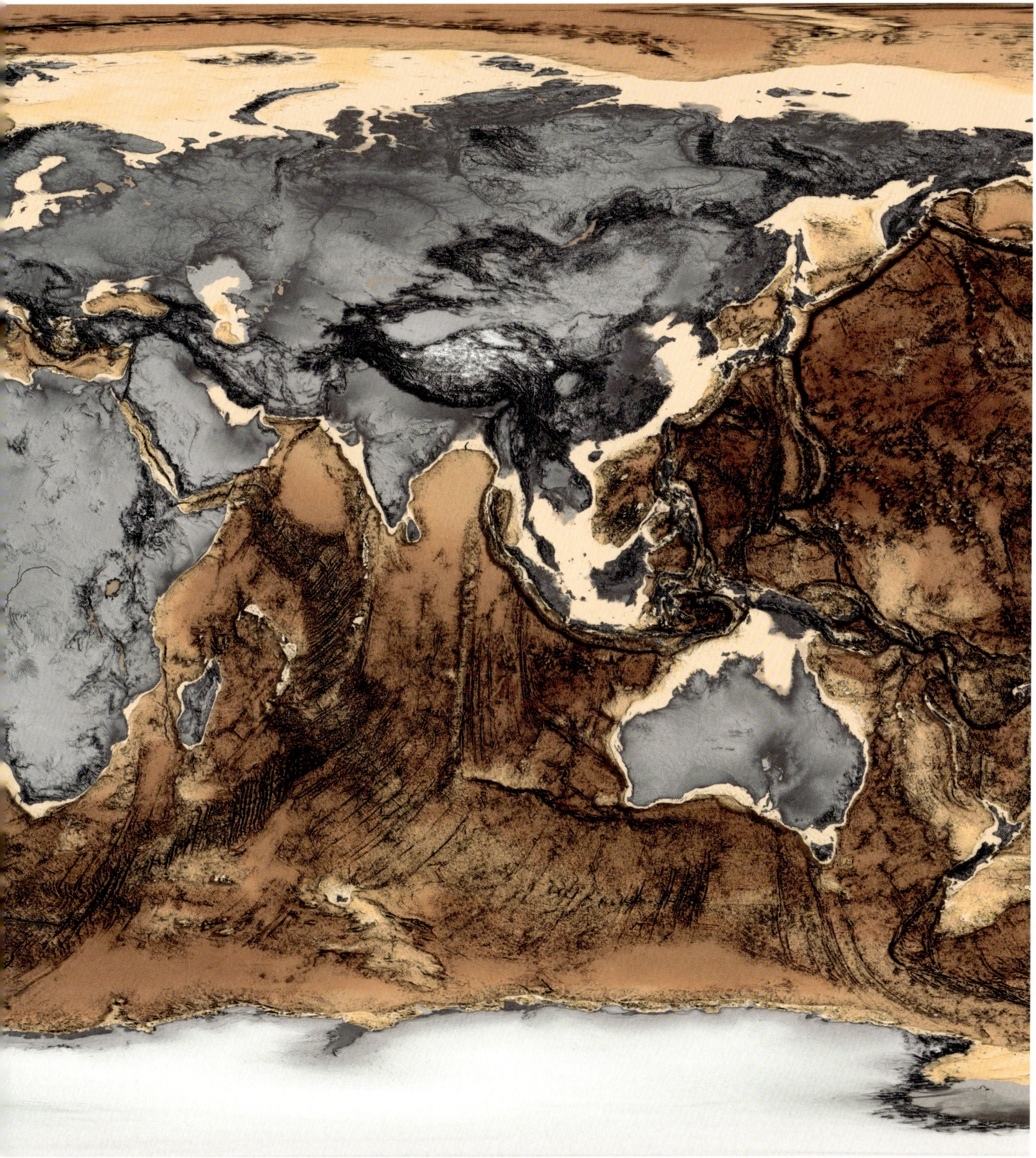

are transformed into one of the most spectacular landscapes on the planet. The bank rises from the abyssal plain in sheer cliffs that stretch unbroken to heights of over three kilometers. On a visit to the Bahamas, you'd never suspect you were exploring the steepest and largest cliffs on the planet, more spectacular than the sheerest rock faces of the Himalayas.

But the Great Bahama Bank is nowhere near the most spectacular ocean landscape. Drain the water from the oceans across the world and, before long, the peaks of an enormous mountain chain begin to appear. These are the mid-ocean ridges that encircle the entire globe like the seams on a basketball. Nothing on land even approaches the scale of this global mountain range. The ridge marks the junctions of the Earth's tectonic plates, where molten rock wells up from deep below. This lava creates the mountains, but it also creates new sea floor. The plates to either side of the mountain range are moving apart as new rock wells up along the center of the chain. These mountains are where new ocean is born.

In the Atlantic Ocean, the mountains rise spectacularly from the abyssal plain, but the extraordinary scale of ocean landscapes is even more evident in the Pacific Ocean, demonstrated, perversely, by its sheer lack of spectacular scenery. Part of the mid-oceanic ridge system in the Pacific is known as the East Pacific Rise, but the landscape here could not be more different from that in the Atlantic. Approaching the ridge here would be like driving across Kansas—many times over. The abyssal plains of the ocean cover about the same area as all the dry land on the planet, a vast featureless landscape on a scale impossible to imagine, especially when it comes to the Pacific. When you view one hemisphere of our planet from above the Pacific, virtually no land is visible at all. We really do live on Planet Ocean. The Pacific abyssal plains are so vast that, drained of water, we could drive across a featureless, pancake-flat landscape for many thousands of miles. The distances are so vast that we simply wouldn't notice our approach to the mid-ocean ridge. In addition, the spreading process is so fast in the Pacific that there is less time for rugged mountains to be built up. In the Atlantic, North America is moving away from Europe at about two to five centimeters a year, while the rate along the East Pacific Rise is one of the fastest recorded—at nearly fifteen centimeters a year.

The landscapes of the mid-ocean ridges are like nothing on land, although there is one place where these submarine mountains rise above the surface to permit a closer view. The island of Iceland is a small section of the Mid-Atlantic Ridge that has been thrust above the surface, offering the tiniest glimpse into what is hidden in the pitch black under several kilometers of water. The island is split by a wide rift valley where the eastern and western halves of the ridge are moving apart. It has one foot in the old world and one in the new. Iceland's landscapes are undeniably spectacular, but below the waves, this rift valley achieves the proportions of Arizona's Grand Canyon—although in reality it's an even grander canyon. This one runs for thousands of miles—and this is still not the most astounding landscape down there.

The spreading process is not entirely uniform, and the mountain range sometimes takes a sharp step sideways. Think of drawing a linear mountain chain across two sheets of paper and then siding one piece of paper sideways. The two offset sections of the mountain chain are linked by a transform fault where the plates must push past each other laterally since they are moving in opposite directions along the fault. Such transform faults also exist on land, where we can examine them more closely, the most famous being California's San Andreas Fault. In some places the fault is »sticky« and doesn't move until enough tension has built up to unstick it. In such circumstances, the abrupt lurching of two plates past each other has devastating results, such as the 1906 earthquake and the 1989 Loma Prieta quake, which both hit San Francisco. In other places, the fault is more lubricated and moves continuously in a process geologists call *aseismic creep*.

In these places, the fault still shapes the landscape, although in a much gentler way than powerful earthquakes and on a smaller scale. Hollister, in Central California, sits astride a lubricated section of the San Andreas Fault, so part of the town is slowly moving in the opposite direction to the other. Along the fault itself, curbstones and walls are offset by a few inches, a curious feature that draws geologists from around the planet. Not far away, on the Carrizo Plain, the landscaping effects of the fault are more obvious. Wallace Creek crosses the fault and, where it does, the creek takes a sharp right angle turn before turning back again to resume its original course. Over the last four thousand years, the San Andreas Fault has offset the creek by over one hundred meters.

The same process is operating along transform faults on the mid-ocean ridges but on unimaginably larger scales. The seafloor mapping data reveals one especially large transform fault along the Mid-Atlantic Ridge, known as the Romanche Fracture Zone. The extent of this landscape made it one of the hardest to try to model. We had real data from seafloor mapping, but the question was how to visualize structures on such an enormous scale. The Romanche Fracture Zone consists of a vast canyon — about as wide and as long as the Grand Canyon but four times as deep. It is so deep that the fissure connects the deepest waters of the eastern and western Atlantic basins. The amount of water flowing through this canyon between the two basins is the equivalent of about ten Amazon Rivers, each at peak flow.

Creating this landscape and then viewing it from the perspective of a human would give us no idea of its true scale. When the first Spanish conquistadors encountered the Grand Canyon, the river at the bottom looked so small, they assumed it was a stream only a few meters wide. In reality, the Colorado River is a mighty torrent some one hundred meters across, but the size of the canyon was so far beyond anything any European had previously seen that the conquistadors had no conceptual framework with which to understand its true scale. We faced the same problem multiplied many times over. In the end, all the scientists we worked with to create these ocean landscapes came to our rescue in a way that really helped bring the series to life.

They all enthusiastically agreed to visit our drained landscapes, to explore a world they could only ever see secondhand in the real world. This involved carefully planned and choreographed shoots in green screen studios, which allowed us to drop each scientist into the landscapes they had been studying remotely and, in some cases, also required some complex physical model building to recreate small parts of the ocean landscape in the studio. This certainly helped create a sense of scale on, for example, the Great Bahama Bank, but we still had problems getting any sense of the true size of the Romanche Fracture Zone. If we were to film such a landscape for real, we would film it from the air, so we decided to do the same for our *Terragen* landscape.

Fellow filmmaker Adrian Warren was also an intrepid pilot and, more importantly, owned a small Cessna aircraft. We didn't need him to fly anywhere, we just needed to build a green screen studio around his plane and for him to »fly« our expert along the Romanche Valley. Gillian Foulger, Professor of Geology at the University of Durham with whom we had already filmed the terrestrial portion of the mid-ocean ridge in Iceland, readily accepted the challenge of taking us on an aerial tour of the virtual landscape of the Romanche Fracture Zone. She turned out to be a consummate actor, and viewing this vast oceanscape from a tiny plane provided the perceptual handle we needed to imagine the unimaginable.

Other scientists were equally enthusiastic about being able to walk around landscapes they had only ever seen remotely. Gregor Eberli, of the Rosenstiel School of Marine and Atmospheric Sciences in Miami, walked the dizzying tops of the cliffs along the edge of the Great Bahama Bank. Peter Talling, of the National Oceanographic Centre in Southampton, strolled over the vast flat landscapes of the abyssal plain. John Smith, of the Hawaii Undersea Research Lab, revealed Hawaii's Big Island as the world's largest mountain. Above the waves, the peak of Mauna Loa, a large shield volcano, is impressive enough, but draining the ocean reveals that the volcano extends all the way down to the abyssal plain. Measured from its base, this mountain rises a thousand meters higher than Everest. Putting people into these landscapes emphasizes their scale in a way that CGI alone cannot. But it's not all about scale.

Some areas of the ocean floor have been mapped in much higher resolutions than the global Smith and Sandwell dataset. These areas have been mapped with a different technique—using sonar instead of the satellite altimetry of ocean surface topography. This allows us to visualize structures that are normally hidden in the dark depths in much more detail, data which reveals yet more surprises. Monterey Bay, near San Francisco, is home to the Monterey Bay Aquarium Research Institute (MBARI), an extraordinarily well-equipped oceanographic facility. Just offshore in the bay, beneath the waves and the surfers, is a huge underwater canyon, equivalent in scale to the Grand Canyon. The Monterey Canyon runs for five hundred kilometers, is a kilometer and a half deep in places, and has been mapped in great detail by MBARI scientists. A few small sections have also been visited by manned and remote submersibles, and are monitored by remote sensing stations that feed their data back to MBARI from tethered buoys above. So, we know far more about Monterey Canyon than we do about most of the ocean floor. Indeed, as David Sandwell pointed out while we pored over maps of the ocean floor in his office at Scripps, we know more about the surface of Mars than we do about four-fifths of our own world.

Detailed knowledge of Monterey Canyon has answered the most basic question about this hidden wonder: What is it doing here so close to the shore? And it turns out that it's not just ocean landscapes that lie beyond our everyday perceptions. So too do the events that created them—events that are often on such a cataclysmic scale that they are far more impressive than anything we can witness on land. Sampling the floor of the canyon reveals a jumble of gravel and sand, suggesting the area is subject to occasional sand avalanches of such power that they have carved the canyon, much like the Grand Canyon has been carved by the erosive power of the Colorado River. As in many bays, waves wash sand and gravel toward the center of Monterey Bay, where it slowly accumulates until it reaches a tipping point, triggering a powerful underwater avalanche.

Using the detailed scans of the canyon, together with information on how the avalanches form and travel through the canyon, we were able to recreate these hidden events in CGI. Who would have thought that the gentle rippling of sand through your toes as you splash along the bay's shoreline could produce such grand scenery or that, as you pause at the water's edge with the Pacific stretching off to the horizon, you are standing at the head of a vast hidden canyon? Recreating these ocean landscapes illustrates how blinkered our normal view of the world is. And the Monterey Bay sand avalanches are just a drop in the ocean compared to what has happened in other areas.

Peter Tallin has sampled sediments from the abyssal plain off the coast of northwest Africa and discovered evidence of a landslide of a different order of magnitude. This landslide began in another offshore canyon but, within a few hours, transported more sediment than all the world's rivers do in a whole year. Traveling at around twenty meters a second, the avalanche erupted from the canyon mouth and spread into a wall of debris over 150 kilometers wide. It was so powerful that it swept over the abyssal plain, as flat as a pool table, for 1,500 kilometers, until it washed up against the flanks of the Mid-Atlantic Ridge. All this happened sixty thousand years ago, but similar events still occur today, although they are completely invisible to us with our normal view of the world. In fact, the more we explore remote sensing data and visualize these hidden worlds, the more common such mind-boggling events seem to be.

It's not just the sheer scale of the ocean landscapes or the powerful events that created them that make them difficult to visualize. Even more intimate landscapes, at scales we are more comfortable with, have no counterparts on dry land. Indeed, we had no idea that one of these landscapes even existed until 1977. In February of that year, Robert Ballard and a team from the Woods Hole Oceanographic Institute (WHOI) were part of an expedition aboard the research vessel *Knorr* visiting the Galapagos Rift, located in the Pacific Ocean, six hundred kilometers off the coast of South America. They were reviewing videos from a remote camera platform called ANGUS that they had been towing 2,500 meters below the surface when they noticed shimmering water emerging from the top of a tall chimney. They were the first people to see a hydrothermal vent on the sea floor.

Two years later, again in the Pacific, more vents were found, and, this time, instead of shimmering hot water, the vents were gushing out super-heated water, black with minerals. For this reason, these strange structures are often called black smokers. And more startling still, the vents are crowded with life, from crabs and clams to huge tube worms, all living perilously close to water that reaches four hundred degrees Celsius. Scientists had long suspected something interesting was happening on the ocean floor, close to the mid-oceanic ridges, because they had discovered broad plumes of mineral-rich water drifting in the ocean. However, discovering their source rewrote both geology and biology. The animals living down there depended on bacteria that harvested the minerals for energy, meaning that the whole food chain was independent of the sun and the process of photosynthesis that supports the rest of life on our planet. These vents are a little like the geysers that draw so many tourists to places like Yellowstone National Park in Wyoming.

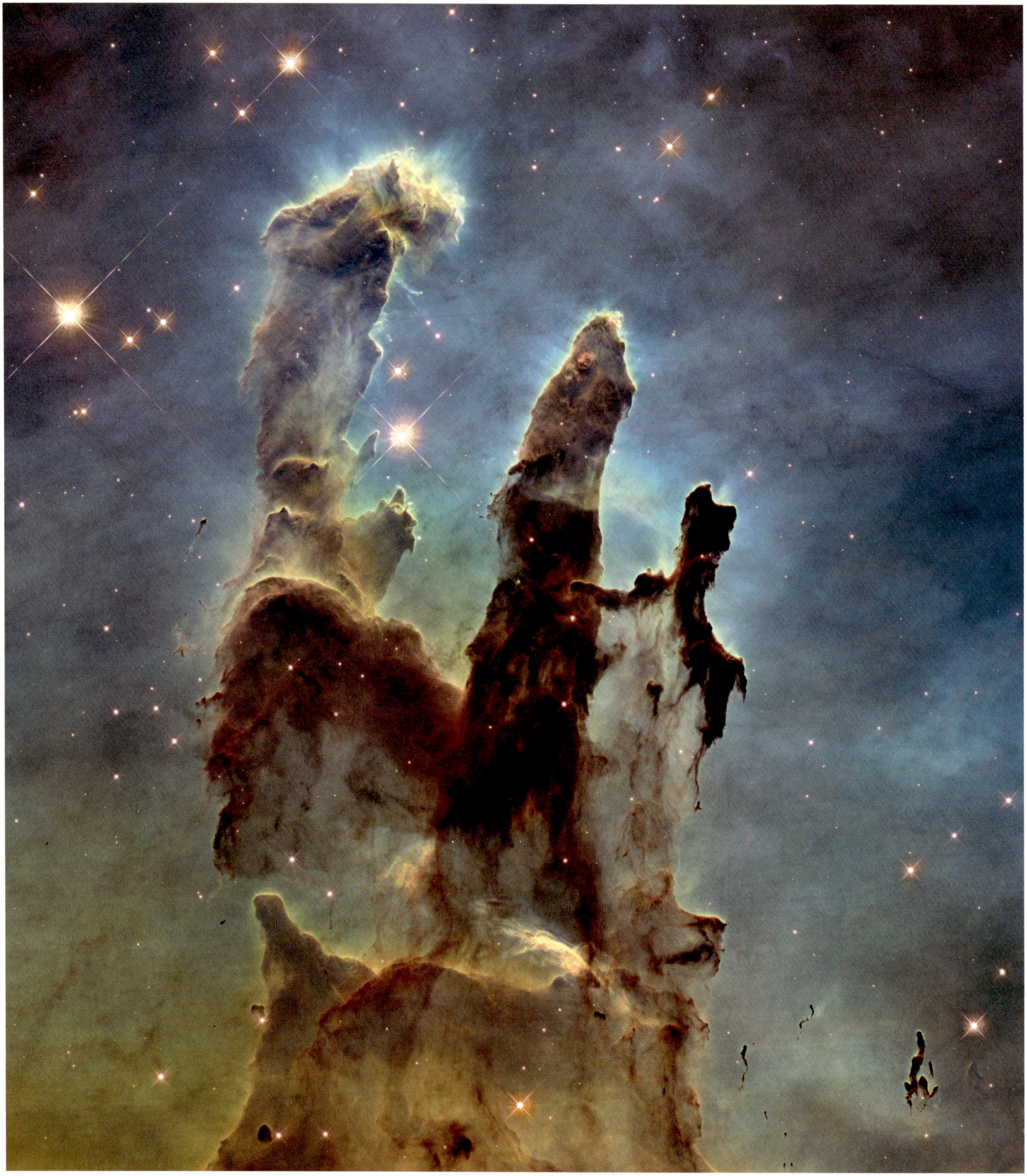

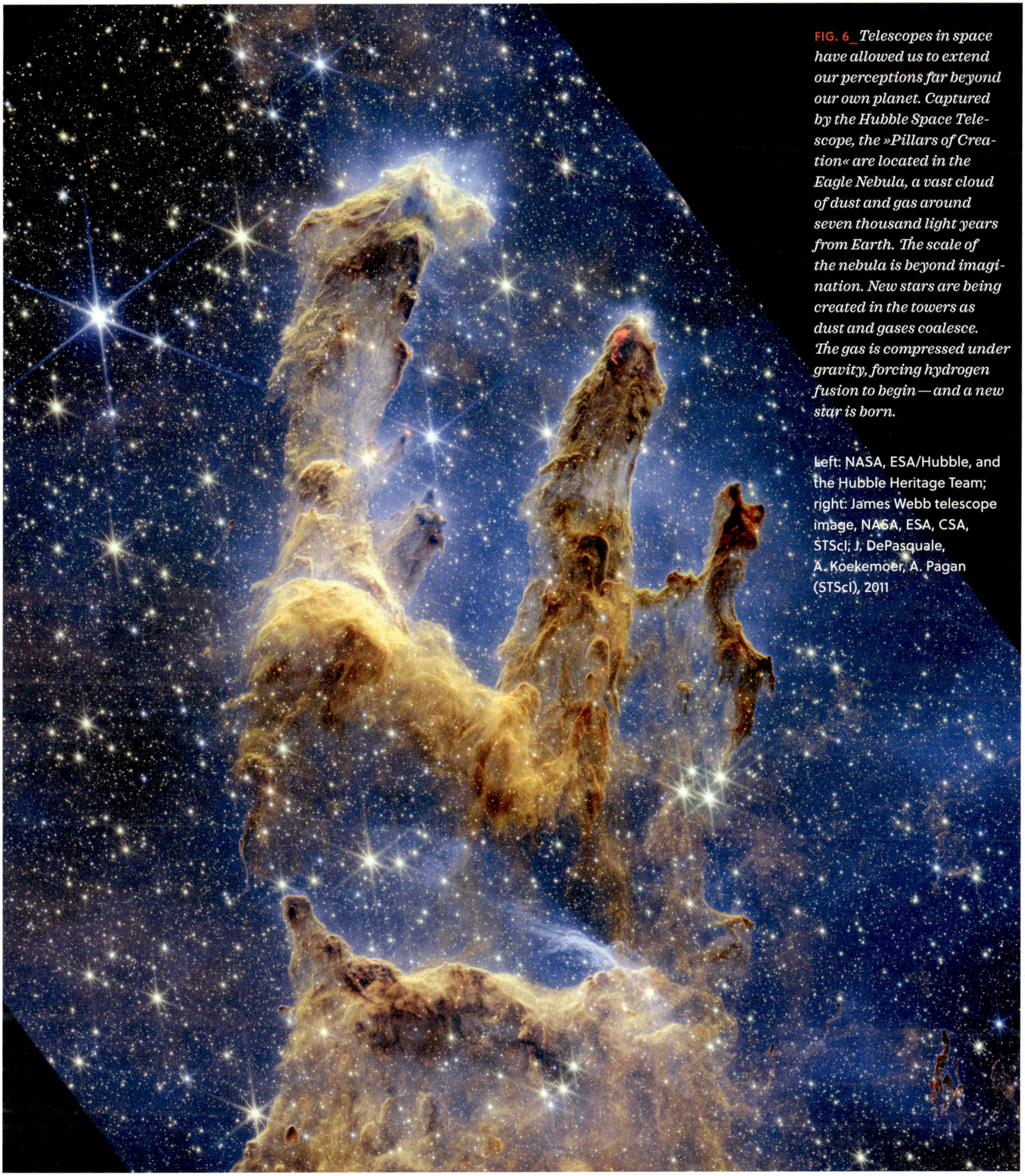

FIG. 6_ *Telescopes in space have allowed us to extend our perceptions far beyond our own planet. Captured by the Hubble Space Telescope, the »Pillars of Creation« are located in the Eagle Nebula, a vast cloud of dust and gas around seven thousand light years from Earth. The scale of the nebula is beyond imagination. New stars are being created in the towers as dust and gases coalesce. The gas is compressed under gravity, forcing hydrogen fusion to begin — and a new star is born.*

Left: NASA, ESA/Hubble, and the Hubble Heritage Team; right: James Webb telescope image, NASA, ESA, CSA, STScI; J. DePasquale, A. Koekemoer, A. Pagan (STScI), 2011

Magma below the surface here heats ground water, which erupts from fissures in spectacular fountains. On its journey to the surface, the hot water dissolves minerals from the rocks, and these precipitate out of a solution around the base of the fountain, gradually building up a mineral cone around the geyser. Over time, these cones grow into more substantial chimneys, although erosion by wind and water prevents them from getting too large. At least, that's the case on land. Below the waves, where erosion is much reduced, these chimneys can quickly reach the height of a twenty-story building, impossibly narrow and towering above the sea floor, with what looks like thick black smoke gushing from the top.

Oceanic hydrothermal vents are powered in a similar way to Yellowstone's geysers. They are found along the mid-ocean spreading ridges, where molten rock rises to the surface. Seawater seeps into the flanks of the mid-ocean mountains, where it is heated to high temperatures. This hot water travels back toward the surface, dissolving rocks as it rises, until it emerges black with sulfide minerals that begin to build the vent's chimneys. The chimneys grow quickly but don't last long. They fur up on the inside as minerals are continuously deposited and, once cut off from fresh supplies of building material, they collapse, while fresh ones rise around new fissures in the ocean crust.

Scientists have visited many of these sites in deep-dive subs, such as WHOI's *Alvin*, and, when I met with Cindy Lee van Dover, an *Alvin* pilot and biologist, she showed me a collection of the most extraordinary creatures I have ever seen. Seeing these extraordinary ecosystems firsthand, as in Cindy's case, or secondhand as preserved specimens, as in my case, is a huge privilege that forever alters your perception of our living planet. It's a world that can only be glimpsed with some very serious technology, and even then, our perspective is limited by how far the sub's lights can penetrate the pitch black of the deep ocean. But we do have detailed maps of some of these sites, which have been mapped using high-resolution sonar. One such site is the TAG (Trans-Atlantic Geotraverse) mound, on the Mid-Atlantic Ridge. The mound is some five kilometers across and built entirely of sulfide minerals of collapsed chimneys. The scale of this substantial undersea hill emphasizes just how much geothermal activity exists at this one site alone. We now know of hundreds of hydrothermal areas along all the mid-ocean ridges, but it's an absolute certainty that there are many, many more hidden in the inhospitable (to us, at least) deep sea.

Perched on top of the TAG mound are active chimneys, and the sonar data allows us to plot their location with high accuracy. But to get a real sense of what it would be like to visit this world without having to be surrounded by a pressure-resistant metal skin or being forced to peer out through tiny ports of thick acrylic into the small spotlight of the sub's lights, we wanted something more interactive than a CGI model—no matter how accurately we could render that model. So, we turned to model-builders who normally construct sets for feature films. After viewing footage of real vents and speaking with experts on hydrothermal vents, we were able to build a replica of the chimney field on top of the TAG mound in our green screen studio, which allowed one of our experts, geologist and biologist Crispin Little, of the School of Earth and Environment at the University of Leeds, to saunter through this strange landscape at his leisure.

Since our life-size model was constructed inside a green screen studio, we were able to add the broader landscape details in CGI. Likewise, we were also able to add the black smoker fluid, and CGI allowed us to model this as if it were hot water gushing into cold water. So, it appeared as it does on real vent fields but, in our vent field, Crispin could walk between the chimneys as he explained how they worked. The overall effect was bizarre—exactly what we wanted, since it illustrated just how strange these normally unseen environments really are.

So much of our world is beyond our everyday perceptions, but technology, coupled with the endless human imagination, has allowed us to extend our natural perceptions from the invisibly small to the inconceivably large. Two decades ago, in *Limits of Perception*, the goal of Alfred Vendl and myself was to take our audience on a journey from one extreme to the other and to show that we have been able to expand our perceptions even to the largest scale that we know—that of all of creation.

Back in the mid-eighteenth century, philosopher Immanuel Kant speculated that we lived in a rotating disk of stars which he called an *island universe*. Furthermore, he thought that objects in the night sky called *nebulae* were actually other island universes. We now know that such island universes or galaxies exist in numbers beyond imagining. When the Hubble Space Telescope was pointed at a patch of what scientists thought was empty space and took a photograph with an exposure time of four days, the blackness was found to be filled with innumerable galaxies. The universe is around fourteen billion years old, and since we depend on light to observe it, we will only ever know our universe out to fourteen billion lights years, as that's as far as light has traveled since the dawn of the universe. But that's plenty to keep us going. Scientists now estimate that there are 350 billion large galaxies out there—and an astounding seven trillion dwarf galaxies. That adds up to thirty billion trillion (3 × 1022) stars.

The universe has been mapped in some detail out to about two billion light years and reveals that, even beyond the scale of galaxies, there is structure. Although the universe is expanding, gravity draws neighboring galaxies together. Our own Milky Way galaxy is part of a cluster of six galaxies, including Andromeda and the Magellanic Clouds. But our Local Group is moving toward the center of a much bigger cluster, a supercluster, called the Virgo Supercluster, of around a thousand galaxies. There are about ten million such superclusters in the observable universe, strung together to make the largest structure we can ever know. We finished our journey in *Limits of Perception* with a 3D map of the universe, up to date at the time, showing the absolute limits of our perceptions.

What lies beyond the edge of the observable universe is forever unknowable, although, where technology fails us, human imagination keeps on going. Our explorations into the strange world of quantum mechanics have raised the possibility of many more dimensions than those of the space and time with which we are familiar, and of multiple universes existing in parallel to our own. This multiverse lies far beyond anything we can ever hope to perceive, although that hasn't stopped many science-fiction films and TV series from exploring the idea in various ways. But even if these ultimate layers of reality lie beyond the limits of our perceptions, we've come a long way from simply relying on our natural senses to inform us about the world. Our extended senses give us a much truer sense of our place in the universe, of our insignificance, even at the scale of our own planet. Opening the doors to these hidden worlds is the perfect antidote to the arrogance of a species that thinks itself the master of all it surveys.

The Serpent and the Dragonfly: Into the Unknown

Ina Conradi, Mark Chavez

Now then, throw it_to the demented creator,_to the demented darkness.

Beginnings

We met in 1987 while doing our Master of Fine Arts (MFA) at UCLA in Fine Art and Film Studies. Ina's works were large-scale hand-woven sculptures for immersive and walk-through environments. As a recipient of the Japan Foundation Fellowship, Ina continued her studies in Japan's traditional hand-crafted and contemporary fiber and textile art.

Mark arrived in Los Angeles before Ina, completing a BFA in Drawing at Arizona State University in 1980. After he arrived in LA, he started working as a freelance animator with a company that made large-scale immersive projections for special venues and concerts with laser light projections. There he developed a technique for vector-based animation that made coherent animation possible. His MFA thesis premiered in laser light during the 1984 Olympics and was projected onto the side of the Westwood Federal Building. It was an applied thesis that consisted of a laser-animated short film. His UCLA academic studies blended well with industry, where computer animation techniques for live-action films were just beginning. This allowed him to contribute ideas and techniques that supported computer animation in many different media, including vector-based laser animation techniques at LaserMedia Inc., raster graphics and early polygon modeling at Symbolics Incorporated, and early interactive techniques at Phillips Interactive. He worked in broadcast television in Japan at Tokyo Broadcasting System and helped to make early PlayStation games at Acclaim Entertainment in New York. He was recruited into the first group of artists hired by DreamWorks SKG. He later worked at Rhythm and Hues Studios in Playa Vista and has credits on numerous big-budget, award-winning Hollywood films. While working in production, Mark designed real-time chatbot avatars, where spontaneous discourse generated within a given framework would enable a somewhat free-flowing conversation. A web-based talking 3D avatar used a chatbot software language called AIML or Artificial Intelligence Mark-up

Language. This technique used pattern matching to associate input phrases with topical output. The AIML language, created by Richard Wallace, is an extension of an earlier chatbot, ELIZA, made in 1966 by Joseph Weizenbaum. It is a *multi-targeted intelligent computer interface* for chatbot technology that simulates conversation by using a pattern matching and substitution methodology. The project interfaced with Haptek Inc.'s live character animation technology under a start-up called Clone3D LLC. The web-based animation used a unique animation technology called a *Chaos Engine*, which enabled the AI-trained chatbot to move so that it appeared to respond emotionally to questions. The animated characters were hosted on a website in 2002.[1]

Journey to Southeast Asia

In 2005, Mark and Ina were invited to Singapore to help found a new School of Art, Design, and Media (ADM) at Nanyang Technological University (NTU). Mark initially worked in the Foundations Art Area, focusing on drawing, and later in the Digital Animation Area. (The school uses the term »area« separate from »department« because of the unified concept on which the school is founded. The meaning is essentially the same.) Ina joined a couple of years later to work in the Foundation Studies program and joined the Visual Communications Area. Currently, Ina is involved in the Interactive Media Area. When we moved to Singapore, NTU/ADM was the only art school within a research university in Southeast Asia.

At the university, Mark's research working with real-time chatbots hosted on the web continued. He branched into working with the Singapore Prison System, where he taught storytelling using video game toolsets. This research led him to a larger project called Cinematics and Narratives (CAN). CAN, funded by the National Research Foundation (NRF) under the Media Development Authority (MDA), focused on intelligent cinematic design. The research attempted to use real-time audience input to guide the visuals as determined by three preset visual styles. To do this, we created the short film in three styles that would morph depending on the scene's tone in the animated movie. These styles would best suit the desired audience's affective response. We considered tracking the audience's reaction to the visuals using various inputs. We ultimately created a system where we could use a slider to adjust the stylistic look of the film. The film was designed in cute, normal, and extreme styles.[2] The team created a director-driven system in the Unreal game engine, where the movie's visuals would morph according to the scene's desired emotional impact. As a case study, *[Vengeance+Vengeance]* was created, a nineteen-minute short film rendered in a game engine that combines science-fiction styles, a melodramatic science-fiction-leaning Hollywood style, and a generic Hong Kong action style to form the narrative.[3] The tagline was, »In a world where scientific achievement has merged the boundaries between man and weapon, one woman stands apart.« The short film won numerous awards at film festivals. Mark continued with the Multi-plAtform Game Innovation Centre (MAGIC) in 2012, where he established a content development think tank in a project titled Game Design for Entertainment. The MDA also funded this project under the NRF and a new agency called the Interactive Digital Media Programme Office (IDMPO).

In 2012, we collaborated on projects combining art, science, and technology. With a few short experimental stereoscopic 3D (S3D) films, we established a long-lasting relationship with the Ars Electronica Center Deep Space 8k Theater and the Ars Electronica Festival. Their large-scale immersive theater enabled us to create one-of-a-kind artistic spaces featuring art-science inquiry. With projected animation, dance, and music in an immersive interactive performance, visuals are displayed in a big theatrical black box with an interactive projection surface on both the wall and the floor. Our animated films were installed in gallery spaces as *mises-en-scène* at international exhibitions such as SIGGRAPH Asia 2013 Art Gallery in Hong Kong, where audiences could experience a film narrative in a walk-through, immersive installation. Rather than being embedded in a pedantic, direct, and readable narrative, in *Elysium Fields*, the main goal was to provide the viewer with an experience reflecting Ina's feelings toward her late father.[4]

While working on these projects, we were deeply fascinated by the ability of the creative arts, e.g., abstraction and surrealism, to recognize and capitalize on one's subjective perception, evoking an emotional response. As with *pareidolia*, we observe images of animals, faces, or mysterious objects in, e.g., clouds and cliff formations, gravel surfaces, wall textures, or floors. Alternatively, we might think we hear indistinct voices in random noise produced by air conditioners or fans. These are part of what our imagination induces naturally. Artists have always looked to these phenomena for inspiration to create works of art in representational and subjective art, where the meaning is left to the viewers' interpretation.[5]

Artists intentionally evoke the essence of things they depict without being representational. For example, to stimulate a creative style of expression, Leonardo da Vinci recommended that his apprentices »[…] look at walls covered with many stains or made of stones of different colors, with the idea of imagining some scene […].«[6] Similarly, the abstract expressionism movement of the 1940s strived to express personal feelings by interpreting random patterns, utilizing a graphic language that leveraged intuitive and creative visual development. Our work attempts to provide the viewer with immersive experiences on an intuitive, cultural, and emotional level rather than embedded in a direct and readable narrative. Many artists use an informed, intuitive approach to describe the ideas and concepts necessary to their biases and interests. For example, to determine the emotional effect of shape and form, Picasso conducted experiments in cubism where he used minimal lines to suggest the essence of a bull. Picasso worked like a technologist.[7] He used experimental cubist methodologies to visually present objects as we experience them in space and time, much like Heisenberg, one of the pioneers of quantum mechanics, who experimented in physics to express the intangible and invisible. Picasso was not interested in merely imitating reality when he adopted the cubist style; instead, he attempted to challenge how we understand reality. During a prolonged period of concentrated experimental labor, he extracted the essence of a bull.

In 2010 we launched the *Emotion Study* project.[8] In this collaboration, we shifted our focus to experimentation in a project where we used data to determine the emotional impact of abstract images. First, we developed a series of paintings and photos and compiled them into a library of non-representational abstract and semi-abstract forms. Next, we ran an online study to see how respondents evaluated the images. The work centered on a standardized set of emotions and examined emotions that abstract images evoke in pursuit of a definition of »meaning« in imagery. We found evidence that emotions and emotional intensities or their

»affective dimensions« interrelate and affect one another. The Circumplex Model of Affect is a theoretical framework that organizes emotional experiences into a circular map. It is a graphic chart that is tagged with named emotions in the structure of a circular illustrated spatial model.[9] The basic idea was to create a slider that uses imagery attached to named emotions on the graphic: sad/gloomy, happy/pleased, and similar opposite emotions. Using the slider, the imagery would change accordingly to evoke these emotions.

The *Emote* project (2017), created for the Media Art Nexus platform, further incorporated the psychology of pairing images and audio.[10] Our study included clinical psychologists from the Nanyang Technological University, Singapore and observed test subjects in the lab (theater). In addition, we tracked brain waves to determine the impact of images and music. Our question was, »Can we measure the emotional impact of abstract animation on the viewer?« *Emote* is a fifty-minute-long animation comprising twenty chapters using directive presets and audio-cued and timer-based settings. The imagery was generated using Derivative TouchDesigner, Open GL Shaders (GLSL), and audio-reactive animation. The artwork was displayed on the Media Art Nexus, a 15 × 2-meter LED video screen at the North Spine Plaza, Nanyang Technological University, Singapore.

FIG. 1_*Media Art Nexus,* QUANTONIUM *by Mark Chavez, experimental animation, LED, 15 × 2 m, Nanyang Technological University Singapore*

Photo by Quek Jia Liang, 2016[11]

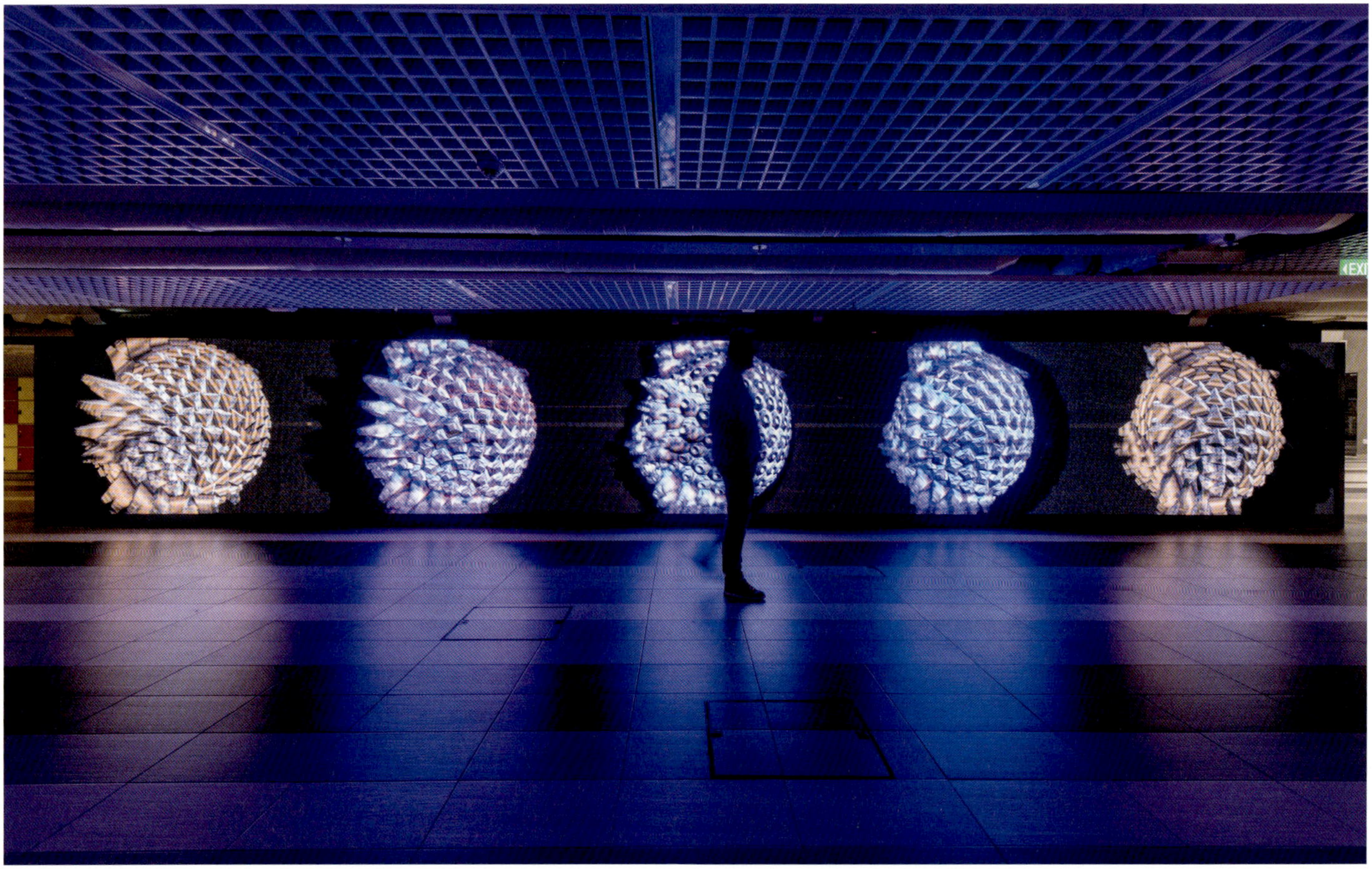

Media Art Nexus Singapore: A Curatorial + Educational + Public + Urban Media Art Platform

Media Art Nexus is an urban public media art installation permanently hosted at the North Spine at Nanyang Technological University in Singapore, dedicated exclusively to video art. This installation was conceived in 2016 and formally launched in 2018 as part of the NTU Museum's public art initiative. The installation has been integrated into teaching and has enabled us to jointly create new artworks and co-curate content with international universities and art institutions such as the UCLA Art|Sci Lab, the Science Visualization Lab at the University of Applied Arts Vienna, Urban Screen Productions (UTV), Melbourne, the Public Art Lab, Berlin, Fraunhofer MEVIS: Institute for Digital Medicine, and the Play and Civic Media Institute at the Amsterdam University of Applied Sciences. We also co-organized an international exhibition and art-sci symposium titled *On/Off 100101010 Colliding and Surrendering: Chaos and Freedom Where Art and Technologies Meet* together with the Web3D Art Gallery, a collaboration between the Queensland University of Technology (QUT), Brisbane, the University of New South Wales (UNSW), Sydney, and the Nanyang Technological University.[12] The symposium examined artistic expression in large-scale immersive displays based on interdisciplinary and hybrid approaches toward art and science in media art. It brought together fourteen international research institutions and media labs.

Art and Science Research Projects

Quantum LOGOS (vision serpent)

In 2019 we delved into animated, interactive, immersive art-science film intended for the immersive theater at the Ars Electronica Center Deep Space 8k in collaboration with science communication experts Bianka Hofmann and Bob Kastner, and leading Austrian physicist Rupert Ursin. Dr. Ursin is Deputy Director at the Institute for Quantum Optics and Quantum Information of the Austrian Academy of Sciences. This research explores quantum mechanics, with the visuals and ideas behind the counterintuitive concepts presented in mythic archetypes. In the film, our challenge was to find an aesthetic approach toward the abstract and invisible knowledge of quantum mechanics by looking at ancient artistic design archetypes. The resulting immersive film *Quantum LOGOS (vision serpent)* was our attempt to use intuitive design, pictorial metaphor, and analogical juxtaposition to explain complex scientific concepts to the layperson. The work explores the basics of quantum theory using intuitive, artistic design archetypes to shed new light on natural phenomena. We examine the nature of existence through scientific observation and illuminate it using familiar design archetypes in such a way as to impart meaning to the untrained nonscientific observer.

Quantum LOGOS (vision serpent) explores the visual iconography of ancient cultures, connecting them with concepts described in contemporary quantum science. Prompted by the notion that humankind has always searched for the meaning of existence, we used the underlying conceptual and visual designs evident in mythic concepts and art to explore ideas behind concrete and mathematical concepts. Though this work uses familiar motifs from ancient cultures, it does not appropriate them. Instead, we researched iconographic graphic design that complements concepts explored in quantum mechanics, creating a new design template that harkens back to ancient ideas and explores the nature of existence using contemporary findings from quantum science. Our pictorial metaphors illuminate the quantum universe beyond ornamental illustration to communicate scientific ideas.

The Greek term *logos* (meaning *word, reason,* or *plan*) has been used for over two thousand years. It means reasoned discourse. It is the divine reason implicit in the cosmos, ordering it and giving it form and meaning. Its meaning has consistently been associated with the idea of reasoned discourse, rational thought, and the underlying order and reason that governs the world.[13] We found that it worked well as part of the basis for our design strategy, using artworks that suggest divine reasoning in traditional cultural beliefs.

Although the film project started in January (2019), Mark had begun research on the topic much earlier, influenced by books like James Maffie's *Aztec Philosophy: Understanding a World in Motion*[14] and Alexus McLeod's *Philosophy of the Ancient Maya: Lords of Time*.[15] These books offer insights into Indigenous Mesoamerican civilizations' metaphysical, epistemological, and ethical wisdom. In addition, the essence of creation myths and themes in these other great cultures provides a rich tapestry from which we derive design inspiration.

These early cultures have often turned to poetry to express ideas for which there are no equations. Science and poetry often do not go together. Nevertheless, poetry helped people feel what they could not understand in ancient times. The following poem was an incantation written by a Maya Shaman after being passed down through the ages from the golden age of Maya civilization.

Ritual of the Bacabs, translated by David Bolles

Can Ahau, they say, is the creator.
Can Ahau, they say, in the darkness.
Coming from the fifth level of the sky,
the head of the dragonfly,
The head covering its worms.
It bit the hand of the unfettered creator
the unfettered darkness.
It licked the blood in the sweat bath,
it licked the blood in the stone hut.
Now then, throw it
to the demented creator,
to the demented darkness.[16]

Our work did not exclusively reference designs from ancient cultures. We also studied the work done by physicists specializing in various areas investigating quantum phenomena. It was only with the generosity of scientists who publish their talks online that we could even begin to understand the complexity of the subject in a minor way. We attended in-person talks at the UCLA Art|Sci Center, with artist Professor Victoria Vesna and nanoscientist Dr. James Gimzewski at the helm, including talks by professors from the California NanoScience Institute, as well as other casual discussions on the subject with academics in the field. Mark also studiously reviewed lectures from the World Festival of Science and other sources on quantum physics, gravity, time, and biology.

SPACETIME STRUCTURE: AN ANALOGY—CONDENSED MATTER AND QUANTUM INFORMATION_While researching the issues related to visualizing quantum phenomena, we found an exciting area of mathematics called *lie theory*. This area of mathematics describes the enigmatic E_8, the most complex shape in our universe: an eight-dimensional mathematical pattern with 248 points, first theorized in 1887. The E_8 Theory, also known as An Exceptionally Simple Theory of Everything, tries to explain the greatest mysteries in physics, such as how particle physics and gravity can be combined in one model. Studying the partially assembled puzzle of the standard model of physics and gravity, we see that all particles' charges fit into a pattern of arguably the most intricate structure known to mathematics, E_8. Theoretical physicists have noted that the structure has geometrically symmetric properties that could bring about a uniform theory in physics. Mathematicians claim that the E_8 theory contains the standard model (of physics) plus the symmetries belonging to gravity. The patterns they have identified are similar to mandalas, making them attractive to the aesthetic eye. We noticed that mandalas share design similarities with mathematical patterns observed in quantum mechanics. By utilizing intuitive forms with significant meaning for less technically informed early cultures, we found that many visualizations made by early cultures shared common design traits with graphically represented math in contemporary quantum physics. E_8 theory asserts that there is a crystalline form that models an 8-dimensional crystal. Quantum theorists describe this as the basic underlying structural form of the universe. Some math involved describes the golden ratio, a well-known artistic design form.[17] Can we take graphical shapes created by early civilizations that describe natural phenomena, such as spirals and twisting forms that arise from celestial observations, to describe quantum mechanics visually?

As stated, rather than stylistically represent the quantum world in an informative, graphic way, we decided to leverage design construction used in the ancient past. Likewise, as in the *double-slit experiment*, exciting natural patterns emerge when we take waveforms and force them to interfere with each other. We also looked at the spiral nature of galaxies, spiral phenomena surrounding black holes, and ideas graphically described by early societies that take a similar form. Finally, we looked at ideas imagined and described by early humans to find forms that resemble observations that our sophisticated instruments now allow us to see. Admittedly, other artists sometimes take this approach to the extreme. We take special care not to align with designs used by pseudo-science.

VIBRATIONS: ERNST CHLADNI AND THE ACOUSTIC MANIPULATION OF MATTER_Ernst Chladni (1756–1827), the father of acoustics, found a simple relationship between sand and various vibration modes on a smooth surface. On polished steel plates, he dusted fine sand and created patterns of nodal lines with the subtle vibrations of a violin bow. The audible vibrations caused by the violin bow stroking the side of the metal plate formed patterns, while the sand formed linear patterns corresponding to the audio waveform. Audio-generated visual patterns create beautiful geometric figures with sound in many areas of nature, such as in snowflakes and even in the colossal weather systems on Saturn. The tortoiseshell and other natural designs also resemble patterns generated by cymatics[18] techniques at specific frequencies. Similar frequencies exist at the subatomic and cellular levels. Intricate, star-like patterns exist in a tiny cross-section slice of our cells' structural geometry. These structures are called microtubules, which are cellular building blocks. Microtubules are tiny cylindrical structures made up of protein molecules called tubulin. They are essential in various cellular processes, including cell division, movement, and intracellular transport. They form the structural framework of the cell, known as the cytoskeleton. The cytoskeleton is responsible for maintaining the cell's shape, providing mechanical support, and organizing the cell's contents. Microtubules act as tracks along which cellular materials, such as organelles and proteins, are transported to their respective destinations. Physicist William Brown proposed a theory suggesting that microtubules may play a role in consciousness and the nature of reality. Brown's theory, known as the Resonant Recognition Model (RRM), suggests that microtubules may be able to process and store information based on variations of resonant geometry. In essence, regardless of media, geometry leads to unseen networking by generating vibrations and amplifying results to an entirely new level.[19]

Design and Concept Approach: Exploring Science with Design Archetypes

Our approach in our project was to use notable mythic Mesoamerican figures. Aside from studying the books mentioned above, Mark began to learn Nahuatl, a Uto-Aztecan language spoken by a subset of Indigenous people in Mexico, to understand Indigenous American ideas better. In particular, it helped him to understand how Aztecs used metaphors to express themselves in poetry. They depicted their universe and the events that occurred in it in a beautiful and evocative way. We saw this as an opportunity to expand research by using collective themes and ideas in a design created intuitively by our ancestors to illustrate ideas reflected in aspects of quantum theory. The most challenging aspect of this project was researching Mesoamerican design archetypes and equating them with accepted concepts in quantum theory. In addition, much effort was put into understanding Mesoamerican thought and searching for elements of quantum theory that would match up. Though there is no direct connection between metaphysical naturalism that scientists or philosophers can discuss and document today, considering what we can learn with advanced technology, there are uncanny intuitive matchups with Mesoamerican philosophy.

Archaeological Findings and Mesoamerican Thought: Through the Smoking Mirror, Visions of the Past, Present, and Future

Scrying was a common practice in ancient Mexico.[20] It involved looking into a black obsidian mirror used due to its highly polished, water-like reflective surface. The Aztecs used it as a divination instrument to see the past, present, and future or to get answers. A scryer, gazing into the mirror, would see clouds of smoke as part of a ceremony, which would reveal a vision of a mystical blur as though through a smoky cloud.[21] The Mexica used obsidian mirrors for divination and as symbols of their power, often wearing them as pendants. Obsidian is associated with the

Mesoamerican Empire. Mexica royalty called mirrors made from obsidian *Tezcatlipoca* or *Lord of the Smoking Mirror* after the god Tezcatlipoca. The name alludes to the deity's connection to the obsidian mirrors used in shamanic rituals and prophecy.[22] Tezcatlipoca, the main creator god, ruled over the first world or *Tonatiuh* and was known as »he who goes forth gleaming.« *Tezcatlipoca* is often portrayed with a stripe of black paint across his face and an obsidian mirror in place of one of his feet.[23] The *tlamatinime* saw reality as something conceived within a dreamy, ephemeral, or illusory stage of terrestrial existence. Nahua *tlamatinime* were the learned people of ancient Mexico, knowers of things, sages, philosophers. The *tlamatinime* claimed everything earthly is dreamlike. They conceived of metaphysics, epistemology, the theory of value, and aesthetics in conceptually overlapping, if not equivalent terms.[24]

TUDOR MYSTICS JOHN DEE AND SIR EDWARD KELLEY_Moctezuma used this or a similar mirror to foresee the approach of the Spanish. It was plundered from Mexico by Cortes and brought to Europe. John Dee, the English Renaissance scholar, astronomer, and adviser to Queen Elizabeth I and her court, used it in the 1580s for magical experiments and as a scrying mirror.[25]

Together with alchemist and medium Sir Edward Kelley, John Dee explored the mysteries of sixteenth-century science in Europe. During the last three decades of his life, Dee sought help to solve unexplained natural philosophy.[26] Edward Kelley convinced Dee that angels in the mirror would open spiritual realms[27] and reveal the universe's greater secrets.[28] Through the scrying mirror, they saw a shadow-filled world where angels gave them information in Enochian. John Dee had one of the largest libraries in England at the time. Their inquiries and those of their peers led to the beginning of chemistry as a science. Careful comparisons between scientific and humanistic sources of information will always reveal a clearer picture.

Mexican Vision Seekers and the Entheogenic Connection

XŌCHIPILLI—MESOAMERICAN DEITY_According to Ronald A. Barnett, Emeritus Professor of Higher Education at the University of London, »It is generally assumed that the idea of other universes is the unique product of ›post-modern‹ thinking based on the theory of relativity and quantum mechanics. However, the ancient Aztecs and Maya probably got there first, albeit for different reasons.«[29] God, Xōchipilli, was the patron of art, games, dance, flowers, and songs in Aztec mythology. His name contains the Nahuatl words *xōchitl* (*flower*) and *pilli* (either *prince* or *child*), and thus means *Flower Prince*.[30] Plants and flowers were associated with mystical experiences. The classic Mexica sculpture of Xōchipilli shows him in the throes of entheogenic ecstasy. American author, ethno-mycologist, and former Vice President of J.P. Morgan & Co. Robert Gordon Wasson says that in the statue's representation of Xchipilli, he is absorbed by *temicxoch*, or *dream flowers*, as the Nahua call the experience that follows the absorption of an entheogen. Devotees would consume *temicxoch* to investigate reality and have an entheogenic experience. There is nothing like it in European art's long and rich history.[31]

K'AWIIL—VISION SERPENT_The vision serpent is one of the essential Mesoamerican deities, also known as Och-Kan and associated with the Maya deity K'awiil, a pre-Columbian god-force connected to royalty, lightning, serpents, fertility, and maize. The vision serpent was an intermediary between the world of the living and the deities. The bloodletting ritual invoked ancestral spirits for guidance, protection, and blessings. Part of this invocation resulted in visual hallucinations. During the vision serpent ritual, participants from the ruling class would cut their bodies, resulting in severe pain and blood loss, stimulating the production of endorphins, the body's natural pain relief.[32] In this trance-like state, participants would burn blood-soaked ceremonial papers. The resulting smoke served as a vehicle for the appearance of the vision serpent. In Maya cosmology, the vision serpent is a portal to the realm of the ancestors. This ritual generates a psychedelic epiphany, manifesting as forms generated randomly in the burning smoke.[33] The world tree and the vision serpent (representing the king) formed the central axis, communicating between the spiritual and earthly worlds or planes. The king could bring the central axis into existence in a temple ritual and create a doorway to the spiritual world with its power.[34] The Maya civilization was founded on chronovision, or the total immersion of individual and collective life in natural rhythms, which was mapped into a mathematical system with numerous cyclical counts operating concurrently.[35]

Metaphysical Naturalism and Cultural Design

The nature of the physical world and our perception of reality are the subject of intense philosophical debate. The Western view of reality has traditionally supported the idea of physicalism/materialism and a duality between the physical world and the realm of dreams. This perspective has been challenged by global Indigenous views of reality, which offer alternative metaphysical frameworks. These alternative frameworks often view the physical world and spiritual or non-physical realities as integrated and interdependent. This perspective can lead to a different understanding of the relationship between humans and the natural world, and challenges the Western view of reality as separate from nature. However, the implementation of these frameworks can lead to challenges in terms of placement and location, as the boundaries between the physical and non-physical worlds are not always clear., i.e.,

> *[t]he problem being addressed is the challenge of identifying entities within a physical world that does not seem entirely physical or natural. This issue has led to debates about metaphysical doctrines, which typically focus on resolving placement problems. A failure to solve these problems may result in the elimination of the entity in question or the rejection of the broader doctrine. Other debates revolve around the proper formulation and understanding of these doctrines, such as clarifying what is meant by calling an entity »physical.« Additionally, scholars discuss whether and how these doctrines can be justified, and examine their implications for science and the treatment of placement problems. For instance, does physicalism mandate that all sciences reduce to physics? These debates highlight the challenges of reconciling metaphysical doctrines with scientific observations and understanding, and illustrate the complexities of grappling with questions about the nature of reality.*[36]

Debates about these metaphysical doctrines often focus on placement problems, where a failure to identify the thing in question may force a rejection of the entire canon. Other debates focus on the proper formulation and understanding of the principles, i.e., whether and how a perception of reality might be justified. What are its implications for science and the adequate treatment of placement problems? Does physicalism require all sciences to reduce to physics?

The viewpoint known as epistemological naturalism holds that scientific investigation is the most reliable way to acquire knowledge (possibly: the only viable method). The ontological claim that metaphysical naturalism makes is as follows: like physicalism, it asserts that everything is natural, but unlike physicalism, it places significant emphasis on providing a precise definition of »natural.« The non-naturalist worldview considers the naturalist worldview to be nothing more than a physicalist one with a new name.[37]

Epistemological naturalism and metaphysical naturalism: both doctrines have significant consequences for our perception of the world, especially human aspects of the world and the nature of mentality. *Quantum LOGOS (vision serpent)* utilizes our current understanding of how Mesoamerican cultures have interpreted nature and existence. This artwork dwells on a contemporary view of natural phenomena reflecting collective mythic memory. In quantum physics, the sun is a source of energy whose gravitational force warps fields of space around it. The quantum effect evident in quantum biology sustains life just like a tree absorbs light waves to grow and absorbs energy through photosynthesis. The animation used the sun as a visual metaphor to describe this effect from a contemporary perspective. The short film presentation ends with an interactive animated interlude. The design of this approach is based on the double-slit experiment's light wave phenomena. Specifically, it builds on the idea that waveforms create distinctive interference patterns when intersecting. This approach expands upon these patterns to explore the forms that emerge when different waveforms intersect and interact. In the contemporary world, we assert that the idea of other universes is the unique product of »post-modern« thinking based on the theory of relativity and quantum mechanics. However, the ancient Mesoamericans may have gotten there first with their intuitive understanding of zero as expressed in mathematics. Religion in ancient Mexico took many different forms. With evidence of human settlement as far back as before the Last Glacial Maximum (26,500–19,000 years ago), proof of human dispersal into the current Mexico region is evident as early as 33,000–31,000 BP.[38]

We assume that humans have always sought out the meaning of life. A desire to understand these matters is at the heart of many ancient religions. »Why am I here? What is this all about, and what happens to us when we die?« Mesoamericans also desired to understand these matters and the enveloping cosmos. For the Aztecs, theirs was a pantheistic worldview, with evidence of it dating as far back as the Olmec civilization at San Lorenzo (3,500 BP) and early Maya and Teotihuacan cultures. They maintained these beliefs until the Spanish invasion in the sixteenth century, when a unique form of Mexican Catholicism was adopted.

Pantheism in the Western world first appeared in Spinoza's *Ethics*, finished in 1675, two years before his death.[39] It is one of the world's oldest forms of spirituality outside the West. The Maya developed a highly sophisticated and complex belief system that posited time as an integral part of their understanding. For the Mexica

(Aztecs), Gods, or expressions of spirit, are an ever-dynamic part of everything, interwoven throughout every aspect of life. (The Indigenous peoples of Mexico had no word for god or goddess, so we interpret these as archetypal forces.) When the priest understands that the spirit body does not exist separately from the physical body, the human in the middle seeks further understanding of the nature of their world. Through an intuitive understanding of the nature of their world, as perceived through mathematics and embedded into their calendar ritual observances, a Mexica or Maya priest would have little difficulty accepting »spooky action at a distance.« In contemporary Chicano commentary, the term *nepantla* is often used. It is a concept that symbolizes the state of »in-betweenness,« reflecting the idea of being in the middle of two distinct realities. The term *nepantla* comes from the Nahuatl language, meaning »in the middle of it« or simply »middle.«

Considering that the Maya and Mexica written documentation, religious or otherwise, was destroyed by the Spaniards during and after the conquest of Mexico, assuming that there is no understanding of complex metaphysical naturalist concepts in documented evidence is shallow. This understanding is evident in Maya and Mexica architectural design and philosophical musings constructed from their mythic narratives.

Scientific Inspiration and Quantum Mechanics Used in the Film

The Copenhagen interpretation states that quantum systems do not possess specific properties before measurement, only probabilities that reduce to certainties on analysis. As a result, quantum physics is as difficult to understand now as it was for Einstein and the scientists of his day. Despite being a difficult concept, artists have been using quantum physics as inspiration since the early 1900s. They have incorporated this inspiration into designs that reflect modern society's concerns. These efforts help us to understand the counterintuitive concepts of space, time, and design dissonance in the physical world.

THE OBSERVER EFFECT_Is it possible that reality is composed of diverse potentialities, much like the visual illusions that our minds create from a cluster of clouds, where shapes can appear in various forms? Recent research by physicists at the

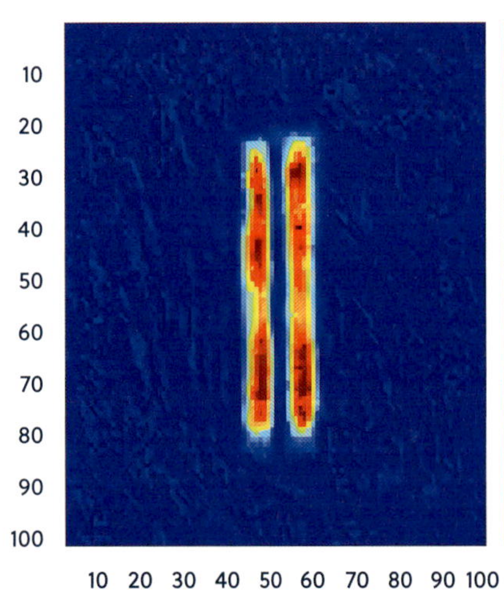

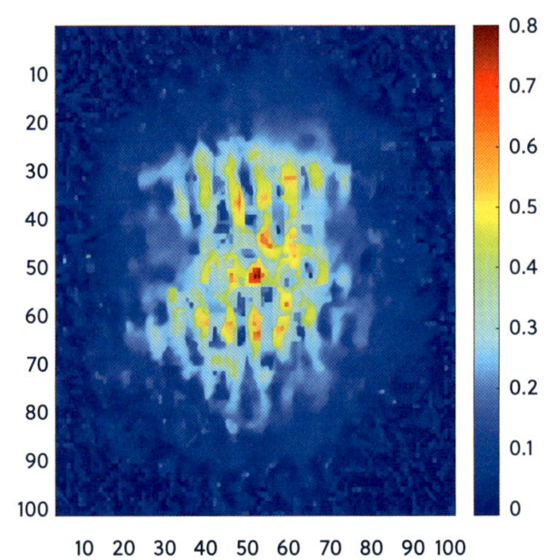

University of Queensland in Australia provides further evidence supporting one of the most puzzling yet essential features of quantum physics: the inherent uncertainty of the quantum wave function. Moreover, this theoretical framework implies the existence of overlapping realities and the coexistence of multiple truths.[40]

The scientific method is built on empirical evidence and objective measures to test hypotheses. The critical aspect of the scientific method is that it does not depend on who observes the results as long as the evidence is reproducible and widely accepted. In contrast, what constitutes objective evidence is not as straightforward in quantum mechanics. The Wigner experiment, a classic thought experiment in quantum mechanics, highlights the challenges to objectivity in this field. The experiment involves two observers, Wigner and »Wigner's friend,« each of whom witness the same event from different vantage points. According to quantum mechanics, observation can influence an experiment's outcome, and this idea is taken to its extreme in the Wigner experiment. The paradox of the experiment is that both Wigner and his friend can witness different realities of the same event, suggesting that multiple realities can coexist simultaneously.

The concept of multiple realities challenges long-held assumptions about the nature of reality and the idea of an objective, observer-independent reality. The implications of the Wigner experiment and other similar thought experiments in quantum mechanics continue to be the subject of intense debate and exploration among physicists and philosophers as they try to reconcile the apparent paradoxes with our understanding of the world. This paradox raises fundamental questions about the nature of science and the nature of measurement. Is it possible for objective facts to exist? Can evidence be something that defies our understanding? Scientists rely on empirical evidence to discover concrete truths, but how can they agree on those truths if they face contradictory realities? These are some of the profound questions explored at the intersection of quantum mechanics and philosophy.[41]

The film employs a design that mimics the *double-slit experiment*, which demonstrates the wave-like properties of light and the duality in nature. The behavior of both light and matter is characterized by particle and wave properties, depending on the type of interaction they undergo. Light would only produce two lines on the screen without interference and diffraction. Interestingly, elementary particles like atoms, molecules, protons, neutrons, and electrons, which are material objects, behave like waves (non-material entities) when they are not observed. However, when an observer attempts to perceive what is happening, the waves collapse into particles. It remains unclear whether the particles are aware of being watched or if the observer, in some way, creates reality simply by observing, or perhaps both concepts are relevant. This apparent paradox continues to be a subject of debate among scientists and philosophers.

FIG. 2_*Double-slit experiment: the image on the left is of two independent strips of coherent quantum matter, and the image on the right is that of the interference fringes after letting them interfere.*

© Nanyang Technological University Singapore School of Physical and Mathematical Sciences, Division of Physics and Applied Physics, 2022

The design of our exhibit was inspired by the double-slit experiment, which demonstrates how different waveforms can interact and create fascinating patterns. Our goal was to create an immersive experience that would transport the audience to the depths of a sacred Maya cenote, a natural sinkhole in the Yucatan Peninsula filled with water. Through graphic illusions, we aimed to simulate the interference of waveforms and the resulting ripple effects. As visitors move across the floor, the exhibit creates ripples that impart a sensation of being surrounded by dynamic energy fields. Moreover, by using light to symbolize a pool of water, the

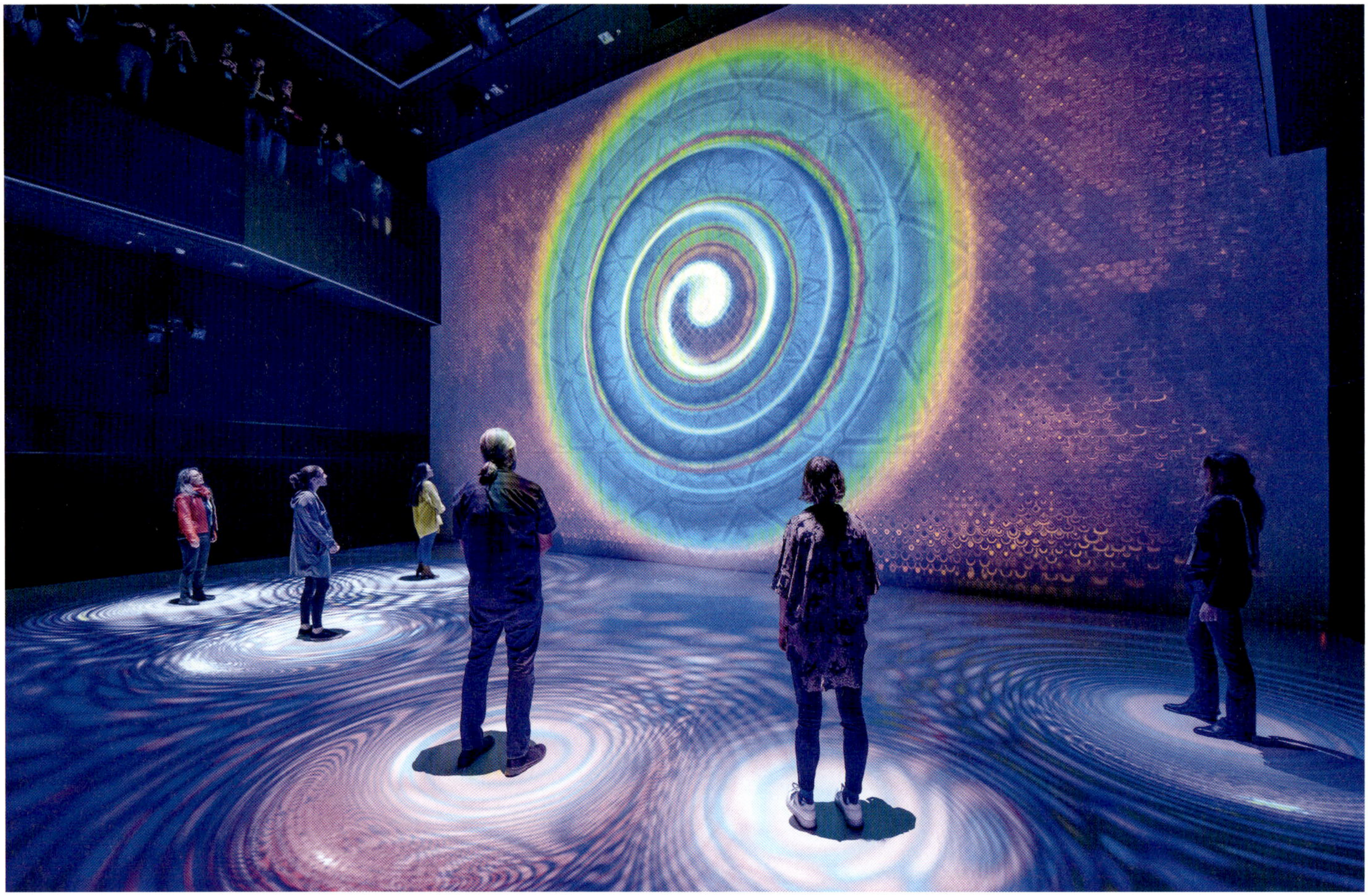

exhibit generates reflections that embody our place within a constantly shifting field of energy.

Quantum LOGOS (serpent vision) and Maya Cenote

The project begins with the sound of a conch shell horn, followed by ten-minute plays that explore different aspects of quantum theory using abstract animation. The audience is then invited to enter the quantum world, where the room is transformed into a pool of quantum energy, and the front screen and floor become interactive. The design is inspired by the depths of a sacred Maya cenote, creating an immersive experience that simulates the rippling of energy fields. Likewise, light is a metaphor for a pool of water, reflecting the audience's place within the ever-rippling waves of energy.[42]

The works premiered at the Deep Space 8K at the Ars Electronica Festival 2019. In 2022, the Guizhou Provincial Museum hosted *Fission: The New Wave of International Digital Art*, a curated exhibition featuring the works of forty-four digital media artists from all over the world, including Zhang Xiaotao and Li Fei. The exhibition showcased fifty-four pieces of art, making it the first international exhibition of digital art held in the Guizhou Province.

FIG. 3_*Mandala,* QUANTUM LOGOS (VISION SERPENT)*, Deep Space 8k Ars Electronica, 2019, immersive installation*

Photo by Wolfgang Simlinger, 2019

VIDEO_ QUANTUM LOGOS (VISION SERPENT) *trailer*

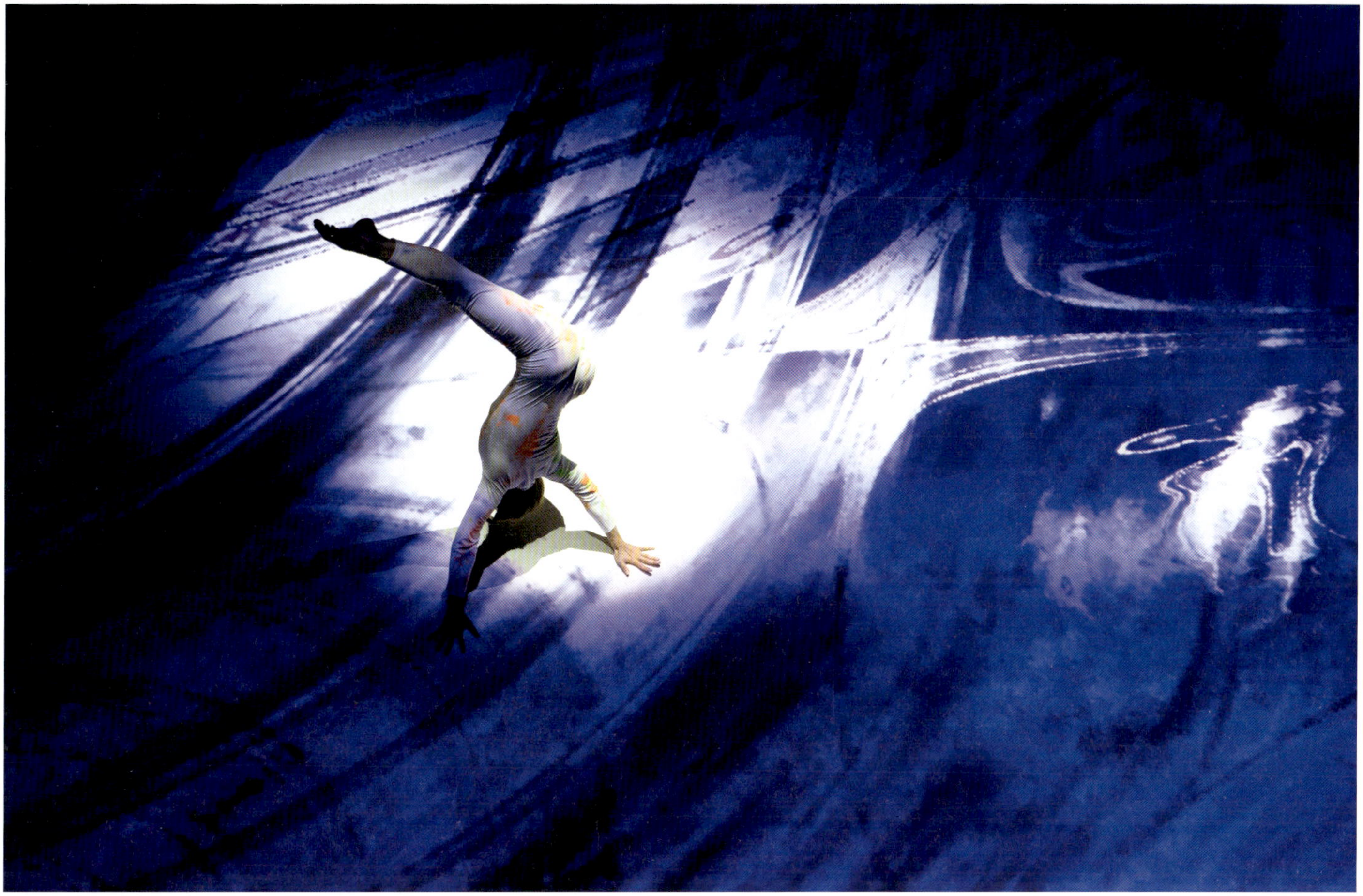

FIG. 4_NOCTURNE, *Deep Space 8k Ars Electronica, immersive animation*

Photo by Tom Mesic, 2021

VIDEO_ NOCTURNE *excerpt*

Reconnecting with Emotive Imagery

NOCTURNE_In 2017, the Singapore Tourism Board Agency (STB) invited Singapore-based artists and us to participate in a *Singapore Inside Out* art exhibition held in Japan at the Bank Gallery, a prominent exhibition space in the Harajuku Shinjuku area on Cat Street. As part of the project, a nineteen-minute musical piece was commissioned for the large-scale installation of *Crepuscular Rays of the Moon*. Tate Egon Chavez created the musical composition. Although the initial vision was for a light composition inspired by the anime *Sailor Moon*, the resulting piece, *Nocturne*, took on a more introspective and moody character, featuring a range of emotive stages that culminated in a lifting interlude.

During the early stages of the COVID-19 pandemic, *Nocturne* was submitted to several film festivals for further feedback. It garnered attention from a few minor film festivals as a stand-alone abstract animated artwork, pleasantly surprising us. The piece was further revised, and we attempted to incorporate a dancer into the work. At the 2019 screening of *Quantum LOGOS (vision serpent)* at the Ars Electronica Center Deep Space 8k theater, Mark met dancer and choreographer Victoria Primus. She had shown interest in collaborating with us. We decided to utilize the isolation of lockdown and our introspection during that time to attempt a remote exhibition in that space. The relevant parties were contacted, and we were delighted to be invited to exhibit the piece at the Ars Electronica Festival 2021 in the Deep Space 8k theater. Nocturne was presented as an immersive, reactive movie projected onto the floor and wall of the AE Deep Space 8k theater. The artwork acknowledges the ongoing global pandemic while recognizing our challenges. It offers the audience a contemporary artistic interpretation of our struggles and a fresh perspective on otherwise adverse events. Obstacles arose during the production of this piece in October 2020, aside from the COVID-19 isolation imposed in Singapore and worldwide. Ina was diagnosed with stage 4 Non-Hodgkin's lymphoma large B-Cell aggressive cancer. Fortunately, her type of cancer has been curable for the past twenty years, even at stage 4. Ina's cancer is now in complete remission.

The challenges encountered required a reevaluation of our artistic design choices. Our initial approach was highly focused on design. However, we realized that tying our imagery to specific scientific meaning led to an exciting yet aesthetically forced outcome, lacking the visceral and emotive impact we felt aligned more with our artistic expression and interpretation. We have a background in fine arts, with Mark formally studying fine art drawing and animation, and Ina studying fiber arts, sculpture, and painting. For our large-scale dancer-performed installation, we incorporated immersive, interactive, and audio-reactive visuals featuring emotive-abstract animation that mapped emotions and design in real time with the flow of the dance performance. The installation is a tribute to ancient rites of spring, utilizing imagery, music, and motion to evoke the past and signal hope for the future amidst uncertain times.[43] *Nocturne* premiered at Ars Electronica Center, Deep Space 8k for the occasion of the Ars Electronica Festival—A New Digital Age, on September 11 and 12, in Linz, Austria.[44]

Quantum Theory and Southeast Asian Textile Archetypes

***MOIRAI, THREAD OF LIFE*_**Our artistic works utilize observations, experiences, and narratives to elucidate our understanding of the world. We pondered the possibility of the quantum mechanism governing all aspects of existence and wondered whether we could observe its manifestation in our everyday experiences. We considered whether quantum mechanics could be the key to comprehending the universe's workings and interconnecting all its facets. Ultimately, we contemplated the potential for humans to uncover a final truth that could answer all of our inquiries.

This project focuses on Southeast Asian fabrics as the primary source of inspiration and interest. We use fabric, the process of fabric-making, and its material as metaphors to explore quantum theories and their relationship to the everyday person. The quantum superposition principle describes that quantum states can be represented as a sum of distinct states, similar to the warp and weft at the start of a weaving. Quantum systems, such as atoms, photons, or spins, can exist simultaneously in two distinct states. The wave-particle duality principle states that electrons behave as both particles and waves.[45] The fabric's structure illustrates how the whole cloth, both front and back, reveals its source, phase, and pattern between every warp and weft.

Textile patterns are produced by weaving and interlacing every warp and weft. The complementary elements of the warp and weft create the whole and reconcile the vertical and horizontal forces, resulting in the opposing structures on the horizontal and vertical axes. These structures can be likened to the binary's 0 and 1 states. As such, we view the horizontal and vertical sequence as binary. We draw a parallel between the textile creation process and the paradox of quantum superposition. Just as every choice between warp or weft could relate to a quantum decision, a cat in Schrodinger's experiment remains in superposition until observed. Once observed, the superposition collapses into one of the possible definite states. Therefore, any warp or weft option could be considered a quantum decision, leading to a wide range of potential outcomes. In our imagination, the textile creation process involves a multidimensional space.

Each fabric contains essential information about the world, whether parallels, spacetime, or dimensions, in this tesseract of fabrics. The culmination of all of them in this space is a visualization of what is presumably a whole truth, where the seen and unseen worlds are together and appear to make sense. For this, we borrowed from the many worlds interpretation (MWI), where all possible outcomes of quantum measurements could be physically realized in some parallel world or universe.

Moirai, Thread of Life is a metaphor for origins, the life span, the link between past and present, and human destiny. The film is based on the many worlds interpretation, which says we are some small part of the whole (fabric of the universe) and connected, entangled with our other invisible destinies through threads. Moirai is the goddess of faith and destiny in Greek mythology. She weaves these destinies

FIG. 5_MOIRAI, THREAD OF LIFE, *Experimental animation 2022*

VIDEO_ MOIRAI, THREAD OF LIFE *trailer*

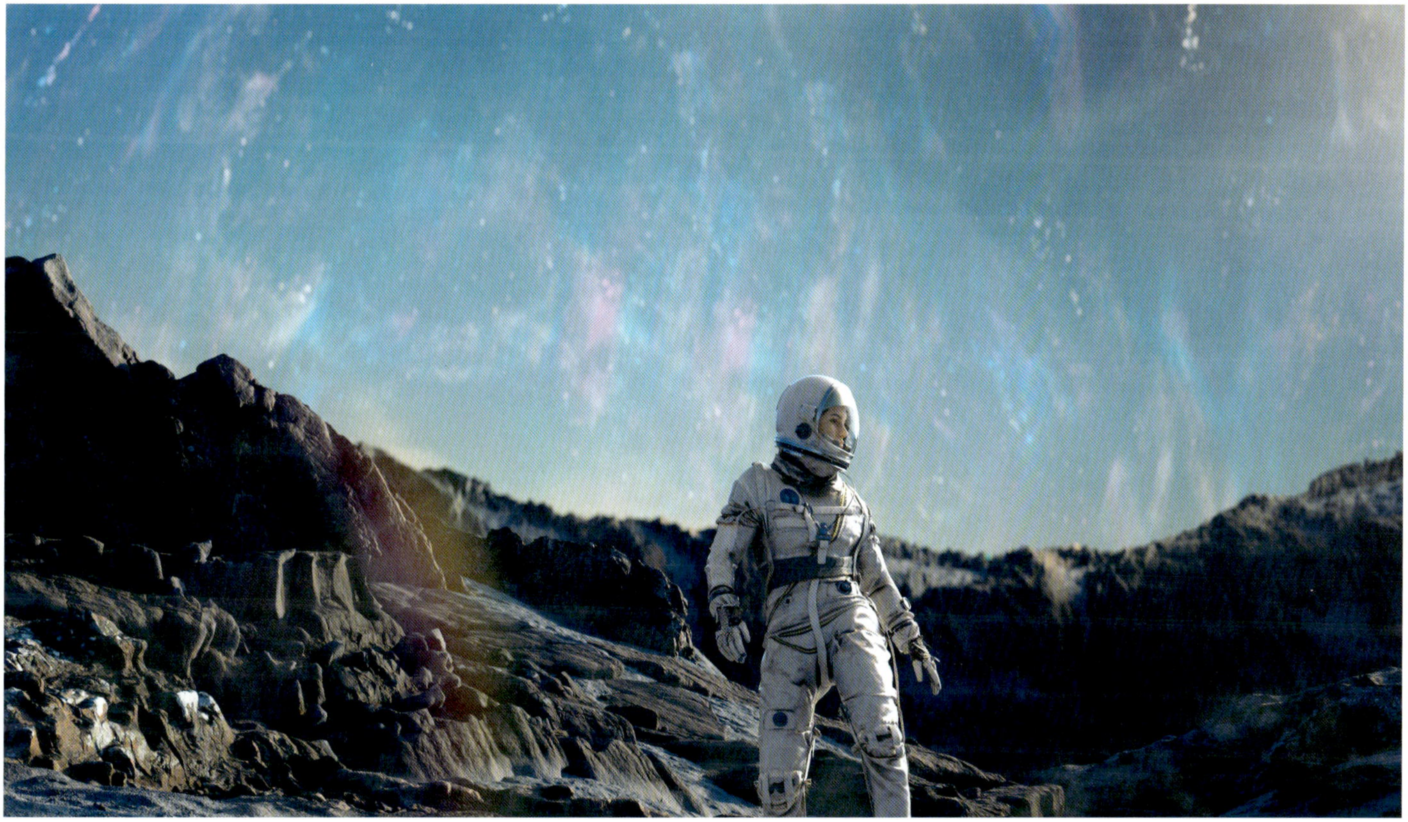

and allows us to exist at several places potentially and in several states simultaneously—in quantum superposition—where all possible outcomes exist. She connects the threads into one observable state and allows threads to be entangled in all possible existences.

In January 2021, we embarked upon our research with Professor Rainer Helmut Dumke of NTU's College of Science, School of Physical & Mathematical Sciences, Division of Physics & Applied Physics, who is a member of the Centre for Quantum Technologies (CQT), Singapore. Thanks to his assistance, we obtained ongoing feedback on the accuracy of our original ideas and graphic storyboards. His feedback was invaluable to the project. During this period, we also studied textiles and the various craft traditions found in Southeast Asia.[46]

The project's visual development coincided with Ina's traditional textile background, from which she had ten years of weaving experience. Ikat textiles from the Indonesian Archipelago inspired the concept of incorporating data photos into AI-generated artworks. AI was used to visualize some of the film's essential concepts, although it was not ultimately used in the project's final stage. However, it added a layer of complexity to the film's textural conceptualization. Being part of this project was uplifting and motivating during Ina's chemotherapy treatments.

Exploring AI-Generated Art

CAN LUCHADORES TRAVEL THROUGH WORMHOLES?_Currently, we are investigating the new and novel field of *generative adversarial network* (GAN)-based AI visual art. Many of these art generation tools leverage word-prompted techniques to generate visual art. One field of image creation has recently found widespread interest among social media platforms and WEB3.0 entrepreneurs. However, these techniques have been around in various forms for many years. The latest iterations leverage the vast libraries of metatagged data that have only recently become available. As a result, large new reference libraries of parseable data can be organized into image-centric database models. These databases also comprise video from live-action sources, traditionally animated motion, and computer-generated animation. Unique to this new reference-based platform is how visuals are called into the image composition space by word-prompted language structures. This allows artists to blend visual art with language using new prompt-based image creation techniques. Our first case study explored quantum ideas through Mexican artistic, ethnic, and cultural tropes. We had recently completed *Quantum LOGOS (vision serpent)* and *Nocturne* and wanted to take a respite from complex technical animation techniques to interpret scientific data and personal feelings. *VQGAN+CLIP* was the first technique we used to generate imagery using word prompts.

VQGAN and *CLIP* are distinct machine learning algorithms that can collaborate to create images based on text prompts. *VQGAN* is a generative adversarial neural network that can generate images that resemble others, but it doesn't generate images based on prompts. On the other hand, *CLIP* is another neural network that can evaluate how accurately a text prompt corresponds to an image. When used together, *VQGAN* and *CLIP* can create images based on text prompts by leveraging the strengths of both algorithms.[47]

Creating images through text input was first introduced in academic publications by OpenAI before the DALL-E software for artists was released. The papers

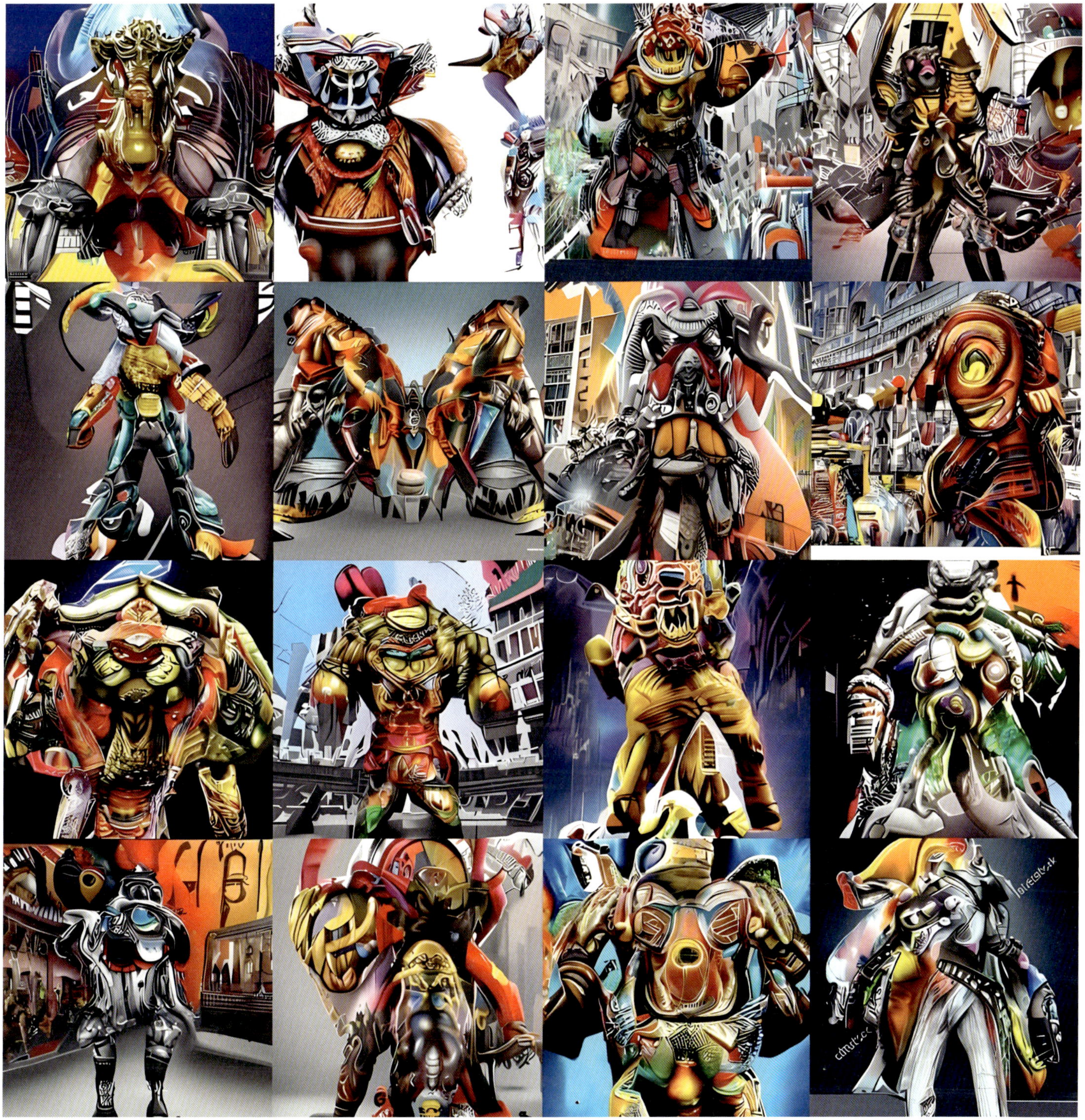

FIG. 6_CAN-LUCHADORES-TRAVEL-THROUGH-WORMHOLES

AI-generated art by Mark Chavez, 2022

presented two algorithms, later adopted by artists such as Ryan Murdock and Katherine Crowson in early 2021. They created versions of the *VQGAN+CLIP* algorithms and made them available as installable software on GitHub. The packages included links to papers explaining the machine learning techniques and clickable links to Google Colab notebooks for running the software in demo mode.

The open-source code generated from this process is available on GitHub for the generative art community to use. Several Python notebooks are accessible on Google Colab, a subscription-based Linux software solution that provides access to various computers. These notebooks utilize different image database models and animation techniques to produce diverse visual styles and approaches. Users can modify and extend the software to suit their artistic goals.

The use of AI toolsets in image creation allows for a unique approach to designing imagery that blends meaning with visuals generated from descriptive language. This technique can result in imagery with unexpected messages embedded in its design. Additionally, combining video-motion tracking with styled expressions overlaid onto the image plane opens up new possibilities for generating looks and means of expression. Overall, this process provides a refreshing and innovative approach to image creation. Mark used quotations from famous scientists, filmmakers, authors, and media theorists as a case study, mixing prompt calls to generate Luchadores or Mexican Wrestlers with folk art masks and an *alebrijes*[48] aesthetic to generate or word prompt the imagery_FIG. 6.

Summary

In the last fifteen years, our work has become more aligned with a shared vision of expressing dynamic content that addresses issues and ideas important to us. We looked at early Maya cultures and their nature, and found many design traits shared with graphically represented math in contemporary quantum physics. When reading about the ideas represented by these cultures and how many of them were searching for a fundamental understanding of existence, we decided that this was a path we could take for discovery. Our latest film *Moirai, Thread of Life* interprets quantum mechanics through the lens of Southeast Asian fabric design. In the piece *Nocturne*, we confront COVID-19 isolation, disease, and the pain of being separated from loved ones in a reactive, interactive, immersive dance movie projection with a position-based laser tracking system, performed by Victoria Primus in a reactive theatrical space in the Ars Electronica Center Deep Space 8k.

We recently began exploring Generative AI. We can learn from our own personal experiences rather than learning and interpreting complex code and forcing it to convey our ideas. The use of generative AI in the creation of artworks makes it possible to combine word phrases and metaphors in a way that is both more powerful and more personal than what is possible using written language or artwork alone. These new methods serve as an inspiration for a new medium of expression. What might generative AI and machine learning teach us about art through adaptive inquiry, and at the same time, how can they provoke us to see our relationship to the world and each other in entirely new ways?

1 M. Chavez and A. Lioret, »Artificial Beings That Look Back at Us,« *Research Gate*, January 2006, https://www.researchgate.net/publication/249012007_Artificial_Beings_That_Look_Back_At_Us.

2 *Leonardo Electronic Almanac* 19, no. 3, July 15, 2013, https://journals.gold.ac.uk/index.php/lea/article/download/87/199.

3 M. Chavez, »[Vengeance + Vengeance] the movie,« *IMDB*, 2012, https://www.imdb.com/video/vi3018827545/; see also Animation-Xpress Team, »[Vengeance+Vengeance] wins the Best Animation Film Award at the Marbella International Film Festival 2012—In Repartee with Mark Chavez,« October 30, 2012, https://www.animationxpress.com/interviews/vengancevengance-wins-the-best-animation-film-award-at-the-marbella-international-film-festival-2012-in-repartee-with-mark-chavez/.

4 I. Conradi, »Elysian Fields,« accessed June 19, 2023, *Vimeo*, https://vimeo.com/83420196.

5 J. Houran and R. Lange, *Hauntings and Poltergeists: Multidisciplinary Perspectives YouTub* (Jefferson, NC: McFarland & Company, 2001), 195–213.

6 L. da Vinci, *Leonardo on Art and the Artist* (Mineola, NY: Dover Publications, 2002), 349, stains on old walls.

7 A. Miller, *Einstein, Picasso: Space, Time, and the Beauty That Causes Havoc* (New York: Basic Books, 2001).

8 I. Conradi, »Emotion Study, Experimental Animation,« accessed June 19, 2023, *Vimeo*, https://vimeo.com/46398056.

9 J. Posner, J. Russell, and B. Peterson , »The Circumplex Model of Affect: An Integrative Approach to Affective Neuroscience, Cognitive Development, and Psychopathology,« *Development and Psychopathology* 17, no. 3 (2005): 715–34, https://doi.org/10.1017/s0954579405050340.

10 M. Chavez and I. Conradi »Emote Portfolio—01_Emerald,« 2018, *Vimeo*, accessed July 1, 2022.

11 M. Chavez, »Quantonium,« January 28, 2017, *Vimeo*, https://vimeo.com/201446339.

12 I. Conradi, »on_off 100101010 Symposium Trailer,« April 17, 2019, *Vimeo*, https://vimeo.com/331113880.

13 »Logos,« in *Encyclopædia Britannica*, vol. 16 (Cambridge: Cambridge University Press, 1911), 919–21.

14 J. Maffie, *Aztec Philosophy: Understanding a World in Motion* (Boulder: University Press of Colorado, 2014).

15 A. McLeod, *Philosophy of the Ancient Maya: Lords of Time* (Lanham: Lexington Books, 2019).

16 D. Bolles, *A Translation of the Edited Text of the Ritual of the Bacabs* (Lancaster: Labyrinthos, 2003).

17 S. Olsen, L. Marek-Crnjac, J. He, and M. El Naschie, »A Grand Unification of the Sciences, Arts & Consciousness: Rediscovering the Pythagorean Plato's Golden Mean Number System,« *Journal of Progressive Research In Mathematics* 16, no. 2 (2020): 2888–931, accessed February 13, 2021, http://www.scitecresearch.com/journals/index.php/jprm/article/view/1850.

18 Cymatics is a subset of modal vibrational phenomena.

19 u/d8_thc, »Fractal Holographic Unified Field Theory,« 2014, accessed June 29, 2022, https://www.reddit.com/r/holofractal/comments/2zong8/slice_of_the_geometry_of_microtubules_the_stuff/.

20 Scrying is telling the future using a crystal ball or other reflective object or surface.

21 D. Anderson, »Origins Of Aztec And Inca Obsidian Mirrors Revealed Through Scientific Analyses,« *Forbes*, July 26, 2019, https://www.forbes.com/sites/davidanderson/2019/07/26/origins-of-aztec-and-inca-obsidian-mirrors-revealed-through-scientific-analyses.

22 S. Milbrath, »The Maya Lord of the Smoking Mirror,« *Tezcatlipoca: Trickster and Supreme Aztec Deity*, ed. S. Milbrath (Boulder: University Press of Colorado) 163–96.

23 E. Baguedano (ed.), *Tezcatlipoca: Trickster and Supreme Deity* (Boulder: University Press of Colorado, 2015).

24 J. Maffie, »Aztec Philosophy,« *Internet Encyclopedia of Philosophy*, accessed December 6, 2020, https://iep.utm.edu/aztec/.

25 G. Olivier, *Mockeries and Metamorphoses of an Aztec God: Tezcatlipoca, »Lord of the Smoking Mirror«* (Boulder: University Press of Colorado, 2008).

26 D. Laycock, J. Dee, and E. Kelly, *The Complete Enochian Dictionary: A Dictionary of the Angelic Language as Revealed to Dr. John Dee and Edward Kelley* (Boston: Weiser Books, 2001).

27 E. Asprem, *Arguing with Angels* (Albany: SUNY Press, 2012).

28 J. DeSalvo, *The Lost Art of Enochian Magic: Angels, Invocations, and the Secrets Revealed to Dr. John Dee* (Rochester: Destiny Books, 2010).

29 R. Barnett, »Mesoamerican Religion and Multiverses: Part Two,« *MexConnect*, May 1, 2008, https://www.mexconnect.com/articles/3383-mesoamerican-religion-and-multiverses-part-two/.

30 Barnett, »Mesoamerican Religion.«

31 R. Wasson, »The Role of ›Flowers‹ in Nahuatl Culture: A Suggested Interpretation,« *Botanical Museum Leaflets, Harvard University* 23, no. 8 (1973), 305–24.

32 M. Coe and S. Houston, *The Maya*, 9th ed. (New York: Thames & Hudson, 2015).

33 M. Coe and M. Van Stone, *Reading the Maya Glyphs* (New York: Thames & Hudson, 2016).

34 D. Freidel, J. Parker, and L. Schele, *Maya Cosmos: Three Thousand Years on the Shaman's Path* (New York: William Morrow, 1993).

35 J. Fraser, *Time: The Familiar Stranger* (Amherst: University of Massachusetts, 1987).

36 D. Witmer, »Physicalism and Metaphysical Naturalism,« *Oxford Bibliographies*, March 30, 2015, https://www.oxfordbibliographies.com/view/document/obo-9780195396577/obo-9780195396577-0258.xml#obo-9780195396577-0258-bibItem-0008.

37 D. Witmer, »Physicalism and Metaphysical Naturalism.«

38 C. Ardelean, L. Becerra-Valdivia, and E. Willerslev, »Evidence of Human Occupation in Mexico Around the Last Glacial Maximum,« *Nature* 584 (2020), 87–92, https://doi.org/https://doi.org/10.1038/s41586-020-2509-0.

39 M. Levine, *Pantheism: A Non-Theistic Concept of Deity* (London: Routledge, 2003).

40 M. Byrne, »New Measurements Show that the Unrealest Part of Quantum Physics Is Very Real,« *Vice*, February 4, 2015, https://www.vice.com/en/article/3dk4nv/new-measurements-show-that-the-unrealest-part-of-quantum-physics-is-very-real.

41 New Mind, »Wigner's Friend Paradox: Is Observation Inherently Flawed?« March 29, 2019, https://youtu.be/5AodzEpvzZw.

42 M. Chavez, »Quantum LOGOS (vision serpent),« 2019, *Vimeo*, https://vimeo.com/355776584.

43 M. Chavez and I. Conradi, »Ars Electronica Festival | A New Digital Deal: NOCTURNE,« *Media Art Nexus*, September 19, 2021, https://www.mediaartnexus.com/?p=3591.

44 Ars Elecronica, »Nocturne,« 2021, *YouTube*, September 12, 2021, https://youtu.be/lAXpEwKN9G4.

45 T. Young, »Bakerian Lecture: Experiments and Calculations Relative to Physical Optics,« *Philosophical Transactions of the Royal Society* 94 (1804), 1–16.

46 M. Gittinger, *Splendid Symbols: Textiles and Tradition in Indonesia* (Singapore: Oxford University Press, 1985).

47 A. Russell, »How to Use VQGAN+CLIP to Generate Images from a Text Prompt—A Complete, Non-Technical Tutorial,« *Medium*, August 15, 2021, https://medium.com/nightcafe-creator/vqgan-clip-tutorial-a411402cf3ad.

48 *Alebrijes* (Spanish pronunciation: [aleˈβrixes]) are brightly colored Mexican folk-art sculptures of fantastical (fantasy/mythical) creatures.

https://inaconradi.com
https://giantmonster.co
https://www.mediaartnexus.com
https://moirai.mediaartnexus.com

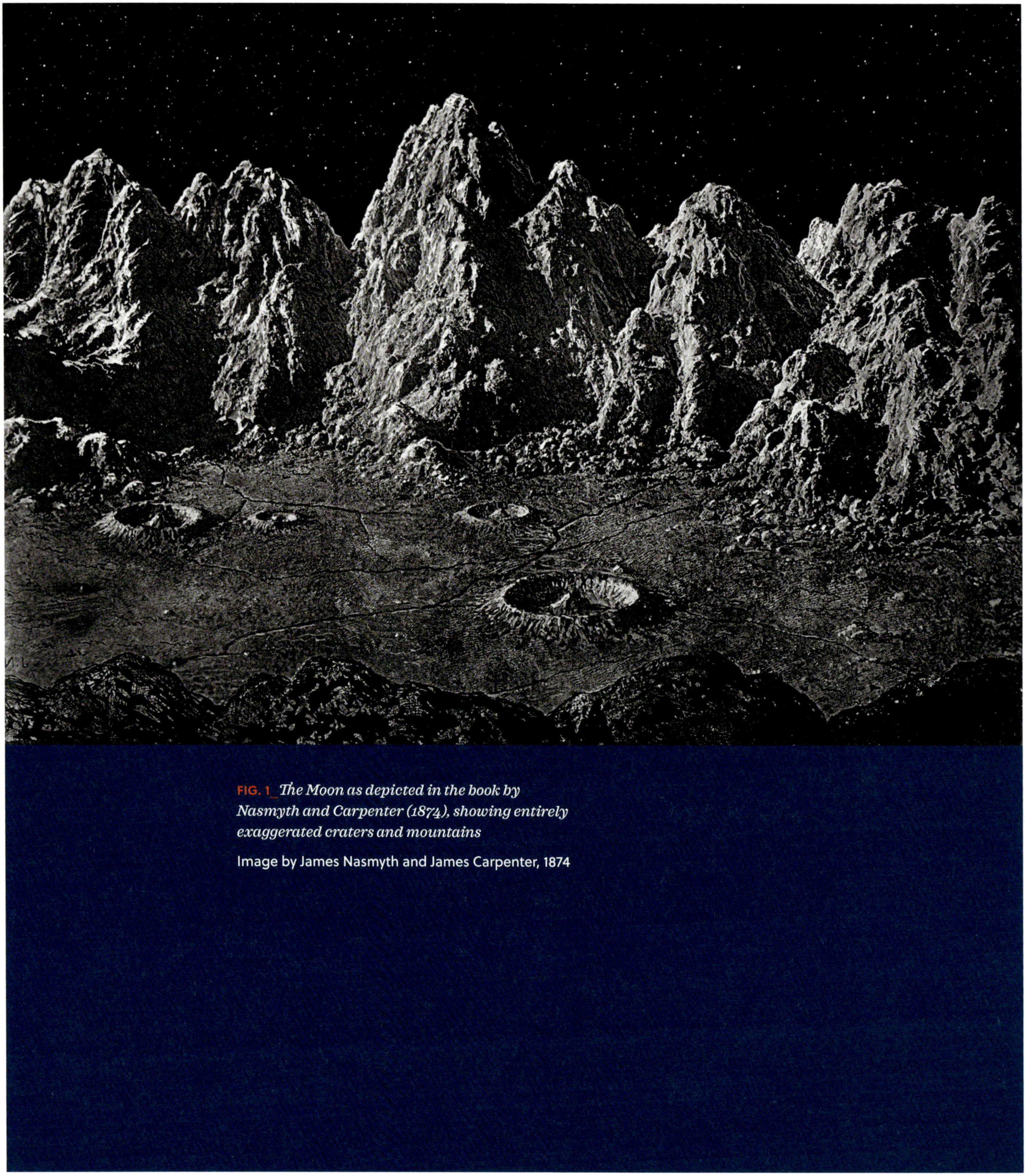

FIG. 1_ *The Moon as depicted in the book by Nasmyth and Carpenter (1874), showing entirely exaggerated craters and mountains*

Image by James Nasmyth and James Carpenter, 1874

Impact Cratering: The Impact of Visualization on Science and Outreach

Christian Köberl

The impacts of asteroids and comets are the most energetic and spectacular geological process currently known to us. The study of impact craters on Earth and the Moon, and astronomical investigations into the orbits of asteroids and other solar system bodies, have allowed the cratering rate for Earth to be determined (how many craters of which size form over a certain period of time).

This chapter describes impact craters on the Earth and elsewhere in the solar system, and the nature and origin of the impacting bodies (asteroids and comets), and gives a short summary of how impact craters can be recognized (mainly on the basis of shock metamorphic effects in rocks and minerals, and/or by searching for the remnants of the extraterrestrial body that are mixed in with the terrestrial target rocks). The physical effects of large-scale impact events are severe, ranging from burning due to the expanding fireball, seismic effects, possible tsunamis, and the ejection and deposition of large amounts of rock and dust from the impact site. In very large events, these effects are global.

One example of an impact event that had global implications is the formation of the Chicxulub impact structure, roughly two hundred kilometers in diameter, in Mexico, sixty-six million years ago, at the end of the Cretaceous. This event led to a severe mass extinction, in which more than half of all of the then living species (fauna and flora) became extinct. Despite the well-documented link between the impact and mass extinction in this case, there have — so far — been no clear links between impact events and any other mass extinctions. Nevertheless, the importance of impact events for the geological and biological evolution of the Earth is undeniable.

Visualization techniques have helped us to not only scientifically investigate impact craters and processes but also to understand their rather spectacular appearance and formation. These techniques have been used in scientific studies — often providing important information and allowing for interpretations and conclusions that would otherwise not have been easy to generate or draw — but also in the popularization of science, as well as in art and culture, all the way to entertainment, as in (for example) science-fiction films.

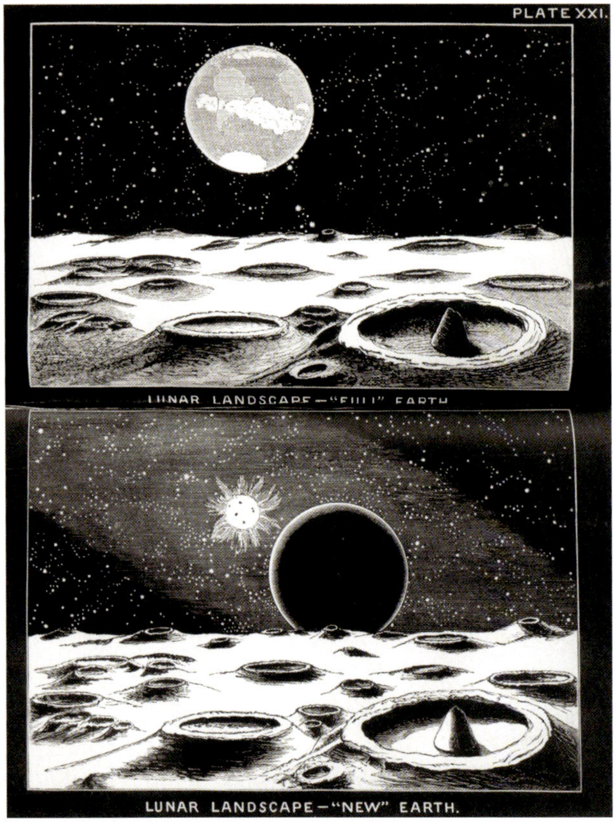

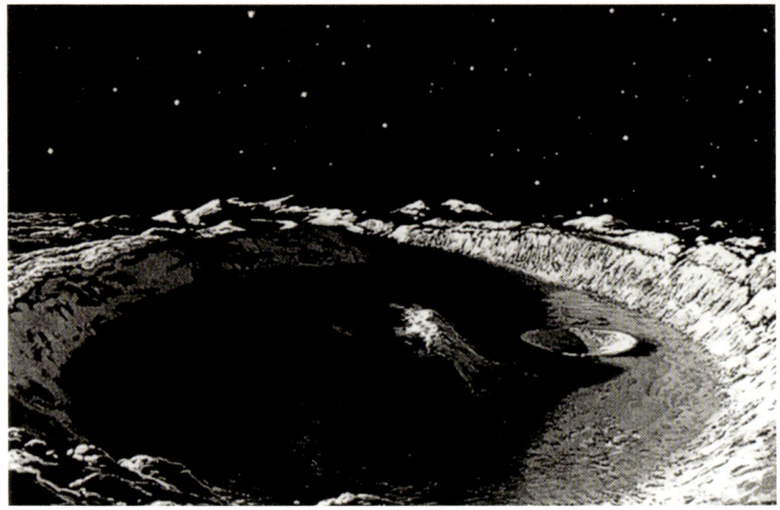

FIG. 2_*Another completely unrealistic lunar surface.*

Image by Richard Procter, 1873

FIG. 3_*In contrast to most other space artists in the first two-thirds of the twentieth century, Lucien Rudaux painted realistic lunar landscapes, the dimensions of which he deduced from geometric considerations.*[4]

Image by Lucien Rudaux, 1937

Introduction: Impact Craters on the Earth, the Moon, and in the Solar System

Impacts of extraterrestrial bodies (asteroids, cometary nuclei) are among the most spectacular, high-energy geological processes on Earth. Of particular interest are the effects and dangers of typical impact events for the geological and biological evolution of the Earth. For example, a huge impact, sixty-six million years ago, marking the transition from the Cretaceous to the Tertiary period, had a catastrophic influence on the biosphere. Impacts of extraterrestrial bodies on Earth carry a risk that should not be underestimated.

Even in the early twentieth century, the study of meteorite impact craters was seen as a topic for astronomers, not geologists.[1] Even though, since the invention of the telescope, craters were known to be present on the surfaces of many bodies in the solar system (mainly the Moon), following detailed investigations by Galileo Galilei and his successors from about 1610 onward, there was a multitude of hypotheses trying to explain their formation. Galilei, in his famous book *Starry Messenger*, published in 1610, included drawings of craters on the Moon that he made on the basis of his first telescopic observations—maybe one of the earliest »visualizations« of craters. However, the predominant hypothesis, which prevailed until the beginning of the twentieth century, was that basically all lunar craters were formed by volcanism. Until the middle of the last century, geologists were of the opinion that impact craters do not exist on Earth, as the formation of such a feature had never been observed by humans.

One of the earliest researchers to speculate about the origin of lunar craters was Robert Hooke, who came up with two alternative options in 1665: first, he dropped solid objects into a mixture of clay and water and found that these experiments resulted in crater-like features. However, he rejected the possibility that lunar craters could have been formed by such »impact« processes because it was not clear from »whence those bodies should come,« as interplanetary space was, at that time, considered to be empty. After all, Hooke conducted his experiments 135 years before the first asteroid, Ceres, was discovered by Piazzi in Palermo. Thus, he preferred a second hypothesis, on the basis of which he carried out experiments with »boiled alabaster,« concluding that lunar craters had been formed by some kind of gas — rejecting a perfectly correct explanation because the »boundary conditions« were missing.

The eighteenth century was dominated by the volcanic theory, as expressed by, e.g., Herschel, Schröter, and Beer and Mädler. In 1787, astronomer William Herschel (1738–1822), who discovered Uranus, claimed to have observed a volcanic eruption on the Moon. On the other side, the slightly eccentric German astronomer Franz von Paula Gruithuisen (1774–1852) was a strong advocate of the impact hypothesis, but his position was not taken too seriously as, a few years earlier, he had announced that, through his telescope, he had seen populated cities on the Moon, with cows grazing on lunar meadows and a star-shaped temple.

Not much progress was made in the nineteenth century. In the early 1870s, two important books on the Moon were published, both of which were titled *The Moon*. The more influential of the two was Nasmyth and Carpenter's 1874 volume,[2] in which they recreated lunar features in plaster casts on the basis of visual observations and photographed these models under low-angle illumination, resulting in spectacular and somewhat exaggerated lunar landscapes_FIG. 1.

Nasmyth and Carpenter were firm believers in the volcanic theory of the formation of lunar craters and cleverly explained even the formation of central peaks. In contrast, Richard Proctor, the author of the other book published in 1873,[3] rejected any resemblance between lunar features and terrestrial analogs, and seriously advocated the impact theory for the formation of lunar craters_FIG. 2.

These illustrations/visualizations got it quite wrong by showing rugged peaks with exaggerated relief. This is all quite surprising as it is not too difficult to derive the actual height/slope/depth relationships between mountains and craters from the geometry of shadows — any astronomer would have been able to point out that the craters and mountains on the Moon are smooth and more like rolling hills than ragged mountain peaks. An interesting exception was Lucien Rudaux, who was already painting realistic lunar landscapes in the 1930s (e.g._FIG. 3),[5] while many other space artists (e.g., Chesley Bonestell [1888–1986]) continued to depict dramatic moonscapes. By the mid-1960s, spacecraft were providing close-up views of the lunar surface, revealing a landscape that was significantly different from the ragged mountains painted by most space artists and depicted in most movies. Those illustrations are examples of visualizations that put effect before scientific accuracy — knowingly or not. As a side note, it is even more surprising to see exaggerated mountains and craters persisting in some films until the present day.[6]

At the end of the nineteenth century, in 1892, Groove Karl Gilbert, chief geologist of the US Geological Survey, concluded—partly based on experiments conducted in his hotel room during a lecture tour—that the formation of lunar craters can be best explained by impact theory. In contrast, he rejected the hypothesis that Meteor Crater in Arizona had been formed by an impact. This is odd because fragments of iron meteorites were actually found around that crater.

If one wants to intuitively understand the importance of impacts for the Earth, the best place to start looking is the Moon. The lunar surface is totally covered, to the point of saturation, by impact craters and basins of all sizes. The reason why they are so well preserved on the Moon but not on Earth is because the Moon is a geologically dead (inactive) object that lacks any atmosphere or hydrosphere and has thus recorded impact activity for millions and billions of years. Due to its larger gravitational cross section, the Earth is subject to an even larger impact flux per unit area than the Moon, but the results of that bombardment are barely preserved. The reason why there are so few impact scars visible on the Earth's surface is related to the active geological processes that occur on our planet: erosion, sedimentation, subduction, volcanism, plate tectonics, etc. These processes cause impact craters on Earth's surface to be eroded, covered by younger sediments, or otherwise destroyed.

Detailed geological studies have confirmed about two hundred impact craters on Earth so far. Planetary scientists recently came to the conclusion that impact events have been a lot more significant for the origin and evolution of our solar system than previously assumed. Impacts have been important since the formation of the solar system, when collisions between small bodies, the planetesimals, led to the accretion of the larger planets that we know today. On the surfaces of almost all bodies in the solar system that have a solid surface (planets, moons, and minor bodies), impact craters are the dominant landforms.

Astronomical studies of minor planetary orbits and the dating of known impact craters allow us to derive the frequency with which impacts of certain sizes have occured on Earth. Such research has indicated that, for example, bodies with a diameter of one to two kilometers, which would cause craters in the order of twenty to forty kilometers in diameter (such as the Ries Crater in southern Germany), have hit the Earth on average about once every million years. But smaller impacts, resulting in craters of about one kilometer in diameter (e.g., Meteor Crater in Arizona), occur much more often, namely every few thousand years. Really large impacts, which cause craters with diameters of more than one hundred kilometers, are fortunately not as common and happen only every fifty to one hundred million years. But such large impacts may result in global catastrophes and can even lead to mass extinctions of life on Earth.

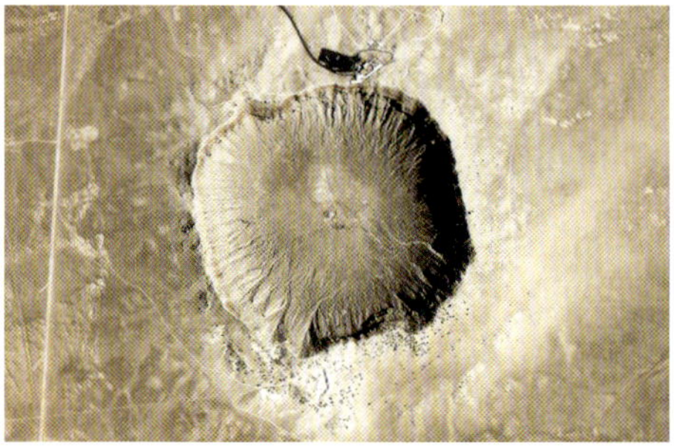

A

FIG. 4_*Examples of terrestrial craters in space images*
A_*Meteor Crater (Barringer Crater), Arizona, U.S.A., diameter approx. 1.2 kilometers*[8]
B_*Tenoumer crater, Algeria, diameter approx. 1.9 kilometers*[9]
C_*Gosses Bluff, an impact crater sandwiched between the MacDonnell Range to the north and the James Range to the south in Australia's Northern Territory—about 160 kilometers west of Alice Springs. This is one of the most studied Australian impact craters. The impactor, an asteroid or comet, was probably about one kilometer in diameter and crashed into the Earth about 142 million years ago. The isolated circular feature within the crater consists of a central ring of hills about 4.5 kilometers in diameter. The grayish feature surrounding the inner ring probably marks the original boundary of the outer rim, which was approx. 22 kilometers in diameter.*[10]
D_*Manicouagan impact structure in Canada; in this image the eroded central uplift is visible with a diameter of approx. 65 kilometers.*[11]

Images by USGS, 2009; NASA 2001–3

How to Recognize an Impact Crater?

From a morphological point of view, we need to distinguish between an impact crater, i.e., the feature that results from the impact, and an impact structure, which is what we observe today, i.e., long after the formation and modification of the crater. On Earth, we are aware of two distinctly different morphological forms: simple craters (small bowl-shaped craters) with diameters of up to ≤2 to 4 kilome-

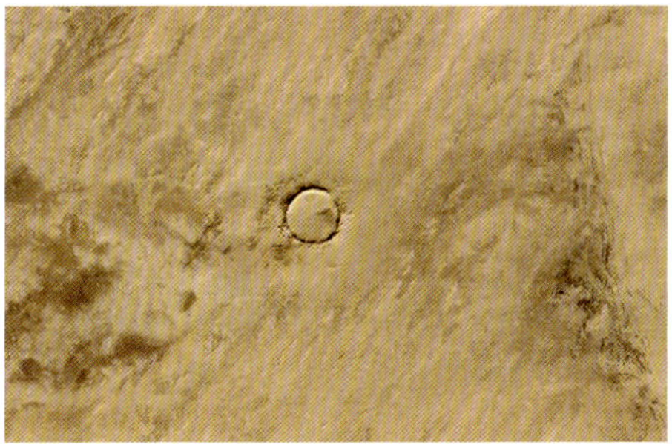

B

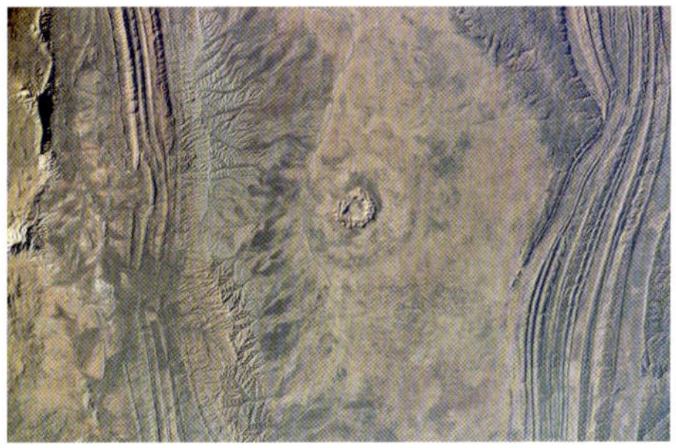

C

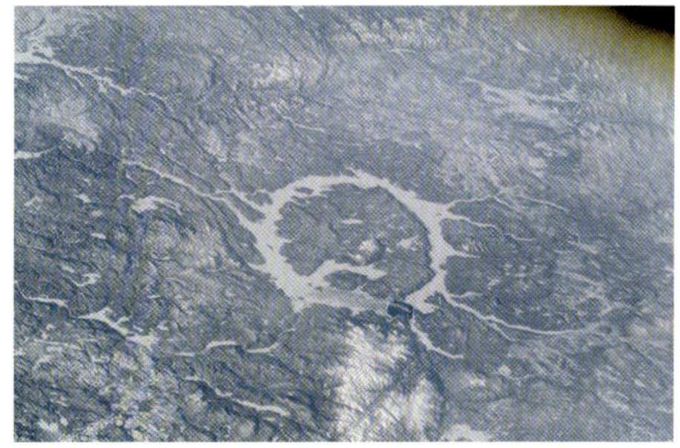

D

ters, and complex craters, which are larger and have diameters of ≥2 to 4 kilometers (the exact diameter at which a simple crater becomes a complex one depends on the composition of the target). Complex craters are characterized by a peak or peak ring that consists of rocks that have been uplifted from greater depths and would not normally be exposed on the surface. The stratigraphic uplift amounts to about 0.1 of the crater diameter.[7] Craters of both types have an outer rim and are filled by a mixture of fallback ejecta and material that slumped in from the walls and crater rim during the early phases of formation. Such crater infill may include brecciated and/or fractured rocks, and impact melt rocks. Fresh simple craters have an apparent depth (measured from the crater rim to the present-day crater floor) that is about one-third of the crater diameter, whereas the value for complex craters is closer to one-sixth. On Earth, basically all small craters are relatively young because erosional processes obliterate small (0.5–10 kilometers in diameter) craters after a few million years, causing a severe deficit of such small craters. _FIG. 4 shows examples of young, bowl-shaped (»simple«) craters on Earth. _FIG. 4 also shows several »complex« craters with a central uplift or a central ring of hills, which formed from the collapse, and later selective erosion of the central uplift. These images are examples of the simplest form of »visualization,« namely »normal« photographs—but from a perspective that is normally not accessible to us, i.e., from space. Such images have helped to identify geological features that may have been formed by impact processes.

However, it is necessary to point out that simple aerial or space photographs alone cannot provide evidence that confirms that a feature was formed by impact. But space photographs together with other information, such as topography, can assist in more detailed geological studies of confirmed impact structures. _FIG. 5 provides some examples of this for the Bosumtwi impact crater in Ghana. This is a relatively young, complex impact structure (ca. 1.1 million years old), about eleven kilometers in diameter, which is mostly filled by a lake_FIG. 5A. The lake sediments obscure the central uplift that, in more deeply eroded structures such as those shown in_FIGS. 4A and D, is exposed. Here the crater is barely eroded. The images in _FIGS. 5B and C show a topographical visualization with some geological information, such as the drainage pattern.

Planetary geologists are often confronted with the question of how to identify and confirm that a certain geological feature was formed by a meteorite impact. Here it is necessary to reiterate that it is not possible to confirm the impact origin of

FIG. 5_*Bosumtwi impact structure, Ghana*
A_ *Perspective view of Aster satellite image over SRTM digital elevation model (ten-fold vertical exaggeration), looking north*
B_ *Projection of the SRTM data close to Bosumtwi (north points toward the upper left), showing the crater rim, the elevation of the Obuoum Range, and the faint outer ring feature*
B *same as* A *but with the drainage pattern (creeks from the Survey of Ghana topography regional maps) superimposed; the drainage pattern emphasizes the circular drainage in the slightly depressed annular zone outside the crater rim.*[12]

Images by Christian Köberl and Wolf Uwe Reimold, 2005

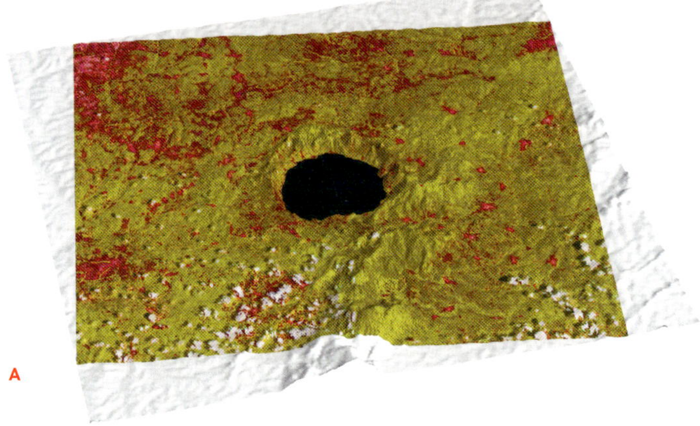
A

a structure from remote sensing (e.g., aerial or satellite images) and geophysical data (e.g., gravity, magnetics) alone. The geological community knows only two definitive criteria that make it possible to identify impact craters on Earth, and both require the study of actual rock samples: first, the presence of meteorite fragments, or chemical traces thereof, in breccias, glasses, or melt rocks; or, second, evidence of the shock wave interacting with terrestrial rocks and minerals (which leads to shock metamorphic products). The high pressure of a shock wave, which results from the high-velocity impact of an extraterrestrial body, causes irreversible changes in the crystallographic characteristics of minerals or even the shock-induced melting of the target rocks and minerals (see more detailed discussion below).

Thus, confirming the impact origin of a suspected geological structure is a long and tedious process that involves time-consuming, complicated analytical geochemical and mineralogical studies. Unfortunately, in the age of Google Earth, any circular feature is all too often designated an impact crater without any of the necessary geological research being conducted.

The Impact Process

The formation of a crater by hypervelocity impact is—not only in geological terms—a very rapid process that is customarily divided into three stages: 1) contact/compression stage, 2) excavation stage, and 3) post-impact crater modification stage. Schematically, the formation of an impact crater can be summarized as follows: first, a relatively small extraterrestrial body, traveling at a velocity of several tens of kilometers per second, hits the surface; this marks the beginning of the contact and compression stage. Almost immediately, a small amount of material is ejected from the impact site during a process called jetting, with velocities that can approach about one-half of the impact velocity. The jetted material is heavily contaminated with projectile material. When the projectile hits the surface, a shock wave is propagated hemispherically into the ground. Because the pressures in the shock waves are so high, the release of pressure (decompression) results in the almost instantaneous melting and vaporization of the projectile—and of large

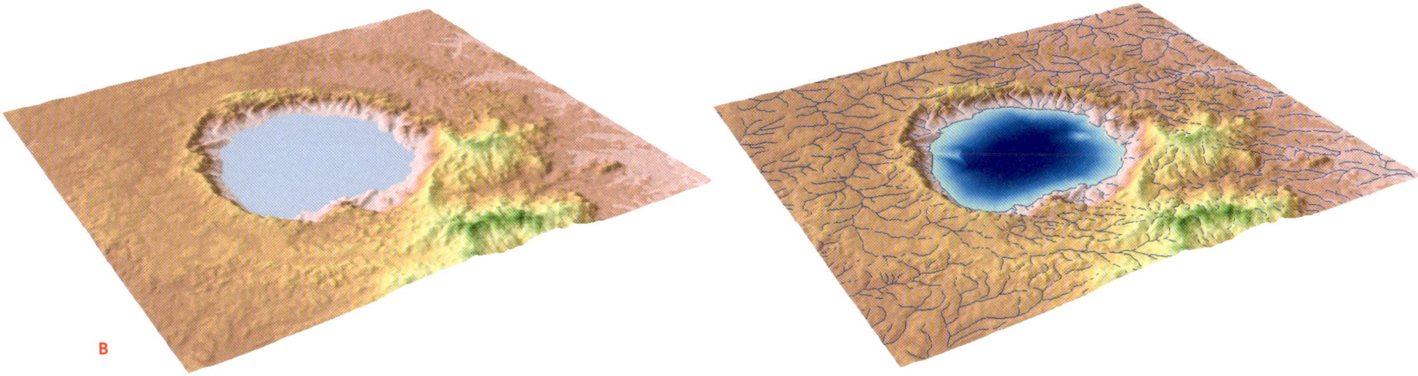

B

amounts of target rocks. The results of the interaction between the shock wave and matter can be observed in various forms of shocked minerals and rocks, all of which originate during the contact (or compression) stage, which only lasts up to a few seconds, even for large impacts. After the passage of the shock wave, the high pressure is released by a rarefaction wave (also called a release wave), which follows the shock front. The rarefaction wave is a pressure wave, not a shock wave, and travels at the speed of sound in the shocked material. The rarefaction wave leads to the creation of a mass flow that opens up the crater, marking the beginning of the excavation phase. Important changes in the rocks and minerals occur upon decompression, and the material follows a release adiabat in a pressure-versus-specific-volume diagram. Excess heat appears in the decompressed material, which may result in phase changes (e.g., melting or vaporization).

The actual crater is excavated during this stage. Complex interactions between the target and the shock wave(s), and then the release wave(s), lead to an excavation flow. In the upper layers of the target, material mainly moves upward and out, whereas in lower levels, material mainly moves down and outward, which results in a bowl-shaped depression, the transient cavity. This cavity grows in size as long as the shock and release waves are energetic enough to excavate material from the impact location. At this point, it is necessary to sound a note of caution: for a crater of about two hundred kilometers in diameter, the transient cavity can easily reach a depth of sixty kilometers. However, only about one-third of this is excavation; the rest is simply material that has been pushed down. Afterward, gravity and rock-mechanical effects lead to the collapse of the steep and unstable rims of the transient cavity, widening and filling the crater. Compared to the contact stage, the excavation stage takes longer, but still only up to a minute or two, even in large craters of more than two hundred kilometers in diameter.[13]

Obviously, nobody in recent history has observed a large-scale impact event —fortunately! Thus, the study of what happens during such events, resulting in descriptions like the one given above, is carried out using physics and chemistry in (often very complicated) computer models, and these calculations can be visualized in a variety of ways, from the schematic to the hyper-realistic. _FIG. 6 (for a gif image, see the online version) shows a simple computer model that uses a vertical

FIG. 6_ *Short video sequence showing a computer model of the formation of the Bosumtwi impact structure, Ghana*[14]

Image by Ludovic Ferrière, Christian Köberl, Boris Ivanov, and Wolf Uwe Reimold, 2008

VIDEO

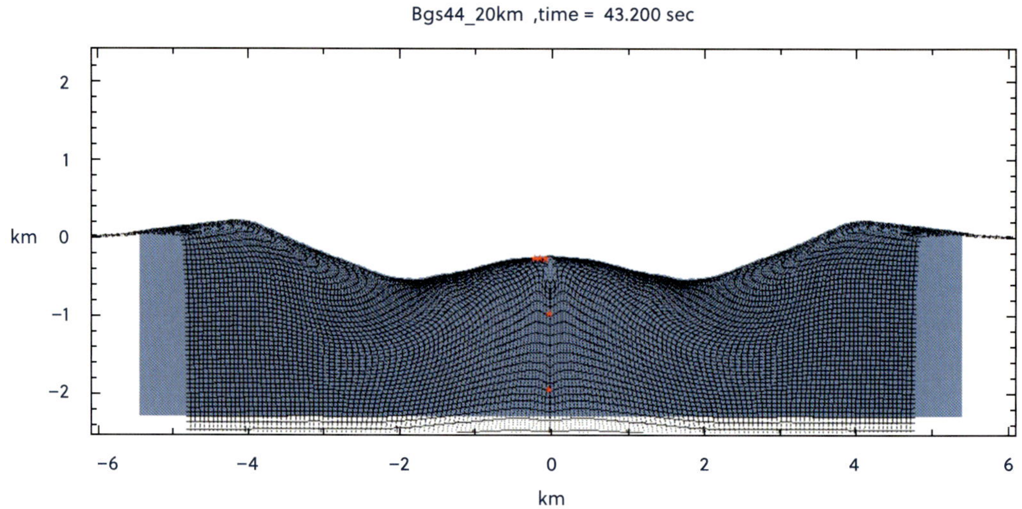

impact to schematically visualize the formation of a crater about the size of the Bosumtwi impact structure in Ghana.[15]

A meteorite impact releases enormous amounts of energy, but, of course, there are huge differences between the impacts of small and large bodies. For small bodies, the composition of the impacting object is also important. Small stony meteorites behave differently in the Earth's atmosphere than iron meteorites, mainly due to the much higher internal strength and cohesion of iron meteorites. Smaller objects are slowed down in the Earth's atmosphere and can, as a result of internal stress and strain, break apart. For example, Meteor Crater in Arizona was formed by the impact of an iron meteorite that was on the order of fifty meters in diameter, although the largest part of the meteorite melted or was vaporized upon impact. And only a minor part of the original meteorite remained in the form of meteorite fragments outside the crater rim. The energy released during this impact was equivalent to an explosion of ten megatons of the explosive TNT. In comparison, the energy released during the explosion of the atom bomb over Hiroshima was the equivalent of just twenty kilotons of TNT—five hundred times less! Stony meteorites of a similar size (but lower mass, because of the lower density) suffer a different fate. Calculations show that such objects can explode in the atmosphere as a result of inner strain during deceleration in the atmosphere.

One example of such an explosion is the famous Tunguska event: on June 30, 1908, a huge explosion over the woods of central Siberia destroyed an area of about two thousand square kilometers. The explosion was felt fifteen hundred kilometers away, and several hours later, the barometric pressure wave was registered as far away as in Europe. The most commonly accepted explanation is that a stony meteorite with a diameter of somewhere between twenty and fifty meters (depending on the entry velocity and angle into the Earth's atmosphere) exploded at an altitude of about ten kilometers above the ground with energy equivalent to five to ten megatons of TNT. Such small atmospheric explosions are actually fairly common.[16] Even though such explosions of Tunguska size happen in the Earth's atmosphere every hundred years or so, they can cause enormous local destruction. The Tunguska event devastated an area of about two hundred square kilometers; in comparison, the area of Vienna is just over four hundred square kilometers.

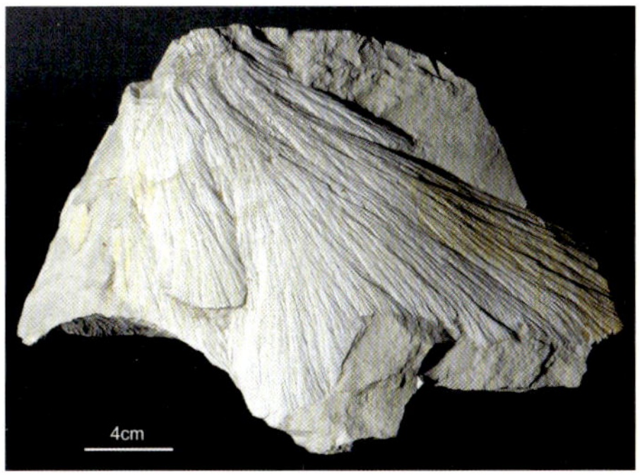

FIG. 7_*Shatter cone from the Steinheim impact structure in Germany, providing unambiguous evidence of an impact origin*

Image by Christian Köberl, 2014

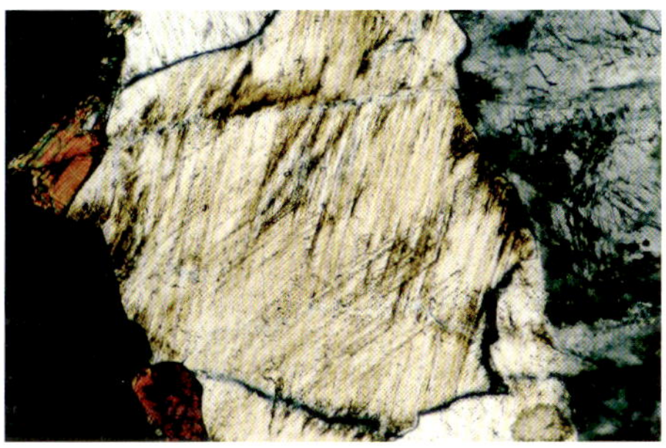

FIG. 8_*Shocked quartz crystal in the optical microscope under a crossed polarizer (providing a »visualization« that is different from plain light). Particularly evident are the planar deformation elements (»shock lamellae«), which are planar (straight), permeate the whole crystal, are parallel to each other, and occur in more than one orientation (more than one »set«). Such features are uniquely characteristic of impact and are used to confirm the impact origin of a geological structure. The image shows a quartz crystal from an impact breccia of the Woodleigh crater (Australia).*

Image by Christian Köberl, 2003

Recognition Criteria for an Impact Crater

Important witnesses of the characteristics of the impact process are the affected rocks. As mentioned above, crater structures are filled with melted, shocked, and brecciated rocks. Some of these are in situ, while others have been transported, in some cases, considerable distances from the source crater. The latter are called ejecta. Some of that material can fall back directly into the crater, and most of the ejecta end up close to the crater (<5 crater radii; these are called proximal ejecta), but a small fraction may travel much greater distances and are then called distal ejecta.[17]

Coming back to the topic of how to actually recognize an impact crater, we have to consider that, on the Moon and other planetary bodies that lack an appreciable atmosphere, impact craters can commonly be identified on the basis of morphological characteristics; however, on Earth, complications arise as a consequence of the obliteration, deformation, or burial of impact craters. This problem has made it necessary to develop diagnostic criteria for the identification and confirmation of impact structures on Earth.[18] The most important of these characteristics are a) crater morphology, b) geophysical anomalies, c) evidence of shock metamorphism, and d) the presence of meteorites or geochemical evidence of traces of the meteoritic projectile. Morphological and geophysical observations are important in providing supplementary (or initial) information. Geological structures with a circular outline that are located in places with no other obvious mechanism for producing near-circular features may be of impact origin and at least deserve further attention. Geophysical methods are also useful for identifying promising structures for further studies, especially in the case of subsurface features. In complex craters, the central uplift usually consists of dense basement rocks and contains severely shocked material. This uplift is often more resistant to erosion than the rest of the crater, and thus, in old, eroded structures, the central uplift may be the only remnant of the crater that can be identified. The geophysical characteristics of impact craters include, among others, gravity, magnetic properties, reflection and refraction seismics, and electrical resistivity.[19]

Of the criteria mentioned above, only the presence of diagnostic shock metamorphic effects, and, in some cases, the discovery of meteorites or traces thereof, are generally accepted as providing unambiguous evidence of impact origin. Shock deformation can be expressed in macroscopic form (shatter cones_FIG. 7) or in microscopic form. These same criteria apply to distal impact ejecta layers and make it possible to confirm that material found in such layers originated in an impact event at a possibly still unknown location.

The shock deformation leads to irreversible changes in the rocks and minerals of the rock into which the impact occurs. In particular, the crystal structure of the minerals is disturbed by high pressures. The high-pressure forms of some minerals (e.g., coesite from quartz, stishovite from coesite, or diamond from graphite) can also form. Crystallographic changes typical of shock metamorphism (also called shock wave metamorphism) are referred to as planar deformation structures

(PDF).[20] These are due to changes in the crystal lattice of minerals (such as quartz or feldspar) that can only be seen with a microscope. These are regular, parallel, and generally equidistant, extremely straight (planar) lamellae running through a quartz crystal (hence the often informally used term »shock lamellae«). These lamellae are very thin (ca. one micrometer), parallel, and regular, and appear in the form of one or more intersecting systems with well-defined crystallographic orientations. A micrograph of a shocked quartz crystal is shown in_FIG. 8.

The shape and type of planar deformation structures are characteristic of, and naturally only occur in, impact rocks in this way. The presence of such shock structures is therefore clear evidence of impact crater formation.

In addition to the petrographic criteria for crater identification, it is also possible to use geochemical characteristics, namely the enrichment of typical extraterrestrial elements in impact breccias, especially in impact melt rock.[21] It is extremely difficult to detect such small amounts of meteoritic matter in the normal upper crustal compositional signature of the target rocks. Only elements that are in high abundance in meteorites but low in terrestrial crustal rocks are useful. It is also necessary to take into account the fact that different meteorite groups and types have different compositions. When compared with target rock abundances, elevated siderophile element contents in impact melts can be indicative of the presence of either a chondritic or an iron meteoritic component. Problems may arise if siderophile elements are in high abundance in the target rocks or if the siderophile element concentrations in the impactites are very low. In such cases, the use of the osmium and chromium isotopic systems can help to establish the presence of a meteoritic component in impact melt rocks and breccias. An example of a characteristic isotope system, the Re-Os system, is shown in_FIG. 9, where the abundance ratio of the isotopes of two elements, Re and Os, has been plotted, and clearly show the difference between terrestrial and extraterrestrial compositions. Confirming that a geologically interesting and conspicuous structure is an impact crater requires relatively lengthy and often technically complex geological, geophysical, petrographic, and geochemical investigations.

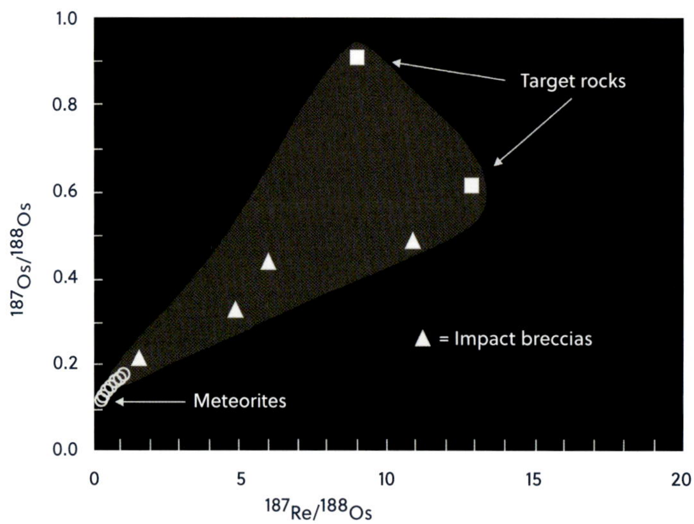

FIG. 9_*Re-Os diagram. The decay of radioactive Re-187 to Os-187 causes an increase in the abundance of the latter isotope; terrestrial crustal rocks contain high abundances of Re and have low Os contents, whereas in extraterrestrial rocks the situation is reversed. Thus, meteoritic material has low isotopic ratios and terrestrial rocks have high isotopic ratios in this diagram, allowing us to easily distinguish an extraterrestrial component in predominantly terrestrial rocks.*[22]

Image by Bevan M. French, Christian Köberl, 2010

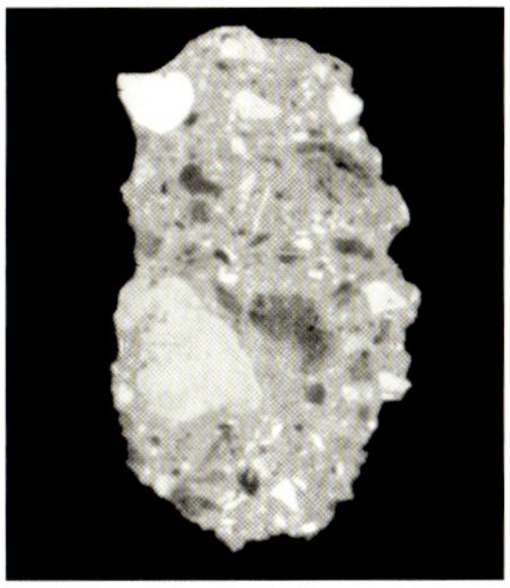

A

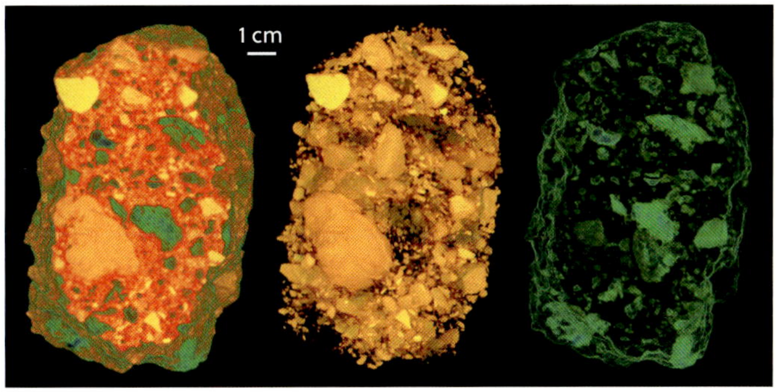

B

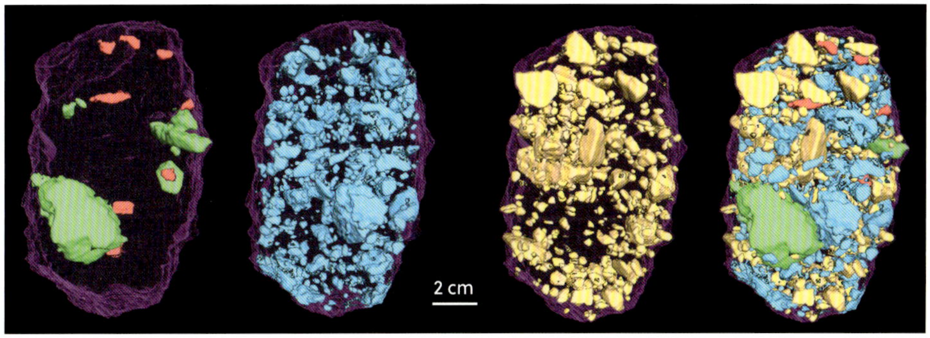

C Granite + schist Melt glass Aplite Pegmatite

FIG. 10_*High-resolution X-ray computed tomography (HRXCT) images*
A_*High-resolution X-ray computed tomography (HRXCT) image, showing a slice through a suevite from the Bosumtwi crater, Ghana. Width of sample approx. ten centimeters. Dark gray inclusions are of low density; light inclusions are of high density. The polymict breccia nature is evident.*
B_*Three-dimensional reconstruction of the clast population in the Bosumtwi suevite. The image on the left shows the sample with color-coded clasts, with the matrix (groundmass) rendered partly transparent. The center image shows only the high-density clasts, with the low-density clasts (and the groundmass) rendered transparent, and the image on the right shows only the rock edges and the low-density clasts (i.e., frothy impact glass).*
C_*Three-dimensional distribution of the four different types of rock clasts in the Bosumtwi sample that were identified by petrographic analysis and discerned in the X-ray images based on their density differences. The pervasive and small-scale distribution of impact glass is especially obvious.*[24]

Images by Christian Köberl, 2002

Detailed Studies of Impact-Derived Rocks

Various imaging techniques are used in detailed mineralogical, petrographic, and geochemical studies of impactites. These studies apply a variety of techniques, using sophisticated analytical methods ranging from X-ray scans to electron microscopy and analytical electron microprobes, to name just a few, some examples of which are discussed here. All these methods aid in understanding the composition and formation of impact-related rocks.

An interesting new technique that has just recently been applied for the first time to the study of impactites is high-resolution X-ray computed tomography (HRXCT). Köberl et al. used this method to image the interior of suevites from the Bosumtwi and Ries craters in order to determine the three-dimensional distribution of clasts within the matrix of the suevites, to test this technique with respect to its suitability for the recognition of different clast types of different densities, and to determine the textural characteristics of the tektite.[23] Their results show that HRXCT makes it easy to discriminate between the relatively frothy inclusions of glassy melt in the suevites as they are darker than the matrix in the raw X-ray scans and can be traced through the whole sample_FIG. 10A. Color or gray scale applications allow us to distinguish between at least four different clast types based on density differences_FIG. 10B. This method enables us to determine the 3D distribution of the various clast populations in the impact breccia using image processing techniques and to quantify their relative abundances_FIG. 10C. However, due to the highly specialized equipment required, it remains to be seen if this technique will become routine in use.

FIG. 11_*Backscattered-electron (BSE) and* A_*cathodoluminescence (CL)* B_*images of Libyan Desert glass do not exhibit an inverse relationship between the brightness of the BSE signal and the corresponding cathodoluminescence intensity. The CL-dark (with a relatively low SiO2 content) areas (with a relatively low SiO2 content) show rims as zonations (or layers) with variations in the CL intensities in an otherwise CL-bright (silica-rich) background. The images are approximately 600 × 600 micrometers.*[25]

Image by Arnold Gucsik, Christian Köberl, Franz Brandstätter, Eugen Libowitzky, and Ming Zhang, 2004

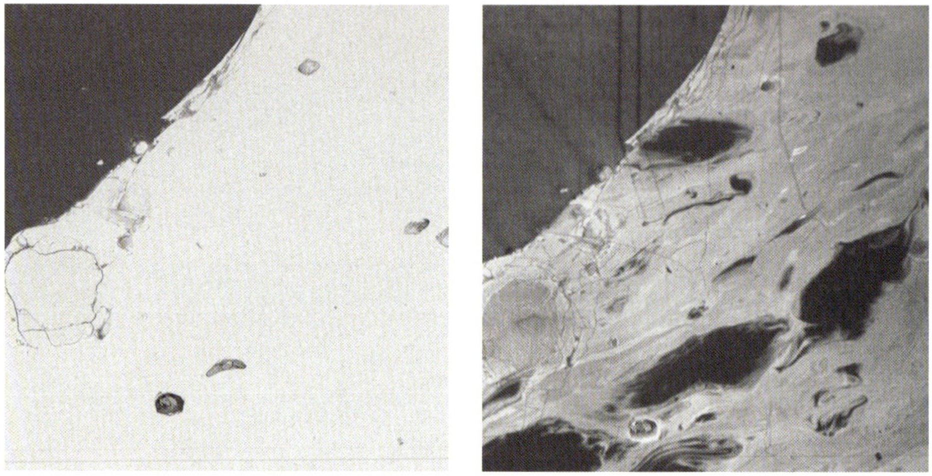

A B

Another method that is applied generally in the study of rocks and, of course, impact-derived rocks is electron microscopy. This method is most often used to obtain high-resolution and high-magnification images of rocks and minerals, and their constituents, but can also be used in some more specific applications, such as cathodoluminescence (CL). In this method, electrons are used to bombard the surface of rocks and minerals, causing light to be emitted. This method helps us to understand minor differences in the chemical composition of the material studied. For example, FIG. 11A shows a normal electron microscopic image of the surface of a rare impact glass (the Libyan Desert glass), indicating rather homogeneous glass. In contrast, the CL image, also taken at the same high magnification in the electron microscope, shows some very minor structural and compositional difference, indicating how the glass was flowing at the time of its formation. Using just normal optical or electron microscopy, these details would not have been visible.

Yet another method used frequently in the study of rocks and minerals, including impactites, is electron microanalysis (with an electron microprobe), where

FIG. 12_*Electron microprobe image and data of spherule layers formed by impact events on early Earth* B.[26]
A_*Backscattered electron image of chromium spinel (gray, euhedral crystals) in shaly groundmass (dark); bright crystals on the right are sulfide (B-SH13, depth: 71.44–71.46 meters).*

Image by Seda Ozdemir et al., 2017

A B

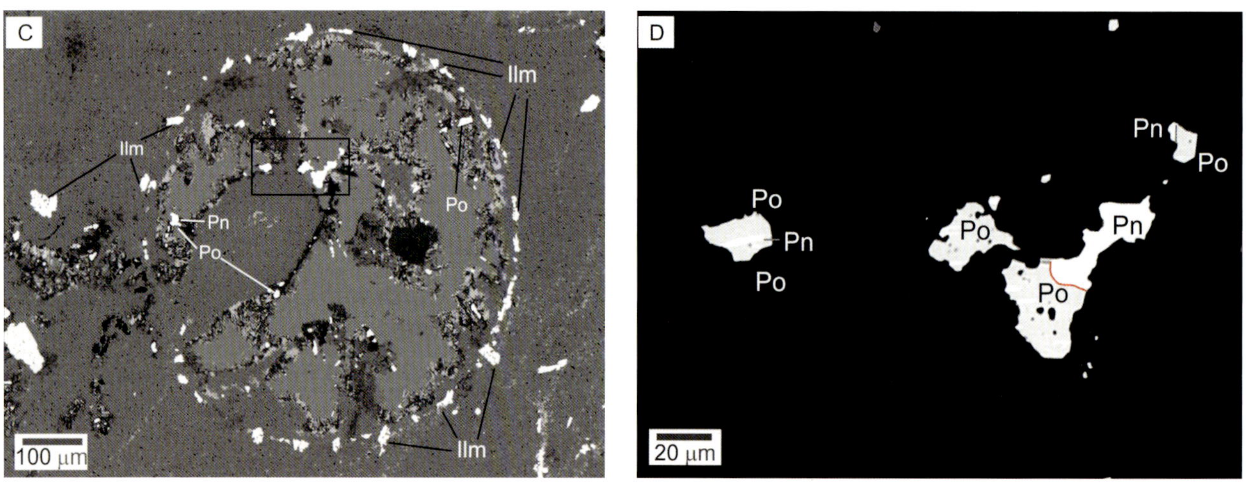

FIG. 13_ *Same samples as the ones shown in figure 12. SEM-EDS element maps for Ti, K, Al, and Si, for three spherule types* A, F, *and* K *are backscattered electron images.*
A–E_ *Al-rich spherule (B-SL5, depth: 65.398–65.429 meters)*
F–J_ *K-rich spherule (B-SL9a, depth: 69.338–69.351 meters)*
K–O_ *K-Al-rich, zoned spherule (A-SL1, depth: 7.780–7.788 meters)*[28]

Image by Seda Ozdemir et al., 2017

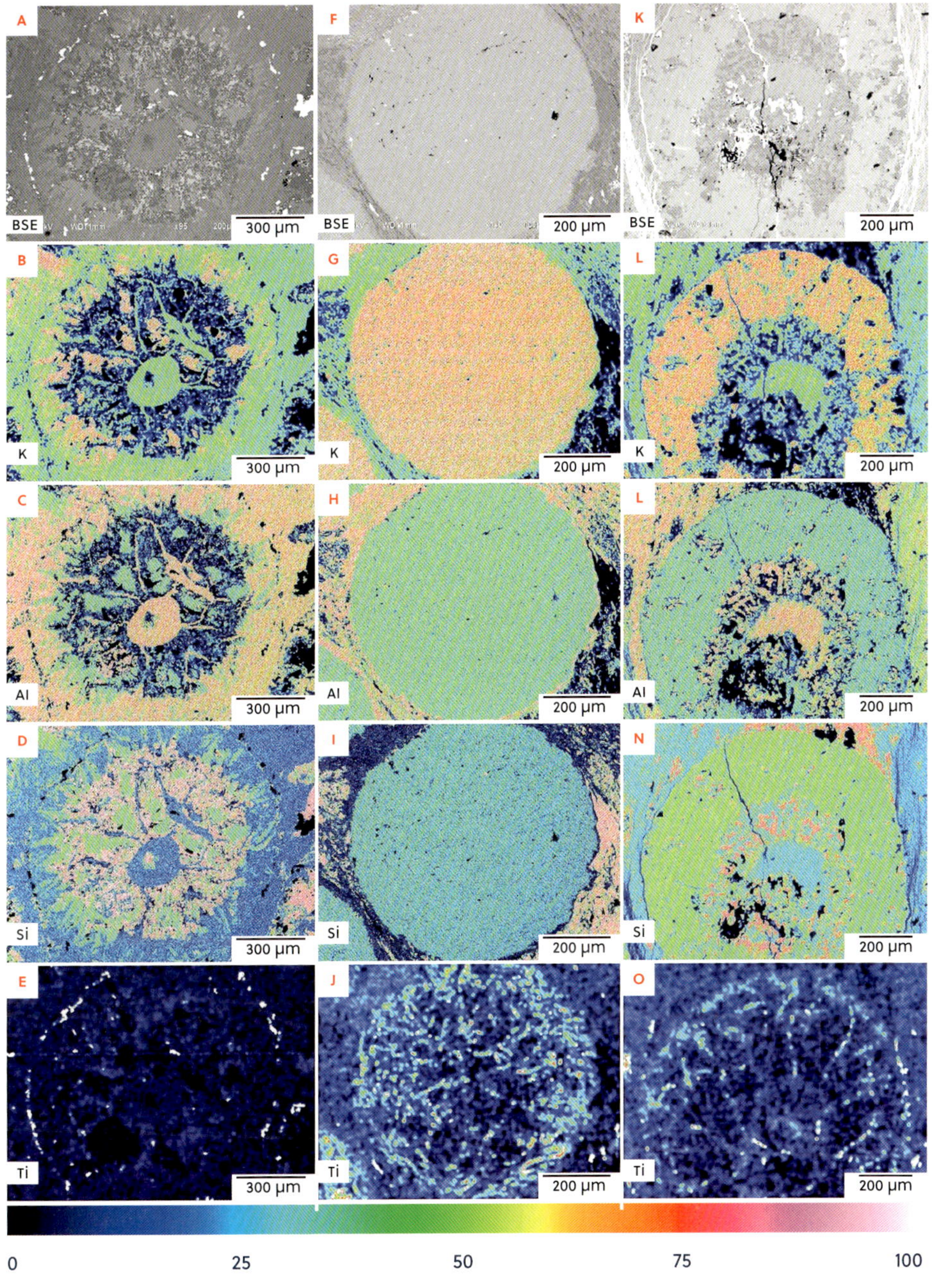

electrons bombard the surface of the sample to release X-rays that are characteristic of certain elements and that can thus be used to map the elemental contents of a sample on a microscopic scale and help to identify the presence of specific minerals, as shown in_FIGS. 12A and B (the examples shown here are spherules that formed billions of years ago in some of the oldest preserved impact deposits on Earth[27]). The resulting element distribution maps, with a color code that indicates the abundance (in weight-percent), are shown in an example in_FIG. 13.

These images (and methods) are just a few examples of what is currently being used in impact studies and the visualizations that are employed in such studies.

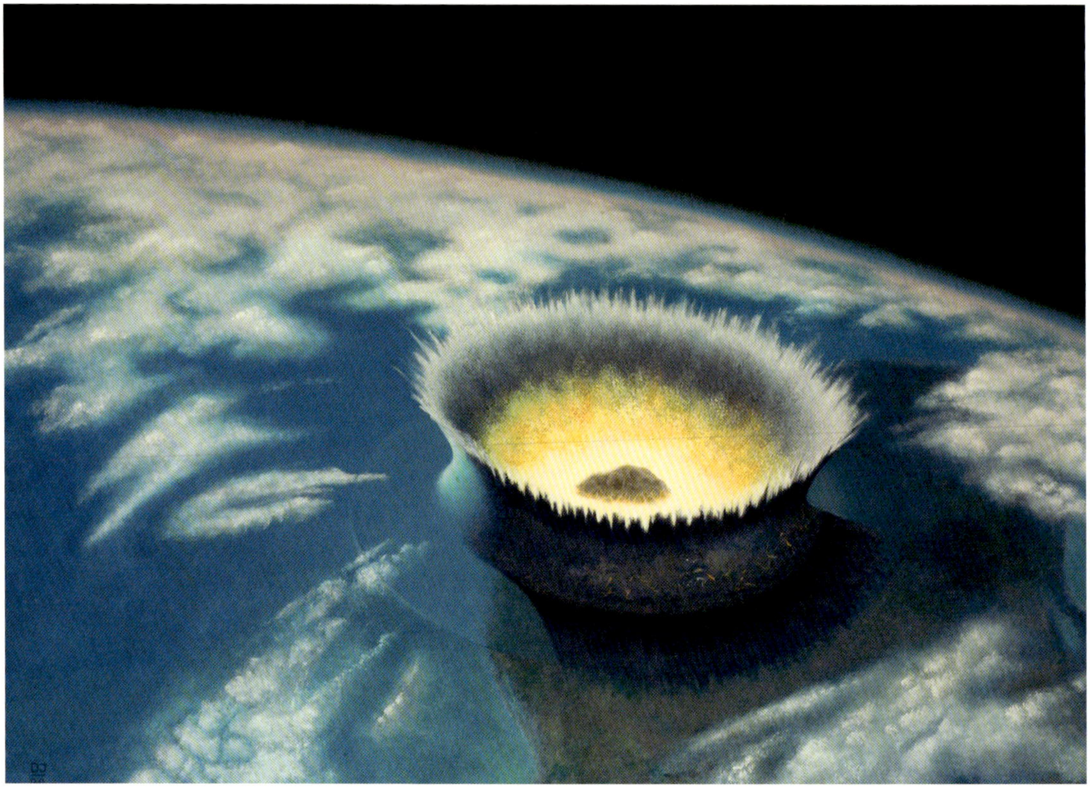

Use of Visualization in Outreach and the Arts

Not only are a variety of images and visualizations used in scientific studies (leading to important scientific discoveries), but such techniques are also of great importance in public outreach, as well as in artistic discourse related to science, such as in a variety of art-science projects. It would go beyond the scope of this short contribution to attempt a comprehensive review of the scientific illustration and art-science aspects of this topic, so I will name just a few examples. Entering the key words »meteorite impact« into various well-known online search engines yields mostly computer-generated images of colorful meteorites impacting the Earth, with rising impact plumes. Many of these images either exaggerate the size of the impacting body or mix up the time scales (e.g., with plumes already on top of the atmosphere when the impacting asteroid is still visible). Not every impact illustration, often made for maximum effect, is also scientifically accurate, which makes it even more worrisome that many news sources, especially scientific news portals, use illustrations of such dubious quality. Journalists often do not have the detailed background knowledge to even notice that there might be a problem or contradiction, and editors often do not know or care, as they just see an eye-catcher, not a scientifically correct visualization. The ideal situation would be scientists being involved whenever possible, which might be difficult for simple news items but would be highly beneficial for longer stories or documentary films.

One topic that has often featured in impact-related visualization with regard to our topic is the impact that caused the demise of the dinosaurs sixty-six million

FIG. 14_ *Oil painting by Dona Jalufka (Vienna, Austria), 1994/2005 (70 × 50 cm). Artistic rendition of a cataclysmic impact, such as the Chicxulub event that caused the end-Cretaceous mass extinction, sixty-six million years ago. The impact of a multikilometer-sized asteroid into a shallow ocean is depicted a few seconds after the initial surface contact. This image was used in a variety of publications and also on the cover of a special issue of the geosciences journal* ELEMENTS *in 2012.*[29]

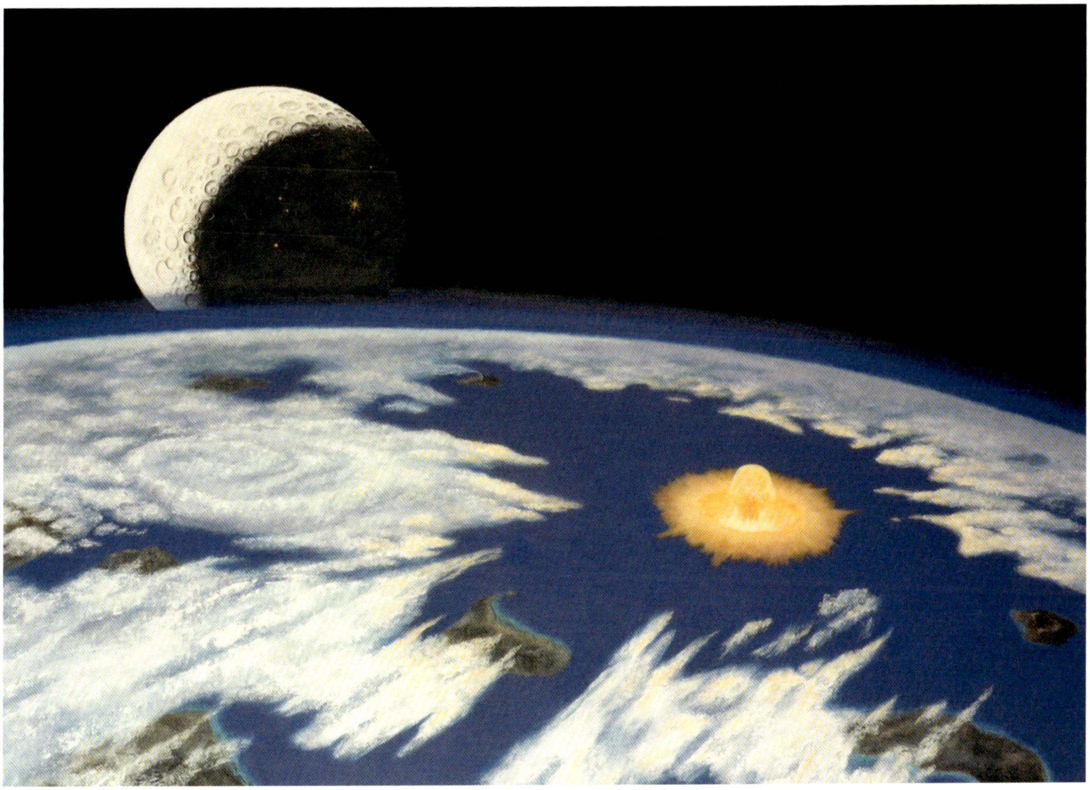

FIG. 15_ *Oil painting by Dona Jalufka (Austria), 2005 (60 × 40 cm). A possible view of the early Earth, sometime between 4.4 and 4.0 billion years ago. There is now evidence of rather clement conditions starting several tens of millions to three hundred million years after the Moon-forming impact event. The young Moon was much closer to Earth than it is today and did not yet have any of the maria that now dominate its Earth-facing side. Toward the end of planetary accretion, a higher impact rate than today led to the barraging of both Earth and the Moon. One medium-size impact event can be seen occurring in the right side of the image, while several smaller events are taking place on the Moon. An ocean dominates the surface of Earth, but island-arc formation and associated intense volcanism indicate the presence of an early continental crust. Earth is covered by a dense, CO_2-dominated atmosphere. This image has been used on the cover of Reimold and Gibson's book*[30] *and in a paper by Köberl.*[31]

years ago. There are many examples of images in which the impacting body is much too large, or where the illustration compresses several stages into one image and is therefore physically implausible or impossible. _FIG. 14 shows an oil painting that was prepared with scientific input and that portrays the early stage of an impact event of such a magnitude. A similar image is shown in _FIG. 15, which depicts an impact event on the early Earth, maybe about 4.2 billion years ago, when the Moon did not yet have its characteristic maria (the large, dark, basalt-filled impact basins that mostly formed around 3.8 to 4 billion years ago), and the Earth's surface had oceans but no large continents yet.[32]

In addition, impact craters and processes have featured in a number of art projects, for example, the Geological Society of America conference held near Ancona, Italy, in 2007 on the topic of Late Eocene impact events[33] was paired with an art exhibition titled *Rimanere Colpiti—Awestruck* by the artists Dona Jalufka and Paula Metallo, which included a variety of pieces that somehow related to the topic of »impact.« In 2009, another exhibition with the title *ImpactArt-ArtImpact* by the same artists was held at the Ries Crater Museum in Nördlingen, Germany.[34] These are just a couple of examples (listed due to personal connections), but there are others all around the world. Another related topic where impact plays a role is films, ranging from films about the Moon, which first appeared in the early twentieth century, with lunar craters depicted in an often exaggerated and scientifically incorrect way,[35] to films that specifically deal with impact events on Earth and how to possibly mitigate any danger. Well-known examples of the latter are *Deep Impact*

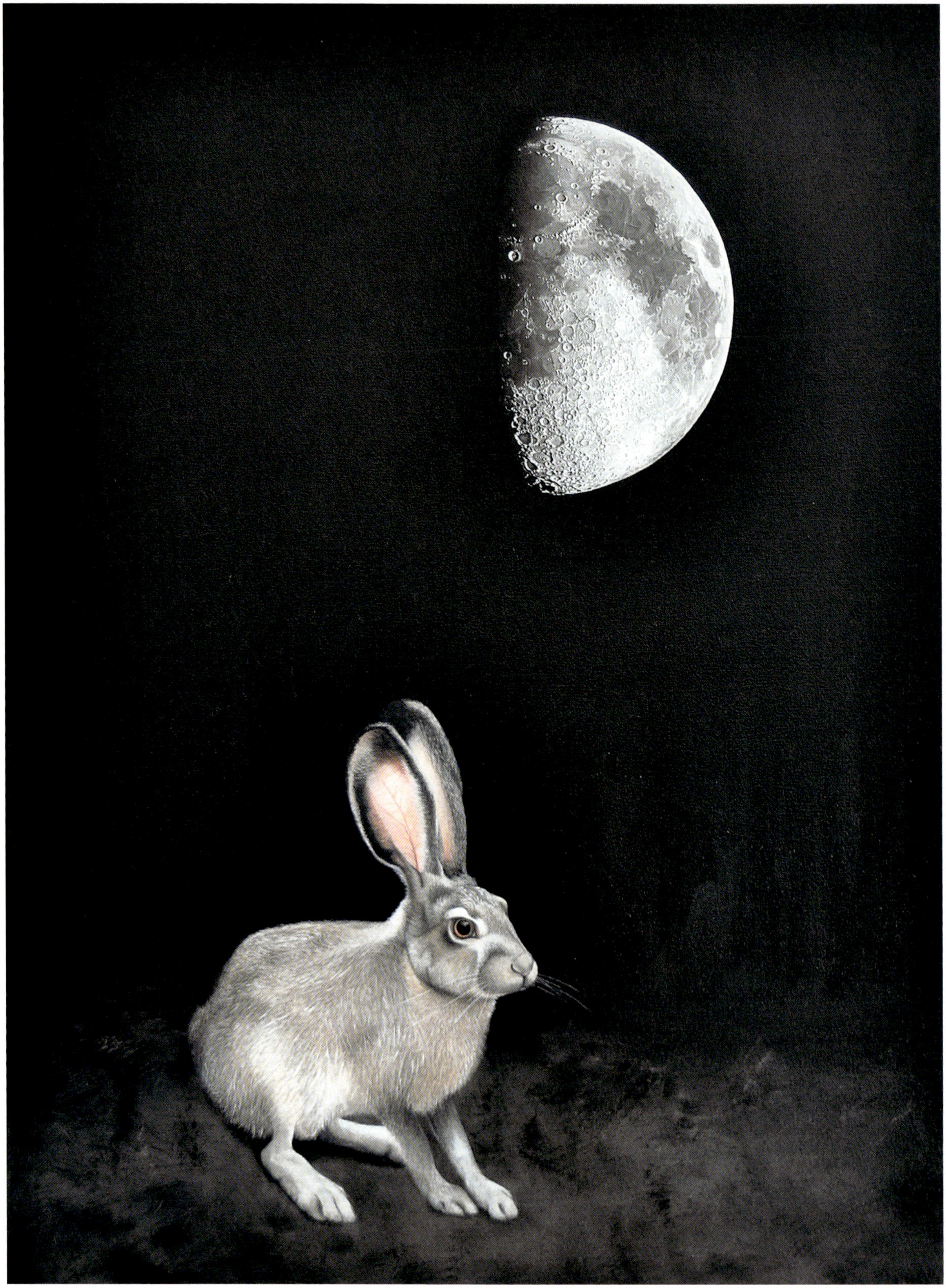

and *Armageddon*, both from 1998, which fall in the category of »science-fiction disaster movie.« There is also the more recent media satire *Don't Look Up* from 2021, which is somehow reminiscent of certain premises of the 1979 film *Meteor* (which had a star-studded cast). None of these films are particularly accurate in scientific terms — although *Deep Impact* at least contains some sequences that are relatively realistic, while others easily give up accuracy for effect and are prone to considerable exaggeration — but a more detailed analysis of this genre would require a separate publication.

Summary and Conclusions

Scientific investigations of the past few years have showed that impact events were much more important in the history of our own planet than had been previously assumed by geologists. Earlier in the twentieth century, the study of impact events was only slowly accepted by mainstream geologists. This picture only changed after irrefutable evidence was found, demonstrating that the large impact at the Cretaceous-Tertiary boundary was the cause of the coeval mass extinction. Impact events can easily cause local, regional, or even global disruption and disaster, and can be responsible for the death of many animals and plants (and even humans). However, if only a few individuals of a species survive, it is anywhere from difficult to impossible to obtain paleontological or chemostratographic evidence of the environmental effects of such an impact. Therefore, the danger of impacts should not be overestimated and can be easily compared with that of other »geohazards,« such as earthquakes or volcanic eruptions. It will be important to continue the astronomical search for more asteroids on potentially Earth-crossing orbits.

The study of impact craters on Earth can teach us a lot about the effects of impact events and the possible consequences of future such events. A large variety of different types of scientific imaging and visualization have been used, ranging from simple photographs (but from positions that provided new and previously unreachable vantage points, such as Earth's orbit) to computer-generated views of large amounts of data (such as magnetic or gravity maps), microscopic images using certain enhancements or data visualization (such as back-scattered electron images or element distribution maps), images using wavelengths other than just plain light, to scientific and artistic illustrations. All of these have been essential in generating previously unknown scientific observations and interpretations, drawing new conclusions, and/or making scientific concepts available to the interested public. This contribution has included some examples, but, of course, these have been limited in many ways, partly also due to personal interest and the involvement of the author.

FIG. 16_ *Oil painting* SISTER HARE AND HER MOON NO. III *by Dona Jalufka (2019; 60 × 80 cm). The Moon is our closest neighbor in space. It is a geological body, but appears prominently in our sky as an astronomical object, and it has had considerable influence, also cultural, on humans. The myths and legends associated with it transcend millennia, cultures, and religions. As a lunar symbol, the hare (related to the rabbit) is renowned for its fertility and shape-shifting powers. The »hare in the Moon,« dating back to ancient times, was predominant in East Asian folklore, and came to Japan and China from India with Buddhism.*

A Personal Epilogue

As noted above, both the text of this contribution and the examples shown to illustrate various points are, of course, a highly personal selection. On the one hand, this is due to my academic history and development, and my interests in scientific research, but also my long-standing interests in film and the visual and fine arts, which have been amplified due to my more than three-decades-long relationship with an artist—my wife Dona Jalufka. My interest in the arts and film date back even longer, to my student years, and I always found it interesting and important to use images, art, graphics, and illustrations (including those of the moving or animated sort) as an important aspect of not only conveying scientific information to the relevant scientific community but also helping to make science understandable to the general (and interested) public. Science is too important to leave all the fun and understanding to just the scientists. In addition, having been in constant contact with artists, I have always found it interesting to see how artists approach certain scientific topics, and informed dialogue between science and the arts has been enriching for both sides. Having been involved in a number of exhibitions, for example during my ten-year tenure as Director General of the Natural History Museum in Vienna, I always found it important to include artistic perspectives (both historical and contemporary) in topics that primarily comprise the natural sciences, and many responses to such exhibitions have noted that these kinds of interactions are quite enriching. The same goes for art exhibitions in galleries and art museums, which can benefit from the occasional interaction with scientists (such as in dialogues between artists and scientists on exhibition tours or in joint lectures). As a final illustration of this point, I offer _FIG. 16, in which a photorealistic Moon meets a hare to illustrate the old folk myth of seeing a rabbit in the light and dark markings on the lunar disk.

Acknowledgments

I would like to thank the editors, in particular Alfred Vendl, for many years of fruitful collaboration in the fields of visualization and the popularization of science. Thanks also go to my many collaborators for the joint research that is cited here and from whom I have drawn some of the example images used here, and to Dona Jalufka for the use of images of her oil paintings.

1 For details, see C. Köberl, »Craters on the Moon from Galileo to Wegener: A Short History of the Impact Hypothesis, and Implications for the Study of Terrestrial Impact Craters,« *Earth, Moon and Planets* 85 (2001): 209–24.

2 J. Nasmyth and J. Carpenter, *The Moon: Considered as a Planet, a World, and a Satellite* (London: John Murray, 1874).

3 R. A. Proctor, *The Moon* (London: Longmans, Green and Co., 1873).

4 From L. Rudaux, *Sur les Autres Mondes*.

5 L. Rudaux, *Sur les Autres Mondes* (Paris: Libraire Larousse, 1937).

6 See the discussion below; see also D. A. Jalufka and C. Köberl, »Moonstruck: How Realistic is the Moon Depicted in Classic Science Fiction Films?« *Earth, Moon and Planets* 85 (2001), 179–200; D. Jalufka and C. Köberl, »Der Mond in klassischen Science-Fiction-Filmen: Kunst kontra Wissenschaft?/The Moon in Classic Science Fiction Movies: Art vs. Science?« in *Ticket to the Moon, Exhibition Catalog—Kunsthalle Krems* (Vienna: Verlag für moderne Kunst, 2019), 142–65.

7 See, e.g., H. J. Melosh, *Impact Cratering: A Geologic Process* (New York: Oxford University Press, 1989).

8 Source: USGS National Map Data Download and Visualization Services, August 13, 2009, https://en.wikipedia.org/wiki/Meteor_Crater#/media/File:Meteor_Crater_-_Arizona.jpg.

9 Source: NASA/Aster Image.

10 NASA, astronaut photo ISS007-E-05697, taken in 2003.

11 NASA, astronaut photo ISS001-E-5739, taken in 2001.

12 All images from C. Köberl and W. U. Reimold, »Bosumtwi Impact Crater: An Updated and Revised Geological Map, With Explanations,« *Jahrbuch der Geologischen Bundesanstalt, Wien (Yearbook of the Austrian Geological Survey)* 145 (2005): 31–70 (+ 1 map, 1:50,0000).

13 Details on the physics and mechanics of the formation of impact craters can be found in Melosh, *Impact Cratering*.

14 For details, see L. Ferrière, »Shock Metamorphism.«

15 See above; see also L. Ferrière, C. Köberl, B. A. Ivanov, and W. U. Reimold, »Shock Metamorphism of Bosumtwi Impact Crater Rocks, Shock Attenuation, and Uplift Formation,« *Science* 322 (2008): 1678–81.

16 See, e.g., J. S. Lewis, *Rain of Iron and Ice: The Very Real Threat of Comet and Asteroid Bombardment* (Reading, MA: Addison-Wesley Publishing Company, 1997).

17 For more detailed information on impact ejecta, see the following reviews: A. Montanari and C. Köberl, *Impact Stratigraphy: The Italian Record* (Heidelberg: Springer Verlag, 2000); C. Köberl, »The Geochemistry and Cosmochemistry of Impacts,« in *Treatise on Geochemistry, Second Edition*, vol. 2, *Planets, Asteroids, Comets and the Solar System*, ed. H. D. Holland and K. K. Turekian (Oxford: Elsevier, 2014), 73–118.

18 See also B. M. French and C. Köberl, »The Convincing Identification of Terrestrial Meteorite Impact Structures: What Works, What Doesn't, and Why,« *Earth-Science Reviews* 98 (2010), 123–70.

19 For a review, see R. A. F. Grieve and M. Pilkington, »The Signature of Terrestrial Impacts,« *AGSO Journal of Australian Geology & Geophysics* 16 (1996): 399–420.

20 See D. Stöffler and F. Langenhorst, »Shock Metamorphism of Quartz in Nature and Experiment: I. Basic Observations and Theory,« *Meteoritics* 29 (1994), 155–81; B. M. French and C. Köberl, »The Convincing Identification.«

21 See, e.g., C. Köberl, »The Geochemistry and Cosmochemistry of Impacts.«

22 For more information, see also B. M. French and C. Köberl, »The Convincing Identification«; C. Köberl, »The Geochemistry and Cosmochemistry of Impacts.«

23 C. Köberl, C. Denison, R. A. Ketcham, and W. U. Reimold, »High-Resolution X-Ray Computed Tomography of Impactites,« *Journal of Geophysical Research* 107 (2002): 1–9.

24 All images have been taken from C. Köberl et al., »High-Resolution X-Ray Computed Tomography of Impactites.«

25 For details, see A. Gucsik, C. Köberl, F. Brandstätter, E. Libowitzky, and M. Zhang, »Infrared, Raman, and Cathodoluminescence Studies of Impact Glasses,« *Meteoritics and Planetary Sciences* 39 (2004): 1273–85.

26 For details, see S. Ozdemir, T. Schulz, C. Köberl, W. U. Reimold, T. Mohr-Westheide, D. Hoehnel, and R. T. Schmitt, »Early Archean spherule layers from the Barberton Greenstone Belt, South Africa: Mineralogy and geochemistry of the spherule beds in the CT 3 drill core.« *Meteoritics & Planetary Science* 52, no. 12 (2017): 2586–2631.

27 See also C. Köberl, »Impact Processes on the Early Earth,« *Elements* 2 (2006): 211–16; C. Köberl, »The Record of Impact Processes on the Early Earth—A Review of the First 2.5 Billion Years,« in *Processes on the Early Earth*, ed. W. U. Reimold and R. L. Gibson (Boulder: Geological Society of America, 2006), 1–22; D. Hoehnel, W. U. Reimold, U. Altenberger, A. Hofmann, T. Mohr-Westheide, S. Özdemir, and C. Köberl, »Petrographic and Micro-XRF Analysis of Multiple Archean Impact-Derived Spherule Layers in Drill Core CT3 from the Northern Barberton Greenstone Belt (South Africa),« *Journal of African Earth Sciences* 138 (2018): 264–88; S. Ozdemir, T. Schulz, C. Köberl, W. U. Reimold, T. Mohr-Westheide, D. Hoehnel, and R. T. Schmitt, »Early Archean Spherule Layers from the Barberton Greenstone Belt, South Africa: Mineralogy and Geochemistry of the Spherule Beds in the CT3 Drill Core,« *Meteoritics and Planetary Science* 52 (2017), 2586–631.

28 For details, see S. Ozdemir et al., »Early Archean Spherule Layers.«

29 *Elements* 8, no. 1 (2012), http://elementsmagazine.org/past-issue-archive/2012-volume-8/.

30 W. U. Reimold and R. L. Gibson (eds.), *Processes on the Early Earth* (Boulder: Geological Society of America, 2006).

31 C. Köberl, »Impact Processes on the Early Earth.«

32 For details, see C. Köberl, »Impact Processes on the Early Earth«; C. Köberl, »The Record of Impact Processes on the Early Earth.«

33 A. Montanari, C. Köberl, F. Hilgen, and R. Coccioni, »Penrose Conference Report: Hothouse, Icehouse, and Impacts: The Late Eocene Earth, Monte Cònero (Ancona), Italy 3–6 October 2007,« *GSA Today* 18, no.1 (2008): 26–8.

34 See https://dona-jalufka.at/biocv/.

35 See the discussion and reviews in D. A. Jalufka and C. Köberl, »Der Mond in klassischen Science-Fiction-Filmen.«

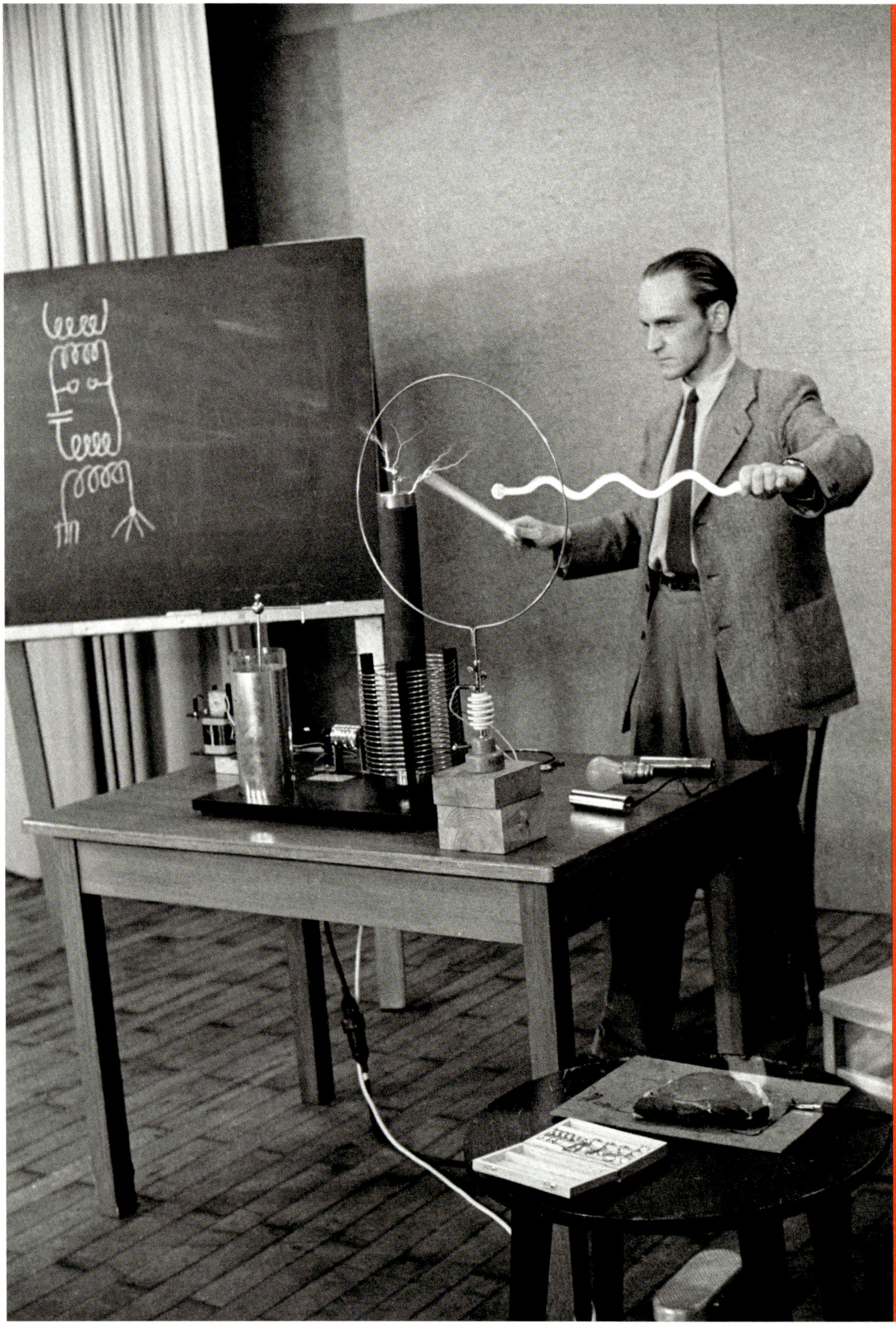

FIG. 1_ *The beginnings of visualization: experiments in an improvised studio — to mark the one-hundredth birthday of Nicola Tesla*

Putting Knowledge in the Picture

A personal account by Thomas Matzek with Silvia Heimader, Barbara Kerb

How visualizations have shaped educational programs on Austrian television since 1955

A charismatic theater star playing twelve Austrian rulers, a bearded biologist in a safari shirt introducing animals in a studio decorated with reeds—these were some of the images that influenced me as a twelve year old. A few years later, I was the one on the road with a semi-professional cine camera: standing in front of historic buildings searching for their characteristic details or on a raised bog in the alps looking for a rare plover species.

Nature, history, science—compellingly presenting the phenomena in these fields became a passion that still grips me today. It was television that took me there—the images on the TV screen that showed us things that we would not have been able to see otherwise. The people whose programming once inspired me are the protagonists of a journey through time that I would like to embark upon here—seventy years of visualizing knowledge on Austrian television.

Since time immemorial, photography and film have helped to convey facts to a wide audience and to make things visible that cannot be seen by the naked eye. Visualizing phenomena is a cornerstone of programming for state-owned television. If something can be demonstrated clearly, it helps an audience without any prior knowledge to understand it. This is the task of education in the original sense—and the basis of the Austrian Broadcasting Corporation (ORF), whose television programming was launched in Austria in 1955. I would like to take a look at the major research topics, such as nature, history, and medicine, to portray the exciting developments that have taken place in the visualization of educational content at the ORF. Typical of the pioneers of the ORF was that the basis of their work was not always the technological opportunities available but their commitment to having an impact on everybody, in line with the idea of public education.

1955 was an eventful year for Austria as the end of the occupation period and the signing of the Austrian State Treaty brought about an euphoric mood—and it was an eventful year for technological developments and inventions as well! In

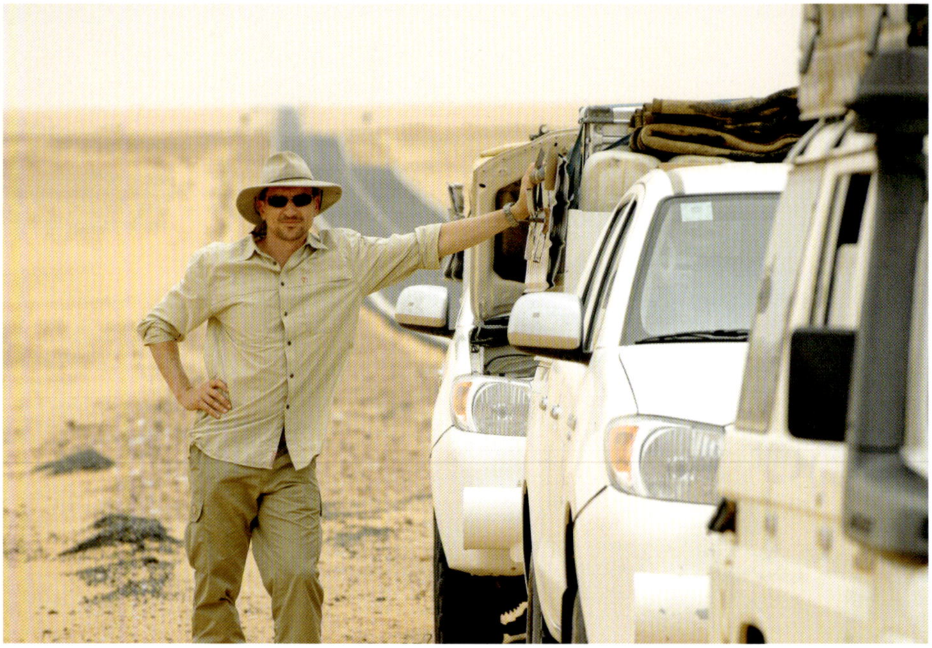

FIG. 2_ *The author on his way to prehistoric rock formations in the Egyptian East Sahara as part of a film expedition*

Photo by Ralf Rabenstein

1955, now independent Austria was to get its own television station. Before World War II, this new medium—based on the electronic scanning of images—had not undergone the same new developments in Austria that it had in places like Germany. Until the signing of the State Treaty, Austria had only had rudimentary experiences with this new technology, especially as the Allied occupational forces had imposed a ban on television technicians in Austria. It was only with the authorization of the Austrian Broadcasting Corporation that experimental programming began on August 1, 1955. Regular programming was launched on January 1, 1957.

The Beginnings of Television—Science as a Point of Reference

Whereas cinemas and community colleges like the Urania in Vienna were already catering to attendees with huge repertoires of both entertainment and documentary films in the early 1950s, television had to assert its place. The ORF's predecessor, the RAVAG, had primarily dedicated itself to public education since 1924, and the television's early programming offerings were also made in this spirit. Alongside news and culture, reports from the fields of science and scholarship were part of its repertoire from the outset. The television broadcaster's first scientific series *Aus der Welt der Forschung* (From the World of Research) was launched in September 1958 with a profile of physicist Karl Auer von Welsbach, who invented the lightbulb. The Kultur und Volksbildung (Culture and Public Education) department responsible for the series produced around sixty hours of programming in 1958, including the *Sendungen aus dem Gebiet der reinen Wissenschaft, der Welt der Forschung und aus dem Volkstum* (Programs from the Field of Pure Science, the World of Research, and Folklore), as they were called back then.

The individual shows in the new *Aus der Welt der Forschung* series were thirty minutes long and consisted of live elements and film clips. The fact that visualization was subject to the dictates of education was unmistakable. The shows began with physics formulas being filmed off a board. Scientists, often older gentlemen in white coats or suits, gave explanations like they would in a lecture hall or classroom. There were no other means of visualizing abstract content available yet. These shows used established film techniques: close-up sequences, medium shots, long shots, montages of visually compelling details, dynamic cuts, and accentuated music. All of this was supposed to make the work carried out in research institutions more »appetizing.« It was joined by graphic explanatory elements such as diagrams, pictures, and formulas.

The chief director of photography at the Austrian Broadcasting Corporation back then was Hans Imber, who had already been a documentary-maker in the First Austrian Republic and a war correspondent during World War II. His cinematographic repertoire is showcased in the half-hour documentary *Die Sprache der Bienen* (The Language of Bees) from April 6, 1959. It presents a portrait of Austrian zoologist Karl von Frisch — who would later win the Nobel Prize — and his research into bees.

This kind of programming was typical of television back then. Conveying knowledge about Austria's researchers (most of them men) was at the top of the agenda for the new medium's education mandate. Imber did not just use conventional cinematic means like macro shots but also recreated experiments in front of the camera — for example, to find out how bees find their orientation. These experiments showed Frisch demonstrating how bees communicate with each other — a simple, effective visualization. This made the research findings clear for the audience.

And this is what early television-makers were interested in: education was to appeal to and entertain the masses, even if its settings do seem rather academic from today's perspective. But it was »hour zero« after all. Television in Austria did not have any competition from cable, satellite, or streaming platforms; it was a completely new medium: in 1955, there were just six thousand »TV connections« in all of Austria. Only a few people had a private television at home. Most people went to pubs or stood outside the electronics shops that had televisions running in their windows. Alongside culture, in particular broadcasts of theater plays, the people in charge of programming focused on science and education but also on vocational training and school television.

One of the masterminds of the time was Helmut Zilk, who would later become Television Director, Minister of Education, and Mayor of Vienna. Zilk, a teacher and educator in the postwar period, had already become a pedagogical television consultant by the mid-1950s and came up with concepts for many TV shows.

One show in the field of science that he helped to develop, *Kleine Welt der großen Wunder* (A Small World of Great Wonders), which aired on September 23, 1962, was about the possibilities of showing what is hidden from the human eye.

»Nature is a spring that never runs dry. Thousands of connections open up to those who search and marvel.« Thus began the commentary that was intended to make the research adventure appetizing to the audience. Zilk and his team introduced the opportunities presented by the microscope — by examining the work of

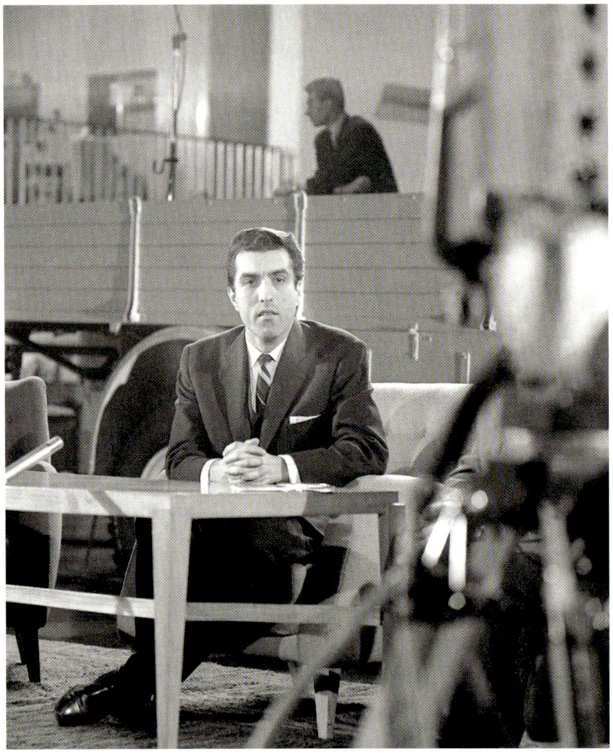

FIG. 3_*Teacher, public educator, and television-maker: Helmut Zilk had a considerable impact on early programming, in particular with school television.*

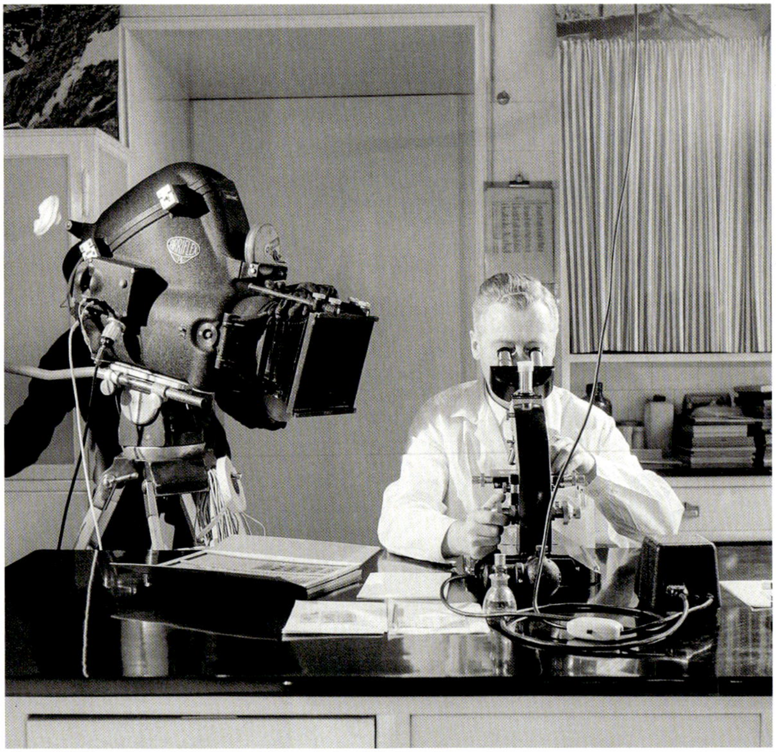

FIG. 4_*Making visible what is hidden to the human eye: microscope images were a fascinating component of early educational programming.*

Austrian microbiologist Ewald Schild. They used the microscope camera to convey insights into areas that had only been possible for a select few back then, with images of rod-shaped blue-green algae, spiral-shaped putrefactive bacteria, and the inner lives of single-celled organisms like amoeba.

Great Hopes for the New Medium

But Zilk & Co. soon recognized the true strengths of television compared with film: being there live, as well as depicting events that lasted longer without breaks—because nobody had to switch out rolls of film. The live character of television would open up very unique creative opportunities to the new medium, which relied on studios, that classic documentary films were unable to provide. Back then, the Austrian Broadcasting Corporation had only modest means—the premises it had at its disposal in the mid-1950s were a school at Singriener Gasse 23 in Vienna's Twelfth District. A studio was set up in a former classroom, where everything—from news shows to dramas—was realized using changing decorations.

Television did not just allow for theater to be broadcast but also for people to observe natural phenomena that were difficult for them to access in real life. Broad-

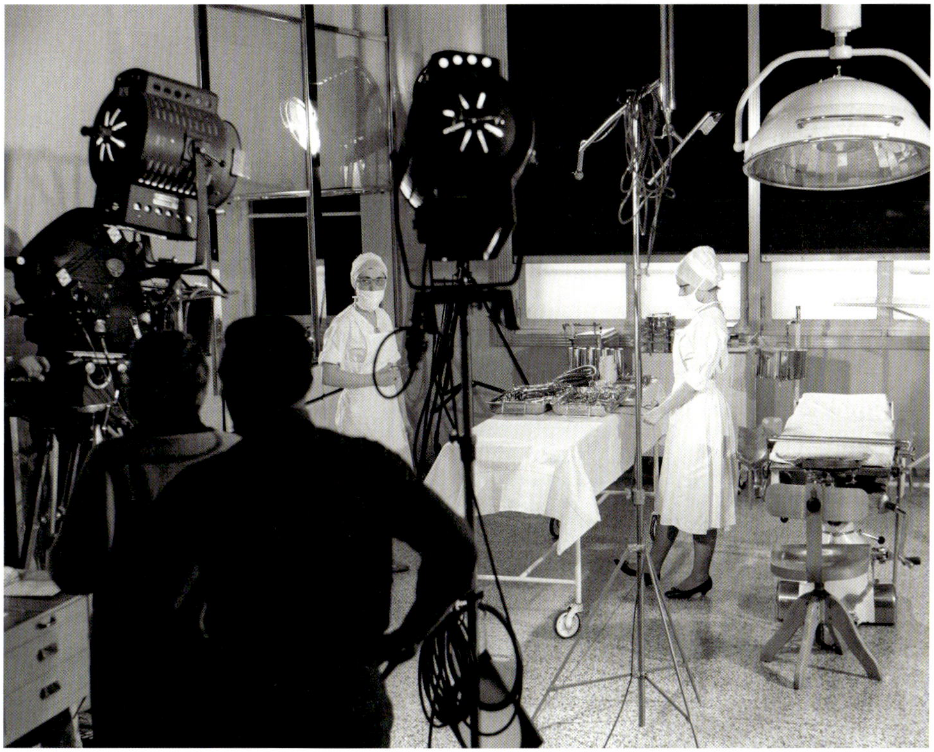

FIG. 5_ *When the operating theater becomes a television studio: program-creators were interested in people and medicine from the very beginning.*

© ORF, 1962

casting technology actually made it possible to capture images without a cameraman on site—today we would say remotely—and therefore to see what was far away, which is reflected in the German verb for watching television, *fernsehen*, which literally means to see (*sehen*) far away (*fern*). An external screen was used to check what was happening.

On September 10, 1955, in an article with the title »Die Fernsehkamera im Dienste der Wissenschaft« (The Television Camera in the Service of Science), *Radio Österreich*, the journal of the Austrian Broadcasting Corporation, enthused about all the new possibilities television presented—under water, in space, but also during medical operations. This led to the formation of a collaboration between the General Hospital in Vienna and the recently established Austrian Broadcasting Corporation. On January 31, 1957, »for the first time in Europe« (quote: *Arbeiter-Zeitung*), an appendix operation was broadcast using a color television camera. The ORF provided the technicians, but the camera itself, which had been developed in the US and was worth one million schillings—about 500 thousand euros today—was made available to the ORF by the American doctor's club in Austria.

»The main goal we are striving for is to show a medical operation and the techniques used while it is being performed to a large number of doctors and students—without disrupting it. These doctors and students sit in a neighboring room and follow the operation live on television, allowing them to observe the operation in peace.« In the US, this broadcasting technology had been established since the late 1940s, but electronic image broadcasting now made its way into the Vienna

General Hospital with the technical installation of an operating studio. No recordings of these shows have survived. Television back then meant »seeing faraway« but not recording. The first magnet coils used to record shows did not arrive in Austria until 1958.

Medical issues and health topics have been part of the fixed repertoire of Austrian television programming since early on—on February 26, 1958, the half-hour documentary *Griff ins Gehirn* (Reaching into the Brain), for example, portrayed an operation in great detail, from the patient's admission to his recovery; the only thing missing was close-ups of the operation actually being performed on the patient's head. The audience was thought to be so interested in these topics that in-house series were developed—such as, for example, *Hüter der Gesundheit* (Guardians of Health).

Evidently as a result of the 1957 Asian Flu, another pandemic that haunted Europe, and due to the increase in viral diseases during the rise of international aviation, on April 8, 1959, a short documentary was broadcast about vaccine production that used the example of a smallpox infection. The segment was mainly made up of carefully performed reenactments that presented the complete course of the illness. The medical professionals who appeared were real—like then head of the Austrian Federal Institute of Vaccine Production, Franz Puntigam. These experts helped to reenact the chain of information as well: from the hospital via the health department and the social ministry to the World Health Organization, which then informed health authorities all over the world—contact tracing anno 1959. Montages accurate in every detail documented the hunt for the pathogen—from examining the patient, an Asian-looking aircraft passenger, to shots taken by the electron microscope that identified the pathogen. It thus showed how the smallpox vaccine was produced in precise detail: from retrieving infected material from cows to its production in the laboratory right down to the testing of the vaccine on mice and rabbits. Back then, visualization worked not by way of animation and graphics but by meticulously depicting the process—the show could also pass as an educational film for medical professionals in training.

Animal Behavior up Close and Personal—in the Television Studio and under Water

One fixture in the experimental programming on Austrian television was nature documentary-maker Otto König. The animal photographer and ethologist produced the show *Wunder der Tierwelt* (Wonders of the Animal World) as early as in 1956, filmed in an improvised studio at the Biological Station Wilhelminenberg, the research institute that König himself had established shortly after World War II.

König brought his concept of education with him from his family home. His father, Otto König senior, had gained a reputation as an adult educator in the workers' movement during the interwar period. With his writings and the lectures he gave at community colleges in Vienna, the journalist, who had long worked for the social democratic *Arbeiter-Zeitung* (Workers' Newspaper), shaped the climate in which his son grew up. König junior combined these experiences with his enthusiasm for nature and the animal world, which he conveyed to his audience in lively demonstrations—just like today's *YouTube* influencers.

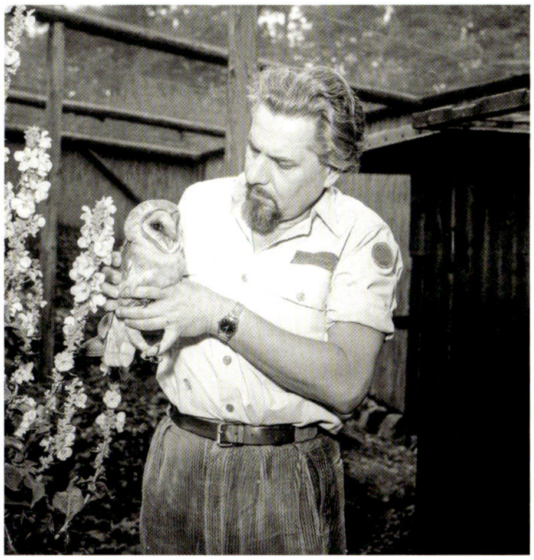

FIG. 6_*Bringing the animal kingdom to the studio: Otto König clearly explained the morphology and behavior of animals using simple tools.*

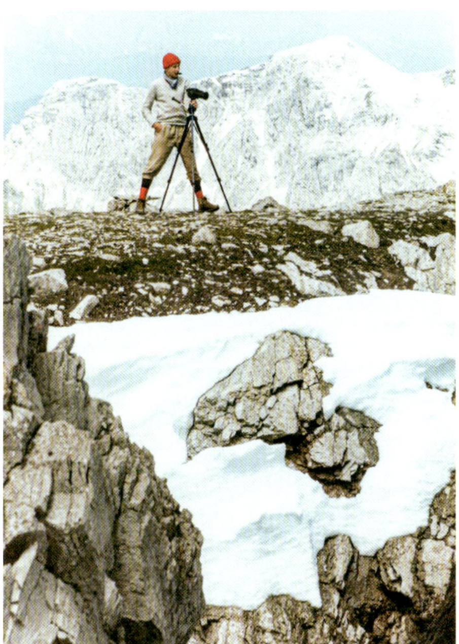

FIG. 7_*The young author—wearing a Cousteau-inspired woolen beanie—in the Alps on the hunt for the Eurasian dotterel*

Although Otto König had gained some experience as a nature documentary-maker on expeditions, he went down a completely different path with his show. Instead of taking expensive, sophisticated technology to animal habitats, he brought the animals to his own habitat—the television studio. Walls of reeds in the improvised studio successfully created an expedition atmosphere in the living room, and the live character did the rest. In order to explain bird migration, for example, König simply used maps hanging on the wall. With the help of these maps and by describing the characteristics of the birds, which he held up to the camera, he was not only able to demonstrate the phenomenon of bird migration but also pointed to the necessary ecological conditions for migration—e.g., intact wetlands. With this mix of tools, the bearded natural history enlightener brought a wide audience closer to the animal world and its behaviors. König was so successful with the compelling demonstrations he carried out using studio animals, objects, maps, studio interviews, and short film clips, that his show become a perennial favorite—and remained on air until the 1990s.

The live character that television made possible was not just effective in the first studio productions but also in the oceanic wilderness. It was not an Austrian production that revealed the possibilities of watching underwater television, it was a French one, but it too would remain a fixture in ORF programming for years to come. *Geheimnisse des Meeres* (Secrets of the Oceans) was the name of the series that I longed to see each week as a middle schooler in the early 1970s. The show was a production by French oceanographer Jacques-Yves Cousteau, who, after the war, had managed to raise enough international funds to go on costly expeditions on his research ship *Calypso*. The technological advancements that Cousteau made in order to research life under water, and that turned each episode into a great adventure for humanity, left a lasting impression on me as an adolescent. Diving and television pioneer Cousteau became a personal idol—I even got someone to knit me a red beanie so I could look like »Capitaine Cousteau.«

The charismatic Frenchman had recognized the potential of television early on and, in the mid-1950s, was already utilizing what was then still nascent broadcasting technology. He used remote-controlled cameras to reach oceanic depths that otherwise posed health risks to divers. This was how he discovered a two-thousand-year-old shipwreck and caught it on television camera—as the journal *Radio Österreich* enthusiastically reported in its August 6, 1955, issue, in the hope that this technology would soon be able to conquer new horizons in Austria as well.

The pioneering spirit of television, with its new means of visualization, did not shy at addressing topics from the past. Early on, the question was raised of what history education might look like for a wide audience. The predecessors to today's Education, Science, and Current Affairs department at the ORF threw themselves into making elaborate, high-quality documentary productions—today we would call them »blue chip«—and, based on that, special versions for school television, which had just been developed and was primarily being pushed by Helmut Zilk.

School television was launched in 1959 with a documentary about Emperor Maximilian to mark the 450th anniversary of his death. *Weißkunig* (The White King)—the title came from an autobiographical volume that »the last knight« had commissioned himself. The original *Weißkunig* book had already been a visualized representation of history: the chronicle of the emperor's life is like a cartoon consisting of 251 woodcuts. The documentary-makers made extensive use of these images as historical sources. Dynamic camera tracking shots, high-quality black-and-white aesthetics, and the voice of a prominent narrator — Fred Liewehr, an actor from Vienna's Burgtheater — ensured dramaturgical variety in spite of the historical subject matter.

Even more so than that of the Middle Ages and early modern period, modern history was at the top of the agenda for the public educators. In particular, it was Helmut Zilk's particular objective to address the events that had shaped Austria in the twentieth century. Under his leadership, a documentary show was produced in 1964 to mark the July Putsch of 1934 using all of the elements available to come to terms with history at the time: archive material, maps, reenactments of historical events — and an atmosphere of dramatic suspense generated with the use of a ticking clock. The script for *25. Juli 1934 — eine Dokumentation über den Dollfuss-Mord* (July 25, 1934 — A Documentary About the Dollfus Murder) was written by Hellmut Andics, and the documentary was directed by Walter Davy (who would later find fame as »Schremser« in the cult series *Kottan ermittelt* [Kottan Investigates]).

The actors playing personalities like Federal Chancellor Engelberg Dollfuß and assassin Otto Planetta were not selected on the basis of how much they resembled the people being portrayed, as whom they were identified by name plates they wore on their chests. The commentary gave the following reason for this: »These three men are not actors; we have deliberately decided that we do not want to make use of external resemblances. The name plates provide guidance. We want to reenact the events that we have learned about from historical records using verifiable eyewitness statements, just like the criminal police would reconstruct a crime.« The narrator thus also played the role of a criminal police investigator giving the actors instructions. He even got them to interrupt and repeat the events when an action was not being carried out the way the records suggested. The murder itself, the shot fired by assassin Otto Planetta that ultimately cost Dollfuß his life, was acted out multiple times. Rather than produce an entertaining illusion, this gave the show credibility.

The early decades of television helped to change the way we looked not just at the past but also the future. In the early 1960s, space travel began attracting huge public interest. On the one hand, people were fascinated by the attempts being made to push into new worlds and to explore what was reality and what was science fiction. But what is more, space travel was an adventure that the world could participate in directly. The discoveries being made in the expanses of space were not just conveyed in newspaper and film reports — the opportunities presented by television turned space travel into something that the audience could experience themselves. Just like new perspectives have always led to new insights, looking at the Earth from space changed the way we look at this planet. As the first shots revealed the clouds moving over South America, the interconnections in our atmosphere became clear to every layperson. It is a sensitive ecosystem we live in.

A Visual High-Altitude Flight with the Apollo Program

In 1969, there was an even greater boost in knowledge — with the reporting on the moon landing performed by US astronauts Armstrong, Aldrin, and Collins.

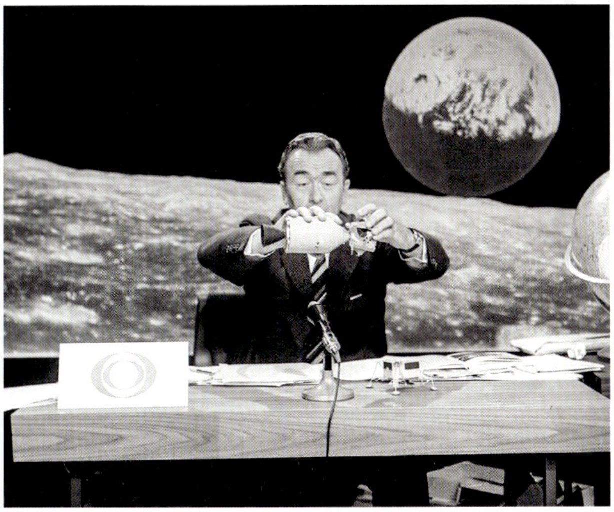

FIG. 8_ *Television history was also written using simple means. Herbert Pichler, also known as »Mond-Pichler« (Moon Pichler), used models to bring life to the many broadcasts.*

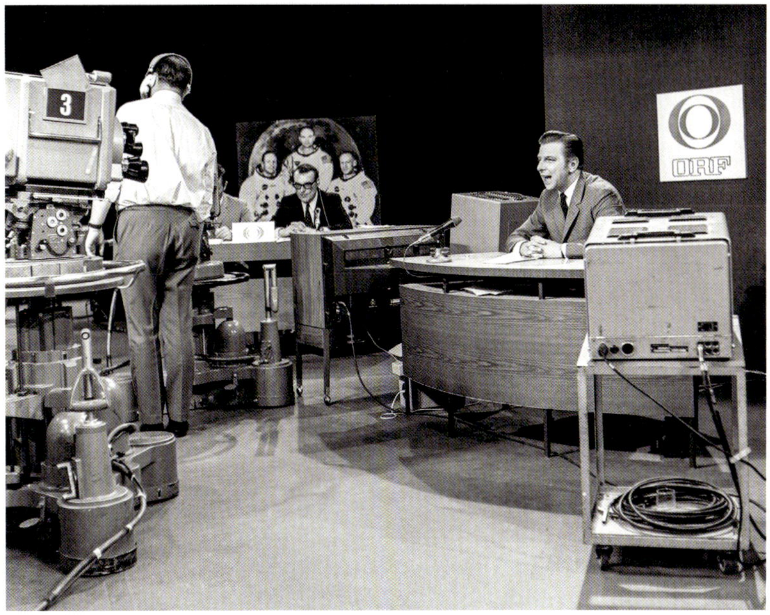

FIG. 9_ *The ORF reported on the Apollo mission for twenty-eight hours. The studio was simple, the passion of chief reporter Hugo Portisch great.*

There had already been detailed reports emerging about NASA's preparations before the flight—from the ORF team too, in particular from chief reporters Hugo Portisch, Peter Nidetzsky, and Herbert Pichler, the latter who was given the nickname »Moon Pichler.« The moon landing was an absolute highlight in the ORF's programming, for which the station mustered all the technology and personnel at its disposal.

The highlight of the broadcast was the images that became visible to an audience of millions directly from space. In order to legitimize its space travel spending, it was important to the US authorities, above all NASA, to allow as many people as possible to take part in the moon landing, which is why the agency spent years investing in suitable technology. It is only when we look at the details that it becomes clear how costly this mission really was. In order to get the camera to weigh as little as possible, they used not the latest technology but a system from 1928 that recorded images in reduced, monochromatic quality, which meant that the camera weighed not even three kilos.

It was the broadcast itself that was difficult. On the lunar module, a screen-shaped antenna broadcast the event through extremely fine gold threads. It transmitted its signal more than 320 thousand kilometers above Earth. Back then, the best receiver was a radio telescope in Australia that transmitted the data it received—still separated into individual components—to the NASA control center. It was not until it arrived there that the image that would be broadcast to the television stations of the world took shape.

Twenty-eight hours and eighteen minutes: that was how long the ORF's broadcast of its live moon landing show lasted. It was a piece of television history—in 1969, only around half of Austria's households had a television set, but that was still ten times as many sets registered as there had been in the late 1950s.Worldwide, 650 million people watched the movements of the Eagle landing capsule as well as the famous first steps on the moon taken by Neil Armstrong. The live images that the television transmitted from the moon into households everywhere changed the idea of »first-hand experience.« There had not been millions of people there on the Moon, and yet millions shared in the experience of what they were seeing—even simultaneously, because in countries in time zones like Austria, viewers got up at 3 a.m. in order to live through the moment that would go down in history with Armstrong's words: »That's one small step for man, one giant leap for mankind.«

Space travel led to a huge technological boost in the field of visualization—from the images from space that made climate phenomena visible in real time to satellite data that allowed viewers to watch objects on Earth from afar. It was not until the development of this technology—which would become available for civil use two decades later as the Global Positioning System (GPS)—that nature documentary-makers gained the ability to track animal migration routes to the exact meter and to be at the right place with the camera at the right time to capture the thrilling images that make nature documentaries today so fascinating.

It was not least the computer screen that enabled the moon landing broadcast to become an icon. Computer images using measurement data would go on to shape the following decade of science visualization. Whether in animation, models, or graphics—computers, computer screens, and even computer fonts would become symbols of scientificity.

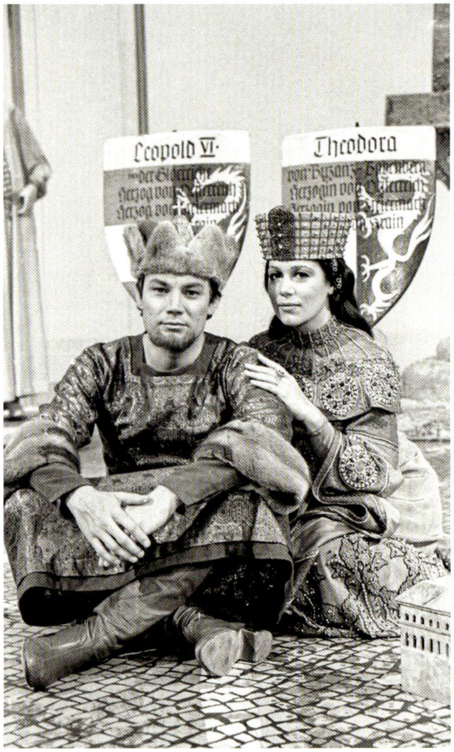

FIG. 10_*Historical reconstruction in the year 1976. Acting star Klaus Maria Brandauer embodies all twelve Babenberg dukes at once. At his side: Hannelore Elsner.*

History Live—Being There in the Past

The station made use of these insignia just a few years after the moon landing, in 1976, to mark Austria's National Day. *Die Babenberger—wir waren 12* (The Babenbergs—There Were Twelve of Us), a two-part show about the history of Austria one thousand years before, combined all of the visual elements available at the time. The focus of this production was the ruling dynasty that, for the first four centuries of its existence, had decided the fate of the country first mentioned in writing as *Ostarrichi* in 996: the Babenbergs.

The home base for this two-part series was a television studio where mobile scenery elements representing the Middle Ages between the tenth and twelfth centuries were set up—a throne, archways, towers, frescoes, model towns. These props could be pushed in and out of the frame. This created an interplay between cohesive scenes and the impression of construction—because the historical portrayals of bygone ages were based on construction and reconstruction. Between these historical decorative elements, there were also video walls that film clips could be played on as well as a computer display that summarized the most important facts. For example, what made the Babenbergs so successful was that this dynasty managed to ensure the continuity of twelve rulers in just three hundred years, while other kingdoms, including the German Empire, were shaped by frequent changes in ruling houses.

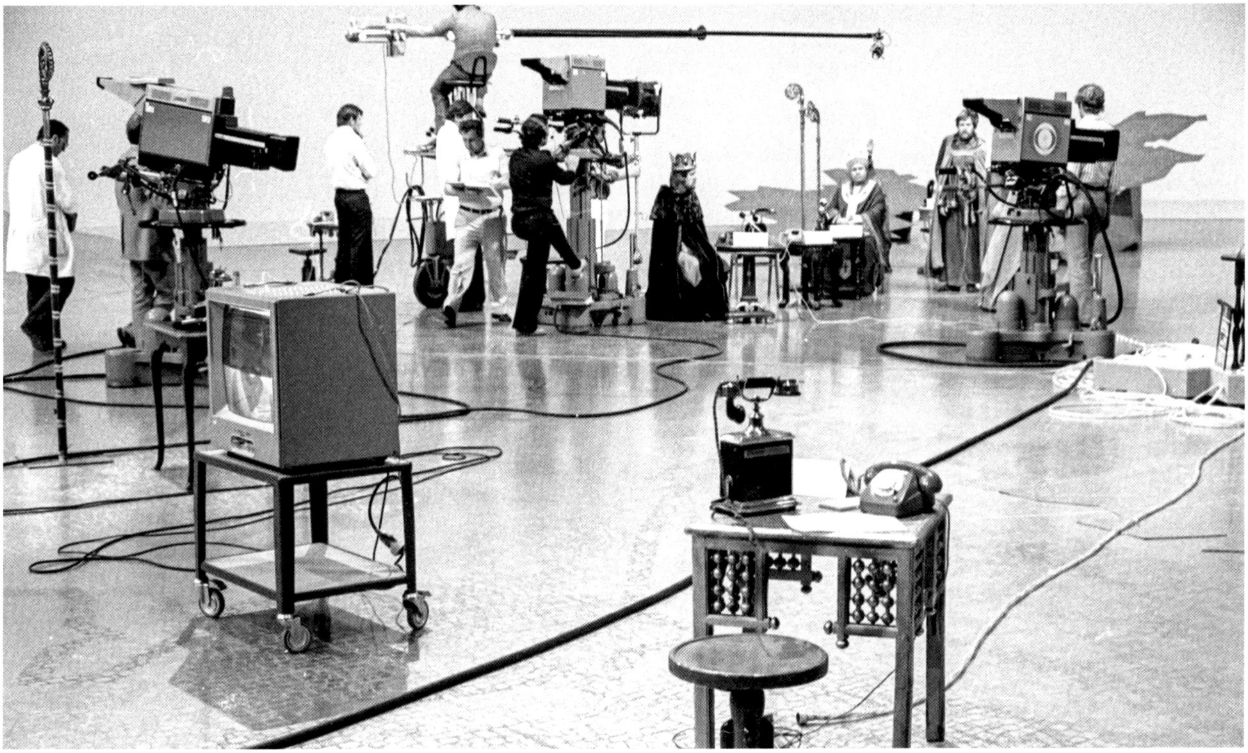

FIG. 11_*The studio as television's multi-tool. Historical props helped to create illusions—close up! Long shots were then used to deliberately break the illusion up again.*

With the computer display, a new visual narrative element focused on the content that was important to documentary-maker Harald Zusanek, who relied on stylistic breaks to make processes clear to viewers. One example of this was a scene about the imprisonment of English King Richard Lionheart, a European scandal in which the Babenbergs played an important role. Here, the rulers of Europe sat in a studio whose floor was decorated with a map of Europe. Each of them sat at a table in their own country—with a telephone that demonstrated communication between the monarchs. Of course, the rulers actually communicated by means of messengers on horseback and secret diplomacy—but that would have required elaborate scenes that would not have been very entertaining. And entertainment was part and parcel of the television concept of visualizing educational content from the very beginning. In this sense, it is not surprising that the show used animation to depict a dispute between a Babenberg and a bishop.

In order to reach a wide audience, the makers of the show also used well-known faces from the ORF: ZIB 2 host Annemarie Berté opened this television event, and other reporters like Klaus Emmerich, Christoph Laschober, and Ingrid Vendl followed. Rising star Klaus Maria Brandauer, who played the roles of the twelve Babenberg dukes, explained the world of this influential Austrian ruling dynasty, sometimes in period costume, sometimes in contemporary clothing. Other members of the cast, playing the emperor, the bishop, his entourage, and his wife, were embodied by up-and-coming TV stars of the time, actors like Hannelore Elsner, Sigmar Solbach, and Günther Strack.

Education Undercover—*Universum* as an Engine of Innovation

From the beginning, nature documentaries were characterized by their ambition to make the invisible visible. Cutting into trees to film woodpecker holes, getting birds used to ultralight airplanes in order to film them mid-flight, and the use of super-slow-motion cameras to catch the world's fastest predators on film mid-chase—the imaginations of nature documentary-makers have seemingly never known any bounds. The English BBC and US channels like Discovery Channel and National Geographic have shown the way. High-quality documentaries with sophisticated visualizations have become audience hits—in Austria too. The *Universum* (Universe) series came into being in 1987 as a result of the great resonance that documentaries had found with audiences. Its motto was unambiguous; as the show-makers crudely put it back then: education undercover.

The initial shows were pilots, and on December 19, 1989, the first real production was broadcast—about an environment that had once been part of old Austria: the Adria. The idea came from Rupert Riedl, who had already caused a furor at the Urania public education cinema forty years before with his underwater films. His research led to the five-part series *Die Gärten des Poseidon* (The Gardens of Poseidon). Alongside Riedl and Walter Köhler, who would later become the *Universum* boss, one of the directors was Alfred Vendl, a chemist who had dedicated his life to visualizing natural phenomena. For the underwater scenes, he had already gained experience using the cameras of his production company. But he also utilized Riedl's know-how.

One of the scenes shown in the first Austrian *Universum* film shows sea urchins moving in the ocean at night. An underwater flash system was built to film it. The devices were installed on the seafloor of a bay near Dubrovnik. Every two seconds, the flash would capture the position of the sea urchins. The movements were then recorded like in a time-lapse. Once the technical requirements had been met, the team endured the shoot in a tent on the coast. But what made them worry about their recordings was that the flash enticed fish, who watched the spectacle in disbelief. The nature documentary-makers were afraid they would be discovered before they had finished filming. But the mysterious phenomenon must have had a deterrent effect—the fish left, and filming continued. The result was a series of unique images. They show the sea urchins »polishing off« rotting algae residue on a rock like little lawn mowers. These were not the only scenes in the film that showed sea life from a new perspective.

The ORF's *Universum* series is renowned for capturing scenes like these and for the Austrian filmmakers who have created ever-fascinating image worlds and prize-winning documentaries with their innovations. Alfred Vendl is one of them and has been a guarantee of innovation for decades. His *Universum* films have tended to be about scientific phenomena like the *Limits of Light*, the *Limits of Perception*, the *Limits of Time*, and the effects of poison. His three-part series about bionics, *Nature Tech Reloaded*, which he realized together with Steve Nicholls, was even awarded an Emmy, the highest award for television productions.

Nature Tech Reloaded utilized modern technology to realize what Otto König had achieved fifty years before Vendl with studio animals, charts, and humorous explanations. For example, in order to explain bird migration, a time-lapse camera,

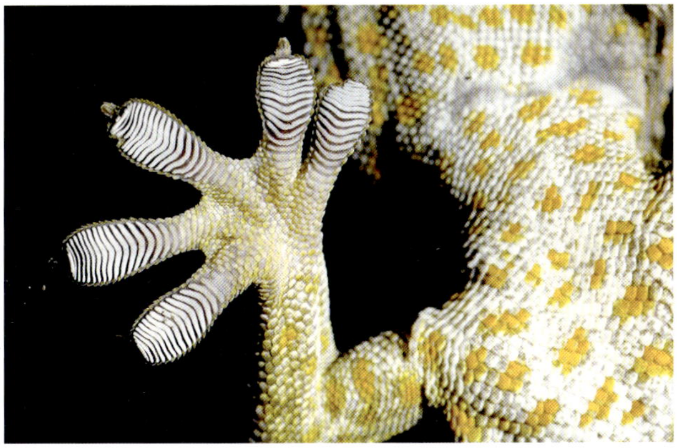

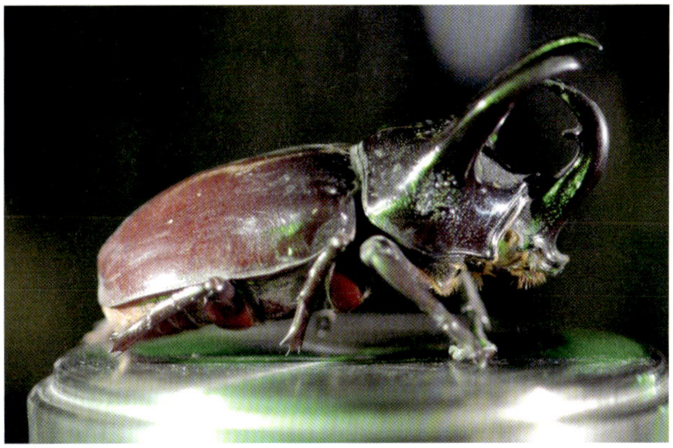

FIGS. 12, 13_*Close-ups of materials analysis. Using innovative macrotechnology,* NatureTech reloaded *became the award-winning highlight of the* Universum *nature documentary program.*

comparative flight studies, and detailed shots of feathers and vibrations made a lavish, innovative overall impression by cinematic means that allowed the audience to learn why bird flight works the way it does.

A project like *Nature Tech Reloaded* is only possible when science and media come together to make knowledge visible. This symbiosis was almost institutionalized in the mid-1990s with the weekly science magazine *Modern Times*, developed by Werner Mück, who would later become Head of Specialist Factual, and moderated by Josef Broukal, an ORF news anchor. The slogan of *Modern Times* was »The future is possible.« New technologies and the internet played a significant role—presented in a defining way during the special show broadcast on Net Night on May 5, 1995, with Broukal and Mück researching various topics live on the nascent internet.

The work I did editing *Modern Times* led to a real boost in my own personal development, as this was where the entire range of the latest camera technologies was being used, such as time-lapses for long-term observations of plants, minicameras mounted on the wing platforms of hailstorm control planes, avalanche animations, and data transfers from the electron microscope—all possible tools were being used to depict scientific phenomena and to help a wide audience to understand them, not necessarily in high-tech quality, but on a weekly basis. This is how they became a stylistic element.

In particular, the reports that I created with the Austrian medical professionals involved in the Austromir space travel project had a huge influence on me. The need to medically monitor astronauts had also led to a boost in medical imaging techniques after NASA's Apollo moon landing program. In 1991, more than a dozen scientific experiments were being carried out within the scope of the Austromir project, a collaboration between Austria and the Russian MIR space travel program.

For the Pulstrans project, Austrian physiologists developed measuring devices that made the strain on the organism visible. Sensors were installed in a vest in order to observe and measure physical strain on the circulatory system by looking at pulse waves and how they changed—presented in the amplitudes of heart frequency variability. Combining this with real images of moving or sleeping people allowed external and internal images to be obtained at the same time. The man heading up the Pulstrans project, chronomedicine expert Maximilian Moser, became a key figure in my own personal attempts to visualize science. Using the measuring devices from the Austromir program, he was able to portray human physiology as a system of vibrations. The dynamically changing heartbeat synchronizes with stress during activity and with regeneration during sleep. Parallel to this, he was able to depict the secretion of hormones.

In his contributions to the *Modern Times* series, Maximilian Moser used vivid experiments to show the impact of vibrations, e.g., the patterns and designs they create when powder or fluid is placed on the outer membrane of an amplifier. This meant that, initially, no special camera technology was required to make the phenomena visible that take place inside us. The graphics of the measuring system then demonstrated the impact on screen.

In the Age of Visuality

After the turn of the millennium, the wheel of technological development began turning faster and faster. In the field of science and nature documentaries in particular, an international competition took shape. Television stations, production companies, and filmmakers tried to utilize technological innovations in order to outdo each other in every production. The development of computer animation and the possibility of combining high-quality shots from reality with digitally generated images opened up previously unknown opportunities for visualization.

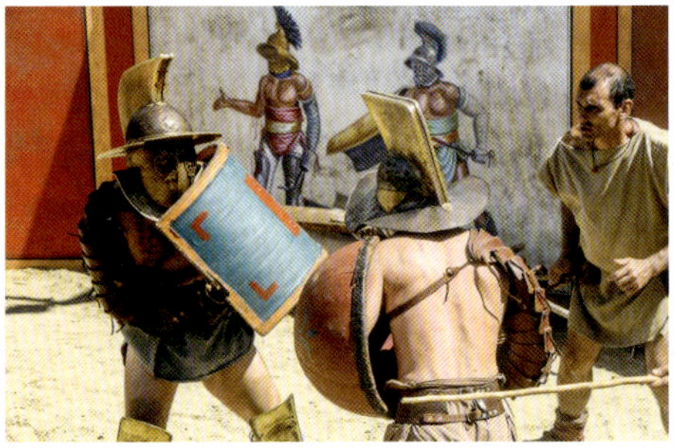

FIG. 14_ *Not a sword-and-sandal in sight: when gladiators fight in* UNIVERSUM HISTORY, *it is based on scholarly expertise.*

Augmented reality, which combines real images and digitally generated animation, has made it possible to bring objects from faraway worlds into the TV studio, such as the different generations of the Mars Rover, the exploratory vehicles that explored the red planet on their missions in 1997 and 2020. For viewers it was a memorable moment seeing vehicles like the Sojourner and the roughly four-times-larger Perseverance rolling through the studio of science anchor Günther Mayr and his weekly *Mayr's Magazine* show. It was a virtual studio, incidentally, where, apart from the moderator, all of the images were entirely computer-generated.

On July 16, 2019, the special show *50 Jahre Mondlandung* (Fifty Years Since the Moon Landing), created by ORF contemporary historian Andreas Novak, made for a special TV experience. Augmented reality technology brought the launcher, originally 110 meters high, into the studio—initially reduced to the size of news anchor Tarek Leitner. This allowed him to show in detail everything that was hidden behind the Saturn V rocket's three layers of casing. Vividly, and close enough to touch, he inspected the command pod, the lunar module, and the engines. One of the five engines, for example, measured 3.2 meters. The ORF graphics were able to turn them all around on their axes, to magnify and shrink them—which is just what moderator Tarek Leitner needed to explain the amazing key facts of the Apollo 11 mission. It was one small step for technology, one giant leap for popular science visualization.

Today's nature documentaries are combining the latest cinematic techniques with methods from data journalism, synchronizing the image and the message. Digital technology has made this possible. Whereas film reels were once required as a storage medium, now all that is needed is a microchip or a mini SD card. These tiny cameras can get closer to animals than ever before. They can be installed in nests,

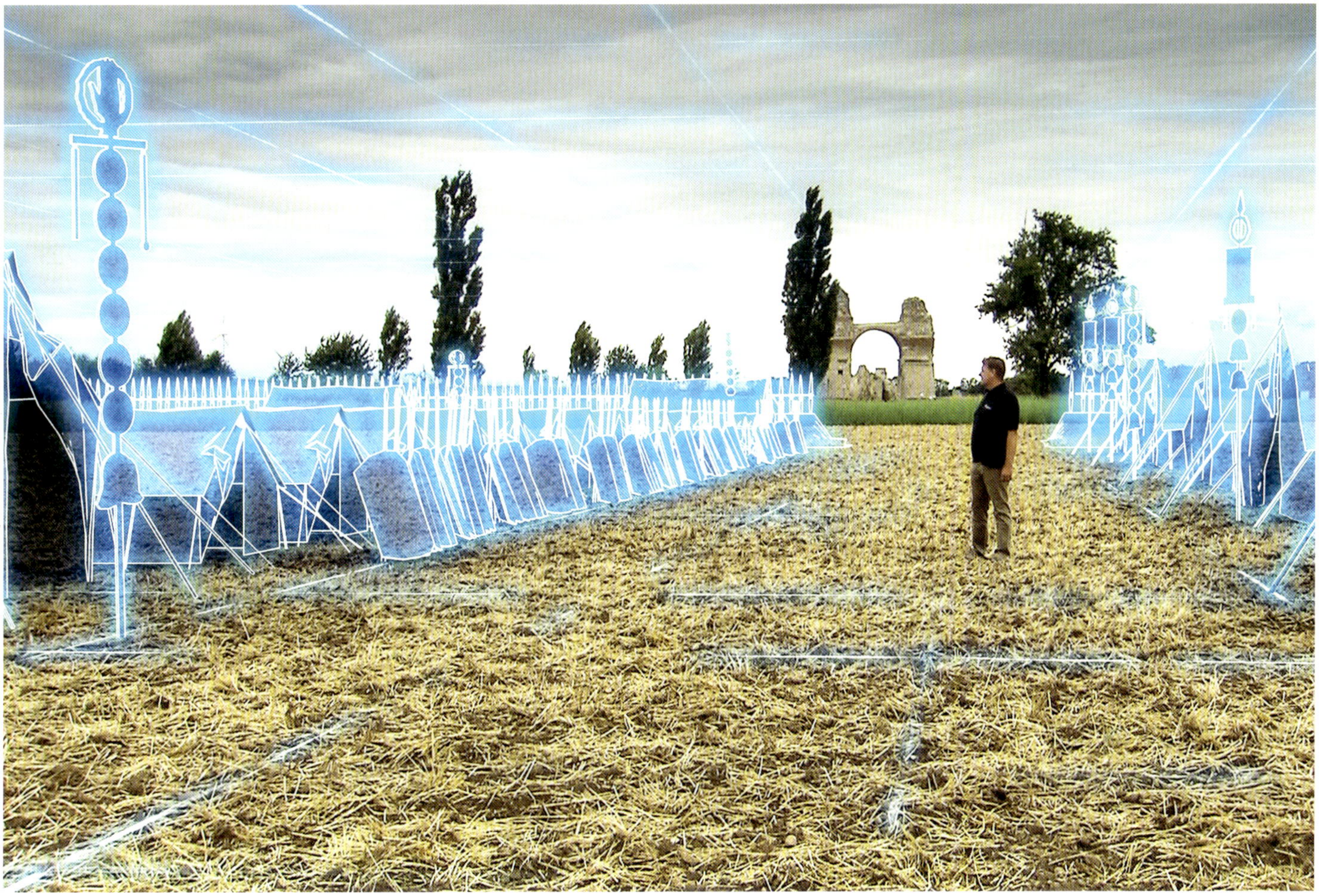

FIG. 15_ *Reconstructing lost cities virtually: with data from archaeologist Wolfgang Neubauer, the Roman city of Carnuntum comes to life.*

© ORF, 2015

caves, right in the ground, or in remote-control robots that have been made to look deceptively realistic. And not just that: mini-cameras are now so small that animals themselves can become filmmakers and reveal the world from a literal bird's eye perspective.

Austrian nature documentary-makers Mario Kreuzer and Leander Khil rely on the value of scientific news. In their film about migratory birds in the age of global change, it is not just their impressive shots of bird migration that count, which they film using a 12K camera (3 × ultra HD); rather, it is about demonstrating the impact of global warming, which reveals itself in changing trigger factors. Some birds are no longer leaving Austria at all because it does not get cold enough anymore — like the greylag geese at Neusiedler Lake, which were still spending their winters in North Africa thirty years ago. The drying up of important saltwater lakes in Burgenland is also a factor in the film — as it has erased an important starting point for bird migration from the map.

In order to make flight paths and their changes visible in their latest *Universum* production, the two bird experts are using incredibly small GPS sensors, which are attached to the birds like bird bands. One of their protagonists is the ruff, a bird

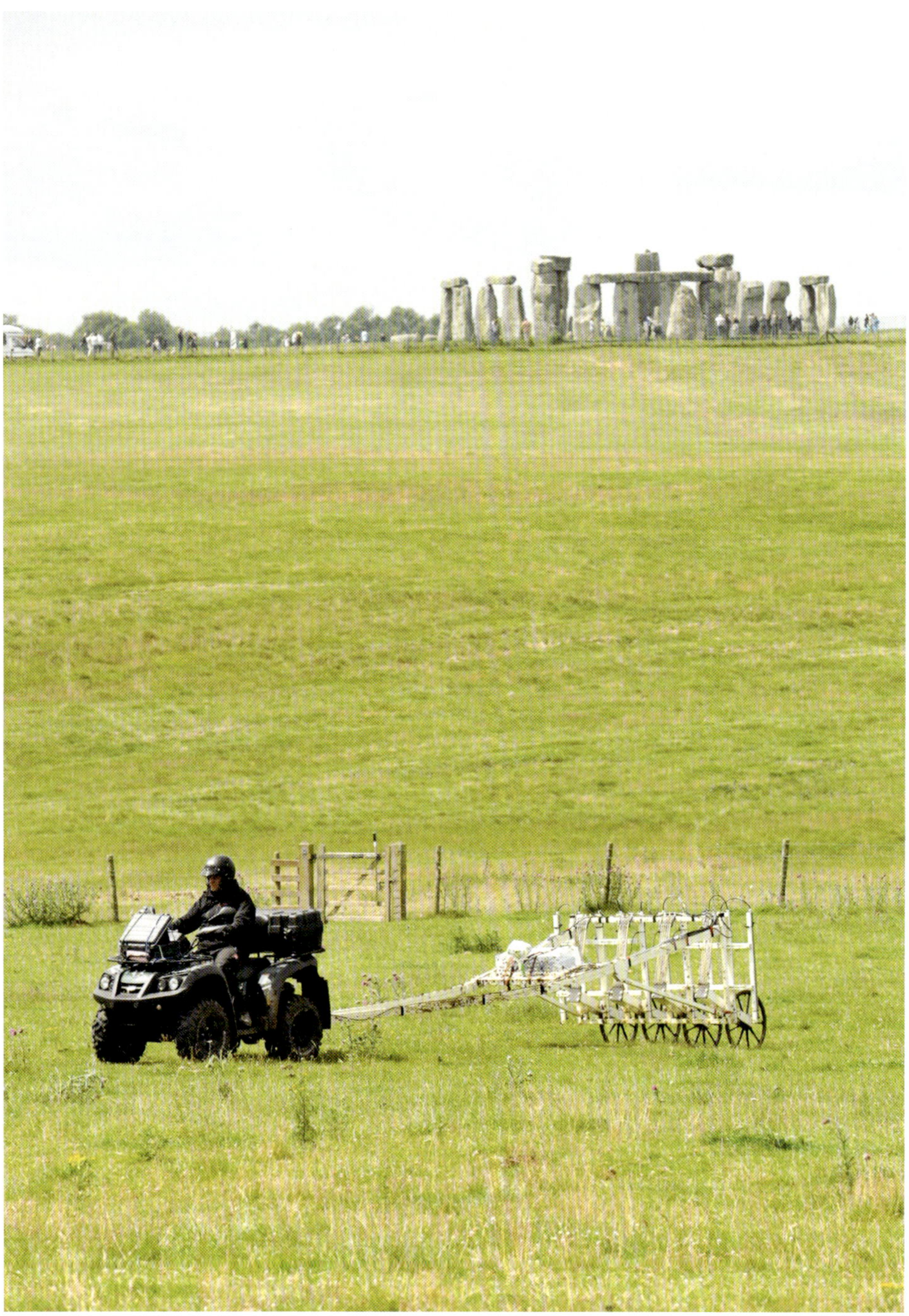

FIG. 16_ *Full speed into the Stone Age: this vehicle with a ground-penetrating radar on its trailer reveals all kinds of things that are yet to be found beneath the earth at Stonehenge.*

species that commutes between the Arctic and Sub-Saharan Africa, stopping off in Austria at the Burgenland Seewinkel. The sensors weigh just five grams and are equipped with a tiny solar panel that supplies them with energy. The filmmakers attached these GPS sensors to seven ruffs. Some of them were lost, but the overall findings are very impressive. One bird flew 1,200 kilometers non-stop—in one night. Another bird going by the name »5K HD« put away 6,500 kilometers from its resting ground at the Seewinkel to its breeding ground on a Russian island in the polar sea. It was only with this ultralight technology—both in the measuring system and in the cameras—that the *Universum* filmmakers were able to gain and convey completely new insights into the behavior and migratory paths of birds.

History education at the ORF has also continued to develop in the last fifty years. The effort required to reconstruct and visualize historical events on the basis of the latest research findings has increased. While actors were still being identified with name plates in 1964, the series *Universum History*, directed by Caroline Haidacher, is now bringing the past back to life by means of careful casting, masks, and costumes. Visualization techniques developed in scientific research are playing an increasingly important role here as well.

One example of this is the ground radar invented by Wolfgang Neubauer at the Ludwig Boltzmann Institute. This noninvasive archaeological technology allows invisible worlds to be made visible—layers deep under the earth's surface that cannot be excavated without wreaking destruction. Neubauer's research has regularly featured in *Universum History* documentaries—in projects about the Roman settlement of Carnuntum and the Viking slave trade, for example.

FIGS. 17, 18_ *Newsroom 2019: the modern television studio is made up of virtual elements—only the host is real. Augmented reality also makes it possible to bring a rocket into the studio.*

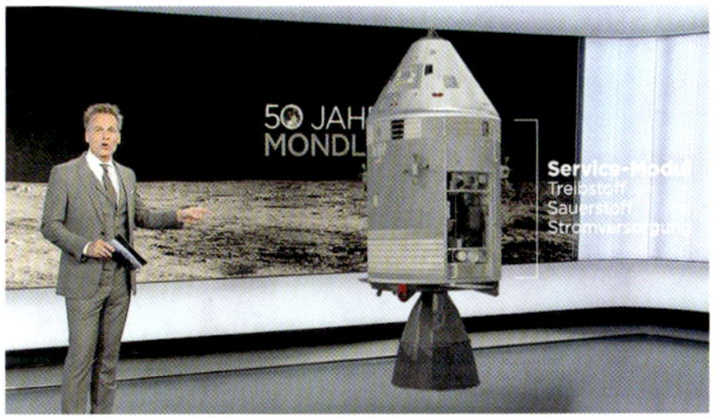

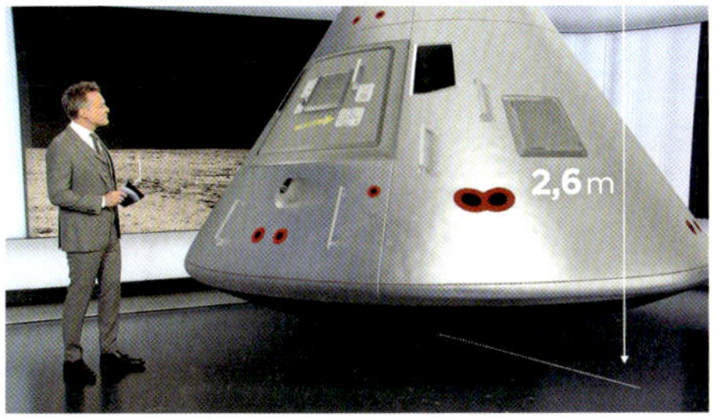

One special question was at the heart of the documentary *Der Aufstieg der Habsburger—die Schlacht am Marchfeld* (The Rise of the Habsburgers—The Battle on the Marchfeld), which aired on January 6, 2022. In 1278, the final conflict between Bohemian King Přemysl Ottokar and Swiss Count Rudolf von Habsburg, the leader of the Holy Roman Empire, took place on the Marchfeld—and was a turning point in the history of Europe. Those who look at the battlefield today might think that Ottokar had set up his headquarters at a strategically unfavorable site: a sloping field. It was Wolfgang Neubauer's ground radar that brought new insights: his images showed that there was once a castle at the site of Ottokar's headquarters—the ideal place to stage a battle. Without this noninvasive method, which provides images of structures beneath the earth's surface, it would not have been possible to obtain this information. Highly specialized visualization methods from research itself played a crucial role here in obtaining new knowledge and conveying it to a wide audience.

In the mid-1970s, when I was hiking around the fields of Dürnkrut and Jedenspeigen with my father to get an idea of the place where the events that had shaped Europe and Austria seven centuries ago had transpired, none of this was known. Everything I knew about that period as a twelve year old came from my school books, the explanations provided by my history teacher, and those supplied by the history show *Die Babenberger—wir waren 12*, broadcast in 1976.

Just like the broadcasters present at the birth of Austrian television, we at the ORF's Education, Science, and Current Affairs department are still trying to convey knowledge in a simple, understandable way using all of the visual means available to us, and we will still be trying tomorrow—together with researchers from Austria and the world. Technology has made so much possible for us, but even more so it is the irrepressible urge to make knowledge available to a wide audience that motivates us—making complex phenomena visible in order to allow people with no prior knowledge to understand them. Putting knowledge in the picture—that's our mission.

Awards in the Time of COVID

Rose Anderson

»Tell Me a Story« is an invitation to step into another world and go on an adventure. Stories can lift our spirits, drive change, and give us hope, Storytellers explore. enlighten, enchant, and engage—make us laugh, make us cry, make us aware, and remind us we are all connected.

Prologue

For over a decade, I have headed the TV & Film Awards and Radio Awards competitions at New York Festivals. Every year, we receive stories from over forty countries across all platforms in five genres: Entertainment, News & Sports, Documentary, Promos, and Brand Image. Hundreds of documentaries come from indie filmmakers, production companies, and networks. And through our award-winning Grand Jury, New York Festivals honors their achievements as storytellers. Everyone wants recognition. We all want to be valued and our work to be appreciated by our peers.

Before I came to New York Festivals, I led the Sports Emmys and was director of production for the TV Academy. My production career included multiple Olympics (Emmy Award), entertainment specials (another Emmy Award), drama series, and arts programs (multiple Emmy nominations).

Nothing much happened in the small New England town where I grew up, but for a few nights in the summer, I could look out a bedroom window over the elm trees and see the night sky light up with fireworks from the fairgrounds. I would focus with all my being and try to remember all the colors and shapes and sounds just as they were so I could revisit them in my mind and keep them in my memory just to recapture that sense of wonder.

But what is the vocabulary used to communicate those images and sounds? How do you describe them to others? What language do you use to bring those ideas into being? How do you bring the story you see in your imagination to life? How do you achieve quality? And finally, how do you quantify quality?

I learned that the scenic elements have very specific terms in the world of film and TV production. The vocabulary includes names for shot selection, camera movement, and job titles: wide shot, close-up, extreme close-up, establishing shot, two-shot, medium close-up, silent audio, graphics, sound effects, audio playback, video playback, pan, tilt, dolly, technical director, senior audio, associate director, director of photography, and many more. I found out that, on set or in a script, they are abbreviated: WS, CU, ECU, EWS, MCU, MOS, GFX, SFX, SOT, AT, VT, 2-shot, pan, tilt, dolly, TD, A1, AD, DP. This is a playbook for a team where visual memory, visual literacy, and creative visualization are the skill set every member excels in. As technology changes, new acronyms come into use as storytellers take advantage of every developmental aid to enhance their creative process. Art and craft and science come together. This is equally true for drama, live event, hard news, documentary, sports, and image films.

»Tell Me a Story« is an invitation to step into another world and go on an adventure. Stories can lift our spirits, drive change, and give us hope, Storytellers explore, enlighten, enchant, and engage — make us laugh, make us cry, make us aware, and remind us we are all connected.

When storytellers tap into our collective memory, we expand our worldview. That magic combination of sound and picture stays long after the last frame fades out. So, here's to all of today's storytellers: the filmmakers, sound artists, journalists, writers, cinematographers, directors, and producers whose passion and imagination transport us.

There is a three-part harmony in successful storytelling: the storyteller, the story, and the audience. The storyteller shapes the narrative and point of view, determines the rhythm and pace. Since this is a collaborative effort with checks along the way at every stage of production, preproduction, postproduction, and air, the characteristics required are patience, focus, organization, long visual memory, and creativity.

The story must have enough material to sustain interest. The framework has a beginning, middle, and end. We meet the characters. They set in motion certain events, moving the action forward until there is a climactic moment. Situations arise and the plot moves forward. We learn more about the main actors and how who they are determines what they do. The effects of those actions play out until the storyline comes to its natural conclusion.

The audience chooses to tune in and is ready to be immersed in the story. With the multiplicity of choices for viewers in today's world, the high bar set for keeping the attention of the audience through imaginative storytelling puts the value of creative visualization in prime position. The mechanics can be daunting, but the theater of the mind is always open.

Happy Birthday, USA

As the most junior member of the ABC News team for the US Bicentennial, I experienced for the first time the amount of planning that went into multi-hour, multi-venue events. The planning was operational, logistical, visual, and editorial. This was before cell phones, Zoom, and the cloud made distant communication so easy. Creative visualization meant long meetings as engineers, directors, and producers talked about getting the signal from one place to another through multiple control rooms with limited input capabilities. Many questions, many scenarios, many possibilities, and many solutions were raised and answered from many perspectives.

I learned the vital importance of specific detail. Every remote location was scouted by teams in advance. We created multipage rundowns describing minute by minute the order of the remotes interfacing with anchor control positions. With as much rehearsal as could be arranged, the central on-site team prepared for what might happen in just a few hours. Scripts were written for anchors who were told by stage managers where the program was going next. Sometimes, the pattern did not follow the written-in-advance formats. And it all appeared effortless to the millions of viewers in the audience.

The nationwide events were taking place in real time all through the day. The phones were ringing non-stop as every remote producer wanted to know when they would be appearing live. Their teams were on standby or loose standby waiting for their cue. The energy of live transmission was fueled by teams of hundreds who had been briefed on the executive producer's creative vision. And the story of the 200th anniversary of the United States of America (July 4, 1776, to July 4, 1976) was reported live from coast to coast. From the majestic tall ships to glorious displays of fireworks, it was a thrilling introduction to live television and a once-in-a-lifetime opportunity.

Today, much is said about bandwidth, meaning the capacity or capability for information delivery. The bandwidth for creative visualization measures the intellectual ability to select and control multiple inputs in a pattern containing multiple possibilities across a timeline. As live events are transmitted by multiple video and audio sources, the creative visualization process requires decision-making at a very fast pace. Staying one step ahead just isn't enough runway to produce an exciting and clean show. Control rooms feature rows of monitors, each with its own individual output: cameras, venues, content playback, graphics. The creative visualization process at this pace means that your brain is ahead of the present (on air) and in the future (preview and all other sources), constantly communicating over headsets to members of the production team at maximum speed and with minimum producer indecision.

When in Rome

The call came from a sports producer who knew that I knew something about music. Was I available for just a few days to work on a one-off concert, to create ancillary material for a future broadcast? We would be recording an introduction on site with our host and a behind-the-scenes interview with the performers. A different production unit would be recording the concert. Fitting in to the performers' schedules, the interview would be taped the day before the concert in a hotel suite. The host stand-up would be recorded backstage at the arena prior to the start of the event.

The behind-the-scenes interview setup had five participants: four interviewees and our host. At the conclusion of the interview, they talked among themselves: José Carreras, Luciano Pavarotti, Plácido Domingo, Zubin Mehta, and Beverly Sills. The arena was the ancient Baths of Caracalla. The performance would take place the night before the 1990 FIFA World Cup Final. This was the start of what would become an operatic phenomenon: The Three Tenors. But that afternoon, they talked about the staging for the concert itself and their creative visualization process as artists and storytellers. Maestro Mehta let them know that the concert impresario had decided the rehearsal that night would be closed: two orchestras on stage and no one in the vast outdoor audience space. After declaring that they could not rehearse without someone to sing to, it was decided that we would be the audience: they would sing to Beverly.

And so, that night, they made their entrance onto the enormous stage with two full orchestras in place. As Maestro Mehta raised his baton, they sang the entire program — including the encores — to just a few of us. The songs were the story, they were the storytellers, and we were their audience. We sat right in front of them. They serenaded us under a huge moon on a sultry July night in Rome with aria solos and trio medleys featuring »Nessun Dorma« and »O Sole Mio.«

The next night, the Baths of Caracalla were at capacity with VIPs, dignitaries, and fans celebrating the World Cup Final. As we made our way to a wing backstage to tape Beverly Sills's introduction, the atmosphere was electric with anticipation. She was perfect. We then took our seats among the six thousand members of the audience as the concert began. It was every bit as dramatic as the artists had hoped for during their rehearsal. They commanded everyone's attention — they were at the height of their artistic powers, and the audience knew this classical concert was something very special — an operatic tour de force.

Perhaps the most significant international venue celebrating excellence is the Olympics. The entire world stops and comes together to watch competing teams give their all for their sports and their countries. As part of the production teams of five Olympic Games for three US networks (ABC, NBC, CBS), my responsibilities were varied. They spanned the creation of athlete profiles and technical content to venue production and editorial supervision at Opening Ceremony and Closing Ceremonies.

Each host country creates a magical spectacle welcoming the athletes and introducing itself to the world. For the creative directors, it is a feat of meticulous organization and a triumph of breathtaking visualization. The Opening Ceremony for the 2012 Games of the XXX Olympiad in London earned the Grand Award from New York Festivals Grand Jury.

Awards in the Time of COVID

Storytellers Gala 2021

When travel became restricted and lockdowns were in force, our usual award ceremony could not take place as planned. For over a decade, our strategic partnership with NAB Show, the world's largest and most comprehensive convention encompassing the convergence of media, entertainment, and technology, had made it possible for New York Festivals to give visibility to storytellers and to salute leaders in the field who bring content to life in new and exciting ways.

New York Festivals honored our commitment to salute the storytellers whose work so impressed the grand juries by streaming a virtual ceremony on October 12, 2021, keeping the date of the canceled NAB Show. The ninety-minute special event featured highlights and acceptances from one hundred teams coming from their homes all around the world.

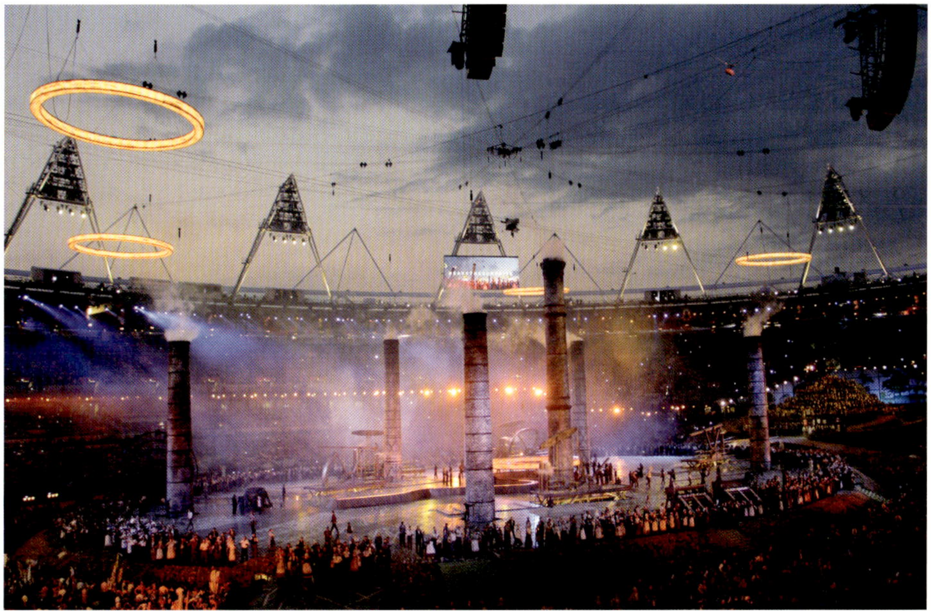

FIG. 1_*2012 Games of the XXX Olympiad Opening Ceremony*

Courtesy of Done+Dusted, 2012

The creative visualization process with accompanying operational and logistical concerns required concentration and communication. What would the plan be? What would the gala be? How would decisions be made?

Together with producing partner PushDesign Creative Director Rob Tobin, I first reviewed previous shows that had been held in Las Vegas. We would have to transition to a prerecorded event with a live feel. We deconstructed the elements of those shows that made the night so exciting for award-winning attendees. Radio and TV & Film award-winning programs were grouped together by genre. Acts would be structured with in-show bumpers and announcer VO.

PushDesign has played an integral part in all the previous live galas for New York Festivals. Founded by award-winning director, producer, and editor Rob Tobin, this video production company blends creative storytelling with technology. With experience in live productions, motion graphics, preproduced packages, interviews, highlights, and postproduction, the end result is always compelling and high-quality.

We visualized minute by minute which of those elements—presenters, billboards, highlight montages, walk-up clips, and acceptances—could be adapted and incorporated into the virtual streaming event. The key was the theme of celebration: a salute to the winning teams all around the world. With cloud technology, our ideas became possible in real life.

To stylize a consistent look, we visualized a clip holder to accommodate the different file sizes of the remarks that would make up so much of the material. Every aspect of preproduction and production was included in a strict timeline. Individual results from the Grand Jury were sent under embargo one month before the gala. Medal winners were invited to send brief video remarks by September 24. Technical specifications were sent to all participants. Files were shared, and the postproduction process for the highlight montages began.

We taped the on-camera introductions one weekend afternoon outside in my garden in New York City. I was the host and Rob was the DP. It was just the two of us. The technical setup was one camera, one monitor, and one mic. I tell everyone that I am living a nineteenth-century life made possible by twenty-first-century technology.

One hundred production teams sent us their remarks, and many included champagne salutes. Sponsors shared files for billboard creation. The graphic packages were shared and proofed. Every acceptance file was color corrected to give the program a consistent look. The show rundown shifted to include all acceptance remarks. Once the acts of the program were completed, screeners were shared for proofing. The final audio session adjusted levels and set a uniform textured soundscape.

We made the decision to post the gala file on New York Festivals' Vimeo channel, available via a landing page on our site. The complete program file would need to be ready early in the day of October 12 and go live for public viewing at 6 p.m.

The 2020 Games of the XXXII Olympiad in Tokyo were postponed to 2021 due to the pandemic. It was the biggest storytelling event of the year. The seven thousand hours of coverage featured across the all the platforms of NBC captivated everyone who tuned in. The drama of athletes competing against the world's best was matched by the emotion of their families who were watching from home.

At the top of that decision-making pyramid was the woman who was the 2021 New York Festivals Lifetime Achievement honoree — Molly Solomon, Executive Producer & President of NBC Olympics Production and Executive Producer for the Golf Channel. NBC Host Mike Tirico made the presentation from the anchor desk at the Ryder Cup, and that segment, which included NBC Sports Olympic highlights, set the heartfelt tone for the exceptional program.

In a year marked by disruptions of every kind, New York Festivals implemented new categories for social justice across all genres and platforms. Filmmakers and storytellers from around the world responded with thought-provoking work, and their acceptance remarks in the program added notes of gravitas and gratitude.

The Grand Award winners included a final championship season, an unlikely school, a portrait of country life, a meditation on meaning, an escape from a war zone, and a once-in-a-generation performance. Four companies earned Industry Awards for their outstanding bodies of work.

With one thousand views and 2,700 impressions, our first Storytellers Gala at home was a success.

Media today is international, and there is an international language of excellence. Over the years at New York Festivals, award-winning entries have been submitted in Arabic, Chinese, Czech, Danish, Dutch, English, Farsi, Filipino, French, German, Greek, Hindi, Italian, Japanese, Korean, Malay, Norwegian, Polish, Portuguese, Romanian, Spanish, Swedish, and Welsh.

Our TV & Film and Radio Grand Juries — award-winners themselves — volunteer their expertise and time for the selection process. We look for wide experience, demonstrated achievement, and geographic diversity in our jury members as a way of quantifying fairness in the selection process. About forty percent of the Grand Jury members are women.

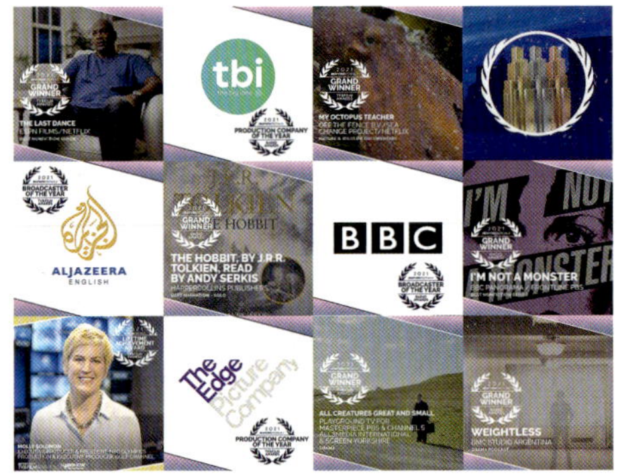

FIG. 2_*2021 New York Festivals Grand Award and Special Award Winners*

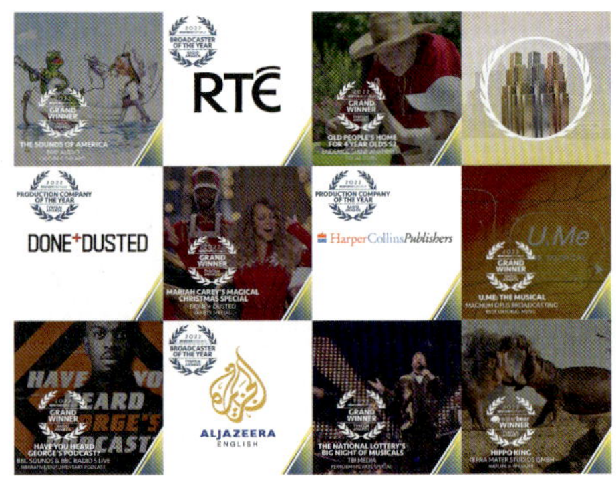

FIG. 3_*2022 New York Festivals Grand Award and Special Award Winners*

Awards do more than keep score. They honor creative excellence and the mastery of craft. They inspire, they encourage, they reward. They help build careers. They can change your life. They mark achievement. They reward passion, courage, and intensity.

For over a decade, New York Festivals has recognized some of the driving forces in the television and film industry—hands-on leaders with a body of work that has made a lasting impression. Their creative visualization prowess is of the highest level.

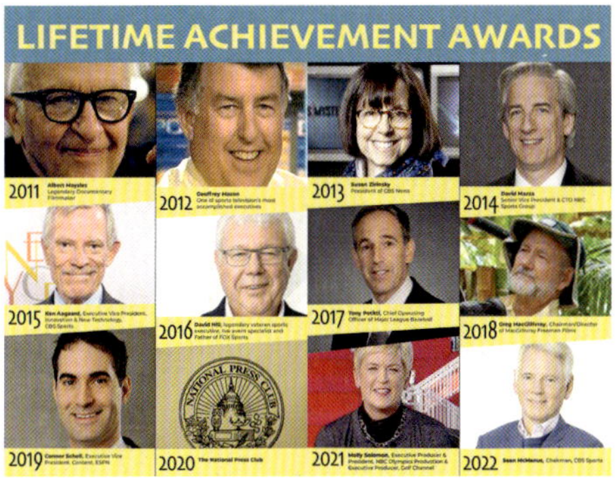

FIG. 4_*New York Festivals Lifetime Achievement Award Winners*

© New York Festivals, 2022

Storytellers Gala Part Deux 2022

We were very excited to plan the 2022 Storytellers Gala, resuming our in-person event at NAB Show in Las Vegas at the end of April. Our ceremony would have two new features: the debut of the iconic NYF Tower in hand-polished Gold, Silver, and Bronze for all trophy-winning teams and the inaugural WaterBear Award for storytellers who are using their talents to save the planet.

Following NAB Show's lead, we sent out a survey to our entrants about their attendance plans. The results showed that there was so much anxiety about the international travel situation; an overwhelming majority of our potential attendees sent their regrets. Only ten percent said they would attend in person. In the face of such an unexpected outcome and with the goal of honoring our prestigious content creators, our team went back to the drawing board to revisit the previous year's alternate scenario of a pre-taped show which we would stream on April 26 at 6 p.m. We really wanted to see each other that year in Las Vegas at NAB Show, but instead we were once again scattered all over the planet, although we did have an opportunity to come together virtually.

This time, we taped indoors with a crew and set for a celebratory evening look. With multiple cameras and full lights, we recorded at »the magic hour« so the iconic Empire State Building would be our backdrop.

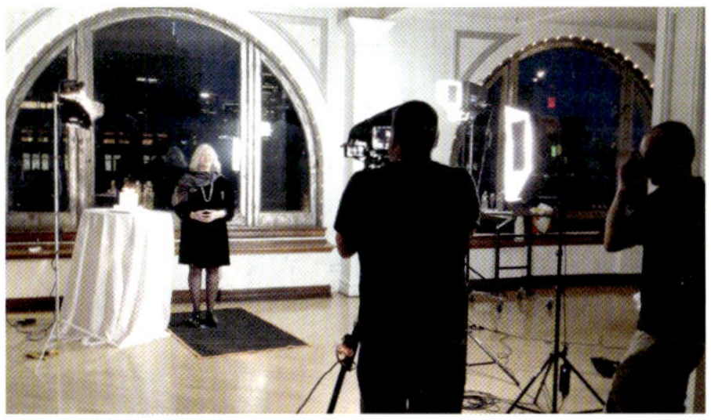

FIG. 5_*2022 New York Festivals Storytellers Gala, Behind the Scenes*

© New York Festivals, 2022

The turnaround timeline was shorter, putting more of a premium on organization and visualization. Winning teams were sent very specific instructions with technical and length guidelines for their remarks. In all, we received almost double the number of acceptances from the previous gala as so many teams wanted to be part of the New York Festivals virtual community.

The companies who took home honors are known for their thought-provoking films and groundbreaking content, their level of dedication, and mastery of their craft. In the face of challenges that impacted every aspect of production and everyday life, they created amazing work, and we included highlights in the ninety-minute gala.

Year in and year out, the most popular shows with the highest ratings are sports broadcasts. From college football and the NFL on CBS to college basketball and March Madness, everyone tunes in for the competitive drama that plays out in real time and goes into overtime for an encore of excitement.

The architect of that programming was our 2022 Lifetime Achievement honoree: Sean McManus, Chairman of CBS Sports. CBS Host Greg Gumbel made the presentation from CBS Sports Broadcast Center in New York.

While trends come and go, excellence never goes out of style.

The Dance of the Spike with Sugar Binders

Yoo Jin Oh, Peter Hinterdorfer, Stefan Mereiter

Peeking into the molecular world of the SARS-CoV-2 spike protein and its sugar coating—a journey toward the real-time visualization of a prospective antiviral therapeutic using atomic force microscopy

The Hidden World of Glycans

Every cell in our body is covered with complex sugar molecules called glycans. These glycans are not as relevant for the metabolism and energy production as the so often-feared carbohydrates that we take up from food, but they are important in defining how our body's cells interact with each other. These glycans are very stable structures and can be found anchored onto proteins on the surface of cells. One important function of glycans is allowing our immune system to discern our own cells from pathogens. For this purpose, immune cells can distinguish the glycans produced by our body's own cells from those that have been produced by, for instance, bacteria, animals, or sometimes even by another individuals (e.g., blood groups are based on different glycans). Unfortunately, pathogens often find ways to outsmart our immune system and sometimes even manage to use our glycans in their favor.

There are still many open questions to be answered with glycans. How are glycans formed? How exactly do they regulate cellular biology? Which changes in glycans occur in diseases, and can we use them to detect or treat those diseases? How do pathogens use glycans to cause diseases? In the research field of glycobiology, we try to find the answers to these questions. However, while these are important questions that could help us to discover new ways of detecting and treating diseases, we are still facing big challenges in the investigation of glycans. On the one hand, glycans are very small, just a few nanometers in size, so they can't be seen, not even using the most sophisticated light microscopes. On the other hand, they are structurally complex and very diverse. A single cell can carry hundreds of structurally distinct glycans on its surface, and even a single protein, of which cells have millions, can be modified by tethering multiple different glycans at the same time.

For decades, researchers around the world have been trying to elucidate the structures of glycans. To do so, we use indirect methods. This means that, since we cannot see glycans directly, we use, for instance, chemical color reactions that only occur if glycans with a certain structural feature are present. Or we use a method called mass spectrometry, with which we measure the mass of a glycan and fragments of that glycan, inferring how the structure most probably would have been. These methods are very informative to glycobiologists as we can add up this evidence and put it together like a jigsaw to understand which glycan structures are present on which proteins of which cell. At least in theory. Sometimes biology can be exceedingly complex, making it almost impossible for us to gather enough information to finish the puzzle.

The Challenge

In 2020, when the coronavirus SARS-CoV-2 became pandemic, researchers around the globe joined forces to understand this new pathogen and to find treatments and vaccines in as little time as possible. Working in the group of Prof. Josef Penninger, who was already well known for his work on the original SARS virus, it was obvious to me (Stefan Mereiter) and my colleague David Hoffmann that we needed to make a contribution to these global efforts. As trained glycobiologists, we were trying to understand the role that glycans play in the infection process of SARS-CoV-2.

All viruses, including coronaviruses, need to get access to a host cell, which they then take control of and change in such a way that they produce and release new viruses. In order to get access to the host cell, the virus needs some sort of molecular key. In coronaviruses, this key is what is referred to as the *spike protein*. The name already gives it away: spike proteins stick out like needles from the otherwise round virus particles. Our immune system acts as a guard and is trying to prevent this from happening. Our immune system is very well-versed in detecting pathogens as something different from anything that should be present in our body. The spike protein of SARS-CoV-2 is therefore a liability for the virus as it is clearly different and therefore exposed on the viral surface. It is an easy target for our immune system to catch in order to clear the virus. Now, this is where glycans come into play once more. The virus has evolved to hide the spike proteins in a cloak of glycans. As mentioned initially, glycans can be used to detect pathogens — this is unfortunately only true if the glycans of the pathogen are different from our own glycans. However, since viruses infect our cells, and our cells are the ones producing the virus, the virus cloaks its spike proteins, of course, with our own glycans. This is a very cunning trick, which not only hides the underlying spike protein but also covers the virus in glycans that look like they are human — a wolf in sheep's clothing.

Aware of the importance of glycans for coronaviruses, we asked: What if we could find something that detects not the spike protein but its glycan cloak instead? In other words, can we spot the wolf by its sheep's clothing? For this completely new attempt at treating a viral disease, we needed to find a molecule that could specifically bind those glycans with which the virus is coated. We took nature as a template and produced more than 140 known glycan-binding molecules, so called lectins, in the lab. After screening all 140 of these lectins, we were indeed able to identify not one, but two new lectins that specially bind to the glycans of the SARS-CoV-2 spike. It was an exciting moment: were there two potential next-generation therapeutics that could be used against the virus?

In the next step, we were faced with the challenge of characterizing the interactions of our two lectins with spike proteins. This was important in order to understand why the lectins bound the glycans of the viral protein, which glycans of the virus were bound, and how strongly they were bound. These were important questions that would help us understand if the lectins we had discovered could really be used as a therapeutic. Which brings us back to biology sometimes being an exceedingly complex puzzle to put together. It turned out that each spike protein forms a complex with two other spike proteins, forming a *trimer*. And since the spike protein is, as explained before, coated in glycans, this protein complex doesn't just carry four or five glycans, but sixty-six glycan sites to which our lectins could bind.

This was a nut too hard to crack indirectly. And this was when we needed to find a new way to analyze the nanoworld — not indirectly, but directly.

The Solution

The Biophysics working group (Prof. Peter Hinterdorfer, Dr. Yoo Jin Oh) uses the cutting-edge technology of atomic force microscopy (AFM), which can practically touch a single molecule to image or to measure the interaction force between two single molecules. During this pandemic situation in particular, the SARS-CoV-2 virus is of key interest in all fields of science. So, we decided to contact Prof. Penninger's group to establish a collaboration between medical scientists and biophysicists working at the nanoscale. Together we wanted to measure which binding forces and how many bonds occur between lectins and the spike protein, and also film these bonds. Lectins can bind to sugar molecule structures like the ones attached to the spike protein. Our collaborative research transforms the nanoscale of a striking biological phenomenon, the relationship between SARS-CoV-2 virus and human lectins, into an interactive visual signal in unprecedented detail.

The Atomic Force Microscopy Technique

The latest advances in atomic force microscopy (AFM) provide us with powerful tools to explore the physical and mechanical properties of cells, viruses, and proteins at nanometer resolution. Most importantly, AFM provides ideal conditions for the nanoscale imaging of biological specimens in their native, physiological environments and for elucidating processes occurring at the interface between viruses and cells. With respect to biological imaging, biomolecular recognition, and the localization of specific binding sites, bio-AFM has now developed into a sophisticated method. It also shows great potential and superior performance for force-probing the strength of molecular bonds and the nanomechanical properties of biomolecules at the single molecular level.

Originally invented to visualize atoms, AFM is an ideal tool for resolving biological structures and is also well suited to analyzing its biochemical and material properties in aqueous media. The measuring principle of AFM can be described in a nutshell as follows: performing like a record player, a sharp probe tip mounted on a micro cantilever scans a surface horizontally, line by line. Due to the interaction between the tip and the surface, the cantilever is deflected and generates a topographic image. Topographical images of surface structures reach submolecular

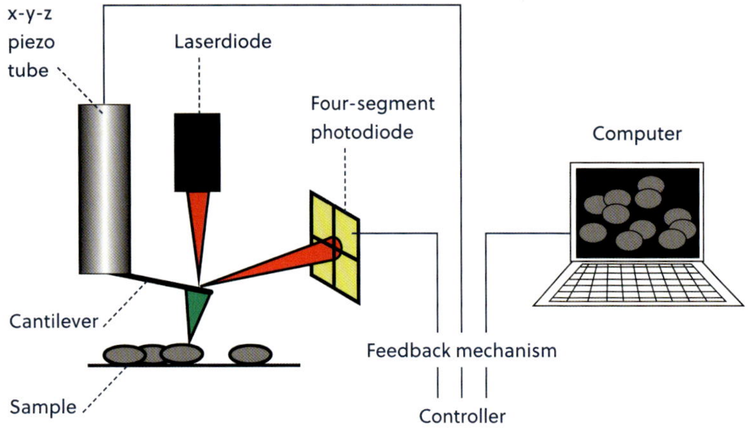

FIG. 1_*Schematic of an atomic force microscope. The working principle of an atomic force microscope is similar to that of a vinyl record player. A cantilever with a sharp tip scans over the surface. Individual molecules (shown as gray oval shapes) are detected as »mountains« by the deflection of the cantilever arm. The deflection is magnified by a laser beam. The computer reconstructs the topographic image from the deflections of the laser beam. Basically, AFM images are generated by measuring forces between a sharp probe (cantilever tip) and the sample surface at a very short distance. Forces between the tip and the sample lead to the deflection of the cantilever. These forces are dependent on the spring constant of the cantilever and the distance between the probe and the sample.*

resolution at the nanoscale in real time and under physiological conditions with minimal sample preparation efforts.

In addition to structural knowledge, forces are of fundamental importance in biology. Many processes on the cellular surface rely on molecular forces, which are a complex interplay of chemical, biological, and physical interactions. Elucidating when and where certain interactions determine cellular processes is key when exploring the forces involved in viral infection. Bio-AFM opens up the possibility of measuring the inter- and intramolecular forces of biomolecules on a single molecular level with a high force-detection sensitivity of a few piconewtons. In what are called *single molecule force spectroscopy measurements*, we visualize molecular interaction forces in force-distance cycles. An AFM tip carrying a ligand is brought into contact with a surface with the respective cognate receptors on it, forming a receptor/ligand bond. This bond is subsequently broken with the application of a measurable »unbinding force« when the tip is retracted from the surface and thereby pulls on the bond. Earlier applications of this technique have included measurements of recognition forces between biotin-avidin (one of the strongest interactions known in biology), complementary nucleotides (the building blocks of DNA and RNA), cell adhesion proteins, and antibody-antigen recognition at a resolution of just a few piconewtons.

Visualized in more detail in_FIG. 2, single molecule force spectroscopy requires one of the binding partners to be immobilized at the apex of the AFM tip (the tip is converted into a molecular force sensor) and the other to a hard surface such as mica or glass. We then repeatedly bring the functionalized tip into contact with the surface and subsequently retract it again while we record the cantilever deflection, which is equivalent to the force acting on the cantilever, as illustrated in_FIG. 2. During most of the approach phase, when the distance of the tip to the surface is sufficiently large (1–3), cantilever deflection remains zero. Upon contact (4) the cantilever bends upward due to a repulsive force that increases linearly (4–5) as the tip is pushed further into the surface. After a predefined contact force (5) has been reached, the cantilever movement is reversed, and the deflection relaxes (5–6). If binding has occurred, the cantilever will bend downward as it is further retracted from the surface (6–7). This process continues depending on the strength of the interaction between the functionalized tip and the surface, and on some other known technical parameters (e.g., cantilever elasticity). When the ramping force reaches a critical value, the binding partners dissociate, and the cantilever jumps back into the resting position (7). The force value at dissociation is called *unbinding force* and is quantified by the vertical distance from the critical bond force to the baseline (7-2). Similarly, the unbinding length is given by the horizontal distance from the contact point to the extension at rupture (6-7, horizontal component). We can clearly see the signature of a single molecular binding event in_FIG. 2.

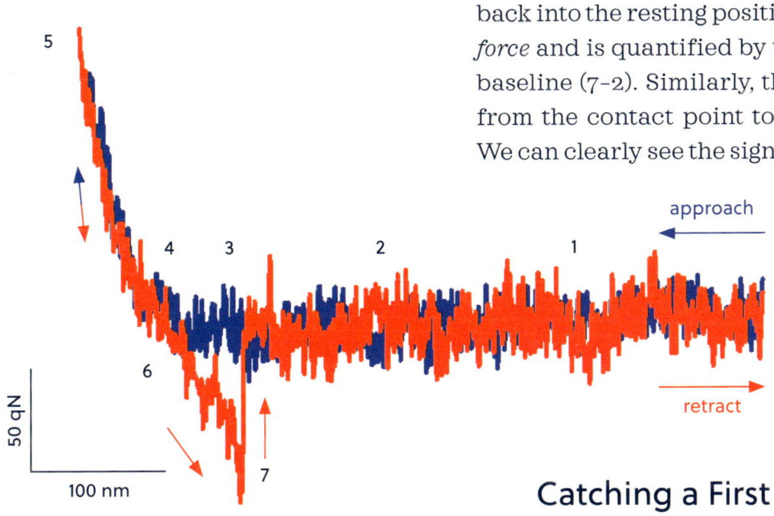

FIG. 2_*Single molecule force spectroscopy: a force-distance cycle (e.g., sweep-rate of 200 nm/s) with a ligand functionalized AFM tip and a dense layer of receptor molecules immobilized on mica. Binding occurs toward the end of the approach phase (1-5) when the tip comes into contact (4) and is pressed into the surface. Under further approach, the tip bends upward, exerting a linear increasing force on the sample (5). During tip retraction, the bending relaxes again. When a ligand-receptor bond has formed, the cantilever progressively bends downward (6), visible as a characteristic parabolic slope in the retract curve, reflecting the extension of the distensible cross-linker ligand-receptor connection. The force increases until the ligand-receptor bond ruptures (unbinds) (7). After it ruptures, the retract part closely follows the approach curve.*

Catching a First Glimpse of Lectins Binding to the Spike

Although the association of soluble lectins with spike trimers had been indicated by conventional methods, the binding steps had not been explicitly resolved. This shortcoming was overcome by studying the interactions on the single molecule level, where monovalent and multivalent binding had been well resolved. For this purpose, the SARS-Cov-2 spike trimer was tethered onto the apex of an AFM cantilever tip, and the CLEC4G and CD209 were similarly linked to an ultra-flat surface. We directly visualized the interaction between the SARS-CoV-2 spike and lectins at the single molecule level by using AFM. Unlike conventional methods, single molecule force spectroscopy can capture every individual binding-unbinding event from spike and lectin entities, thus providing unprecedented details of the interaction. Our single molecule force measurements can observe the binding strength and kinetics, and map the interaction energy landscape between the spike and the lectins.

We performed single-molecule force measurements by moving the SARS-CoV-2 spike-trimer-functionalized tip toward the surface-bound lectins to allow for bond formations_FIG. 3A. After that, unbinding was accomplished by pulling on the bonds, which resulted in the characteristic downward deflection signals of the cantilever whenever a bond was stressed and ruptured_FIG. 3A, RIGHT. The strength of the rupture forces was then determined based on the magnitude of the vertical jumps measured during pulling_FIG. 4B. We saw up to four bonds disrupting sequentially, one after the other_FIG. 3B, and depicted their probabilities for observing the breakage of one, two, three, or four bonds_FIG. 3C. We had anticipated that we would encounter multiple bonds. This is because one spike trimer has sixty-six glycan sites, each site with the theoretical potential to be bound by our lectins. The observed formation of multiple bonds therefore indicated that each spike trimer with its coat of glycans is indeed bound multiple times by the lectins. The formation of multiple bonds also indicates the overall strong binding of the lectins to the SARS-CoV-2 spike trimer. The single molecule force spectroscopy experiments give access to the kinetic and thermodynamic parameters of multiple molecular bonds

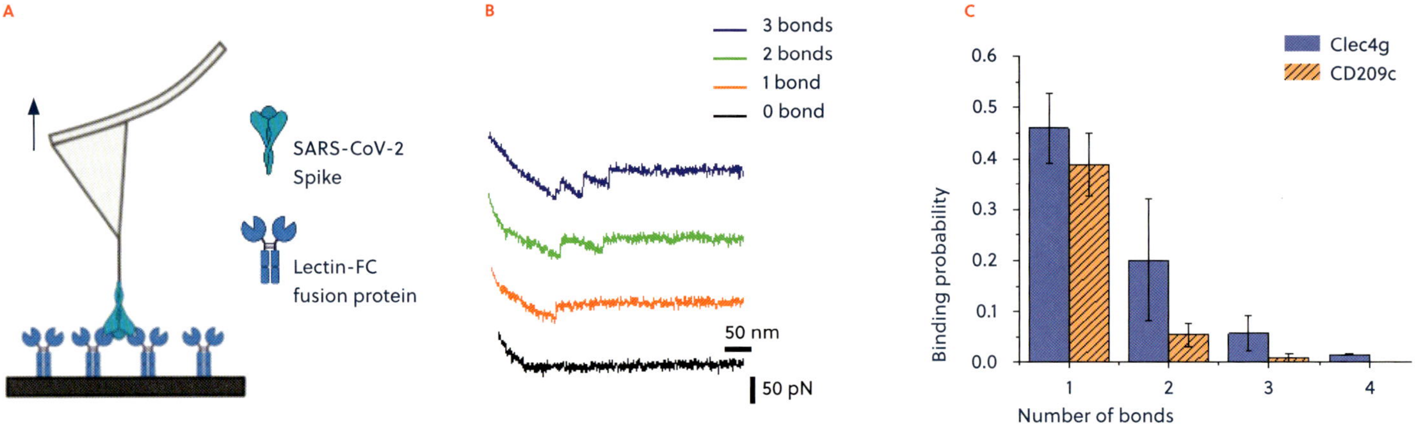

and clarify many aspects of the complex binding interplay between the lectins and the glycosylation sites of the spike trimer. We caught a first glimpse of the nanometer world of spike and lectins. And while this glimpse taught us about binding strengths, one important question was still left unanswered: how many lectins would simultaneously bind a spike trimer if they could engage from all sides, like they would in a therapeutic setting in, for instance, the blood stream of a patient?

FIG. 3

A_*Schematic overview of single molecule force spectroscopy (SMFS) experiments using full-length trimeric spike coupled to an atomic force microscopy (AFM) cantilever tip and surface-coated murine Clec4g-Fc or CD209c-Fc. A maleimide-poly(ethylene glycol) (PEG) linker was attached to 3-aminopropyltriethoxysilane (APTES)-coated atomic force microscopy (AFM) cantilevers. The arrow indicates the pulling of the cantilever.*

B_*Representative force traces showing sequential bond ruptures in the SMFS experiments. Measured forces are shown in piconewtons (pN). Force measurements were performed at room temperature (~25 ºC) with 0.01 N/m nominal spring constants (MSCT, Bruker) in TBS buffer containing 1 mM $CaCl_2$ and 0.1 % TWEEN-20.*

C_*SMFS-determined binding probability for the binding of trimeric spike to CLEC4G and CD209c. Data are shown as mean binding probability ± SD of single, double, triple, or quadruple bonds (N=2). Force-distance plots were acquired by recording at least 1,000 curves with vertical sweep rates of between 0.5 and 10 Hz at a z-range of typically 500–1,000 nm, using a commercial AFM.*

Reprinted from D Hoffmann, S Mereiter, YJ Oh, et al., »Identification of Lectin Receptors for Conserved SARS-CoV-2 Glycosylation Sites,« *The Embo Journal* 40, no. 19 (2021): e108375, https://doi.org/10.15252/embj.2021108375, 2021

Filming the Dynamics of Lectin Bonds

To answer this question, we needed to see more of the interaction. We therefore assessed the structural role of the lectins and their involvement in modulating the conformational dynamics of the spike by recording high-speed AFM films of the SARS-CoV-2 spike alone and of the SARS-CoV-2 spike bound with lectins. High-speed AFM is capable of making recordings of the interactions and the dynamic structural changes of biomolecules in a spatial resolution of nanometers and in a temporal resolution of a few milliseconds. In other words, it lets us observe single molecular interactions in real time! Our data on conformational flexibility, interaction strength, and the binding kinetics of the spike trimer make it easier to understand the dynamics of lectin binding. The SARS-CoV-2 trimeric spike and lectins were applied to freshly cleaved mica. We captured films in aqueous conditions using an ultra-short cantilever or fast imaging_FIG. 4. Areas of 100 × 100 nanometers containing single molecules were selected to capture the high-speed AFM films at a scan rate of ~150–300 milliseconds per frame.

When looking at the spike trimer alone, the incredible flexibility of the viral protein manifested itself. Viruses are masters of evolution; due to their fast replication and rapid acquisition of mutations, viruses are ever-changing. As the virus' entry key to infect cells, the spike protein is the fastest evolving part of the virus. We don't know exactly how long this virus had evolved in nature before it jumped the species barrier to infect humans, but one thing is certain: the spike has been shaped through this evolutionary process into an incredibly optimized tool. Observing its movement and flexibility allows us to only speculate what tricks the virus must have developed to outmaneuver our immune systems and to help it infect our airway cells.

VIDEO_High-speed AFM of the single trimeric spike visualizing its real-time interaction dynamics with lectins

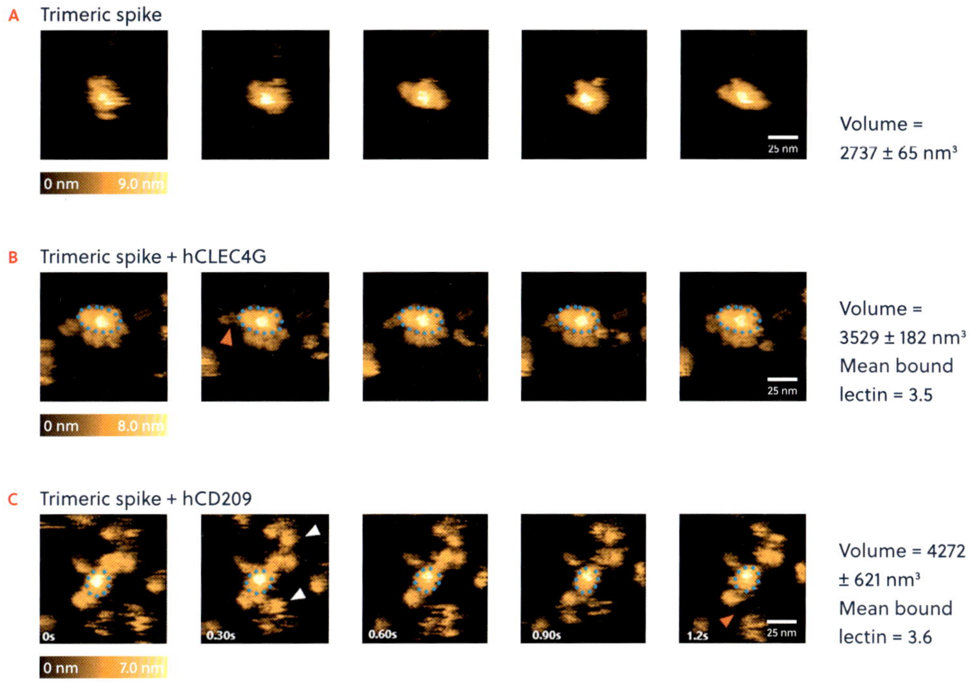

FIG. 4_*High-speed AFM of the single trimeric spike visualizing its real-time interaction dynamics with lectins. The*_A *panel shows five frames of the trimeric spike alone imaged on mica. The panels in*_B *and*_C *show five sequential frames of the trimeric spike/CLEC4G and trimeric spike/CD209 complexes respectively, acquired at a rate of 303 ms/frame. Association and dissociation events between hCLEC4G or hCD209 and the spike are indicated by white and red arrows respectively. The blue dotted ellipses display the core of the complexes, revealing low conformational mobility. Color schemes indicate the height of the molecules in nanometers (nm). The volumes of single trimeric spike and trimeric spike/Clec4g and trimeric spike/CD209 complexes are indicated, as are the numbers of lectins bound to the trimeric spike, averaged over the experimental recording period. The volume of the isolated SARS-Cov-2 spike trimer of 2737±65 nm³*_A *increased to 3529±182 nm³ and 4272±621 nm³ upon the association of CLEC4G and CD209, respectively. Taking volumes for isolated hClec4g and hCD209, the number of molecules bound to the SARS-CoV-2 spike trimer was estimated to be 3.5 for hClec4g and 3.6 for hCD209, which roughly corresponds to the numbers of bonds found in single molecule force spectroscopy measurements*_FIG. 3.

Reprinted from Hoffmann, Mereiter, Oh, et al., »Identification of Lectin Receptors.«, 2022

Next, we analyzed films recorded of lectins forming a complex with the spike. The blue dotted ellipses highlight the core of the spike/lectin complexes displaying low mobility. Dynamic interactions with lectins and flexible morphology were observed at its periphery. Sporadically, lectins would engage with the spike, bind to it, and stay attached for a limited amount of time before dissociating again. This was a dynamic process that repeated itself over and over again. White arrows point to lectins associating with the spike trimer body, while red arrows indicate the dissociation of the lectins from the spike trimer body_FIG. 4. We calculated the numbers of lectin molecules bound to the spike trimers based on the measured mean volumes of the spike, the lectins, and the spike/lectin complexes, averaged over the recorded time periods. Again, we observed the lectins establishing multiple bonds with spike trimers, however, with important differences between different lectins. The binding of the lectin CLEC4G to the SARS-CoV-2 spike led to tighter compaction and less flexibility in the peripheral regions when compared with the binding of the lectin CD209, where CD209 was observed to have considerable fractality and mobility. Thus, by monitoring single molecular events using high-speed AFM, we were able to observe the spike/CLEC4G forming more rigid clusters with lower conformational flexibility as compared to the spike/CD209 complexes in equilibrium conditions. This was in accordance with faster association rates and shorter dissociation paths of CLEC4G as compared to CD209. The association of three to four lectin molecules with one trimeric spike observed in the experiment indicates the formation of high-affinity bonds to one to two glycosylation sites per monomeric subunit of the trimeric spike. This was an exciting finding, as this also meant that the lectins were able to occupy and potentially block all three monomers of a spike trimer simultaneously!

Visualizing the Therapeutic Potential of Lectins

Could lectins really be used to bind to the spike, to glue it all up, and thereby prevent the virus from attaching to our cells? To answer this, we further explored the inhibitory effect of lectins on the binding between spike trimers and receptors on living cells by monitoring their binding affinity in cell force spectroscopy experiments on Vero E6 cells (a standard cell line for studying SARS-CoV-2 infections in the lab). Our comprehensive investigation into the inhibitory effect not only reveals the lectins' potential in therapeutic medicines but also provides a valuable collection of data to decipher the mechanism of the inhibition of the binding between the spike trimer and the host cells. This may provide an experimental basis for designing novel drugs to reliably block the virus from entering the binding interface.

For this, we went back to single molecule force spectroscopy, the method we had used before to measure the binding kinetics. We functionalized the AFM tip with a spike trimer once more, but this time to test the binding of the spike to living cells and to study the effect of the lectins CLEC4G and CD209 on the inhibition of this processor_FIG. 5. For this, Vero E6 cells were first grown on culture dishes. Then, lectins were added in different concentrations to block the cantilever that had been functionalized with the spike. Lastly, with the lectin/spike complex anchored to an AFM cantilever_FIG. 5A, we recorded force-distance curves on living cells with the assistance of a CCD camera to localize the cantilever tip on selected cells. The sweep range was fixed at 3,000 nanometers and the sweep rate was set at one Hertz.

Repeated force measurements on Vero E6 cells reported the binding probability in the absence or presence of lectins in various concentrations. While adding the lectin CD209 caused no reduction in the binding probability over the whole concentration range, CLEC4G was observed to heavily interfere in the binding between the spike trimer and the cell surface at concentrations larger than 70 nanomolar, where it drastically blocked the interaction. An estimation of the inhibitory constant (Ki) revealed that concentrations of CLEC4G as low as 35 nanomolar could block fifty percent of the spike binding. Thus, CLEC4G can functionally impede the spike binding to host cell membranes, thereby providing a rationale of how this lectin can affect SARS-CoV-2 infections.

The Dance of Molecules

Research into sugar molecules and their importance for life dates back to the late nineteenth century. At that time, chemists like Emil Fischer were already elucidating the structural features of sugar molecules. Many years later, around the 1950s, the term *lectin* was coined, derived from the Latin word *legere*, meaning *choosing*. The name was assigned when researchers first discovered the glycan-binding capacities of these molecules and showed that different cells could be crosslinked with lectins. This foreshadowed the very specific nature of lectins and their potential as therapeutics against pathogens. Our knowledge of both sugars and sugar binders has progressed immensely since then and could now fill entire encyclopedias. However, despite this progress, our research (not only glycobiology but in most molecular biosciences) remains in essence indirect. We don't actually observe the molecular processes as they are too small to be seen, but we have devel-

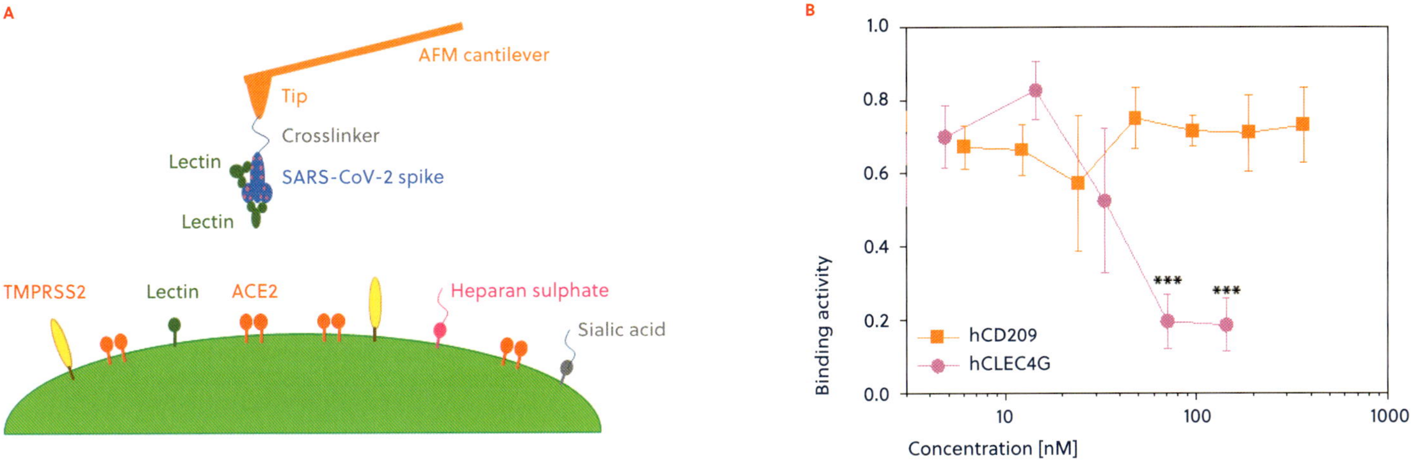

FIG. 5

A_Sketch of binding inhibition experiment. Binding activity of the full-length trimeric spike coupled to an AFM cantilever tip with the viral entrance receptor ACE2 on the surface of Vero E6 cells in the presence of the lectins hCLEC4G or hCD209,
B_probed at the indicated concentrations in single molecule force spectroscopy experiments. A definite decrease in binding was observed at a concentrations of 33–71 nM for hCLEC4G, while no dramatic effect was visible for hCD209. Data have been normalized to the untreated control and are shown as mean + SD (N=4).

Reprinted from Hoffmann, Mereiter, Oh, et al., »Identification of Lectin Receptors.«, 2022

oped fantastic tools that tell us in a short amount of time what might be happening in our test tubes. These experiments provide evidence that confirms or rejects our hypotheses. And while this is really enough to move our scientific knowledge forward and to unravel many aspects of diseases, we sometimes fail to truly understand these molecular processes. Now, as humans we want to visually perceive even that which can't be seen. Using the trick of applying tactile forces instead of light, of feeling the molecules using AFM, we are finally able to peak into molecules in motion and even to measure single molecule interactions. It has truly opened the doors to a hidden world and has allowed us, more than fifty years after their discovery, to finally observe how lectins interact with glycans. It has also enabled us to monitor the spike of SARS-CoV-2 and how it bends and moves. On the molecular scale, completely different forces dominate, yet it appears oddly familiar, perhaps comparable with a tree in the wind. We measured a single trimeric spike to be 2,737 ± 65 nm^3 in volume. After adding lectins, we were able to observe in real time how they engage with the spike protein and how long they stay attached before falling off. Now, the volume of the tree increased to more than 4,000 nm^3. The spike was glued up with lectins. We can watch single molecule interactions, which also allows us to count how many lectins are binding per spike trimer—which brings us back to biology's almost extreme complexity. Observing single molecule interactions has finally allowed us to count molecules that would have otherwise remained unquantifiable. We were able to narrow down how many of the sixty-six glycan sites to which our lectins could bind were actually being bound. We were able to measure how strong these interactions were and finally test the therapeutic potential of lectins.

Science is a journey, and there are still and always will be many open questions to be answered—whether regarding glycans, lectins, viruses, or other diseases. Being able to monitor the molecular world will help us to answer some of these question in the future. But that's not all: opening the door to the hidden world of the biomolecules that form the basis of life has also opened the door to observing a very aesthetic world of moving, bending, associating and dissociating, turning and twisting and turning molecules.

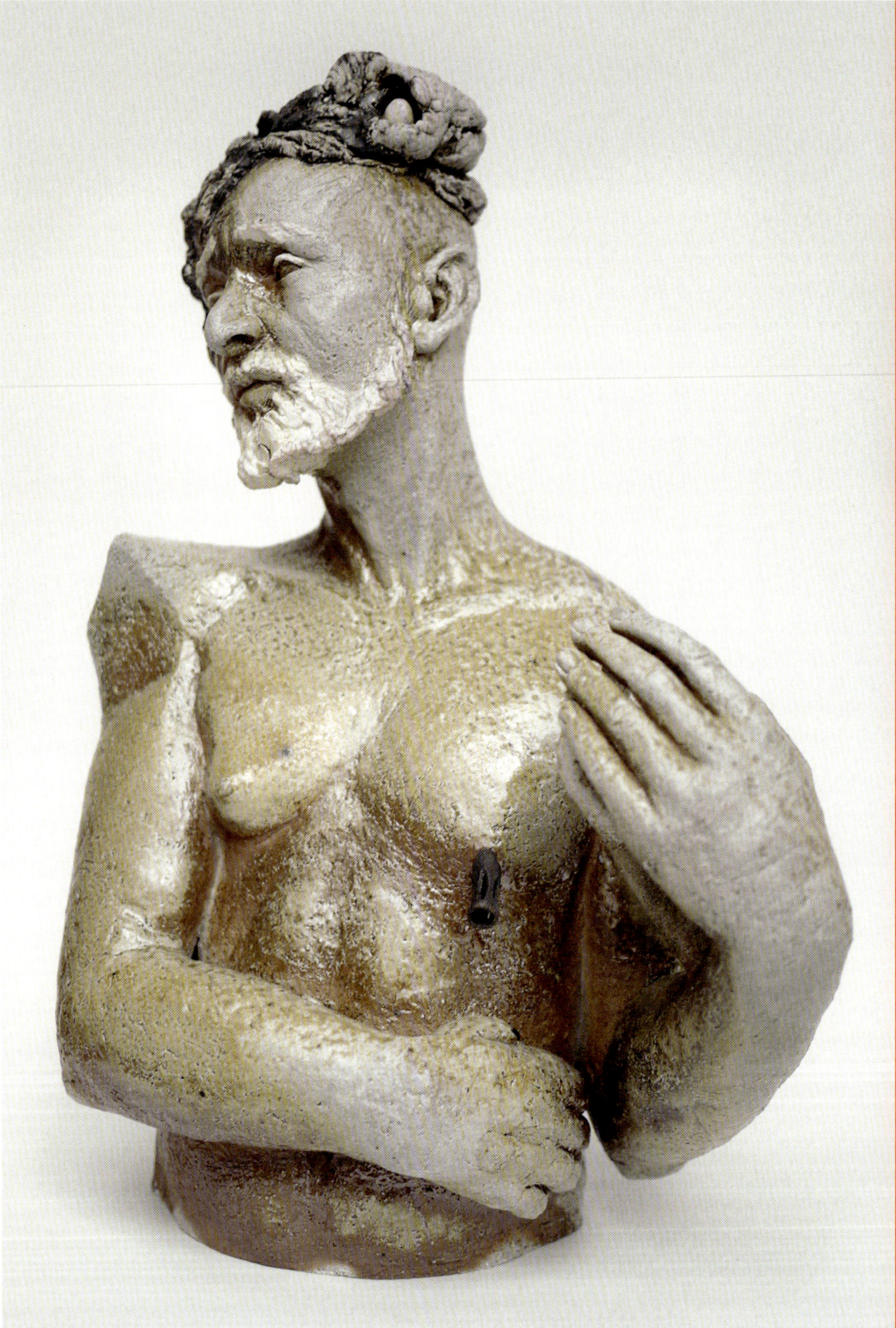

FIG. 1_*Front view of* LA BORNE WORRIER, *Stoneware and porcelain, anagama firing, 40 × 30 × 17 cm*

Ceramics, texts, and photos by Anna Steinhäusler, 2021[1]

VIDEO

© Georg Heimberger, 2021[2]

La Borne Worrier

Anna Steinhäusler

Wood firing provides insights into the earth and activates awareness and community.

This ceramic figure was wood-fired during an advanced training course in the French ceramics town of La Borne, Département Cher. La Borne does not have a supermarket, but it does have an astonishingly high density of ceramicists residing there with a broad variety of kiln forms. The former church is also used as a ceramics museum, presenting the town's history of ceramics and local ceramicists. Beside the church is the Contemporary Ceramics Center. During the aforementioned course, a group of twelve international artists fired the center's kiln, an anagama, which translates as »cave kiln.« The surface of the sculpture is testament to wood-firing that lasted three days. The title of the sculpture, *La Borne Worrier*, emerged as the antithesis to the »warrior.« It represents a figure that worries about others.

A Sculpture Emerges

The »born worrier« began as a media image of a worried father who fears for his child's life. During the modeling process, he grew into a hybrid creature that reflects various aspects of what it is to be human, such as parenthood, power and powerlessness, the protection of life, and threats to life. He has manifold attributes: a worried, pensive gaze, his head crowned by a nest that protects the life growing in it in the form of a porcelain egg. Used at different times, his body has one angular and one soft shoulder. In this section, the structure-giving impact of both geometry and logic, and the mobility of gentleness and flexibility reach a balance. Due to the plasticity and receptivity of the clay and the soft form of the shoulder, the latter sank when it was touched from outside. Being in contact with the surrounding environment leaves traces on people. The chest area changes from a female to a male half. The end of a gun barrel functions as a male nipple. This juxtaposition is an expression of the human potential to nourish and take life. The right hand transforms

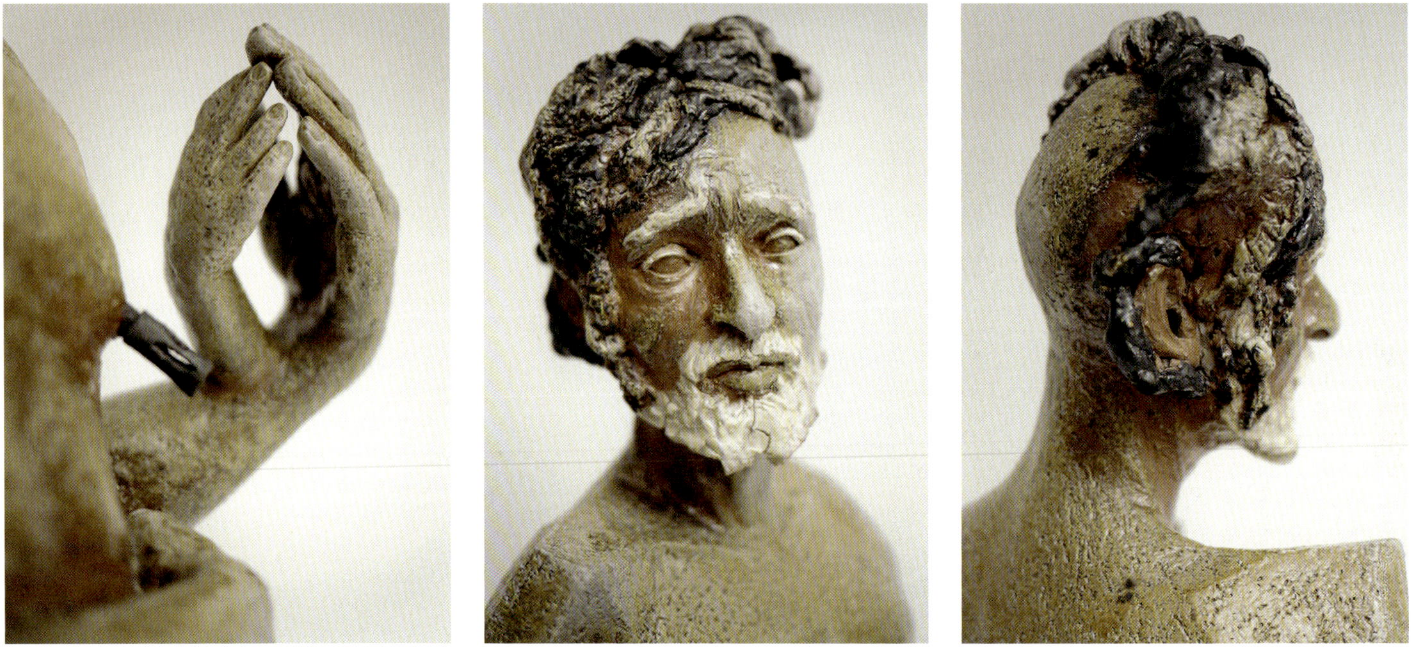

into a lion's paw, which, resting against the stomach in a gesture of protection, lingers without moving. The fingertips of the left hand form a cyclical unity with a child's hand growing out of the forearm, incorporating the coming generation and the father's personal history into the narrative.

FIGS. 2, 3, 4_*Detail on* LA BORNE WORRIER
Stoneware and porcelain, anagama firing
40 × 30 × 17 cm

Taking up the multifaceted appearance of wood-fired ceramics, the figure is not made from one kind of clay but from various colored stoneware clays and porcelain. Porcelain, a white ceramic mass, is valued for its translucency and purity compared with stoneware clays, which are not pure white. The porcelain is pure because, unlike natural stoneware, it has not been carried through erosion and decomposition processes, it has not been redistributed by rivers or winds, and it has not been contaminated with, e.g., iron compounds; rather, it consists to a large extent of white clay, kaolin. In the figure, this material property was suitable for delicate, vulnerable parts like the bird's egg, the soft shoulder, and the woman's breast. In freely modeled areas such as the nest crown, I combined white, black, and red stoneware. The richer in metal the stoneware clays are, the more their metallic character comes to the surface through the high firing temperatures. Clays with a high proportion of metal oxide become unstable because they soften earlier. They appear reactive, eroded, melted, and like they have been attacked. I deliberately placed the worrier in profile to the fire. I wondered if the flames and kiln atmosphere could capture his inner turmoil. The kiln's gift was the painterly bisection of the figure into one light and one shadow side, clearly identifiable in the warm red and cold gray markings on his face.

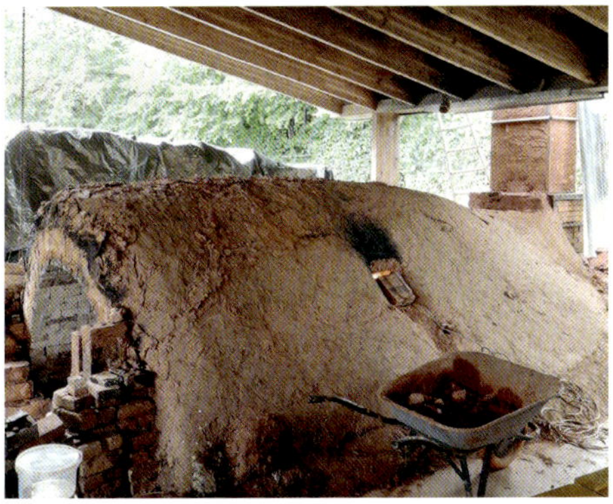

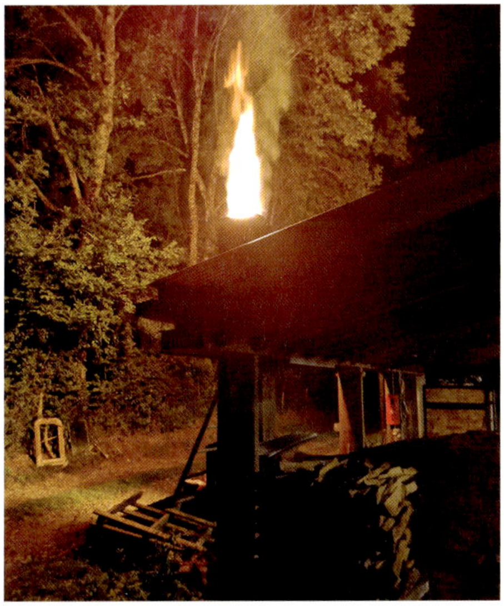

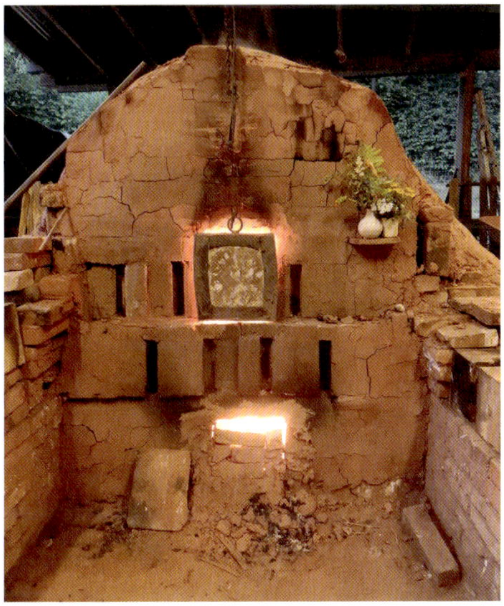

FIGS. 5, 6, 7_*Photos of the anagama, Contemporary Ceramics Center in La Borne*[3]

Encountering a Work of Art—Entering into a Relationship

A sculpture is an idea that has become material—an embodiment. It is the expression of a spiritual world. Through artworks we encounter worldviews, observations, suggestions, questions, problems, and thought-provoking impulses. When viewers contemplate a work of art, their personal history is interwoven with what they see. We are able to encounter human figures by relating to them. By perceiving familiar, our own, and foreign parts, we approach ourselves and learn what we resonate with and what we deliberately distance ourselves from.

Lending an Ear

If I do not distance myself, if I make the problem of the Other my own, it moves me and motivates me to get to the source. In caring, behaving in a way that is sensitive to the situation, and showing empathy lies the potential to foster life. Caring begins with an intention to listen attentively and to thus allow the other to feel that they are being heard—turning to your counterpart as if for the first time—with infinite curiosity, openness, and interest. With this moving definition of empathy, entrenched thought patterns open up. Our body is limited, but our mind can be extended into the lives of others. It takes extension beyond the self, a kind of self-forgetfulness for a problem to become multilayered. By directly categorizing a problem by means of predefined thought structures, potential solutions go unseen.

In addition to its human expression, the sculpture is shaped by the expression of the chosen material. A material imparts a mood through its properties and our memories and conceptual cross-references. This mood can reinforce an idea or shift its meaning. Thus, the artist, the material, and the viewers share in that which a work of art ultimately is. The sculpture is a platform for conversation and exists in a web of relationships and interpretations. It is a memory, a transmitter, and a catalyst for human ideas, inviting us to compare perspectives.

Wood Firing

Wood firing is a form of ceramic firing in which the atmosphere inside the kiln imprints itself on the clay figures. Wood kilns, which are generally built outdoors, come in different shapes and sizes, have a firing chamber of up to several cubic meters in volume, and are fired with wood.

Critical in wood firing are the clays and type of kiln used, the firing duration, and the wood that is burned. The figure shown here was fired in an anagama, a horizontally built, single-chamber kiln from Japan. In a single-chambered kiln, the clay pieces are placed in the same space as the fire and are not protected from it. In other kilns, the clay pieces are deliberately protected from the flames and the fly ash as consistent results are desired and variations will not be accepted. *La Borne Worrier* was exposed to the flame for three days. In this time, it hardened and received its contrast-rich appearance. Firing for several days allows raw materials to melt that would not develop any glass at lower temperatures or during shorter firings. For a firing that lasts several days, a group of people organizes itself into firing shifts to ensure that wood is constantly being added to the expanding fire. At its highest temperature, which is maintained for a long period of time, logs are added approximately every three minutes. Ash is an insignificant gray powder—helpful for cleaning and even considered residual waste in household heating systems. But in wood firing, it is the raw material that provides us with insight into the soil and serves as the basis for glass formation. What is special about long wood firing periods is the possibility of putting clay pieces into the firing unglazed—in the anticipation that a natural glass layer will form. During the firing process, the unburnable part of the wood, the ash, flies through the firing chamber and sticks to the clay pieces until they are covered in a fine coating. Ash contains compounds of metal oxides like potassium and sodium compounds, iron, manganese oxide, and many more, which are absorbed from the soil by trees and are relevant in the formation of glass and its coloring. At a heat of around 1,300 degrees Celsius, ash can melt in reaction with the clay and form glass. This reaction creates the glaze. By looking at these glazes, humans gain an insight into the interconnectedness of trees. The glazes show which minerals plants have absorbed during their growth. The metal oxides stored vary depending on the location of the trees, the type of wood used, and which part of the tree is being burned. Each new tree, in turn, leads to a variation in the glaze. Both the trees and the clays used reveal regional geological features. On the skin of the ceramics, the language of the bones of the earth, the minerals, shows itself.

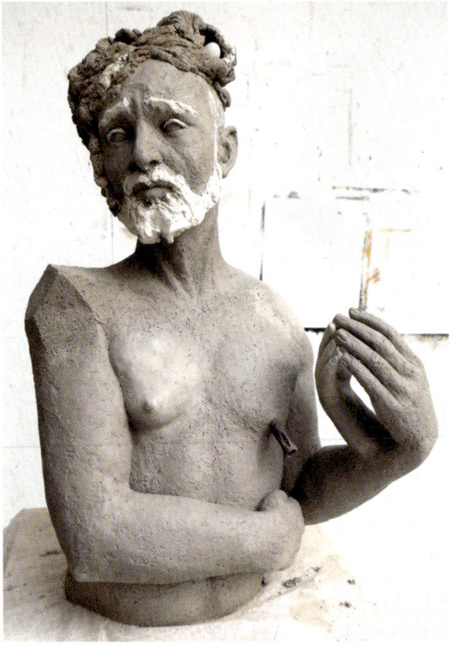

FIG. 8_ La Borne Worrier *sculpture before firing*

The face of the soil takes shape in the firing chamber, a closed-off unobservable space. Ceramicists only have limited insights into this space as the development of high temperatures causes the entire firing chamber to appear in a uniform, orange-yellow annealing color, limiting the usual color spectrum. Technical aids like temperature probes and pyrometers are used to extend our limited human senses but are not consistently reliable and therefore cannot replace them. Wood firing is a practice that requires and produces attentiveness, awareness, and community. Ceramicists do not have the option of distancing themselves by giving the task of firing over to electric controls in an electric kiln. By adding and taking away air and wood, they control the firing curve, which describes by how many degrees the firing temperature is raised in what time period, in order for the ceramics to harden without breaking. The minds and senses of the people firing the kiln must be extended to include the entire kiln and its functioning. The ceramicists lend the kiln their ear. A certain kind of suction sound reveals a long, even flame and good combustion. Within this connected working atmosphere, ceramists become aware of the kiln's needs. Like attentive acquaintances, they change minute details within the kiln, such as by opening small air holes in order to observe how the suction of the kiln changes or how better combustion can conserve the resource wood and release fewer emissions.

Until the cooled kiln is opened, it remains unclear whether all of the pieces have survived; whether a piece has been hit by a log; whether another log has fallen on any other ceramics and they have fused together; whether the forms have warped because the clay has become soft or was modeled too thinly; whether a deliberately positioned ceramic piece has been subject to an intensive burst of ash, forming a jade-green river of glass; whether a glaze has melted too intensely and the ceramic has become stuck to the firing plate, meaning that it can only be removed using a diamond grinding wheel; or whether glass drops began to cool before they hit the

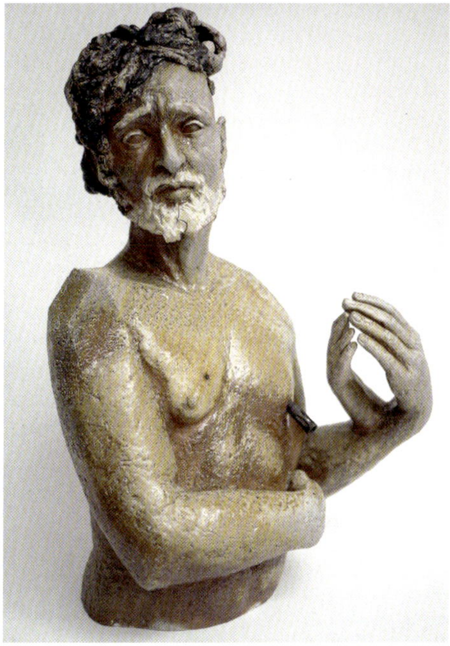

FIG. 9_ La Borne Worrier
sculpture after firing

firing plate and are now decorating the foot of the ceramic like pearls. As it can only be controlled to a certain degree, wood firing is not simple but complex. People have to adjust to this complexity.

Wood-fired ceramics delight us due to their simple beauty, but they are affected by a range of processes that must take place and interlock in order for a multilayered surface to emerge. The image of a natural spectacle becomes visible on them. The originally consistent clay reveals its true colors through wood firing.

A versatile palette of earth tones and structural gradients, from coral-like crusts to perfectly melted glass layers, expand the surfaces of the clay forms. If you turn a wood-fired ceramic around in your hands, it continuously reveals new sides of itself. Ceramicists speak of a »light side« and a »shadow side« to describe the sides facing toward and away from the fire. Ash is more likely to accumulate on the side that is exposed to the fire, making that side more varied and intense. Often, the more protected side does not form glass but reveals the traces of flames in warm tones. Touch—the way the flame moves through the kiln, and how water, choosing the path of least resistance, streaks the pieces—plays just as much a role in the appearance of the ceramics as the position of the pieces to each other. This web of relationships leaves another imprint on their appearance. A larger ceramic piece can deflect fire from the ones next to it, creating a kind of fire shadow on its neighbor's surface in more gentle tones. Wood-fired surfaces outshine perfect, meticulously balanced electro kiln glazes because something has befallen them. They have been designed, shaped, drawn on, they have communed, reacted, exchanged, and survived. They have emerged from this extreme exchange more animated than if they had been fired in a protected atmosphere without turbulence. They can exist in a tremendous variety without visually losing their balance. This diversity, with multifaceted attitudes, can lead to tensions in interhuman relations, but nature exists through them in a balanced interaction in which nothing is without meaning and everything is relevant and has its share. There is no dominant color in the wood firing color palette; all earth colors are represented. Each ceramic is uniquely designed by its firing milieu, but it is only if all the ceramics are viewed together that they explain why they look the way they do. It is specifically this interconnectedness in mutual drawing and coloration that ultimately constitutes their appeal. Glazes and surfaces that are almost impossible for the human mind to imitate are the incentive for focusing on how wood firing works, for the physically taxing work, and the disrupted circadian rhythm. They embody the deviation, unpredictability, the acceptance of variation, and the unexpected. When we engage with them, they allow our thought patterns to regenerate, enable us to question beauty ideals and perfectionism, and to give more space in life to humility and the acknowledgment of interconnections.

1 http://www.annasteinhaeusler.com.
2 http://www.studiosylt.de.
3 http://www.laborne.org.

Insights: Stroke, a Medical History in Neurology and Neuroradiology, and the Spirit of Empathy

Karl Heimberger
with Christian Našel

»I need four technical aids to examine patients: my eyes, ears, hands and my heart.«
Franz Chvostek, MD, 1835–1884

»Healing begins with discovery.«
Motto of the Oregon Health & Science University, Portland, Oregon, US

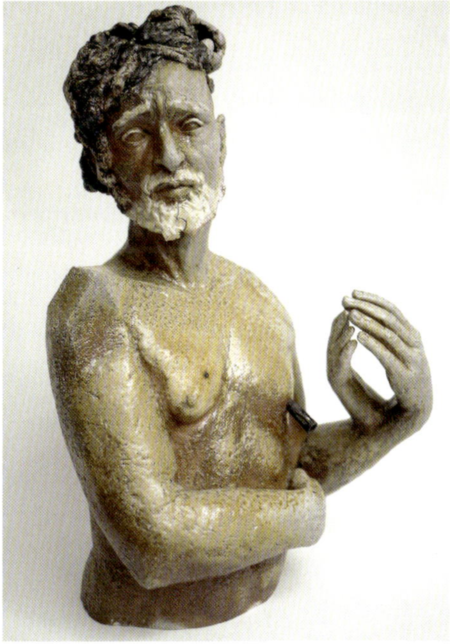

FIG. 1_La Borne Worrier, *stoneware and porcelain, anagama kiln, 40 × 30 × 17 cm*[1]

VIDEO

Simply looking at a sculpture opens up its spiritual inner world. This recognizant process is mainly shaped by the intention of the artist and partly by the viewer's readiness of insight. The artist tries to give an externally recognizable form to what lies secretly hidden deep under the surface of the presented statue's skin_FIG. 1.

In medicine, the rational recognition of what is pathophysiologically occurring in and under the skin of a patient guides the way to therapy. With patient history and physical examination in mind (H&P—including inspection, palpation, percussion, auscultation, screening mobility, muscular force, sense perception, cognition, and... showing empathy), the physician selects the suitable technical assistance, which then confirms or refutes a presumptive diagnosis. Medical technologies serve as gates. In neuroradiology, noninvasive or minimally invasive techniques produce electronic images and films that expose otherwise invisible pathological substrates within and around nervous system structures.

One example of a medical history is presented here with a description of the course of the disease and the recovery of the affected person after correct diagnosis and the application of state-of-the-art therapy, consequently preventing disease recurrence. This medical history is supplemented by discussions of the technical tools that can visualize the underlying causes of this medical condition. Selected imaging methods are also listed in the toolbox section in a compact descriptive format: computed tomography (CT), magnetic resonance imaging (MRI), intra-arterial digital subtraction angiography (IADSA), image-guided surgery, and quality assurance. Readers can access the resulting film documentaries by making use of augmented reality via quick response (QR) codes.

Medical History and Neurological Conditions in Diagnosis, Therapy, and Clinical Outcome

Stroke

A stroke is the abrupt onset of neurological symptoms caused by a central nervous system disorder with cerebral ischemia, an infarction with arterial occlusion, or intracranial bleeding or venous occlusion. Stroke is the major cause of lifelong disability and functional dependency in developed countries. Initiating immediate diagnostic and therapeutic action is the prerequisite for a favorable clinical outcome. A sixty-four-year-old male experienced the sudden onset of complete paralysis of the right arm and leg (hemiplegia) and had extreme problems talking understandably (dysphasia). Within less than two hours after the onset of symptoms, he was taken to the hospital. After rapidly performing H&P (the patient's history was explored by interviewing attending relatives), an MRI of the head was carried out, including diffusion and perfusion sequences. Intracranial hemorrhage was ruled out. A diffusion MRI revealed an infarct in a circumscribed area of the left hemisphere, where a perfusion MRI showed that the blood supply to the brain tissue had stopped (ischemia). But surrounding the ischemia, it also identified critical perfusion in a much larger area of the left cerebral hemisphere_FIG. 2, which urgently needed to be taken care of.

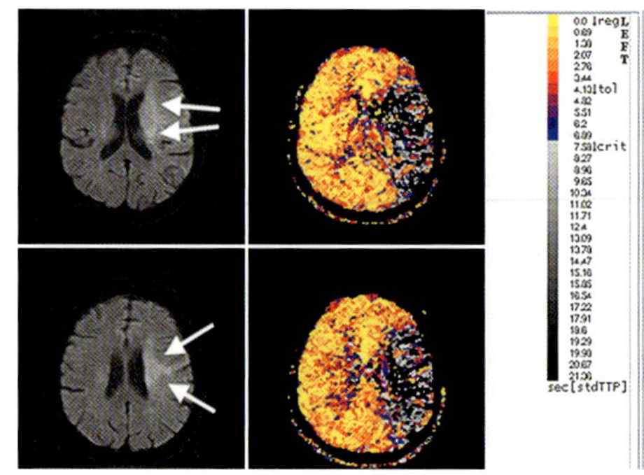

FIG. 2_*Brain MRI. Acute phase of stroke. Viewing axial cross sections from below is the radiological standard. So, the right side of an image represents the left side of the examined person. The black-and-white diffusion MR brain images (first column) show a one-sided infarct (demarcated in shades of light gray to white; arrows), whereas, after the administration of an intravenous contrast agent, the colored perfusion MRI maps (second column; yellow and red show regular perfusion) reveal a much larger critical perfusion area (shades of gray) surrounding the infarct zone (black). This difference in size is called* diffusion/perfusion mismatch *and is important for therapy planning, as we will discuss below.*

Image by Christian Našel, 2020[3]

Perfusion MRI surveys the distribution of an intravenously administered contrast medium in the brain over time. The standardized time to peak (stdTTP) perfusion MRI method presented here calculates an absolute measure of this distribution time, which allows comparisons between different examinations and patients to be made. Furthermore, the calculated stdTTP values allow the examining physicians to reliably quantify the regional perfusion state of the brain at the time of the examination. StdTTP represents the time between the first appearance of contrast-induced signal change to maximum signal change in the various brain regions and is therefore measured in seconds.[4] In order to obtain sufficiently precise measurements, the whole brain is dynamically imaged every 689 milliseconds, which generates more than 1,600 images of the complete brain during a perfusion MRI examination, which usually takes about one minute on the acute stroke setting. Depending on the respective stdTTP time values, various qualities of perfusion are differentiable. These are expressed by a defined color scale that allows rapid interpretations to be made of the perfusion maps of the brain computed using this method. On this scale, regular perfusion corresponding to stdTTP values of ≤3.5 seconds are displayed in yellow to red. Comparably, stdTTP values of >3.5 and ≤7 seconds are displayed in blue. This perfusion state indicates cerebrovascular disease, but immediate treatment is not yet necessary. Conversely, stdTTP values of >7 seconds correspond to critical perfusion with an eighty percent chance of suffering irreversible damage from severe hypoperfusion. These regions are marked in shades of gray and require immediate therapy (e.g., interventional angiography with thrombectomy or thrombolysis). In stdTTP perfusion maps, this method can be used to identify the presence of viable (gray-stained) and salvageable tissue at risk of infarction, predicting the growth of the infarct—if left untreated. Finally, areas without any enhancement in the perfusion MRI are stained black, indicating that the perfusion has stopped, which is equivalent to an ischemic infarct of the brain tissue.

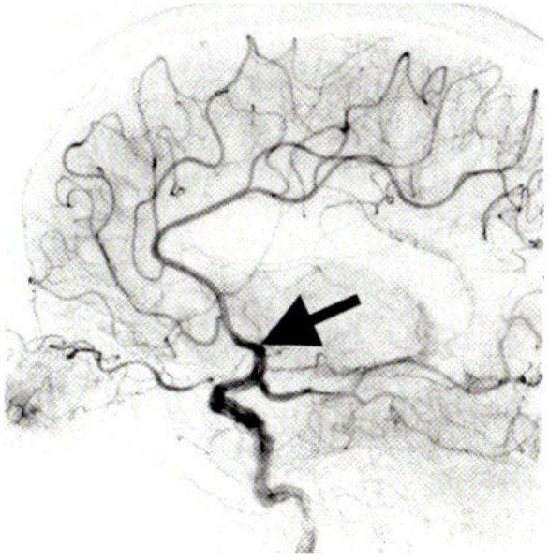

FIG. 3_*Screenshot from video. IADSA. Viewed from the side. Arrow: an occlusion is found at the origin of the middle cerebral artery (MCA)*

Image by Christian Našel, 2020[5]

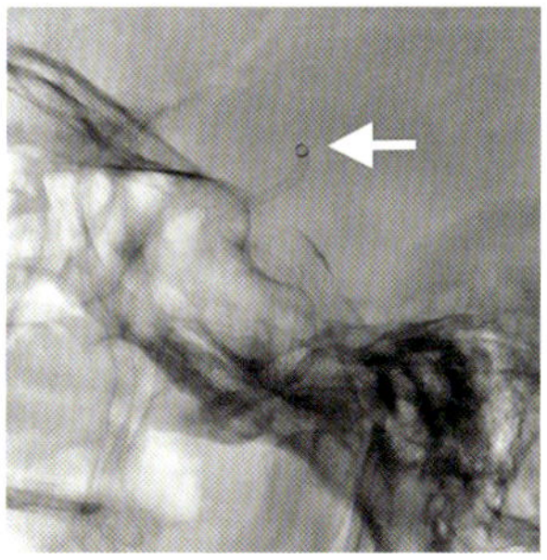

FIG. 4_*Screenshot of cranial fluoroscopy for navigating and positioning an intracranial endovascular aspiration catheter with the tip (arrow) at the thrombus occluding the origin of the left middle cerebral artery. Viewed from the side. The radiodense bone structure stretching from the upper left to the lower right is the skull base.*

Image by Christian Našel, 2020[6]

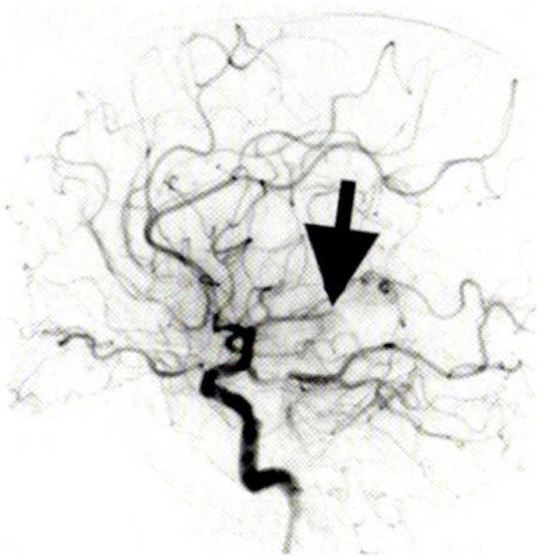

FIG. 5_*IADSA Screenshot. After thrombectomy, a second thrombotic occlusion, possibly a thrombus that had migrated, was found in the posterior part (arrow) of the same artery that had previously been totally occluded (in fig. 3).*

Image by Christian Našel, 2020

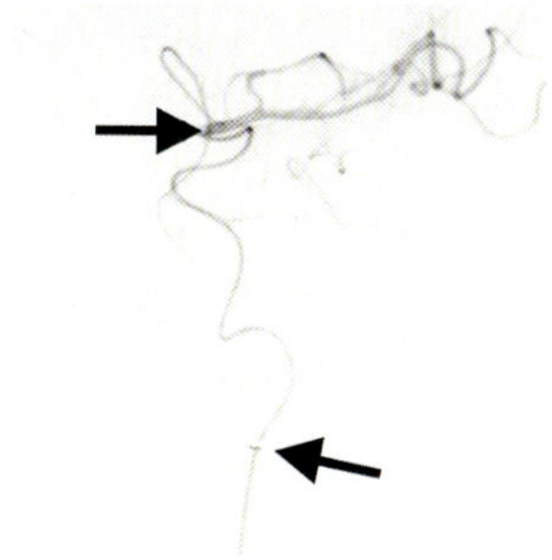

FIG. 6_*Screenshot of superselective IADSA showing patency of the posterior segment of the middle cerebral artery after a second thrombectomy. Upper arrow: tip of a coaxial microcatheter used to inject the contrast agent to control patency. Lower arrow: pointing to the tip of the endovascular guide catheter.*

Image by Christian Našel, 2020[7]

The patient was immediately transferred to the angiography unit. Intra-arterial digital subtraction angiography (IADSA) confirmed the thrombo-embolic occlusion of one of the main brain arteries_FIG. 3.

During the same diagnostic session, a thrombectomy was performed with an endovascular aspiration catheter_FIG. 4.

A second occlusion in the posterior part of the same blood vessel was also spotted angiographically_FIG. 5 and recanalized endovascularly_FIG. 6. The removed thrombotic material is displayed in_FIG. 7.

The angiographic control examination that promptly followed showed the patency of all treated blood vessel segments_FIG. 8.

One week later, a pre-existing narrowing of the lumen of a cervical artery (ICA_FIG. 9) was widened using an intra-arterially expandable tubular wire mesh structure known as a *stent*_FIG. 10.

Fourteen months after thrombectomy and stenting, a follow-up examination showed that the perfusion of brain tissue in the treated hemisphere had been completely restored_FIG. 11.

Clinical outcome: after rehabilitative treatment with OT, logopedics, PT, and the initiation of tertiary prophylactic factors to prevent further atherosclerotic disease and any successive embolism, and by addressing the patient's high blood pressure, putting a stop to smoking, and considering nutritional factors, etc., the patient has been able to lead a self-supported life with a persistent minor speech impairment and mild hemiparetic symptoms.

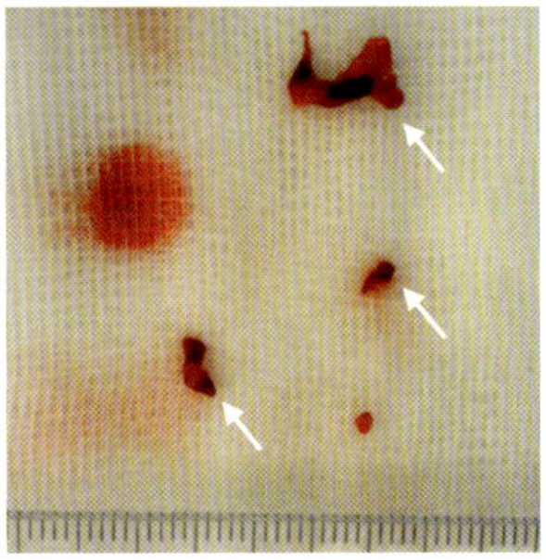

FIG. 7_*Intra-arterially aspirated and removed thrombotic material (arrows). Measuring rod in millimeters.*

Image by Christian Našel, 2020[8]

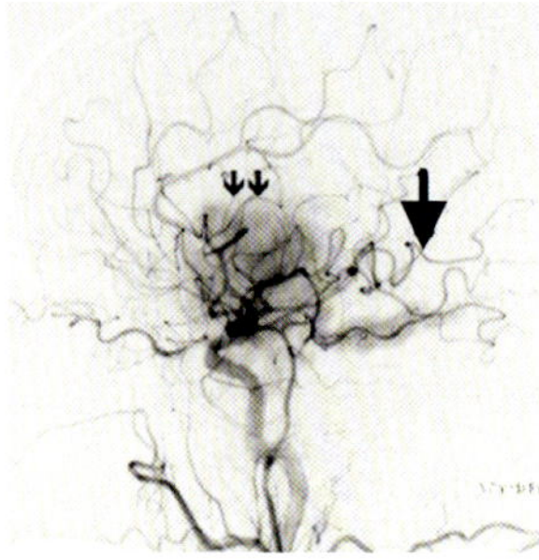

FIG. 8_*Screenshot from video. Control IADSA revealing all re-canalized MCA segments, including dorsal segments (single arrow). Cloudy contrast enhancement constitutes* hyperperfusion *(double arrows), i.e., post-ischemic defunct vasodilation in the core zone of the acute infarct.*

Image by Christian Našel, 2020[9]

VIDEO

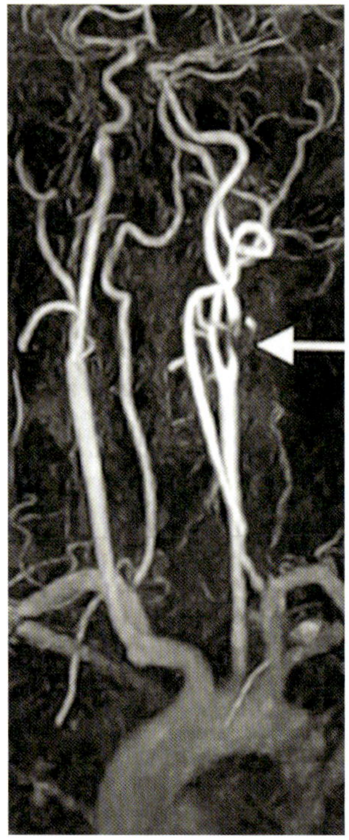

FIG. 9_*MR angiographic segmental, oblique, frontal overview of thoracic, neck, and part of the intracranial arteries showing stenosis (narrowing) in the cervical part of the internal carotid artery (ICA) of the patient's left side (arrow). This region was most likely responsible for the origin of the athero-sclerotic thromboembolism to the brain arteries.*

Image by Christian Našel, 2020[10]

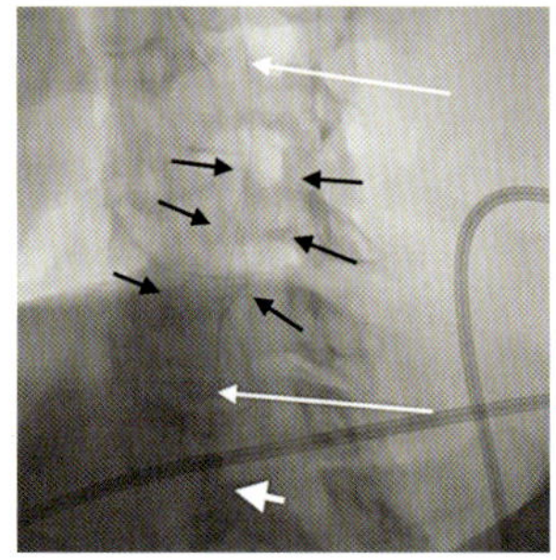

FIG. 10_*Lower cervical spine, viewed from the left side. Screenshot during fluoroscopic angiography after stent positioning and stent expansion to dilate an arterial stenosis in the internal carotid artery. X-ray fluoroscopy is used to visualize the elements introduced endovascularly that are required for the procedure. This is one of the interventional radiologist's viewing methods to immediately check on the progress and impact of his/her demanding, responsible, and ambitious work. Black arrows outline the expansion of the faintly radiopaque tubular stent. The short white arrow shows the distal end of the endovascular guide catheter. A thin endovascular guide wire runs from within the guide catheter via the tip of the lower long white arrow toward the tip of the upper long white arrow.*

Image by Christian Našel, 2020[11]

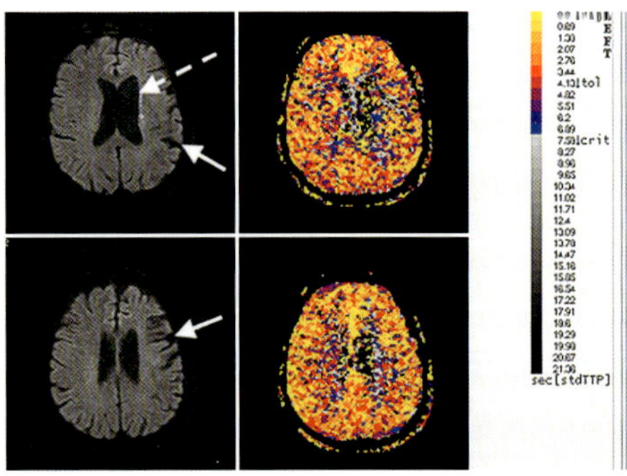

FIG. 11_*Same patient as in figure 2. Follow-up brain diffusion and perfusion MRI, fourteen months post-therapy. The black-and-white diffusion MR images show the discrete diminution of brain convolutions and a widening of brain grooves (linear arrows), as well as an enlargement of the cerebrospinal fluid-containing ventricle (dotted arrow) in the stroke-affected hemisphere. These are manifestations of localized brain atrophy as a long-term biophysical consequence of the acute infarct in figure 2. Colored stdTTP perfusion MRI maps show completely reestablished perfusion in the treated hemisphere. It is certain that immediate therapy prevented the then acute infarct zone from growing in size.*

Image by Christian Našel, 2021[12]

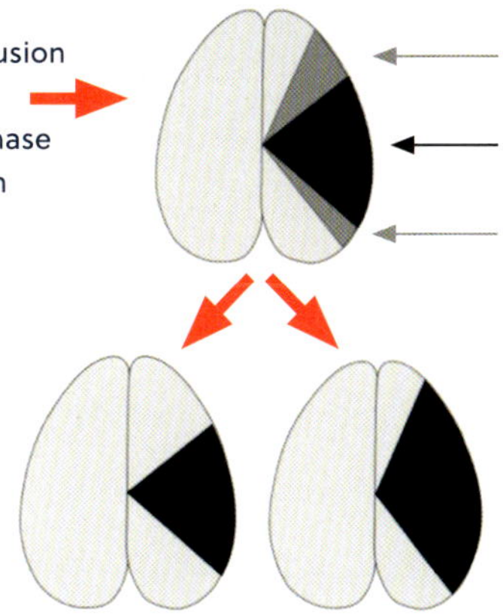

FIG. 12_*Stroke prognosis: the size of an acute infarct grows if regular perfusion is not restored in brain tissue that is salvageable despite critical blood supply — versus if reperfusion is achieved.*

Image by Karl Heimberger, 2022

Discussion: stroke requires emergency diagnosis and treatment within the first few hours after the onset of symptoms (motto: »time is brain«). This ensures the discovery of rescuable brain tissue that would otherwise — without the possibility of consecutive, well-coordinated therapy — be lost to permanent, fast-progressing brain damage (schematic_FIG. 12), leading to a poor clinical outcome for the patient with a more substantial neurological deficit and a life-long, substantially lower degree of self-sufficiency.

During the acute phase of a stroke, diffusion MR images provide information about the core size of the non-reversibly, permanently damaged infarct zone. But the still viable and therapeutically salvageable, adjoining, critically perfused brain area may be much larger and can be detected using a perfusion MRI technique. The initial diffusion/perfusion discrepancy in size between the core infarct and the bulk of savable brain tissue at risk is referred to as a *diffusion/perfusion mismatch*. Together, a recent, sudden clinical onset of neurological symptoms, proven critical perfusion MRI values quantified by the standardized time-to-peak method (stdTTP_FIG. 2),[13] the prognostically important *mismatch* visualization of an acute ischemic infarct_FIGS. 2, 12, and diagnostic IADSA_FIGS. 3, 5 call attention to the need for instant treatment.

Some of the characteristics of stdTTP are:

a) it quantifies and demonstrates the hemodynamic burden of brain tissue immediately after stroke onset and can therefore indicate the necessity for prompt therapy;
b) it is independent of age;
c) it produces absolute time measures by documenting the time span from the first appearance of a contrast-induced signal change to the maximum signal change in all examined brain volume units as a measure of brain capillary perfusion (color-coded time scale_FIGS. 2, 11[14]);
d) it is intra- and inter-individually comparable;
e) it gives results quickly, within seconds for all volumes examined; and
f) there is no need for the input of extra parameters.[15]

Since W.C. Roentgen's innovation of 1895, conventional radiology had been performed by exposing film foils to ionizing radiation. Today, technologies such as electronic detectors and analog-to-digital converters are used to produce images: in CT-scanning, the tube that generates the X-ray more or less spirals around the patient while the table supporting the patient is moved. CT uses the tissue-specific attenuation of X-rays registered with one- or two-dimensional detector orientation for high spatial resolution to discernibly display volumes of less than $1mm^3$. Attenuation numbers (Hounsfield units, value of »zero« calibrated to H_2O) are transcribed to a shades-of-gray scale used universally and documented as cross-section images (multiplanar tomograms) without and/or with contrast agent enhancement, or as volume-rendered ceCT angiograms_FIG. 13 or documented in different projections. A 3D visualization can be reconstructed from ceCT data. The 3D-simulated reconstruction of an intracranial aneurysm (the pouch-like distention of an artery) serves as an example in_FIG. 13.

MRI, however, does not make use of X-rays. To give a very quick and cursory portrayal of how it works: external bursts of variable radio frequency pulses being repeatedly sent into a *magnetic* field result in changes in the energy and magnetization levels of atomic nuclei within the organism being examined. After the radio pulses are cut off, and while the nuclei in various organs return to their original lower levels of energy and their original type of magnetization (i.e., »relaxation«), the nuclei send tissue-specific electromagnetic signals (dependent on the tissue's proton density and conditions like pulse frequency type, repetition intervals) to external antennae. These pickup units receive signals at various predetermined reception times and, in this manner, also determine the tissue contrast in mathematically reconstructed images. The interaction between magnetic field, radio pulse frequencies, and atomic nuclei is called *resonance*.

Different modalities of data acquisition, processing, and projection can be used for different kinds of image reconstruction, also to produce angiography-like *imaging* (magnetic resonance angiography, MRA_FIG. 9). Besides structural information, MRI can also provide functional information, for instance, about brain blood perfusion in stroke_FIGS. 2, 11—or, to be more precise: contrast-enhanced perfusion MRI can predict the danger of an acute ischemic brain infarct growing quickly in size if left untreated_FIG. 12. Color-coded stdTTP perfusion MRI maps document the time span from the appearance of contrast-induced signal change to the maximum signal change as a measure of brain capillary perfusion_FIGS. 2, 11 in all brain volume units.

Those whose interest in neuro-MR techniques does not end at this point may wish to read further.[16] However, within the scope of this article, of greater interest is the question of how deep medical images can provide insights into the hidden causes of disease and how well imaging may help to find specific therapies.

In *IADSA*, X-ray fluoroscopy is employed and connected to electronic detectors, an image-intensifier, and an analog-to-digital converter to exclusively visualize the blood vessels' ramifications and arterial, capillary and venous phases by means of the intra-arterial administration of an X-ray-attenuating contrast agent, and by simultaneously subtracting images of the surrounding bones and soft tissues.

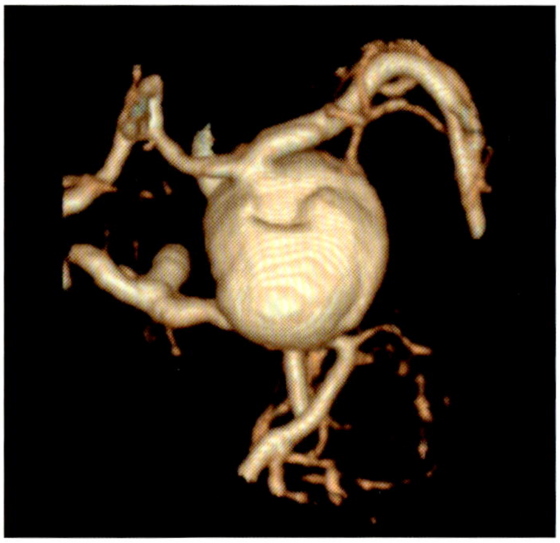

FIG. 13_*Screenshot from video Three-dimensional visualization of an intracranial aneurysm created by applying a volume-rendering technique to contrast-enhanced spiral CT image source data*

Image by Karl Heimberger, 2006

VIDEO

Quality assurance: the reliable diagnostic interpretability of any single imaging method is examined by several independent expert observers in scientific study setups. A loss of image quality can be due to hardware-related artifacts or due to implants (some of them are unsuitable, some are not allowed in MR examinations for safety reasons), or due to software (too low or too high signal-to-noise ratio, etc.), or it can be procedure-related. Regularly performed quality controls ensure the reliability of results. Patient movement can also be a reason for inadequate image quality.

Figures: some of the figures presented here provide screenshots and explanatory legends for videos, which run rather quickly.

Medical Follow-Ups and Personal Remarks

We recommend not losing faith in diligent, ongoing scientific research.[17] Medical research may find ways and means of preventing disease as well as remedies for patients with health impairments that are currently resistant to therapy. Regular check-ups with specialists allow patients to receive information about the latest developments in medical science and their translation into clinical practice, while even expensive, modern, state-of-the-art therapy is made available to all patients regardless of income. This is, fortunately, possible in healthcare systems that are financed on the basis of solidarity. One recent major scientific advancement has been the manufacturing of injectable monoclonal antibodies that specifically target pathogenic mechanisms, like with neurodermatitis,[18] bronchial asthma, or meningiomas.[19] Moreover, the Medical University of Vienna »has defined Medical Imaging as one of its five key focal areas of research« for the Medical Imaging Cluster.[20]

On the other hand, in individual home-based care fields, feedback from clients has been shown to be a major factor in personal job motivation for care staff. The same applies to medical staff. A personal interest in the sustained wellbeing of the patients being treated is the basis of nursing personnel's and medical technical staff's persistent engagement and interest in daily practice and in teamwork in healthcare centers, hospitals, out-patient clinics, etc. even though nursing staff are overwhelmed by workloads,[21] despite a lack of employees, and although nurses and assisting personnel are sometimes not appreciated enough (or at all) by their employers for working so hard. But the fields of medical management and curriculum management at economic universities have turned their growing attention to the beneficial effects of burnout prevention and therapy,[22] and corporate culture (special business management, e.g., commissioned projects and courses with practical training in »organizational culture« and »anti-bullying« at the Vienna University of Economics and Business, Institute of Public Management and Governance, winter semester curriculum 2019/2020; F. Ebinger in cooperation with K. Heimberger, W. Lalouschek, T. Majoros, S. Schwendt, et al.).

For his valuable contribution to the informative content of this article and for providing relevant figures and videos, the authors would like to extend their warm thanks to Prof. Karl Rössler, Head of the Department of Neurosurgery at the MedUni Vienna, Austria.

Acronyms and Reiterations of Physics of Magnetic Resonance

ceCT contrast-enhanced computed tomography

CT computed tomography

Diff MRI, diffusion MR diffusion magnetic resonance imaging, which provides information on the diffusivity or obstructions to the movement of atoms in organisms (e.g., in brain swelling) by using special radio frequency pulses during brief alterations in the equipment's local magnetic fields

H&P medical history and physical examination

H_2O water

IADSA intra-arterial digital subtraction angiography

ICA internal carotid artery

MCA middle cerebral artery

MR magnetic resonance: in a magnetic field, a repetitive external radio pulse sequence produces a state of higher energy and changes the magnetization of a great number of atomic nuclei in the organism being examined. When the pulse finishes, these nuclei return to a state of lower energy and their prior magnetization level. During this time, the nuclei send electromagnetic signals that are typical of various types of tissue and induce electricity in an antenna. The signals that are generated multiple times are used to gain image contrasts.

MRA magnetic resonance angiography

MRI magnetic resonance imaging

OT occupational therapy

Perf MRI, perfusion MRI contrast-agent-mediated perfusion magnetic resonance imaging, providing information on the blood supply to tissue

PT physiotherapy

QRC quick response code; a sort of barcode that electronically connects to a website

stdTTP standardized time to peak; a contrast-enhanced MR-based quantitative measure for brain perfusion

1 http://www.annasteinhaeusler.com.

2 https://www.studiosylt.de.

3 Christian Našel, Head of the Department of Radiology, Karl Landsteiner Private University, Tulln, Austria; secondary affiliation: Department of Biomedical Imaging and Image-Guided Therapy, Medical University of Vienna, Austria.

4 Christian Našel, Nicole Kronsteiner, Erwin Schindler, Sören Kreuzer, and Stephan Gentzsch, »Standardized Time to Peak in Ischemic and Regular Cerebral Tissue Measured with Perfusion MR Imaging,« *American Journal of Neuroradiology* 25 (2004): 945–50.

5 Christian Našel, Head of the Department of Radiology, Karl Landsteiner Private University, Tulln, Austria; secondary affiliation: Department of Biomedical Imaging and Image-Guided Therapy, Medical University of Vienna, Austria.

6 Christian Našel, Head of the Department of Radiology, Karl Landsteiner Private University, Tulln, Austria; secondary affiliation: Department of Biomedical Imaging and Image-Guided Therapy, Medical University of Vienna, Austria.

7 Christian Našel, Head of the Department of Radiology, Karl Landsteiner Private University, Tulln, Austria; secondary affiliation: Department of Biomedical Imaging and Image-Guided Therapy, Medical University of Vienna, Austria.

8 Christian Našel, Head of the Department of Radiology, Karl Landsteiner Private University, Tulln, Austria; secondary affiliation: Department of Biomedical Imaging and Image-Guided Therapy, Medical University of Vienna, Austria.

9 Christian Našel, Head of the Department of Radiology, Karl Landsteiner Private University, Tulln, Austria; secondary affiliation: Department of Biomedical Imaging and Image-Guided Therapy, Medical University of Vienna, Austria.

10 Christian Našel, Head of the Department of Radiology, Karl Landsteiner Private University, Tulln, Austria; secondary affiliation: Department of Biomedical Imaging and Image-Guided Therapy, Medical University of Vienna, Austria.

11 Christian Našel, Head of the Department of Radiology, Karl Landsteiner Private University, Tulln, Austria; secondary affiliation: Department of Biomedical Imaging and Image-Guided Therapy, Medical University of Vienna, Austria.

12 Christian Našel, Head of the Department of Radiology, Karl Landsteiner Private University, Tulln, Austria; secondary affiliation: Department of Biomedical Imaging and Image-Guided Therapy, Medical University of Vienna, Austria.

13 Našel et al., »Standardized Time to Peak.«

14 Našel et al., »Standardized Time to Peak.«

15 For a comparison with other perfusion method concepts, see Christian Našel, Uros Klickovic, Heike-Marie Kührer, Kersten Villringer, Jochen B. Fiebach, Arno Villringer, and Ewald Moser, »A Quantitative Comparison of Clinically Employed Parameters in the Assessment of Acute Cerebral Ischemia Using Dynamic Susceptibility Contrast Magnetic Resonance Imaging,« *Frontiers in Physiology* 9 (2019): 01945; Scott W. Atlas, *Magnetic Resonance Imaging of the Brain and Spine*, 5th ed. (Philadelphia; New York: Wolters Kluver; Lippincot, Williams & Wilkins, 2016).

16 Peter D. Schellinger, Jochen B. Fiebach, Sabine Heiland, and Klaus Sartor, »Perfusion-Weighted MRI,« in *Stroke MRI*, ed. Jochen B. Fiebach, Peter D. Schellinger et al. (Darmstadt: Steinkopf Verlag/Springer Science, 2014), 23–9; SW Atlas, *Magnetic Resonance Imaging*.

17 Medizinische Universität Wien, »Jahresbericht,« accessed September 2022, https://www.meduniwien.ac.at/web/presse-meduni-wien/jahresbericht.

18 Eric L. Simpson, Thomas Bieber, Emma Guttmann-Yaski, et al., »Two Phase 3 Trials of Dupilumab Versus Placebo in Atopic Dermatitis,« *The New England Journal of Medicine* 375, no. 24 (2016): 2335–48.

19 Matthias Preusser and Christine Marosi, »Antiangiogenic Treatment of Meningiomas,« *Current Treatment Options in Neurology* 17, no. 7 (2015): 359.

20 For more information on medical imaging research, see Medical University of Vienna, »General Information,« accessed September 2022, https://cluster.meduniwien.ac.at/mic/general-information.

21 Günter Valda, *House of Fate* (Dortmund: Verlag Kettler, 2021).

22 Wolfgang Lalouschek, »Burn-Out Therapy and Prevention by Coaching Medical Staffs,« personal communication.

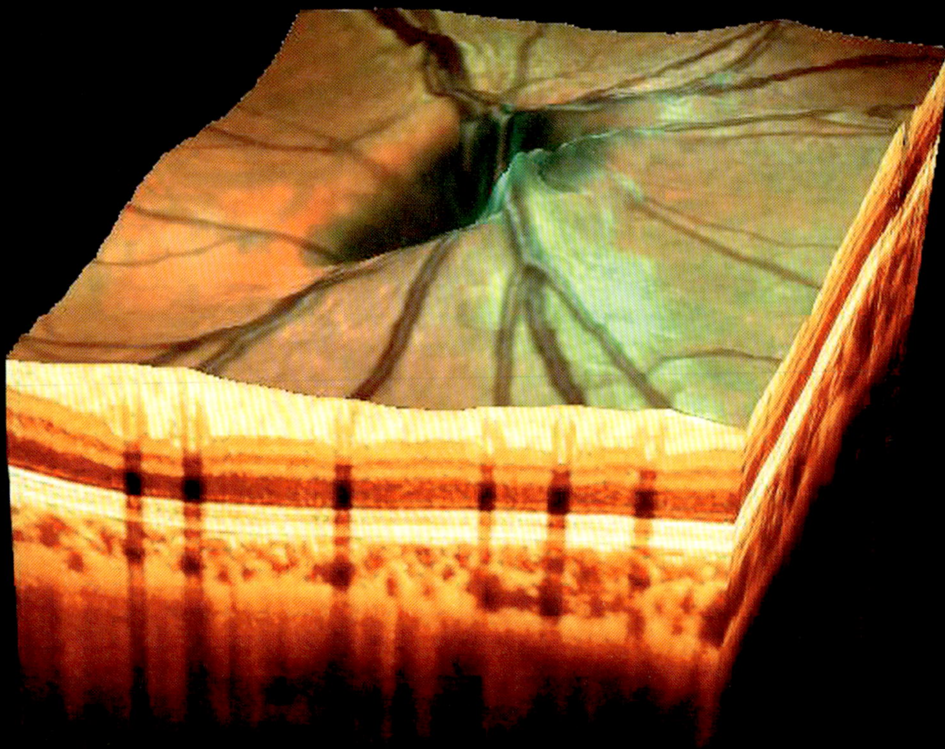

FIG. 1_ *Imaging the human eye by OCT: high-resolution three-dimensional representation of a section of the human retina created using optical coherence tomography (OCT). The image shows the point where the optic nerve enters the eye with the accompanying retinal blood vessels. The different layers of the retina can also be seen in the sectional image.*

Image courtesy of Gerhard Garhöfer, Medical University of Vienna, 2022

Imaging and Medicine

Markus Müller

»Wär nicht das Auge sonnenhaft, die Sonne könnt es nie erblicken.«
(If the eye were not sun-like, it could never see the sun.) ***Johann Wolfgang Goethe***

Vision and Imaging

Medicine and imaging are closely interrelated topics and the medical discipline of ophthalmology provides ample evidence for the interconnectedness of imaging, vision, cognition, wellbeing, and human disease. Vision is a perception of light, an electromagnetic wave, and a bundle of light quanta with a speed of approximately 300,000 kilometers per second, perceived by a healthy human eye at a limited wavelength of 400 to 750 nanometers. Higher-frequency UV light or lower-frequency infrared light eludes human perception. Thus, humans and the visual senses are limited to a narrow »meso-world,« a small sector of our physical world, which extends from the universe to the subatomic level. The development of the eye and thus visual perception was a longstanding riddle of evolutionary theory[1] and is the result of an evolutionary selection process that has taken advantage of the information provided by light quanta. Vision, which is probably the strongest of the human senses, is based on highly complex neuronal processes, which were described in the fundamental work of 1981 Nobel Laureates for Medicine David Hubel and Thorsten Wiesel.[2] The retinal cells and nerves_FIG. 1 serve as sensors for light quanta and pass on this information to specific neocortical regions, which connect to associative areas in the brain.

These areas arouse conscious perception and emotional reactions. Our culture, our aesthetic judgments, and our ability to interpret pictograms, hieroglyphs, letters, numbers, paintings, and pictures is thus wholly dependent on the proper functioning of our visual apparatus. For most of medicine's history, imaging was limited to mere visual inspection, which has a physiological resolution limit in the micrometer range, determined by the distance between retinal cells. Geniuses like Vesalius[3] and Leonardo da Vinci drew many important insights from visually inspecting the human body and its organs, and preserved this information in outstanding drawings. The *Codex Windsor*, for instance, contains a drawing by

Leonardo of a particular vortex formation flow around the aortic valve, a finding that was rediscovered in the twentieth century using modern MR techniques.[4] How important and educational visual inspection has remained up to this day and age is underlined by outstanding artistic artworks by medical illustrators like Frank H. Netter[5] and the controversies surrounding the *Pernkopf Atlas of Anatomy*, an iconic product of immoral science.[6]

Lenses and the Beginnings of a Mechanistic View of the Human Body

The manufacturing of high-quality lenses in the seventeenth century in the Netherlands and Italy led to a revolution in art but also in science and medicine, and pushed the limits of visual detection into the nanometer range. Vermeer and many other artists worked with lenses and the camera obscura to create magnificent paintings, and Galileo Galilei was able to observe sunspots and the moons of Jupiter. Whereas Galileo was looking for the mysteries of the universe, others were looking for hitherto hidden worlds in organisms. For the natural sciences, the development of the microscope by Antoni van Leeuwenhoek was a key achievement _FIG. 2A. Leeuwenhoek was the first to explore microbials from his mouth (»animalcules«), blood cells, and the movement of sperm, and was there at the beginning of scientific reasoning about cells being unique entities and constitutive elements of life.[7] The scientific progress facilitated by lenses is a prime example of the impact of technology on medicine. Technological advances have always led to deeper insights into the mysteries of the human organism. Besides technology, materialistic and reductionistic reasoning provided fertile ground for medical discoveries. The French physician Julien Offray de la Mettrie, a contemporary of Leeuwenhoek, published his book *L'homme machine* in 1748. This product of the Age of Enlightenment pursues a purely materialistic idea of the human organism. La Mettrie's reductionistic concepts and the idea that the human organism can be understood as the ever-smaller layers of parts of a machine still play an important role in the concepts used in medicine and mechanistic biology. Whereas the Leonardian world understood the human machine as the sum of its organs, the Leeuwenhoekian world advanced to a cellular level with deep insights into the cell biology of the human organisms' approx. 3×10^{13} cells, which, spectacularly, are the result of the fertilization of a single original cell. Ironically, Leeuwenhoek and his later contemporaries still adhered to preformistic Aristotelian theories_FIG. 2B despite the visual evidence to the contrary. Later, the twentieth-century world moved along this reductionistic path of reasoning to a molecular level of understanding and describing human biology. The new concepts of cellular and microbial medicine led to an explosion of knowledge about health and disease at the end of the nineteenth century by means of the microscope but also inspired many other disciplines like the arts. Klimt's iconic painting *Danaë* from 1907, for instance, depicts human egg cells and blastocysts, which Klimt studied using a microscope at the Department of Anatomy at the Medical Faculty of Vienna, which was chaired by Klimt's friend, anatomist Emil Zuckerkandl.[8]

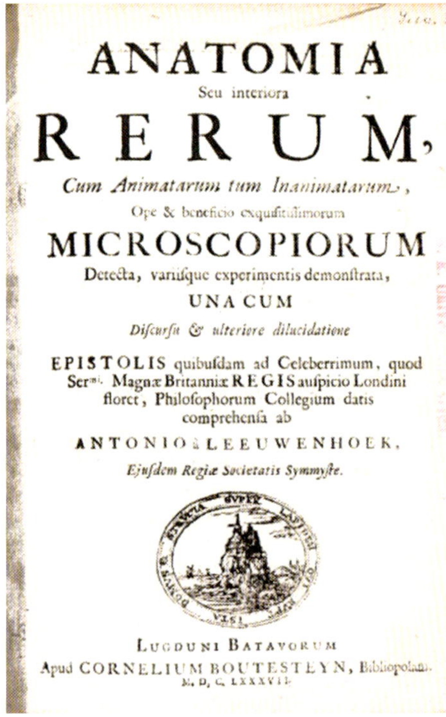
ANATOMIA
Seu interiora
RERUM,
Cum Animatarum tum Inanimatarum,
Ope & beneficio exquisitissimorum
MICROSCOPIORUM
Detecta, variisque experimentis demonstrata,
UNA CUM
Discursu & ulteriore dilucidatione
EPISTOLIS quibusdam ad Celeberrimum, quod Ser^mi. Magnæ Britanniæ REGIS auspicio Londini floret, Philosophorum Collegium datis comprehensa ab
ANTONIO à LEEUWENHOEK,
Ejusdem Regiæ Societatis Symmyste.
LUGDUNI BATAVORUM
Apud CORNELIUM BOUTESTEYN, Bibliopolam.
M. D. C. LXXXVII.

FIG. 2_*Leeuwenhoek's microscopical studies and the discovery of cells*
A_*Preface to Antoni van Leeuwenhoek's* ANATOMIA SEU INTERIORA RERUM, CUM ANIMATARUM TUM INANIMATARUM, OPE & BENEFICIO EXQUISITISSIMORUM MICROSCOPIORUM DETECTA, VARIISQUE EXPERIMENTIS DEMONSTRATE[9] *(Anatomy, or the Interior of Things, of Animate as well as Inanimate Things, Detected with the Help of Clever Microscopes, Demonstrated by Various Experiments)*

Image courtesy of the Collections of the Medical University of Vienna and Christian Schöfer), 2022

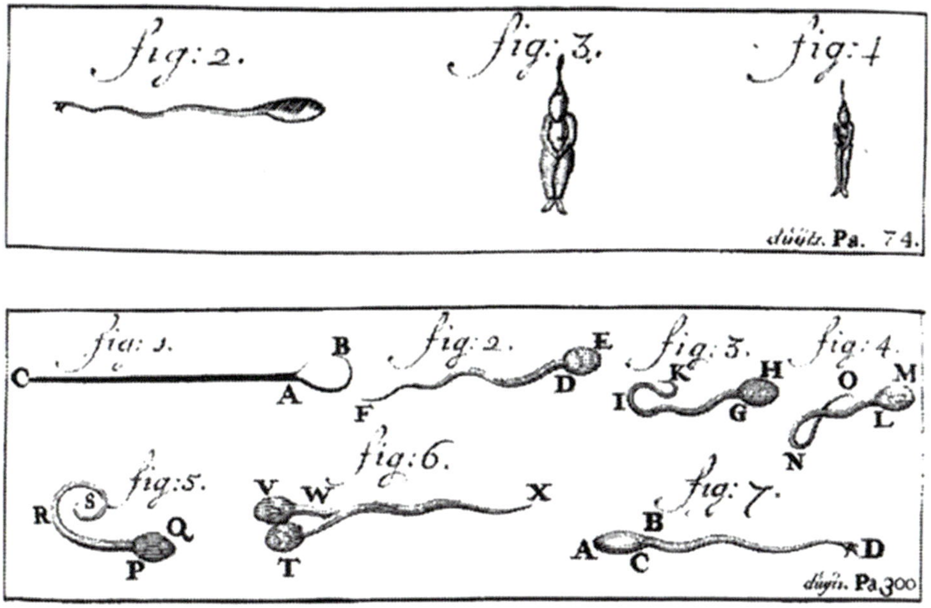

B_*First description of sperm cells from Leeuwenhoek's work, also indicating preformistic theories. The first accounts of Leeuwenhoek's discoveries were published by Robert Hooke and the Royal Society. In 1687, a comprehensive overview was published by Cornelius Boutesteyn, Leiden.*

Image courtesy of the Collections of the Medical University of Vienna and Christian Schöfer, 2022

A Revolution in Anatomy: X-Rays and Radioactivity

In the late nineteenth century, medicine progressed at an unprecedented speed as a direct result of the accelerating scientific understanding of the different layers of the human organism. The catalyst for this process was spectacular advances in imaging technology, most importantly the discovery of X-rays by Wilhelm K. Röntgen in 1895 and of radioactivity by Pierre and Marie Curie in 1898. Both discoveries led to a revolution in the understanding and diagnosis of disease and were honored with the Nobel Price. X-rays provided an in vivo, interior view of the skeleton and human anatomy for the first time and transformed internal medicine and the practice of surgery. Likewise, the discovery of radioactivity led to the development of the discipline of nuclear medicine. However, those early days of unprotected work with ionizing radiation also led to a substantial number of X-ray- and radium-poisoning martyrs, like Austrian radiologist Guido Holzknecht and Marie Curie herself.

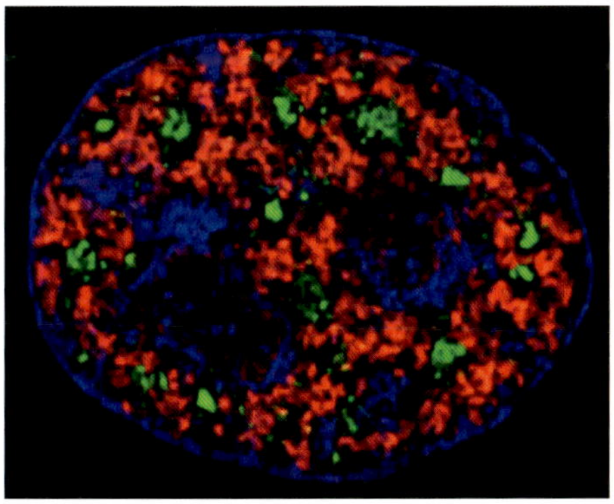

FIG. 3_*Imaging the molecular machinery of gene transcription by SIM: cell nucleus of a HeLa cell depicting the Histon H3.3 variant (red) and the splicing protein SC-35 (green). This picture was produced by structured illumination imaging microscopy (SIM) with a resolution (below the Abbe limit) of about 200 nanometers.*

Image courtesy of Christian Schöfer, Medical University of Vienna, 2022

Pushing the Limits in Microscopy

Besides the X-ray revolution, substantial advances were made by refining optical microscopy technology. The resolution limit was expanded by employing other forms of electromagnetic waves (electron microscopy) and non-optical tools (atomic force microscopy), and by making specific modifications to the microscopic principle, which gave rise to a number of techniques like (immuno-)fluorescence microscopy, confocal microscopy, high-resolution microscopy, and structured illumination microscopy (SIM)_FIG. 3, which offered significant advantages. These techniques made it possible to push the visual limits to study mechanistic cell biology and break the Abbe Limit of approx. 200 nanometers (light). In 2014, the development of super-resolution fluorescence microscopy was honored with the Nobel Prize in Chemistry.

Modern Medicine: Computed Tomography (CT), Photon-Counting CT Scanners, and MRI

The anatomical revolution, which started with Röntgen's discovery of X-rays, was accelerated by the development of tools to obtain sectional images of the human body. In 1979, Sir Godfrey Hounsfield received the Nobel Prize for the development of computed tomography, a technique that, to date, has grown to hundreds of millions of applications annually. Like in the case of microscopy, a vast array of new CT modifications and CT applications, such as perfusion- and photon-counting CT, have allowed us to obtain extremely accurate information about the anatomy and functioning of the human body. Since medical imaging is largely determined by physical constants, e.g., wavelength, another analogy to the development of microscopy is the employment of different sources of energy and fields to the tomographical principle. In contrast to CT, MRI exploits the ability of protons to relax into a resting state after induction by a magnetic field, a now widely used technology with millions of scans worldwide each year, which earned Peter Mansfield and Paul Lauterbur the Nobel Prize in 2003_FIG. 4. Like CT, MRI technology has expanded extensively in scope and application (e.g., MR angiography, functional and perfusion MR, real-time MRI, high-field MRI_FIG. 4).

Functional Imaging: PET, SPECT, fMRI, PET-MRI Fusion

The revolution that started with Curie's discovery of radioactivity led to a number of important imaging technologies based on the in vivo application of radioactively labeled molecules (»radiotracers«) combined with the tomographical principle. Positron emission tomography (PET) imaging describes the distribution pattern of, e.g., 11C or 18F labeled, positron-emitting radiotracers in the human body and is based on the principle of μ-ray annihilation radiation after the collision of a positron with an electron. PET not only provides anatomical images but also allows us to study selected molecules, molecular pathways, metabolism, pharmacology, and behavior_FIG. 5. A related principle, single photon emission computed tomography (SPECT), is based on the application of μ-ray-emitting tracers. Another functional imaging method, functional MRI (fMRI), makes use of functional changes to blood flow_FIG. 6. Today, different techniques are frequently employed in combination, e.g., PET-MRI fusion or MRI spectroscopy, in order to optimize topical, anatomical, and functional information_FIG. 4.

FIG. 4_*Imaging of tumor metabolism by »high-field« 7 Tesla MRI: high-resolution metabolic imaging with 7 Tesla MRI in a patient with high-grade glioma (glioblastoma). In one examination, up to nine different metabolites of the brain can be examined with high resolution on 7T MRT in order to assess the different structures and the extent of the brain tumor in terms of metabolism. High tumor activities can be seen on the choline and glutamine images. The left row depicts the standard contrast MR images on 7T and 3T for comparison. The lower row depicts the ratio images to better highlight the tumor activities. This high-resolution spectroscopic imaging improves the noninvasive assessment of tumor grading, makes it possible to carry out more targeted biopsies of high-grade tumor areas, and facilitates recurrence diagnostics.*

Image courtesy of Siegfried Trattnig, Medical University of Vienna, 2022

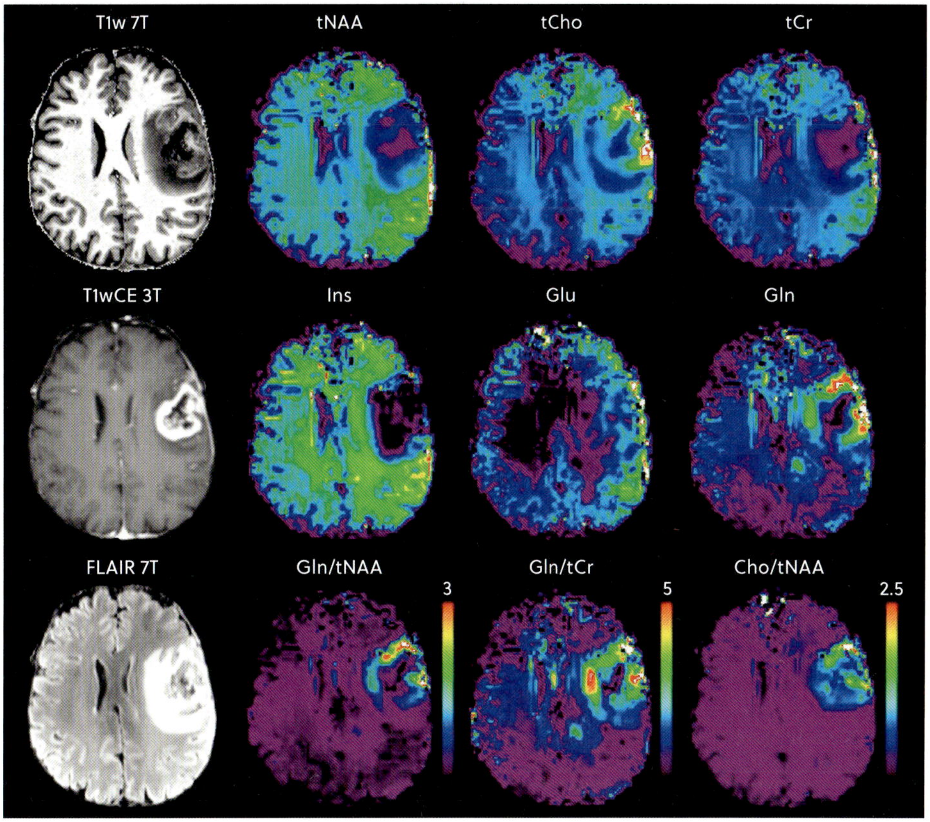

FIG. 5_*Imaging brain distribution of proteins by PET: distribution of the enzyme monoamine oxidase A (MAO-A), quantified by positron emission tomography (PET). MAO-A is important regarding the degradation of modulatory neurotransmitters as dopamine, serotonin, and norepinephrine. Axial (left), sagittal, coronal (right) views.*

Image courtesy of Rupert Lanzenberger, Medical University of Vienna, 2022

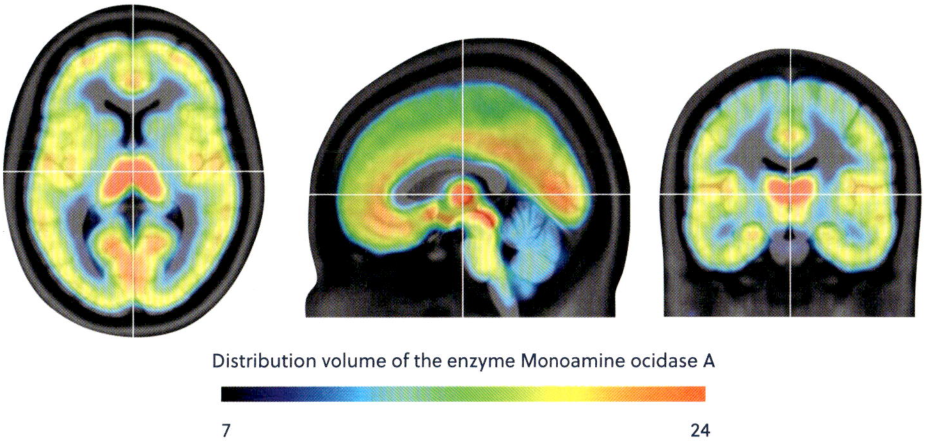

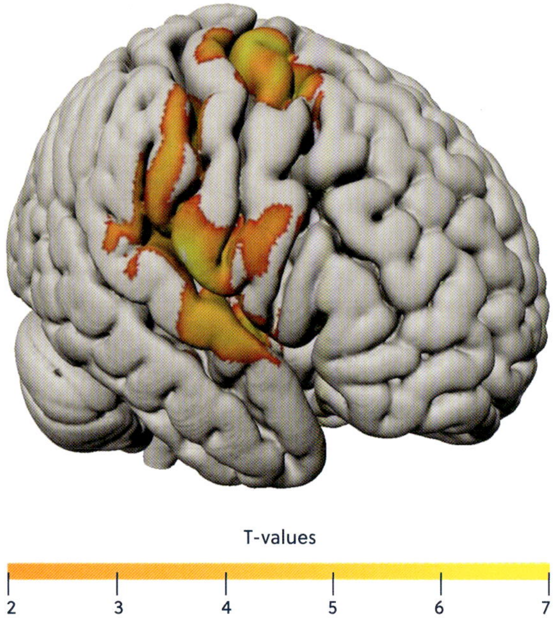

FIG. 6_*Imaging brain activation during cognitive tasks by fMRI: brain activation during a reward task measured with functional magnetic resonance imaging (fMRI, 3T), i.e., monetary incentive delay task (group feedback response to a monetary loss contrasted against gain during a monetary incentive delay task, lateral 3D brain view)*

Image courtesy of Rupert Lanzenberger, Medical University of Vienna, 2022

Other Sources of Waves and Optical Coherence Tomography (OCT)

In recent decades, many different sources of light and energy, and magnetic fields have been exploited for medical imaging purposes. Besides high-frequency ionizing waves like μ-rays and X-rays, lower-frequency technologies have provided infrared imaging, ultrasound, and doppler imaging. Optical coherence tomography (OCT), based on laser light imaging, has proved to be a particularly useful invention as it is a versatile technology that is the basis of a noninvasive method to study the eye_FIG. 7.

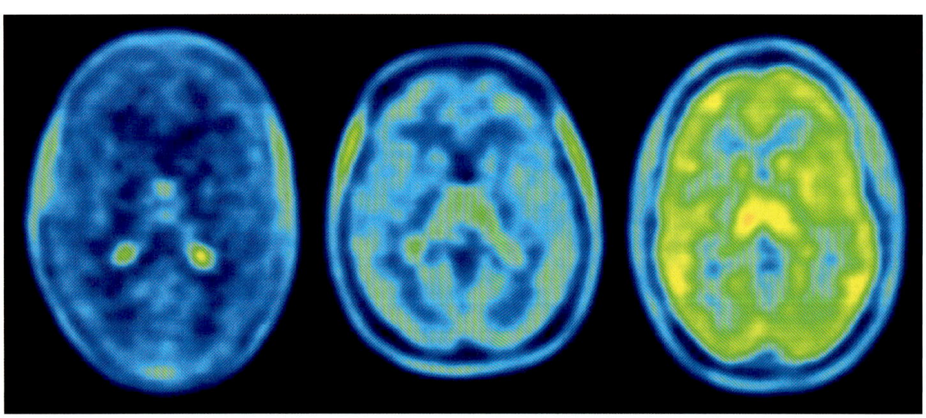

FIG. 7_*Imaging drug delivery by PET: positron emission tomography (PET) imaging of drug delivery to the brain. In this experiment, three times the same amount of a radiotracer was administered to healthy volunteers (baseline left). With another drug, tariquidar (3 [middle] and 8mg/kg/bw [right]), the blood-brain barrier is »opened« and more of the radiotracer enters the human brain.*

Image courtesy of Martin Bauer and Oliver Langer, Medical University of Vienna, 2022

The Future of Imaging: Artificial Intelligence and Machine Learning

In the last 500 years, no discipline in medicine has remained unaffected by the spectacular advances made in imaging. The progress in surgery, internal medicine, pharmacology, obstetrics, psychiatry, and many other disciplines would be unthinkable without imaging tools and is vastly dependent on the availability of a large number of highly sophisticated and methodologically diverse imaging technologies. Many aspects of the physical laws have been exploited to visualize the human body's form and functions at an anatomical, cellular, and molecular level. Most likely, and not unlike in Leeuwenhoek's age, medical imaging will be revolutionized once again by emerging technologies. In the near future, AI and machine learning will change current medical imaging concepts and will open new doors for patients, doctors, students, and lovers of aesthetic images.

1 Richard Dawkins, *Climbing Mount Improbable* (London: Norton, 1996).
2 David Hubel and Torsten Wiesel, *Brain and Visual Perception: The Story of a 25-Year Collaboration* (Oxford: Oxford University Press, 2004).
3 Andreas Vesalius, *De humani corporis fabrica* (Basel: Joannes Oporinus, 1543).
4 Francis C. Wells, *The Heart of Leonardo* (London: Springer, 2013).
5 Frank H. Netter, *Atlas of Human Anatomy*, 8th ed. (Philadelphia: Elsevier, 2022).
6 Markus Müller, Herwig Czech, and Christiane Druml, »The Medical University of Vienna and the Historical Legacy of Pernkopf's Atlas,« *Surgery* 165, no. 5 (2019): 871–2, https://doi.org/10.1016/j.surg.2019.02.011.
7 Paul Nurse, *What is Life?* (New York: Norton, 2021).
8 Erich Kandel, *Age of Insight* (New York: Random House, 2012).
9 Antoni van Leeuwenhoek, *Anatomia Seu interiora Rerum, Cum Animatarum tum Inanimatarum, Ope & beneficio exquisitissimorum Microscopiorum Detecta, variisque experimentis demonstrate* (Leiden: Lugduni Batavorum, Cornelius Boutesteyn, 1687).

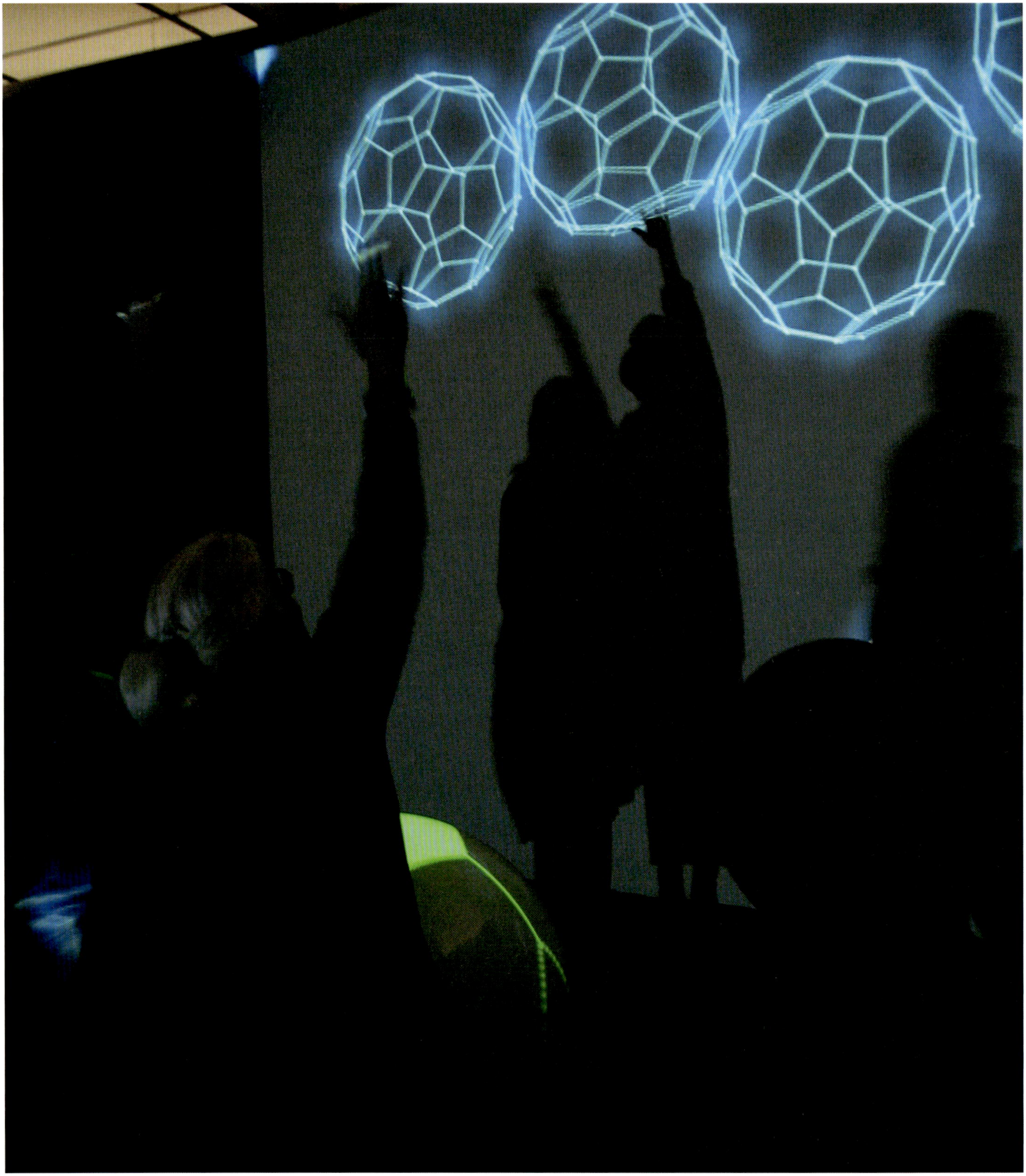

NANO: *Bottom up and in Between*

Victoria Vesna, James K. Gimzewski

This chapter is a retrospective on a long-term collaboration between a media artist and a nanoscientist, focusing on their first large-scale exhibition, *NANO*, which premiered at the Los Angeles County Museum of Art (LACMA) in 2003. Twenty years later, the authors reflect on the inspirations and motivations behind the creation of these large-scale art-science installations.

FIG. 1_ZERO@WAVEFUNCTION: *interactive buckyball molecules being manipulated by Buckminster Fuller's daughter, Allegra Snyder-Fuller (1927–2021), with her shadow, Los Angeles County Museum of Art (LACMA), 2003*

Photo by Victoria Vesna, 2003

VIDEO

NANO involved interconnected immersive spaces that aimed to shift perception in relation to micro-/macroscales. The goal was, and continues to be, to show a wider audience how important it is to understand invisible and inaudible realms. The collaboration continued with the development of further installations that had been part of that inaugural show—*Zero@wavefunction*, *Nanomandala*, and *Quantum Tunnel*—as well as with the creation of additional experiential works delving into the nano/bio realm such as the immersive *Blue Morph* installation. For the past decade, Victoria Vesna and James K. Gimzewski have worked on their retrospective research—Gimzewski using atomic and molecular imaging, and single particle manipulations, and Victoria Vesna with projects focused on environmental and ecological issues—both involving scale and perception from different angles.

Twenty years later, their artistic and scientific paths have crossed again, and they are now collaborating once more on a new work in progress—the *Atomic Gold Standard*—asking the public: What do we value?

FROM NETWORKS TO NANOSYSTEMS: Art, Science, and Technology in Times of Crisis

Less than a year after Victoria Vesna started her tenure as Chair of Design Media Arts in 2000, Gimzewski arrived at the University of California Los Angeles (UCLA) from IBM Zurich, where he had worked in an IBM corporate laboratory for twenty years. Initially, his research focused on a device developed there—the scanning tunneling microscope (STM)—and, later, on different approaches to nanoscale science. At UCLA, he was hired as a Professor in the Department of Chemistry and Biochemistry and was excited to work outside the corporate environment. Around the same time, Victoria Vesna came to the campus soon after obtaining her PhD in Roy Ascott's Planetary Collegium and was envisioning a new department that would cross disciplinary boundaries. Her research centered on the work of Buckminster Fuller and, through her study, she found out about a molecule named after him — the buckminsterfullerene.[1]

Months after their arrival at UC Los Angeles (UCLA), the world experienced 9/11, and shortly afterward, Governor of California Gray Davis announced the creation of the California Institutes for Science and Innovation. The institutes were spread across the University of California (UC) system, were allocated major funding, and were intended to cooperate with California-based industries, with a research focus on biotech and nanotech in bioengineering and medicine.[2]

Together with her colleagues in the UC Digital Arts Research Network (UC DARnet), Victoria Vesna co-organized a multi-campus symposium entitled Networks to Nanosystems: Art, Science, and Technology in Times of Crisis to start a dialogue with the scientists involved in nanotech and biotech research. Each campus was granted a building for an institute with a specific research focus, and nanotechnology was planned for UCLA along with UC Santa Barbara.[3]

While planning the symposium, Vesna was keen to have scientists engaged in nanotech research participating along with artists and humanist scholars. It was at this point that she heard about and invited a new faculty member to attend — nanoscientist Gimzewski, who had just joined the department of Chemistry and Biochemistry as a Professor. She received an immediate positive response from him, and their dialogue was initiated through their common interest in buckyballs — albeit from opposite sides of the pendulum.

Buckminster Fuller — engineer, architect, and visionary — left a legacy that created common ground for artists and scientists to interact. The new carbon allotrope beyond diamond and graphite was discovered only one year after Fuller passed away and was named after him due to its similarity with the structure of his domes. When Harry Kroto, Richard Smalley, and Bob Curl, the experimental chemists who discovered C60, named it buckminsterfullerene, they were honoring the work of Richard Buckminster Fuller, also known as Bucky (1895–1983). Smalley's laboratory equipment could only tell them how many atoms there were in the molecule, not how they were arranged or bonded together. From Fuller's model, they intuited that the atoms were arranged in the shape of a truncated icosahedron with its isolated pentagons giving the hexagonal units a three-dimensional form similar to that of a geodesic dome. Only after a novel phenomenon or concept is named can it be translated into the common currency of thought and speech. H.W. Kroto said that the newly discovered carbon cage molecule would be named buckminsterfullerene »because the geodesic ideas associated with the constructs of Buckminster Fuller had been instrumental in arriving at a plausible structure.«

Victoria Vesna's PhD research centered around the work of Bucky Fuller in relation to natural systems, while Gimzewski had been deeply absorbed in the molecule at IBM Zurich. Gimzewski had already been the first to publish an image of a molecule (copper pthtalocyanine) in 1986 using the STM in a vacuum environment. While it was quite difficult to obtain pure buckyballs at the time, he managed to get a small sample that was sufficient for imaging and published a series of articles that explored the nano-electro mechanical properties of individual molecules. When squeezed, the quantum levels of the molecules' electron levels shifted, suggesting an electromechanical amplifier just one nanometer across. His group was also able to

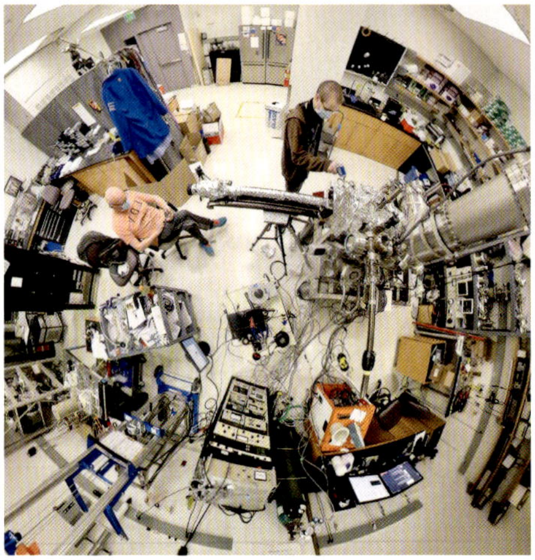

FIG. 2_ *360-degree view of Gimzewski's Lab at the California NanoSystems Institute, UCLA*

laterally move individual buckyballs along a single atomic step of copper. This was used to create an abacus utilizing ten molecules that could be slid to-and-fro along the copper step at room temperature. This creation was entered in the *Guinness Book of World Records* as the smallest calculator in the world and became an early icon of nanotechnology.[4]

Learning about this molecule and how it is manipulated was as fascinating for Vesna as it had been for Gimzewski to learn about the architectural work of Buckminster Fuller. Their exchange moved quickly into philosophical realms as both of them were interested in Buddhist ideas related to empty space and impermanence. Both of them continue to find it fascinating how quantum mechanics views interconnectivity and the nature of electron wavefunctions in a similar way.

Scanning Tunneling Microscopy (STM)— FEELING is BELIEVING/MEASURING INTERACTION/ dynamic VIBRATIONS.

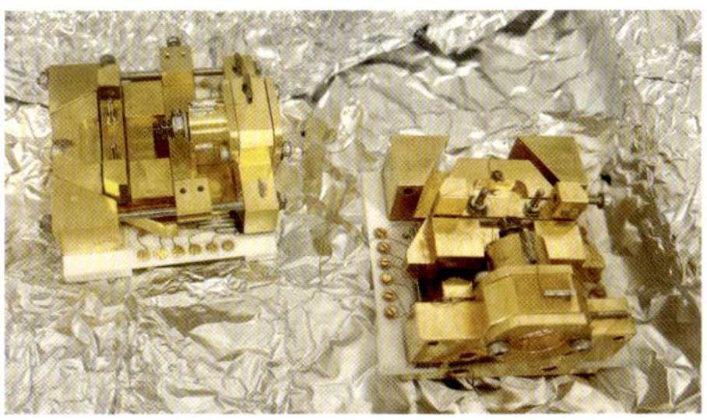

FIG. 3_*Two scanning tunneling microscopes (STM) designed and built by Gimzewski et al., IBM Zurich Research Laboratories, circa 1990*

Image by IBM Zurich Research Laboratories, 1990

One could say that the scanning tunneling microscope (STM) is a bit of a misnomer with respect to what people expect a microscope to contain, such as a lens that acts as a magnifying element to zoom in on a small field of view and make it visible. The STM contains no such element and instead has a sharp metallic point, terminated by an atom, that interacts with a surface and uses the interaction between the two extremities in closest proximity as a sensing element. If we take our finger and feel an object, we can visualize the shape and feel of it. The STM also utilizes an interaction, but on the atomic scale. Consider an atom on a surface and an atom at the end of a probing tip. When we feel something like a table top, the interaction is abrupt: our finger feels through the air until it reaches the solid table interface abruptly. However, the two atoms approaching each other already feel an interaction at a distance due to the fact that atoms are not solid spheres but rather diffuse. As the tip approaches, it dips into an evanescent field of electrons, at first lightly and then more intensely as the atoms come closer. If we define a level of intensity (wave function overlap of atoms) by measuring a small, millionth of a milliampere current (nA) by applying a small bias voltage lower than the one produced by an AAA battery as a constant, we can raster the tip at constant interaction and record the vertical distance (z) required to maintain a constant current when we raster or scan the lateral planes of the surface (x, y coordinates). This results in a topography, so to speak, of the landscape of the surface. One analogy is that of a blind person reading braille, where the bumps on the paper are atoms or molecules.

We may talk about seeing atoms, but that view is very stereotyped and continues to be widely propagated at high schools and universities using ball models of the atom. The ball model is not just misleadingly oversimplified, it is also wrong and, worse still, removes any of the mystery, intrigue, and beauty of its complexity, rendering it sterile and uninteresting. We don't actually »see« atoms; rather, we can visualize them by using image processing tools to create a landscape while we are in fact recording one part of a complex and fascinating interaction. The intensity of the interaction can be changed by using a different constant current or applied bias voltage, and we may see quite a different image reflecting the nature of the new interaction, particularly for a molecule or semiconductor atom. In fact, if we increase the interaction sufficiently, we can actually push and pull atoms around a surface or create new atomic bonds.

Dr. Don Eigler, from the IBM Almaden Research Center in San Jose, California, demonstrated that he could write the company logo (IBM) using individual xenon atoms and, for fun, Gimzewski and his students wrote »UCLA« using just forty carbon monoxide molecules shuffled around on a copper surface. These demonstrations certainly made nanotechnology more tangible, and one could say »seeing is believing,« but like almost everything today, seeing is not necessarily believing in the age of the »deepfake.« In some respects, the images of an experiment show only a scientist's *interpretation* of an atom, which is based on the nature of *interactions*. Science as a whole is based on studying interactions and interpreting the results. Art also involves the interpretation of what one senses and interprets in a highly individual manner. Today, the scientific visualization of data is similar, but the scientific image, for the public, implies a universal truth when, in fact, it is also subject to individual interpretation and bias. The front covers of journals like *Nature* and *Science* are certainly far from art in that they rarely convey the beauty of science or its mystery but rather more closely resemble the characteristics of glitzy industrial, pharmaceutical promotional material. There are several exceptions where artists have attempted to portray atoms and molecules, but rarely in close and direct dialogue with a scientist engaged in this work. Unlike the physical objects we experience every day, there is no visual cue like a landscape or an animal to interpret directly, because such a thing does not exist.

Gimzewski's colleague Franz Giessibl took an image of a silicon atom that was published in *Science* and subsequently republished in a newspaper at a low resolution. Artist Gerhard Richter was so impressed by this image that he made a photocopy from the newspaper article, enlarged it, and turned it into an artwork. This started an ongoing dialogue between the artist and the scientist. This relationship has been discussed in a recent book by Giessibl entitled *First View Inside an Atom: Encounters with Gerhard Richter Between Art and Science*. Their meetings had revealed more than just the science of an atom to Richter and became philosophical as they looked into an atom from many perspectives. Richter himself wrote: »Ich kann über Wirklichkeit nichts Deutlicheres sagen als mein Verhältnis zur Wirklichkeit, und das hat dann etwas zu tun mit Unschärfe, Unsicherheit, Flüchtigkeit, Teilweisigkeit oder was immer.«[5] Which roughly translates as: »I can't say anything clearer about reality than my relationship to reality, and that has something to do with fuzziness, uncertainty, ephemerality, partiality, or something similar.« Franz originally created a visual from his data as a discrete, crisp image; the newspaper printing and photocopying of it paradoxically created a blurred image, and yet, the fuzzy end result portrays the atom in a much more realistic manner in many ways. However one translates it, the aspects of fuzziness, uncertainty, and ephemerality for us constitute the beauty of the atom, which, in the quantum world, extends that atom out to infinity in its connection with the universe and makes it something that creates hope, awe, and mystery in the material, atomic world we live in and are made from. Fuzziness, uncertainty, and ephemerality are also essentially quantum mechanical traits that reinforce the idea that atoms are not just tiny spherical bricks held together with a mortar made of electrons but rather an extended fabric connecting nature and humanity.

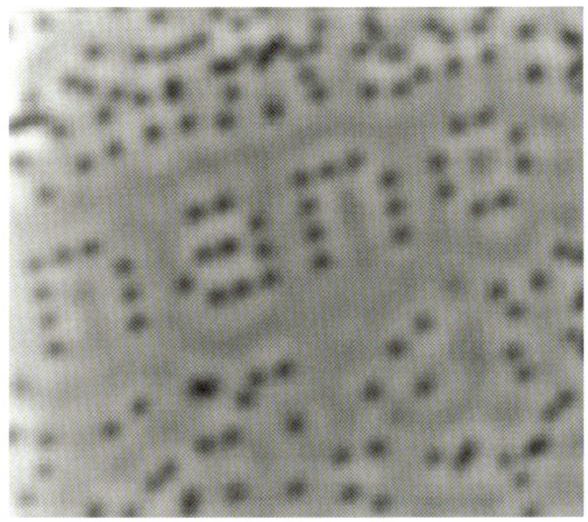

FIG. 4_*NANO written in carbon monoxide molecules by Don Eigler at IBM Alamden Laboratories for the exhibition at LACMA*

Image courtesy of Jim Gimzewski, 2002

VIDEO

Another attempt at the art of the atom worth mentioning can be found in some of the works by sculptor Kenneth Snelson, who was Buckminster Fuller's student at Black Mountain College and helped to develop the tensegrity structure idea. Although it was an impressive deep look into the atom on the part of the artist, one detects that it was not created in collaboration with a scientist involved in this kind of research. Ultimately, the real opportunity in representing the atom lies in the realm of digital visualization as it can better show the transient, ephemeral nature of what we are »observing.« Further, media artists use the same tools — computer imaging — as scientists, and this creates an easy bridge and blur between two worlds.

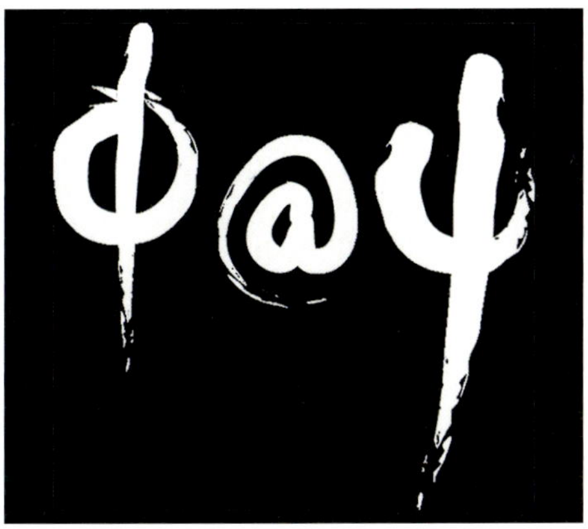

FIG. 5_ ZERO@WAVEFUNCTION *logo*

Logo design by Victoria Vesna, 2002

ZERO@WAVEFUNCTION

It is striking that artist Vesna and scientist Gimzewski, who did not know each other when they attended the Networks to NanoSystems symposium in 2001, showed similar images in their presentations. They were coming to this juncture from opposite sides, with the buckyball as the meeting point. The scientist did not know about Fuller, and the artist knew little of how the molecule behaves or how the STM works. They were learning from each other and started their first collaboration by questioning how molecules are represented in science — as fixed objects — which gives the wrong impression entirely and therefore skews understanding of the underlying principles. Change is constant, and the materiality at the bottom is not easy to grasp...

There was a lot of back and forth, and, during development, there were many iterations of the interactive buckyball that mimics the actual manipulation of the molecule, until they both felt comfortable with the representation. Vesna wanted to project the molecule as large as possible and have the audience members manipulate the shape with their shadows as a metaphor for the way the STM works — from a distance, through a computer, with a tiny tip that »feels« the terrain and sends back data that is translated into an image. The piece was named: *Zero@wavefunction* and premiered a year after their meeting, at the Biennale of Electronic Art in Perth (BEAP), Australia, in 2002.

The name *Zero@wavefunction* is a playful reference to the time-independent Schrödinger equation, which represents the time evolution of a wave function and the quantum-mechanical properties of an individual physical system: $\hat{H}\psi = E\Phi$.

The name also refers to email addresses such as John@home.com, resulting in ϕ@ψ, where the lower-case *phi* is an individual wave function and the upper-case *phi* is the total wave function of everything, a bit like an email address but in a quantum realm. It became the title of the installation, where the gallery observer became an active participant in the total piece, which used people's shadows as a metaphor for the STM tip manipulating buckyballs projected on the wall. The processes, which seriously challenged computer and video technology at the time, included a zero-point energy molecular vibration and nanomechanical deformations of the molecule. These were loosely based on publications by Gimzewski and Joachim et al. on experimental and quantum chemistry calculations of the electronic transparency of a single C60 molecule as well as the electrical resistivity of a single molecule — in other words, of the physics and chemistry of one molecule.

Zero@wavefunction was very much inspired by some of the inherent traits of atoms and molecules that we do not find in the ball-model world. Firstly, molecules vibrate, even at the lowest temperatures, down to absolute zero (−273.15 C). This is called zero-point energy, interpreted in terms of vacuum fluctuations where even space is not vacuous but filled with energy fluctuations. How marvelous that we have this zero-point energy, which to date has not been harnessed as a potentially infinite energy source for humanity! Next, when the atoms of a tip approach a molecule, they experience interactions in the form of forces of attraction; at first, they are followed by repulsion at closer distances. At even closer distances, chemical bonds can be created. Third, molecules are not static rigid structures, and we know from molecular mechanics that they buckle and bend under the forces of the tip interaction.

In order to create an installation that conveys all of these fairly complex but real effects, we decided to create a playful piece where the participant interacts with a visual representation of a molecule and, through this interaction, essentially feels or experiences the physics outlined above. It is important to note that the dimensions and scales of the molecules and interactions had to be adjusted to create a meaningful and reactive interaction between the human mind, the hand that moves the molecule, and its responses. We scaled the data into visual and audio output by around a billion times. The vibrations, sourced on a molecular level, were amplified and sped up for the human ear to hear.

The process started with a transparent line model that is instantly recognizable as a soccer ball or truncated icosahedron or a buckminsterfullerene, depending on your experiences, with sixty apexes forming isolated pentagons and interconnected hexagons. The extension of the electron wave function was added as a glow that was at its highest intensity at the apexes and that gradually decayed into space. The structures also showed a shimmering motion like the vibrational breathing of the molecule in real life. The algorithm included distortions and, when too much force was applied, nothing happened—it would only respond to slow movement and, with patience, could be transformed and moved. In addition, buckyball projections interacted with each other when in proximity and induced mutual distortions.

Soon after the premiere of *Zero@wavefunction*, we were invited by Robert Sain, then director of LACMALab, to create a series of installations at the LACMA Boone Gallery. This was not a planned commission but one that arose due to unusual circumstances. LACMA was in the process of developing a plan for major renovations; the architect who had been selected, Rem Koolhaus, was deemed too controversial, and eventually the whole project was paused. This left a space in the schedule that would normally have been filled years in advance. Sain, inspired by the *Zero@wavefunction* project, was interested in bringing it to a larger audience, giving the artist and scientist a unique opportunity to create a series of experimental, interactive works in a large space. Sain was working in tandem with Carol S. Eliel, Curator of Modern and Contemporary Art at LACMA. What follows is a brief description/walkthrough of the exhibition.

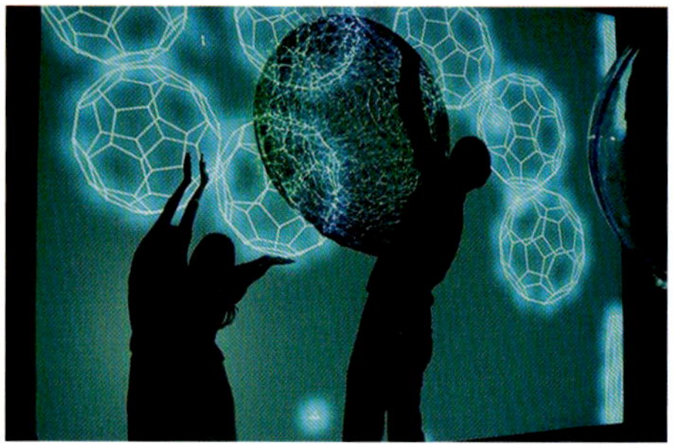

FIG. 6_*Audiences manipulating buckyball projections with their shadows:* ZERO@WAVEFUNCTION, *LACMA*

Image courtesy of Victoria Vesna, 2003

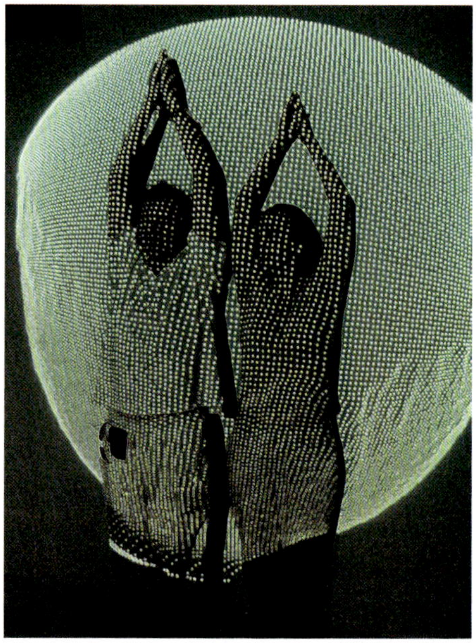

FIG. 7_*Authors in the Quantum Tunnel, LACMA*

Image courtesy of Victoria Vesna, 2003

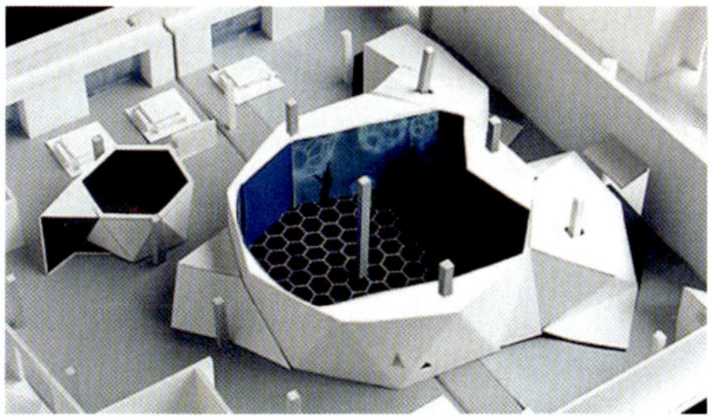

FIG. 8_*Model of the* NANO *architecture for the interconnected installations*

Model by architects Johnston Marklee, 2003

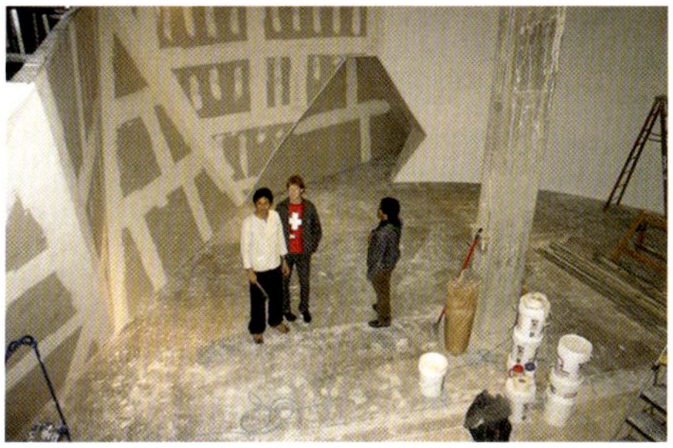

FIG. 9_*Construction phase of the* NANO *sculptural walls, LACMA*

Image courtesy of Victoria Vesna, 2002

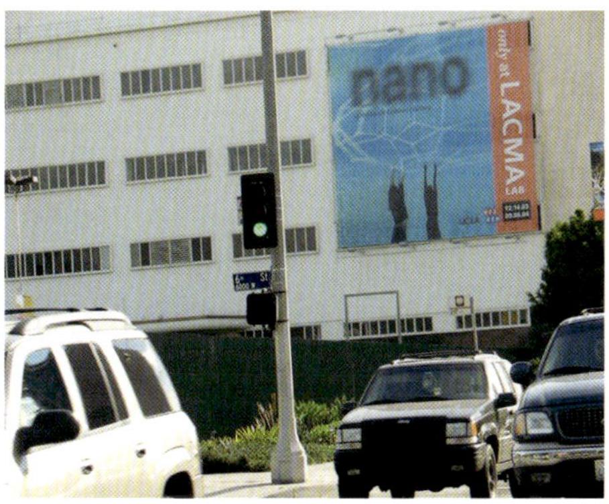

FIG. 10_NANO *banner outside of the LACMA museum, 2003–04*

Image courtesy of Victoria Vesna, 2003

PLANETARY/CELLULAR ARCHITECTURE

Architects Johnston Marklee were commissioned by LACMA to work together with Gimzewski and Vesna to develop the exhibition framework. Together, they envisioned the space that would house the various installations—based on Bucky Fuller's *Dymaxion Map*, with the interactive technology hidden in the structures so that the experience would be seamless and audiences would not get distracted. The complete architecture acted as sculptural structures that were connected to the installations within. As the audience walked in, a group of surveillance cameras captured images of them and projected them in a growing hexagonal pattern. Surveillance was still a major public concern at that point in time, and by showing hundreds of cameras at the very entrance, the audience immediately had a heightened awareness that it was being watched. In retrospect, it is interesting to note that surveillance technology has now been normalized; the idea of swarm cameras is not shocking anymore and is accepted as a way of life in most cities and public spaces.

Once inside, the first space that one entered was the »Inner Cell,« where the sense of scale and perception fully shifted. The floor mimicked a graphene surface that, although it was not yet fully understood at the time, interactively responded to the walking of the audience with lattice distortions. People were made aware that they were sharing the space with large robotic balls that seemed to move around on their own and influence their movement. But as they continued on, they saw the »Atomic Manipulation Table,« which showed a bird's eye view of the »Inner Cell« projected and allowed the movement of the balls to be directed, influencing the movement of the people inside. This not only mimicked the work on a molecular level but also pointed to the social manipulation possible using these surveillance and other remote technologies. The »Atomic Manipulation Table« was a real-time projection of the large robotic spheres in the inner space. Visitors could move the spheres using roller ball mice on the periphery of the table. In real experiments, we also used a computer interface with STM images of atoms and molecules, and the mouse was used to move the tip within interaction distance to reposition them in the lateral X,Y planes. This was more of an engineering-like installation compared to the manipulation of buckyballs discussed later.

Going further along, visitors entered the »Quantum Tunnel.« The process of tunneling is an important part of quantum phenomena where electrons are transported through space by a process that would be forbidden in classical physics. The electron's wave function, or the associated probability of finding an electron, decays rapidly as one moves away from a surface made of atoms. Now, if we bring two metal atoms very close, let's say, several atomic diameters apart with empty space in between, the two atoms will experience an overlap of their wavefunctions and allow electrons to be transported between the two atoms, resulting in a flow of electrons. This invokes the wave properties of the electrons that are reflected and partially transmitted through a classically forbidden barrier. It is still a somewhat strange, mystical process and involves the duality of the electron as a wave and particle. To combine the concepts of wave reflection and transmission, we used sounds and echoes, and two participants on opposite sides of the installation would see fuzzy probabilistic representations of their faces as a particle-like representation.

NANOMANDALA

While planning the exhibition, the curators approached us to look around the contemporary art collection and select a few appropriate works that could enhance the *NANO* spaces. We assumed that this was to make sure people understood that what we were presenting was connected to the art world. But we were not able to identify any artwork connected to the concept, so we turned our attention to the East Asian collection, as ideas of empty space were much more present there. This suggestion was greeted with outrage from some in the organizing committee — mixing spirituality with science seemed sacrosanct, and one member quit in protest over this idea. But we persisted with our ideas and happily got a positive response from the director, who made a statement that, in his view, the Tibetans were the first nanotechnologists. And it just so happened that a sand mandala creation was being planned at the same time, so we contacted the monks and started several interesting philosophical conversations. We met with the Tibetan lama and monks several times to make sure that they agreed with the concept, and, in the end, they did agree that we were working toward the same goal — to show that everything is interconnected. Four monks worked all day for six weeks until the most magnificent sand mandala of the Chakrasamvara[6] took shape. The process of building the sand mandala was carefully observed by us — starting from the center with a few grains of sand and moving out to a very complex imagery.[7]

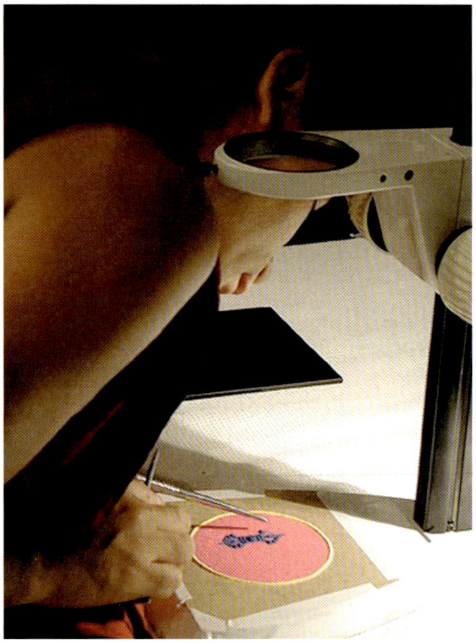

FIG. 11_ *Tibetan monk in Gimzewski's UCLA chemistry lab recreating the center of the Chakrasamvara sand mandala under an optical optical microscope*

Image courtesy of Victoria Vesna, 2002

The *Nanomandala* consists of views that go beyond the powers of ten — from the photographic view, to the optical, to beyond the visible realm — into a grain of sand. Using a crane, high resolution photographs of the nine-foot mandala were taken at LACMA. The photography consisted of 300,000 images that sequentially zoomed in closer and closer on the center. Once the photographic limit was reached, one of the monks came to Gimzewski's Nano Lab to recreate the center that had been imaged using optical and scanning electron microscopy (SEM). Three points of view — photographic, optical, and electron — were edited into a movie of the surface of the grain of sand, moving into the complex details of the entire Chakrasamvara sand mandala and back in a cycle of creation and destruction. The project entailed photographing the sand mandala in several stages. Using a crane, the whole process took well over two months, with the rendering alone consisting of six hundred gigabytes of data, taking one week of rendering time on thirty machines. The sand mandala is a cosmic diagram and ritualistic symbol of the universe used in Hinduism and Buddhism, which can be translated from Sanskrit as »whole,« »circle,« or »zero.« The STM rearrangement of atoms bears some resemblance to the methods monks use to laboriously create sand images particle by particle. However, Eastern and Western cultures use these bottom-up building practices with very different perceptions and purposes.

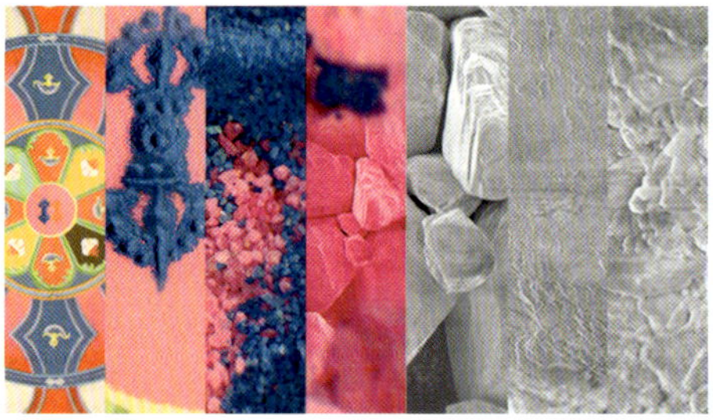

FIG. 12_ *Three different zoom viewpoints of the sand mandala: scanning electron microscope (SEM), optical microscope, digital photography*

Image by Victoria Vesna, 2003

The *Nanomandala* was projected onto a bed of sand — starting with one grain and slowly emerging into the mandala's complex structure, with the sound starting with the ocean and ending with Tibetan chanting mantras recorded while the monks were building it. The audience did not passively look at the projection but was able to touch the sand and thus change the surface terrain.

Visitors were able to watch and touch the sand *Nanomandala* that was projected as a loop of evolving scale, from the molecular structure of a single grain to the recognizable image of a pile of sand, and out into a full view of the image. The

Chakrasamvara mandala slowly emerged from the bottom up. Sand has a particle size of between 0.06 and 2 millimeters. The grains of sand in the mandala are approximately one millimeter in diameter and made of silica (silicone dioxide). One can easily estimate that, in the mandala at LACMA, there were up to 100 million grains of sand. A calculation shows that there are approximately ten to the power of eighteen atoms, which far exceeds the number of grains in the mandala by a massive factor of 10,000,000,000, which is around 10 billion. However, the sand forms a simple crystalline structure, whereas the mandala has a much more complex and deeper form in terms of its cultural and spiritual significance. Indeed, just the number of atoms on the top surface layer of a grain of sand still exceeds the number of sand grains in the mandala.

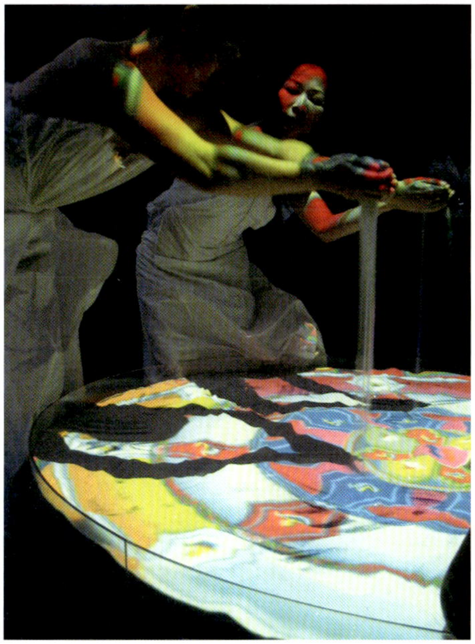

FIG. 13_ANY / NANO / BODY *performance, interacting with the sand mandala, choreographed by Norah Zuniga Shaw*

Image courtesy of Victoria Vesna, 2004

RITUAL Destruction / the Impermanence of Everything

The Chakrasamvara sand mandala was on view for several months, followed by a large dissolution ritual, during which all the beauty and hard work of the monks were ceremoniously destroyed, the sand put into bags and thrown into the ocean at the Santa Monica Pier. In parallel, the center of the mandala that was recreated was subjected to similar ritualistic destruction by the monks. During the exhibition, the monks had commented that turning off the projector had the same effect, the same message — everything appears and disappears on the cosmic, human, and atomic/molecular level.

Another installation entitled *Fluid Bodies* captured live video of audiences passing by with their shadows projected in tiny particles onto the wall dissolving with a slight delay — confronting the idea of time, space, and the inherent impermanence of everything, including our own mainly fluid bodies. Walking out, visitors had an opportunity to look inside the installation from all sides of the structure through large kaleidoscopes. All of this was experienced from another point of view, shifting in multiple mirrors that reflected the spaces within as different viewpoints of the same reality.

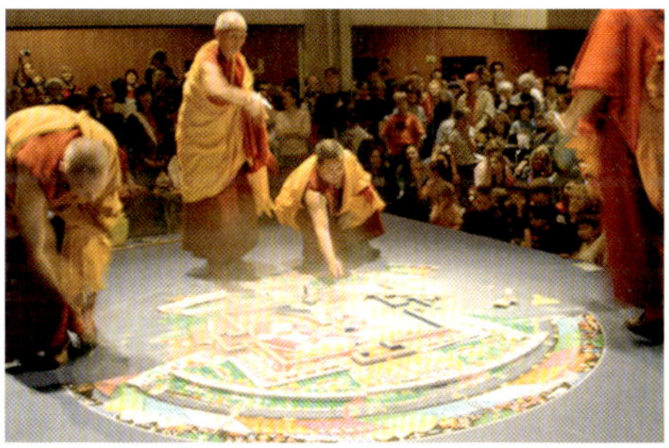

FIG. 14_*Dispersion ceremony at the LACMA: the sand that had been collected was ritually returned to the ocean at the Santa Monica Pier*

Image courtesy of Victoria Vesna, 2004

VIDEO

The exhibition ran for nine months at LACMA, and we opened it up to other disciplines — responding from their perspectives. We worked closely from the beginning with the well-known post-literary critic Kathryn Hayles, who involved her students in developing wall labels, and ultimately a book: *Nanoculture: Implications of the New Technoscience*. The entire exhibition was described in detail and accompanied by essays written by her then graduate students, who unpacked ideas of hybridity, scale, and how all this became part of active cultural construction.[8] The preface was written by Vesna's mentor, Roy Ascott, who referenced a similar bridging of art and science, which also resulted in an exhibition and a book — *Aspects of Form*, edited by Lancelot Whyte, which had accompanied *Growth and Form*, an exhibition shown at the ICA, London, in 1951. He observed that »D'Arcy Wentworth Thompson stood then in relation to overarching ideas of form and pattern, very much in the way that Richard Feymen was positioned, bottom-up as it were, in the debate around nanotechnology today.«[9]

During the exhibition, the spaces were used by Norah Zuniga Shaw, a choreographer who created a piece, *any/nano/body*, with a group of dancers interacting with the artworks. They wore white clothes whose forms were inspired by protein

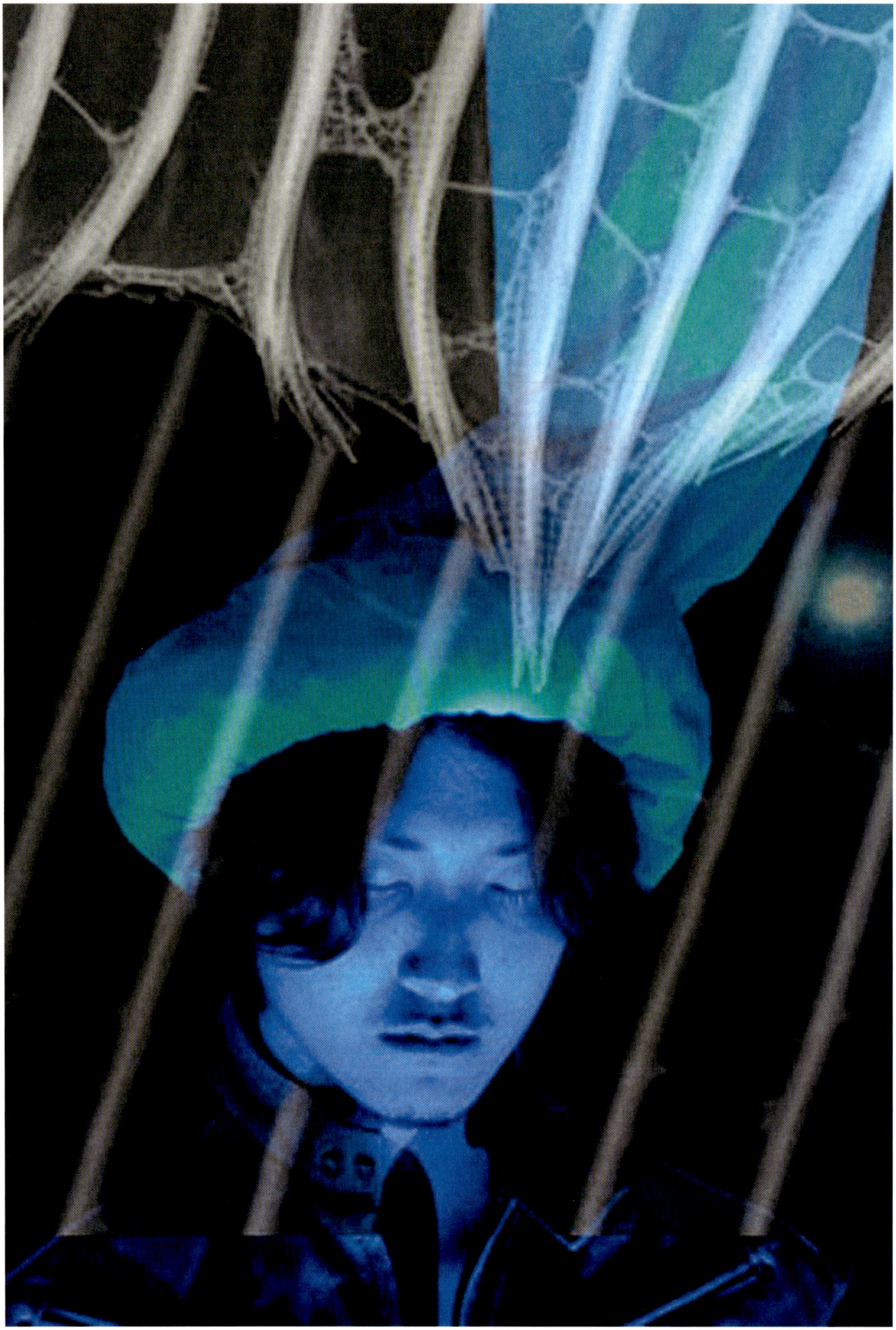

FIG. 15_ *Close-up of a participant in the* BLUE MORPH *installation, Integratron, Landers, CA*

Image courtesy of Victoria Vesna, 2008

VIDEO

folding and designed by Elizabeth Toledo, a Cuban-American fashion designer from New York, and students at the Otis College of Art and Design in LA — inspired by the exhibition's message of the scale and wonder of the nano realm. The clothes were also presented at a fashion show at the Beverly Hilton Hotel. To capture the younger K-12, we also organized workshops on tensegrity systems using models and even staged a rave for the teenagers that lasted all night at the museum.

Moving into NANO/BIOmimicry—Micro/Macro Worlds

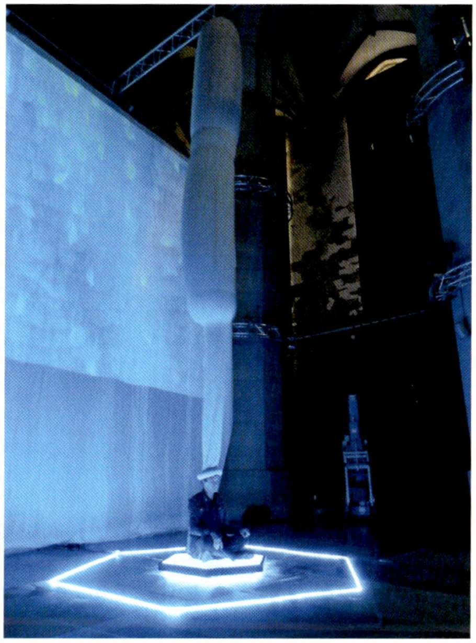

FIG. 16_*Ryszard W. Kluszczyński*
BLUE MORPH *at St. John's Cathedral, Baltic Cultural Center, Gdansk, Poland*

Image courtesy of Victoria Vesna, 2011

During the development of *NANO*, parallel groundbreaking research was taking place in Gimzewski's lab. He was interested in exploring the vibrations of living cells and used a simple yeast cell to measure their movement. While working on the media art projects, he had been inspired to ask his graduate student Andrew Pelling to convert the data into sound. This was then amplified so that it could be heard by the human ear. The resulting sound was incorporated into the installations — namely in the »Inner Cell,« and Vesna went on to develop a site-specific piece using these sounds — *Cell Ghosts*.[10] The research generated a lot of interest and some controversy, and many people from various disciplines contacted Gimzewski about the cell sounds.

But one moment determined the direction of the next piece that was to emerge: Gimzewski was in his office when a phone rang, and he answered. The call was from Ana Castello, who had read an article about the movement of the cells in the *Smithsonian* magazine and was inquiring if it would be possible to measure the vibrations of the metamorphosis of a chrysalis to a butterfly. Being an experimentalist, he was open to her suggestion to send in some chrysalises and give it a try. But it was not easy, as the chrysalis is very large, and it was impossible to work with the microscopes. So, they came up with a solution — to place a micromirror on the back of a chrysalis and reflect a laser off it to measure the movement. When the resulting signal was sped up and amplified for the human ear to hear, the result was surprising. Oscillations occurred in bursts and did not gradually build up as had been anticipated. The sonification of this data inspired a fully immersive installation that put the audience at the center of an interactive artwork that included SEM images and sonification. They would experience the piece fully only when completely still on the interactive pedestal. Conceptually, this was a shape-shifting work and was adapted to the environments it was shown in — from the Integratron, where it premiered, to galleries, festivals, and, most impressively, the Cathedral of St. John the Divine in Gdansk, Poland (2011).

The *Blue Morph* project moved Victoria Vesna to completely dedicate herself to creating work that addresses ecological problems. She continued to focus on sound and creating immersive installations such as *Water Bowls: moon ~ drop ~ oil ~ sound*, going under water and exploring noise pollution created by fossil fuels. Her more recent works have continued scaling and projecting microworlds, drawing attention to planktons and micrometeorites. At the invitation of Dr. Alfred Vendl, Director of the Science Visualization Lab at the University of Applied Arts Vienna, Vesna conceptualized the *Noise Aquarium* installation by working with existing visualizations of plankton, scaling them up to be as large as whales, with interactions that revealed destructive underwater noise. These visualizations on their own were

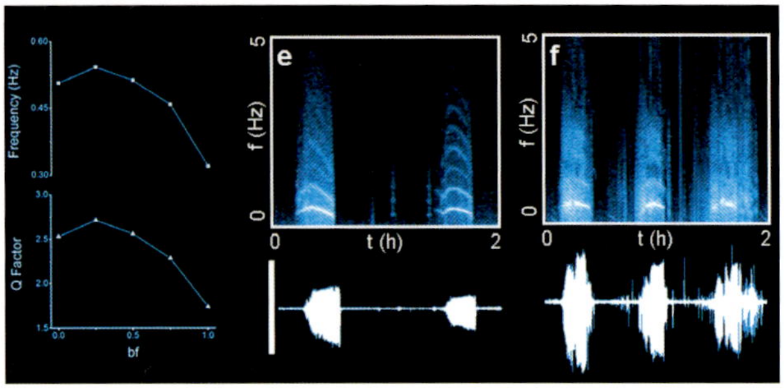

FIG. 17_*Sonogram and waveforms of the metamorphosis of a monarch chrysalis into a butterfly*

Image by Gimzweski Lab, 2007

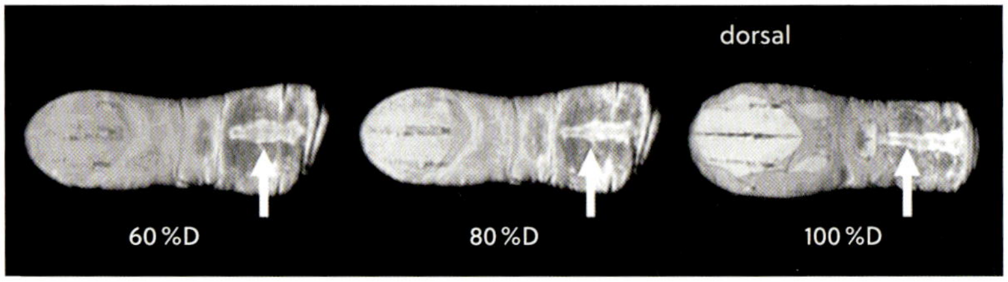

FIG. 18_*8 Tesla MRI imaging of the monarch chrysalis at different stages of metamorphosis*

Image by Richard Stringer, 2007

a fantastic testament to high-level technological achievements as well as excellent animation. But it took an artist's vision of scale, frequency, and perception to bring these endangered micro-creatures to the attention of the public. Additionally, micro- and even nanoplastic particles were added to the mix—the ultimate invisible effect of the fossil fuel industry that is also largely responsible for underwater noise pollution. She employed a similar installation strategy to that of *Blue Morph*, with the audience standing on an interactive pedestal and only enjoying a full view of the creatures when they were completely centered and still. As it happened, most of them moved around wreaking destruction—we are all implicated.[11]

From the depths of the ocean, Victoria Vesna turned her attention to the cosmos when she was invited by geologist and Director of the Museum of Natural History in Vienna, Dr. Christian Köberl, to develop a piece for the meteorite gallery. The meteorites were, of course, fascinating, but what caught her attention was learning about micro-meteorites—approximately one hundred tons of stardust that falls on Earth daily. This dust from space is mixed with natural and human-made pollution that was brought to the attention of the public by a jazz musician from Norway, Jon Larsen.[12]

Created and premiering at the beginning of the pandemic, these dust particles became an important message to humanity, caught in climate, political, and economic disruptions. The invisible and inaudible worlds perhaps offer another way to bring to the forefront, raise awareness of, and remind the larger, global public about the fact that we are all made of stardust. *[Alien] Star Dust: Signal to Noise* evolved into a meditation and augmented reality (AR) piece, in which audiences are put at the center of everything falling apart, a perfect storm with star dust mixing with anthropogenic pollution. Once again, we are observing and experiencing nanoparticles, from nanoplastics to fire ash, gold, silver, platinum, and other elements that arrived on Earth as dust.[13]

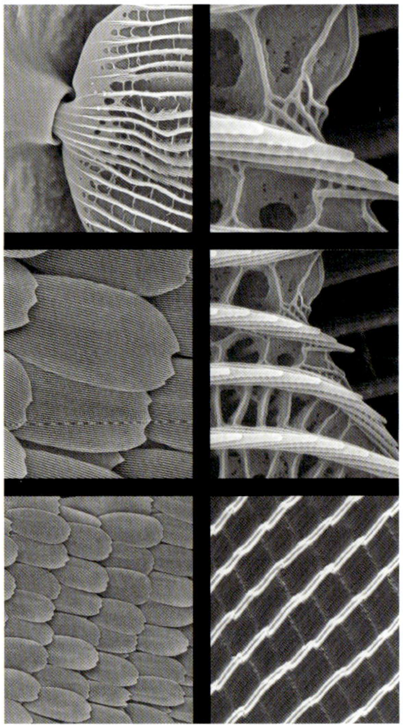

FIG. 19_*Scanning electron microscope (SEM) and atomic force microscope (AFM) images of blue morpho wings at different magnifications*

Image by FEI Co, Hillsboro, Oregon, 2011

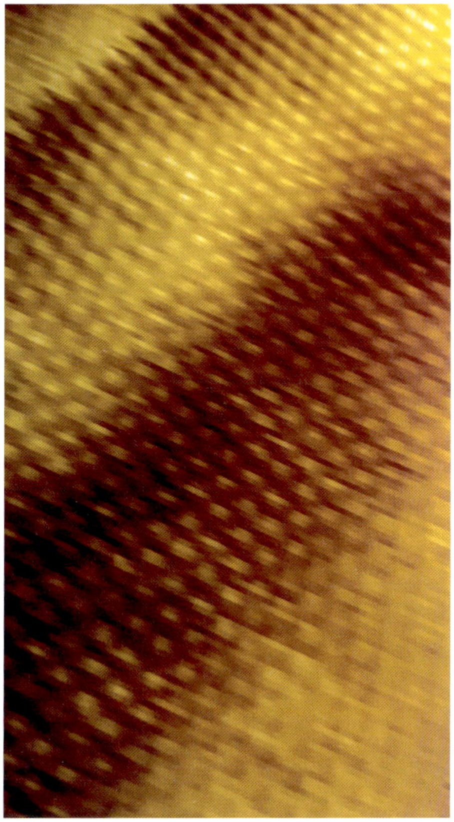

FIG. 20_*Low-temperature scanning tunneling miscroscope (STM) image of gold atoms on a single, atomically clean gold crystal*

Image by Gimzewski Lab, CNSI, UCLA, 2022

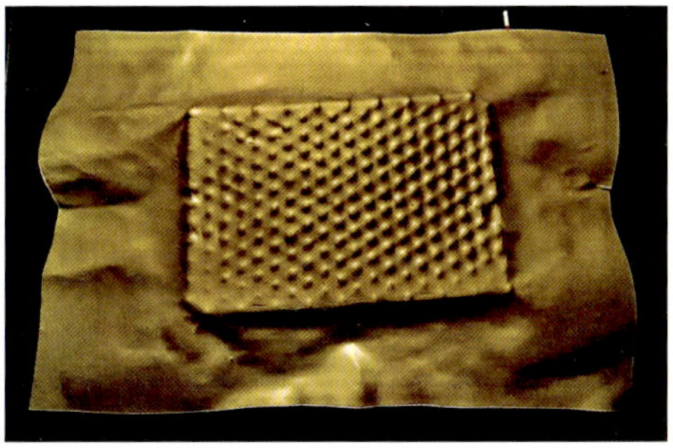

FIG. 21_*3D topographic artwork of gold atoms*

Image by Gimzewski Lab, CNSI, UCLA, 2023

ATOMS & GOLD

After our parallel journeys into the complexity of the micro-/macroworlds, we joined forces again to ask: What do we value? What is value? To address this question, we are using gold as a metaphor and as an actual material that arrived on Earth from space.

Gold is believed to come from a process known as supernova nucleosynthesis as well as from neutron stars colliding, which is a newer astrophysical model that suggests that a single neutron star merger could have created masses of gold multiple times heavier than the Earth's mass. Gold was part of the dust when our solar system formed and clumped together. Many of the heavier elements sunk, but gold also arrived during subsequent asteroid impacts with Earth. The atomic mass of gold is ~197 grams as determined approximately by the total rest mass of its protons, neutrons, and electrons. Its atomic mass unit (AMU), also called the dalton (Da), is $1.67\ 10^{-24}$ g, which means that the mass of a gold atom is simply that constant multiplied by 197, equaling $\sim 3 \times 10^{-22}$ g, which is almost nothing. Put another way, one gram of gold contains $\sim 10^{21}$ or 1,000,000,000,000,000,000,000 atoms. This is, of course, a massive number of tiny atoms, and the STM can image each individual one.

During the pandemic, while Vesna was working on the *[Alien] Star Dust* project, Gimzewski adapted his lab work to remotely scan atoms from almost anywhere on Earth. The focus of the research involved designer molecules on a gold surface that is a small, approximately one-centimeter disk of very pure single-crystal gold. The aim of the project was in fact to transfer a single gold atom on the tip of the microscope to a special molecule that has a set of »hands« to hold it. However, one part of the procedure was to image gold atoms on the surface of the crystal and record the data. These images formed the initial inspiration for the current piece entitled *Atomic Gold Standard*, a new work in progress that, in many ways, connects our art and science research, which has developed jointly and in parallel over the past two decades. We scale the invisible and inaudible to produce another way to perceive reality—understanding empty space, complexity, nano-, bio-, quantum worlds—in relation to our human condition.

The word *atom* invokes a whole range of reactions. The atom may conjure up many images—whether images in chemistry textbooks, science fiction stories, images of Hiroshima, Chernobyl, and of so many other things, many of them not particularly positive. Democritus, in the fifth century BCE, proclaimed that everything is made from *átomon*, ἄτομον, physically indivisible and indestructible units of matter. Moreover, he proposed that they were in perpetual motion, with space between them, and that such atoms were infinite in number and kind. Indian atomism also proposed similar ideas of the atom, some perhaps closer to scientific views of atoms today but dating back to the eighth century BCE, such as the Jainist concepts that include subtle multiple states, reactions, and vibrations. Later, in 1808, Dalton provided experimental evidence to propose that atoms are indeed indivisible constituent parts of matter, and Einstein, in 1905, explained the details of Brownian motion that result in the buffeting of pollen particles previously discovered by Brown studying the particles under an optical microscope. What he initially thought was the buffeting of a living particle was a manifestation of atomic motion.

Gold, on the other hand, is deeply connected to the human psyche, and this has even revealed itself in the recent space research and exploration of 16 Psyche.

According to scientists, there may be trillions of dollars' worth of precious substances such as gold, platinum, and diamond buried deep in asteroids and pieces of space debris. And now, humans seem to be getting one step closer to mining some of those riches. In relation to cryptocurrencies and NFTs, this has some resemblance to the inequalities of the »Gilded Age,« a term from Mark Twain and Charles Dudley Warner's 1873 novel *A Tale of Today*, and the serious social problems of the decade 1920–30, hidden by a thin gold veneer. Gold gilding is around one hundred nanometers thick, but in the *Atomic Gold Standard*, we have reduced that thickness to one atom or around 1/300th of that gilding. One atomic diameter is the ultimate limit of thickness physically possible, and we are rapidly approaching such dimensions with today's computer chips.

Conclusion

While writing this retrospective narrative, we were struck by our shared recurring interest in figuring out ways to bring the atomic/molecular realm to the attention of the wider public. Starting from the bottom up—from the zero point to the series of installations at LACMA—we laid the foundation for our approach to our respective research and collaborations.

Despite some rough early encounters, objections, and even resignations, we integrated spirituality and connectionism into our work from the start. While science has traditionally been concerned with minimizing the effect of the observer, or removing their influence entirely, we acknowledge the observer in the creation of our art-science projects. To us, the observer is always central and participates. Rather than isolate research in a test tube with one variable (closed system), we acknowledge the connection of all things around us, focusing on engagement and being an active part of the experience. At the cutting edge of research, science itself is exciting and mysterious, but any emotional traits associated with the process are usually sanitized out and presented as scientific facts. Increasingly, for real-world issues such as climate change, the financial markets, wars, and so on, the laws of physics are at best very limited when it comes to describing today's realities. We must also be cognizant that the next phase of science is in the field of complexity in which we attempt to unravel nature's collective and emergent capacity to do the unexpected.

It is not by accident that we have become more and more involved with environmental issues and, as time has passed, have become increasingly convinced that this is critical in this age of a major paradigm shift. We started our journey with the Networks to Nanosystems: Art, Science, and Technology in Times of Crisis symposium, a gathering of artists, scientists and humanities scholars, and this topic has increasingly been amplified by world events. The works we have presented have created a basis for our ongoing art and science research, and have gotten us to the point where we are now pointing to gold atoms and asking: what is value? By reconnecting art and science, we have an opportunity to envision a more just world by inspiring an awe for nature, bottom up and in between.

1 Edgar J. Applewhite, »The Naming of Buckminsterfullerene,« in *Culture of Chemistry: The Best Articles on the Human Side of 20th-Century Chemistry from the Archives of the Chemical Intelligencer*, ed. Balazs Hargittai and István Hargittai (Boston, M. A.: Springer, 2015), 21–3, https://doi.org/10.1007/978-1-4899-7565-2_3.

2 Governor Gray Davis initiated an ambitious project to develop California Institutes for Science and Innovation across the entire University of California (2000), combining cutting-edge research with training for new scientists and technological leaders. Two of the funded institutes focused on the Networking and NanoSystems conference dialogues: Cal-(IT)2, a joint effort between UC San Diego and UC Irvine, and the California NanoSystems Institute (CNSI), a joint effort between UC Los Angeles and UC Santa Barbara.

3 UC DARnet—UC Digital Arts Research Network (UC DARNet), funded as a Multi-campus Research Unit by the UCOP, convened an international conference and series of exploratory workshops over a ten-day period. These events took place in several physical locations throughout the state of California in collaboration with the Centre for Advanced Inquiry in Interactive Arts, University of Wales, and the Center for Science, Technology and Art Research, in the School of Computing, University of Plymouth (CaiiA-STAR), and the Banff Centre for the Arts. The conference organizers were Victoria Vesna (UCLA), Sharon Daniel (UCSC), Robert Nideffer (UCI), Roy Ascott (CaiiA), and Sara Diamond (Banff Centre).

4 Smallest calculator, 0.000001 nanometers: »James Gimzewski and a team of scientists at IBM Research Division's Zurich research Laboratory, Switzerland, created a calculator with a diameter of less than one millionth of a millimeter. The molecular abacus consists of ten molecules of carbon 60, that can be moved along a microscopic groove on a copper surface with the tip of a scanning tunneling microscope.« *Guinness Book of World Records* (London: Guinness World Records, 2000).

5 Franz J. Giessibl, *First View Inside an Atom: Encounters with Gerhard Richter Between Art and Science* (Cologne: Walther König, 2001).

6 Chakrasamvara is a powerful god (wrathful deity) of Buddhism, and he is immensely popular in Tibet, Mongolia, and Nepal. This ritual diagram (mandala) is conceived of as the cosmic palace of the wrathful Chakrasamvara and his consort Vajravarahi at its center. These deities embody the esoteric knowledge of the Yoga Tantras.

7 N. Katherine Hayles (ed.), *NanoCulture: Implications of the New Technoscience* (Bristol: Intellect, 2004).

8 Hayles, *NanoCulture*.

9 Lancelot Whyte, »Aspects of Form, Accompanied by Growth and Form,« exhibited at ICA, London in 1951.

10 *Cell Ghosts* premiered at »Crash and Flow,« Seoul-Shinchon Art Festival, Seoul, South Korea in 2004 and was later shown in Apeejay Media Gallery, New Delhi, India, and the Red Gate Gallery, Beijing, China.

11 *Noise Aquarium* premiered in 2017 as a featured project at the WEB3D Art Gallery ON | OFF: 100101010 in Brisbane, Australia, for the 22nd International Conference of 3D Web Technology.

12 Jon Larsen, *In Search of Stardust: Amazing Micrometeorites and Their Terrestrial Imposters* (Minneapolis: Voyageur Press, 2017).

13 *[Alien] Star Dust: Signal to Noise* premiered March 20, 2020, at the Natural History Museum in Vienna.

Journeys into the Hidden Microscopic World

Stephan Handschuh, Thomas Schwaha

Technological advances in microscopic imaging and their impact on biology and science visualization

Microscopy opens doors to hidden worlds. Therefore, microscopic techniques have been key to a huge number of scientific discoveries in biology and medicine carried out in the last few centuries. Microscopy allows us to study tissues and cells, and their pathologies. Many biological and biomedical disciplines such as cancer biology, neurobiology, developmental biology, and evolutionary biology have benefited from recent advances in microscopic imaging, and image visualization and analysis tools. In this chapter, we summarize the technical advances of the last two decades and highlight some novel emerging microscopic imaging techniques. We also provide a personal perspective on how modern computer-aided 3D reconstruction and visualization techniques have had an impact on our own everyday work as biologists in academia with respect to both research and teaching. Finally, we summarize how collaborations with Alfred Vendl's Science Visualization Lab have given us a unique chance to introduce unfamiliar microscopic creatures to a broader public.

Background

We first met in a comparative vertebrate anatomy class at the University of Vienna in 2005. At that time, we were both undergraduate students of biology specializing in zoology. Sharing a fascination for the morphology and evolution of small and lesser-known creatures, we soon became friends taking our first steps toward professional science. It quickly became evident that we were about to launch our academic careers at a turning point. Many of our academic teachers had backgrounds in comparative and functional animal morphology. They repeatedly told us that »these new imaging and 3D reconstruction methods« would have a huge impact on the field, heralding in a new era of animal morphology. Our supervisors Hans Leo Nemeschkal and Manfred Walzl, in particular, were consistently pointing out that it was up to us—the »youngsters«—to get familiar with these new techniques. They repeatedly told us that in-depth technical training in these new techniques would give us a solid foundation for careers in science. We followed their advice and—at the same time—our own interests, as both of us were fascinated by the technical tools available.

We learned tissue preparation and microscopy techniques, which hadn't changed much over almost a century until the late 1990s or early 2000s. Studying the internal anatomy of microscopic animals was usually conducted by cutting specimens into thin sections and investigating them by means of light or electron microscopy. This is indeed a very powerful approach but comes with several limitations: animal morphology is highly complex and three-dimensional, but sections provide only two-dimensional information. To compensate for this limitation, serial sections of whole animals (or parts of them) were cut and then reconstructed in 3D by means of camera lucida drawings or physical wax or wood models. These procedures were used for over a century and led to fascinating insights, but they were prone to error and highly laborious, and it was difficult to communicate the acquired data, especially in the case of physical models. Computer-aided 3D reconstruction was a game changer in this respect. During our studies, the University of Vienna already had a license for a piece of commercial 3D reconstruction and visualization software, which gave us an early opportunity to learn how to digitize serial sections, how to align consecutive sections in the software, how to annotate (segment) specific anatomical features, and how to create highly detailed 3D models from these annotations. This kind of manual, anatomical reconstruction was the starting point for our scientific work and gave us a solid foundation to build upon.

Ten years ago, our careers went in different directions. Thomas Schwaha became a permanent senior scientist at the University of Vienna, working mainly on the morphology and phylogeny of aquatic invertebrates by utilizing a variety of microscopy techniques, particularly confocal laser scanning microscopy (CLSM). Stephan Handschuh got a permanent scientific staff position at the imaging core facility of the University of Veterinary Medicine, Vienna (Vetmeduni Vienna), with a strong focus on microscopic X-ray computed tomography (micro-CT). Despite the different routes we took, our work still shared many similarities, and we collaborated on numerous projects over many years, with our main working area strongly focusing on high-end microscopic imaging, 3D image visualization, and image processing and analysis. All of these fields have undergone tremendous and exciting technical advances over the last decade, which have also had a big impact on our work.

This chapter summarizes how modern imaging and visualizations allow us to travel into hidden microscopic worlds. On the one hand, we aim to outline the technological advances that have taken place in imaging equipment, computer hardware, and image processing and visualization software in the last two decades. On the other hand, we want to provide very personal insights into how these developments have changed our day-to-day routines in many areas of biomedical research and, more specifically, how they have changed the way we investigate the microscopic anatomy of animals. Finally, we want to highlight how state-of-the-art imaging and visualization techniques have also given us new tools for communicating our science to a broader audience.

2000–2020: Inventions and Developments in the Field of Microscopic Imaging and Image Visualization and Analysis

Around 2000, a remarkable portfolio of microscopic imaging techniques was already available to biologists, including, most prominently, traditional (diffraction-limited) light microscopy techniques, as well as scanning electron microscopy (SEM) and transmission electron microscopy (TEM). In retrospect, none of us could have anticipated the incredible technical advances that were about to take place over the course of the next two decades. Summarizing all these advances in microscopy would go far beyond the scope of the present chapter, but we think that it is crucial to highlight some of the most significant contributions and advances as some of them have also had a significant impact on our own work.

Image Data Acquisition

Much progress has been made in various areas of imaging. In light microscopy, huge improvements have been made with regard to the tagging and visualization of proteins and in the development of new sensors.[1] Confocal laser scanning microscopy (CLSM) has continuously improved, providing biologists with a solid and easy-to-use technique for acquiring 3D volumes of biological samples. Due to absorption and scattering, depth penetration and thus specimen thickness is generally limited to fifty to one hundred micrometers unless tissue-clearing is used.[2] New imaging modalities such as light sheet microscopy (LSM) have been introduced and have opened up new areas of application.[3] Together with novel methods of tissue clearing, the invention of LSM offered exciting new possibilities for imaging fluorescent tags in comparatively thick biological samples. Thus, LSM is a key technique for imaging samples up to ten millimeters in diameter at mesoscopic resolutions.[4] Most importantly, the Abbe diffraction limit has been circumvented by a set of new imaging techniques that can be summarized as super-resolution microscopy. These techniques include structured illumination microscopy (SIM),[5] stimulated emission depletion microscopy (STED),[6] singled molecule localization microscopy (SMLM),[7] and Airyscan microscopy.[8] In 2014, the invention of super-resolution microscopy was awarded the Nobel Prize in Chemistry.[9] Since then, the field has continued to develop rapidly. Super-resolution microscopes have become more affordable and easier to use, and the most recent development of MINFLUX/MINSTED has yielded even more improvements in spatial resolution.[10] An alternative way to circumvent the diffraction limit is expansion microscopy (ExM), which physically enlarges biological samples instead of improving the spatial resolution of the imaging system.[11]

A second area that has undergone major improvements is electron microscopy (EM). Various approaches have been developed to automatically or semi-automatically collect large 3D EM datasets. These approaches include serial block-face EM (SBEM),[12] focused ion beam SEM tomography (FIB-SEM),[13] and automated tape-collecting ultramicrotome multiple SEM (ATUM-multiSEM),[14] and they have opened the doors to new research areas such as connectomics.[15] Another improvement has been the development and refinement of methods for investigating sub-cellular structures in vitrified tissue under cryogenic conditions instead of using traditional chemical fixation and heavy-metal contrast enhancement. Cryo-EM, with its roots back in the 1980s and 1990s, avoids fixation artifacts and thus allows subcellular structures to be depicted much closer to their *in vivo state*[16]; it can be utilized to image molecules down to atomic resolution.[17] In 2017, the development of cryo-EM was awarded the Nobel Prize in Chemistry.[18] Major improvements have also been made in the field of correlated light and electron microscopy (CLEM).[19] This powerful approach is capable of linking molecular information in living cells to the ultrastructural details of cell organelles. One specific CLEM approach performed on serial sections is array tomography.[20] More recently, CLEM has been increasingly performed under cryogenic conditions (cryo-CLEM).[21]

Tomographic techniques such as microscopic X-ray computed tomography (micro-CT) have not undergone equally fundamental technological advances in terms of their imaging procedure. Here, the game changer has been setups for microscopic tomographic imaging becoming much more affordable and widespread in the community. Around 2000, micro-CT setups were only very scarcely used in biology, and lab-based micro-CT systems were rare and not easily available to the majority of scientists. Over the last two decades, micro-CT has gradually evolved into a routine technique for the structural imaging of biological samples at micron resolution. Initially, micro-CT was used to image hard (mineralized) tissue such as bone.[22] A breakthrough for imaging soft (non-mineralized) tissue was the introduction of staining protocols that use high-Z compounds to increase their contrast.[23] Lab-based scanners are now widespread and available to most research institutions, making micro-CT a routine tool in many research fields, including vertebrate morphology,[24] invertebrate morphology,[25] paleontology,[26] and developmental biology.[27] Micro-CT nicely complements the wealth of light and electron microscopy techniques available because it is able to acquire 3D image data from dense and non-transparent samples ranging from one millimeter (and less) in diameter up to comparatively large samples with diameters of around three hundred millimeters. Small samples can be scanned in true spatial resolutions below one micrometer. Another key feature of micro-CT is its non-destructive nature. Specimens stay intact, meaning that micro-CT is easy to combine with other modalities in correlative imaging pipelines.[28]

Image Data Visualization, Processing, and Analysis

The development of new imaging techniques has had a huge impact on biological and biomedical research. However, improvements in computer hardware and software, and the development of new image processing, segmentation, and analysis tools have had an at least equally high impact on the everyday work performed by researchers. Commercial software packages such as Amira (Thermo Fisher Scientific), Imaris (Bitplane), Arivis Vision 4D (Zeiss), and VGStudio MAX (Volume Graphics) have improved and have been continuously enriched by new modules. Furthermore, an increasing number of free software tools have become available, including FIJI ImageJ,[29] Drishti,[30] 3D slicer,[31] and Dragonfly (ORS, free of charge for academic use). These tools are available to a large number of scientists and have made it easy for students in particular to enter the field of visualization and analysis using 3D image data. Improvements to computer hardware have been equally important. In 1999, the first single-chip volume rendering system for consumer PCs was released,[32] which allowed researchers to conduct high-end volume visualizations of tomographic data in their own labs. Since then, the computational power of consumer PCs has consistently increased with regard to CPU power, GPU power, and memory (RAM). Since the scientific visualization of image data uses the same set of algorithms and techniques as other computer applications — including computer games — state of the art GPUs allow for high-end visualization at a comparatively low cost. Modern GPUs are so powerful that they are increasingly being used on a wide scale for image processing tasks as well.[33] Now, a workstation that costs between three thousand and four thousand euros can easily process and visualize 3D image datasets of between twenty and sixty gigabytes in size.

Images are important for visual communication. However, there is another aspect of using images in science that is at least as important. Images, if properly acquired, are a rich resource for quantitative data, including information about factors such as the size and shape of objects, the density of molecules, or motion. In current imaging-based research, the processing and analysis of image data is often much more time-consuming and challenging than initial data acquisition. One essential step in almost every image analysis pipeline is image segmentation (annotation). Segmentation is the process of isolating specific image regions (objects, segments) for subsequent visualization and analysis. Segmentation can be performed on every length scale and every level of biological organization. It can be used, e.g., to isolate a human lung from a medical CT dataset, or a specific jaw muscle in a micro-CT scan of a tiny insect's head, or a cell population in a confocal microscopy image volume of a thick tissue slice, or specific cell organelles such as mitochondria or the Golgi apparatus in an FIB-SEM volume. Traditionally, segmentation has mainly been performed either by hand (manual segmentation) or using intensity-based segmentation tools. These approaches are indeed powerful and have distinct advantages. Manual segmentation can take advantage of the experimenter's knowledge together with the human eye's amazing capacity to detect even subtle features. However, it is very subjective and needs to be performed by trained specialists in the field, and it is very slow and time-consuming. The effort required to complete the segmentation of a volume dataset based on single serial section may account for more than seventy-five percent of the whole visualization and analysis workflow.[34] Intensity-based segmentation tools such as threshold segmentation are an objective and very powerful approach. However, they require sufficient contrast between the structure of interest and other regions in the sample/image. Sufficient image contrast is provided by many structures and modalities, such as, e.g., bone in X-ray CT data[35] or fluorescently labeled components in CLSM data.[36] However, many other datasets do not provide enough contrast for intensity-based segmentation, including EM data, stained soft tissues in micro-CT data, and serial section data. More recently, several new automatic and semi-automatic segmentation methods have been introduced, including random walk segmentation,[37] machine learning pixel classification,[38] and deep learning convolutional neural networks (CNNs).[39] These approaches offer exciting possibilities for the fully automated segmentation of, e.g., EM datasets.[40] In particular, deep-learning-based approaches promise to speed up and improve routine segmentation tasks. Although they initially required advanced Python programming skills, many software packages now include a deep learning environment and allow pre-trained models to be shared among scientists, making deep-learning-based segmentation accessible to a large number of scientists.[41] Another field in image processing and analysis that is especially crucial in correlative imaging is image registration. Image registration allows the user to align and merge data from different modalities into a single coordinate system.[42] The integration of powerful registration tools into commercial software such as Amira provides scientists with an easy-to-use tool for image analysis across different modalities and length scales.[43]

The Role of 3D Imaging and Visualization in Our Academic Careers

Research

Microscopic 3D imaging and computer-aided visualization, reconstruction, and image data analysis are crucial tools in our everyday professional lives. They are considered indispensable tools in many preclinical biomedical research areas, including the phenotyping of animal models, the study of molecular mechanisms in healthy cells, and the study of diseases both at the organism and cellular level. They have also revolutionized the field of comparative animal morphology and have become standard tools for the investigation of both comparatively large vertebrates and microscopic invertebrates. This chapter reviews the role played by these tools in our everyday work from two very different perspectives. Thomas Schwaha has worked for a decade as a principal investigator (PI) in the field of comparative animal morphology and evolution, while Stephan Handschuh has spent a decade working as a staff scientist in a scientific service facility.

Since the very beginnings of his PhD thesis, Thomas Schwaha's work has focused on the morphological study of small invertebrates. The animal group that he has focused on the most are bryozoans (moss animals)_FIG. 1. The motivation for his research as a PI at the University of Vienna has come from different directions. On the one hand, studying the morphology of these animals yields insights into the function of specific organs and the ecology of these animals. On the other hand, morphology is often used to find similarities or dissimilarities with other animal groups, and these traits can then be used to infer phylogenetic relationships to other groups within the tree of life. Thomas Schwaha's fieldwork includes regular stays at marine stations all over the world, where he collects specimens for his research. Back in the lab, he utilizes a rich portfolio of imaging techniques, including *in vivo microscopy*, the microscopy of serial sections, CLSM, SEM, TEM, and micro-CT. While some of these techniques (serial sections, micro-CT) are particularly well-suited to reconstructing general models of animal anatomy, CLSM can take advantage of fluorescence labels that stain specific tissues such as the nervous or muscular system, which are especially interesting in the context of studying phylogenetic relationships. In most cases, it is the combination of several of these modalities that yield the best results. As one example, micro-CT is ideally suited to imaging minute invertebrates inside their mineralized skeletons, while other microscopy techniques deliver more anatomical details of specific organ systems.[44] Taken together, the work of Thomas Schwaha has provided insights into the relevance of soft-body morphology in the evolution of this phylum and has already described several new species in recent years.[45] Thomas Schwaha has also worked

A B C D

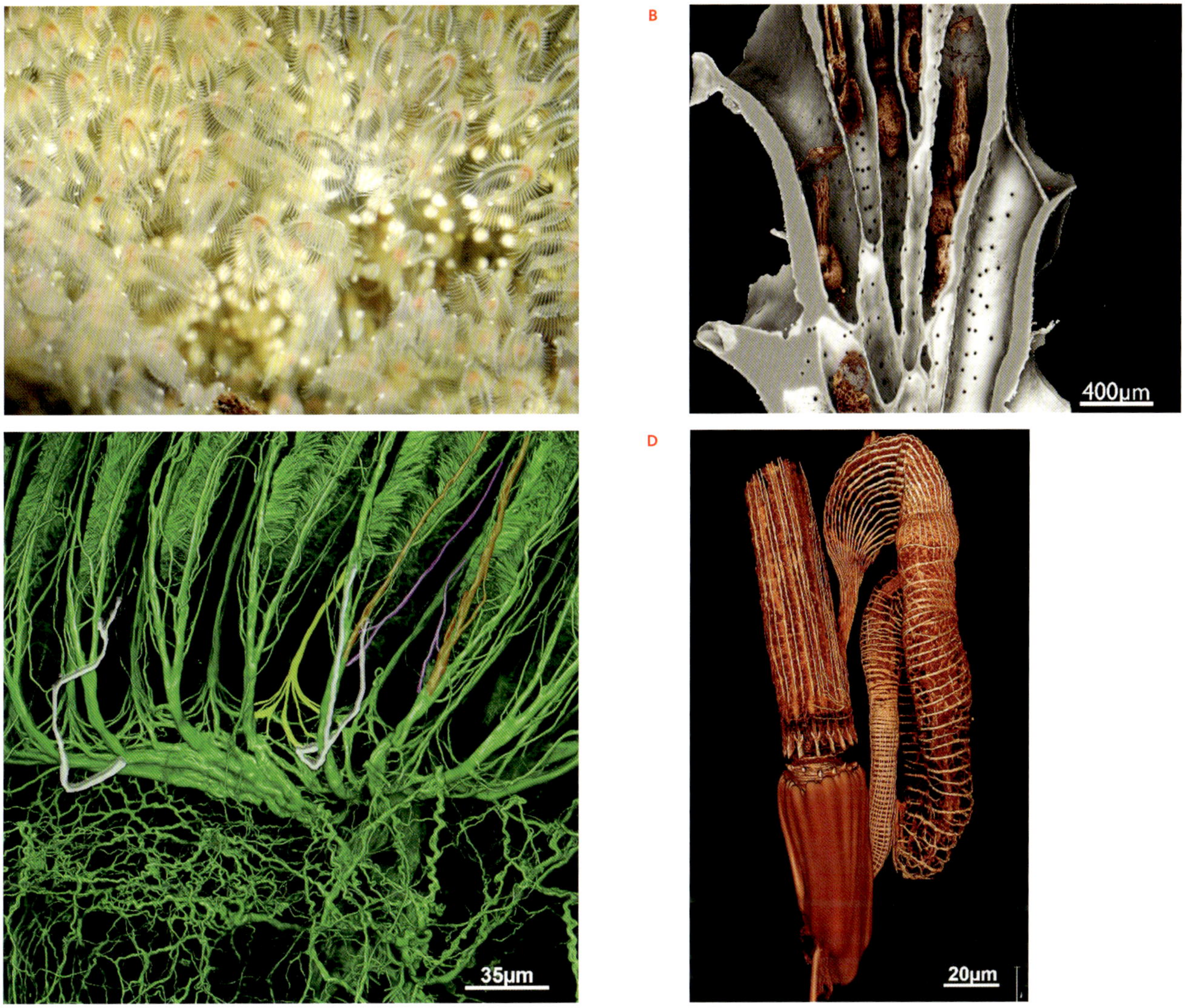

FIG. 1_ *Microscopic imaging for the study of the morphology of small aquatic invertebrates*
A_ *Stereomicroscopy image of the freshwater bryozoan* PECTINATELLA MAGNIFICA

Images by Thomas Schwaha, 2018

B_ *Micro-CT image of a colony of* CINCTIPORA ELEGANS, *simultaneously depicting the mineralized skeleton and the soft body parts*[84]
C_ *CLSM image of* STEPHANELLA HINA *stained for acetylated alpha-tubulin, depicting nerves in the tentacles*[85]
D_ *CLSM image of* PALUDICELLA ARTICULATA *stained for f-actin, depicting muscles of the tentacular apparatus and digestive tract*[86]

For Creative Commons Attribution, see the footnotes.

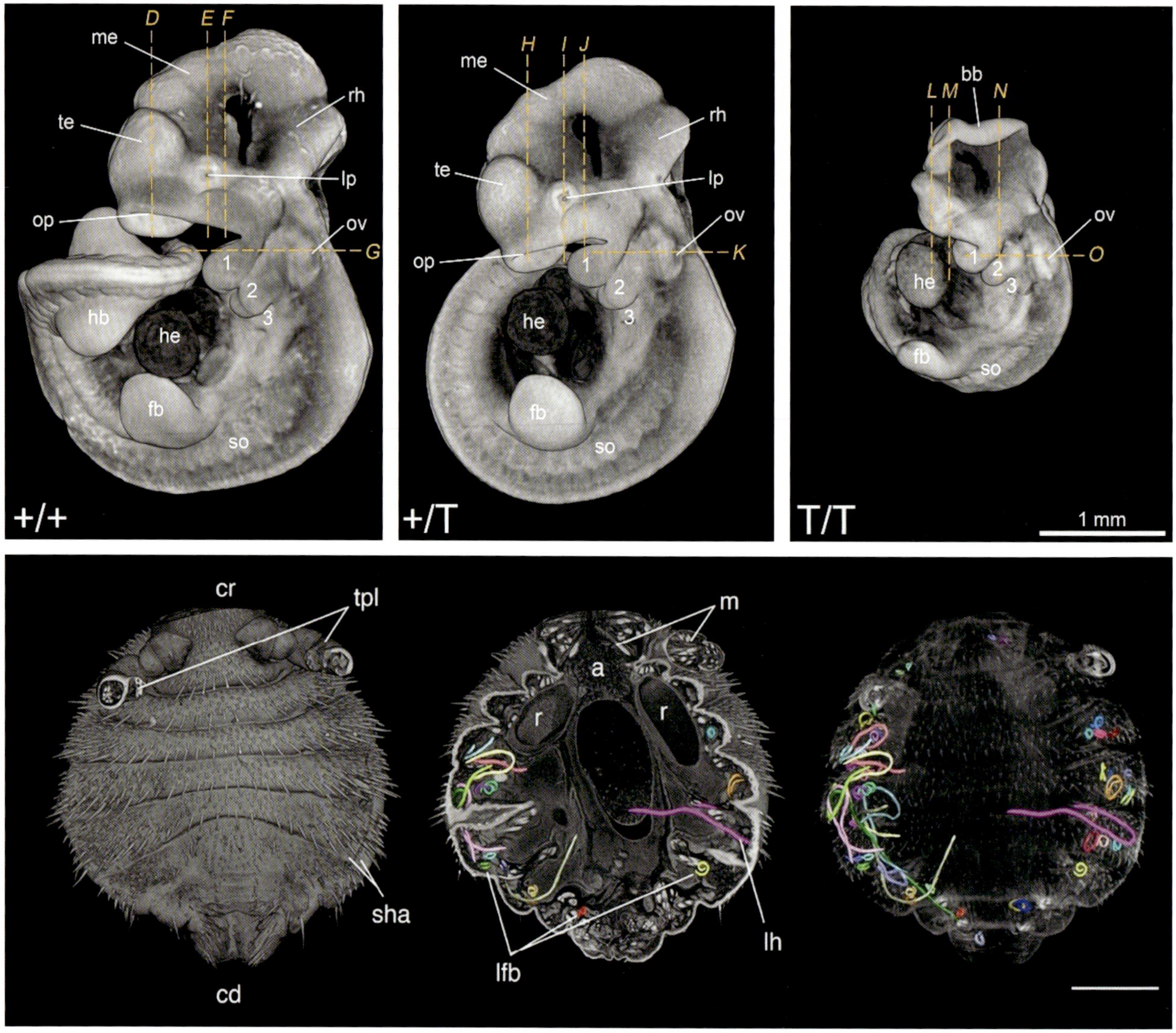

FIG. 2_ *X-ray imaging applications in a scientific service facility*
A_ *Micro-CT in cancer research. Imaging of a wildtype E10.5 mouse embryo (+/+) next to heterozygous (+/T) and homozygous (T/T) Ing3 knockout embryos. Loss of Ing3 expression in homozygous embryos causes growth retardation and defects in several organ systems including the central nervous system, resulting in embryonic lethality.*[87]

B_ *Micro-CT imaging in parasitology. Imaging of seal heartworm* (ACANTHOCHEILONEMA SPIROCAUDA) *larvae in the seal louse* (ECHINOPHTHIRIUS HORRIDUS) *reveals the size and location of parasites in the louse abdomen. Scale bar = 0.5 mm.*[88]

For Creative Commons Attribution, see the footnotes.

in scientific collaborations examining many other animal groups including mollusks, arthropods, and vertebrates.[46]

Since 2012, Stephan Handschuh has worked as a staff scientist in the core imaging facility at Vetmeduni Vienna, where he provides support in imaging-related projects to researchers using the facility infrastructure. One main part of his work focuses on providing training for both imaging systems (widefield microscopes, CLSM, micro-CT) and image processing, visualization, and analysis software. Another aspect of his work is performing the image acquisition of micro-CT image data as a scientific service_FIG. 2. Furthermore, Stephan Handschuh continuously works on developing new methods for sample preparation and X-ray imaging (see section below). In the last decade, Stephan Handschuh has supported many projects of both Vetmeduni Vienna researchers and researchers from other national and international institutions. With a main technical focus on micro-CT imaging and 3D image analysis, he has contributed to projects in various research fields, including developmental biology,[47] neurobiology,[48] cancer research,[49] bone physiology,[50] parasitology,[51] vertebrate morphology,[52] invertebrate morphology,[53] botany,[54] and paleontology.[55] In contributing to these projects as a facility staff member, his main aim is to uphold the highest standards in data acquisition, and data visualization and analysis. Stephan Handschuh's work also has a strong focus on developing, providing, and advocating for multimodal correlative imaging workflows, which are increasingly gaining in importance in different areas of the life sciences.[56]

Method Development

Over the years, both of us have been interested in not only using existing techniques and workflows but also actively develop new methods. We are both trained biologists and do not have any profound background in optical physics or computer programming. As a result, our efforts to develop methods have focused either on sample preparation or on ways to combine available imaging systems and software in order to make new workflows possible. Our first joint method development project focused on the volume visualization of serial sections. Together with Brian Metscher from the University of Vienna, we published a set of protocols for the high-quality 3D visualization of serial section data, focusing in particular on image processing pipelines and rendering settings.[57] Traditional segmentation-based surface models show only simplified geometry models from tissues, while volume rendering is capable of transferring the complete information contained within histological sections to a 3D space. As such, 3D visualizations are enriched by depicting subcellular details such as cell nuclei_FIG. 3A.

In another project, we collaborated with Bernhard Ruthensteiner and Nathalie Bäumler from the Bavarian State Collection for Zoology (ZSM) in Munich to develop a correlative microscopic imaging pipeline that combines micro-CT, light microscopy, and electron microscopy to examine a single specimen.[58] Using image registration tools to merge micro-CT, LM, and EM data in the 3D software, users can zoom in from the tomography of a whole animal onto its organs, tissues, cells, and subcellular structures_FIG. 4A. This approach allows animal specimens to be investigated across multiple length scales and levels of biological organization. The workflow does not require any modifications to be made to existing sample preparation protocols and can be used for samples that are routinely aldehyde-fixed,

A

VIDEO

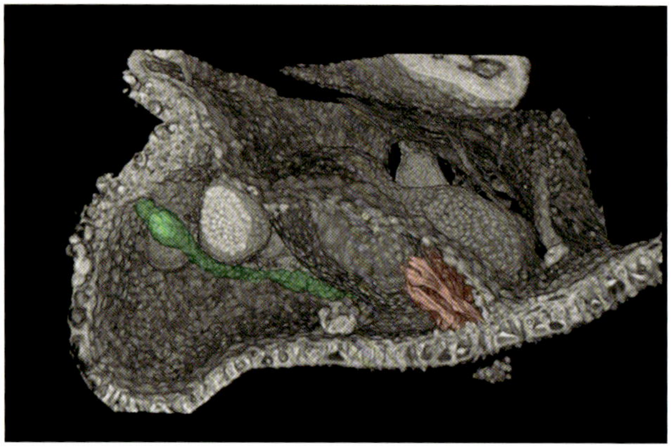

B

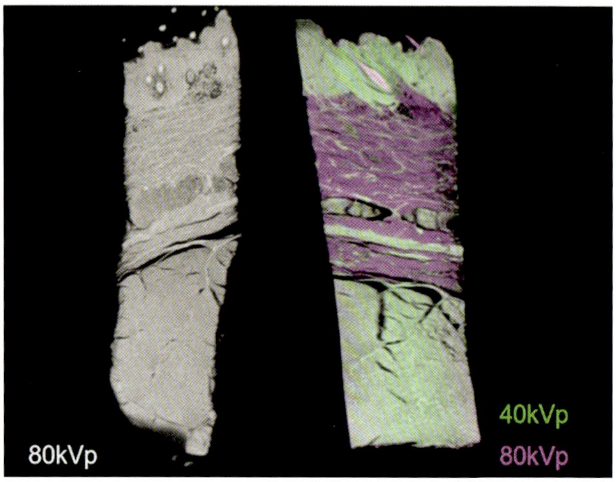

C

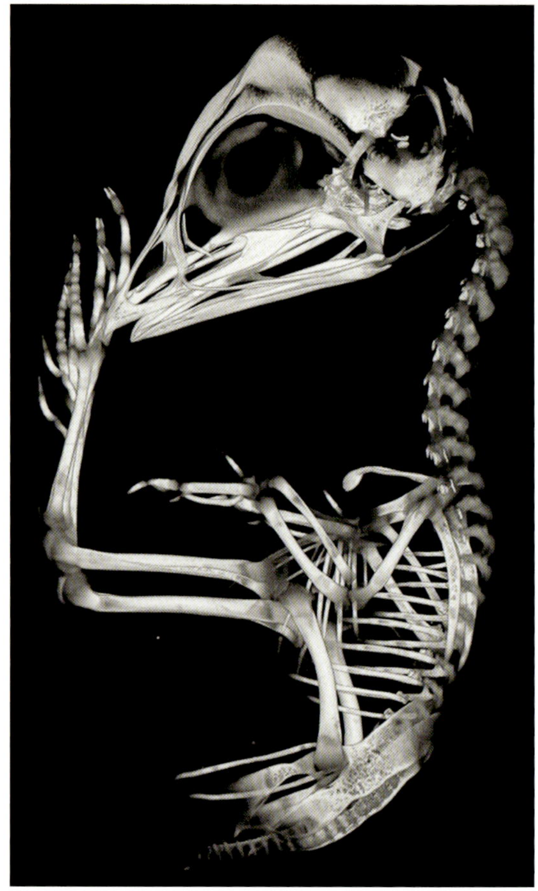

FIG. 3_ *Method development*

A_ *Volume visualization of the budding process in the freshwater bryozoan* CRISTATELLA MUCEDO *based on serial semithin sections, delivering a much higher level of micro-anatomical detail compared to traditional segmentation-based surface models. See also video.*

B_ *Microscopic dual-energy CT (microDECT) of a double-stained porcine skin biopsy (right) compared to a conventional single energy 80kVp scan (left). The iodine stain predominantly binds to fat tissue and hairs, while the tungsten stain predominantly binds to muscle and collagenous connective tissue. MicroDECT reveals spectral differences between the two stains and delivers color information on tissue components.*

Images by Stephan Handschuh, 2011

C_ *Visualization of the entire developing skeleton of an E14 chicken embryo. Bone has high intrinsic X-ray contrast, while cartilage is made visible based on staining with ruthenium red.*

VIDEO

A

VIDEOS

B

C

VIDEO

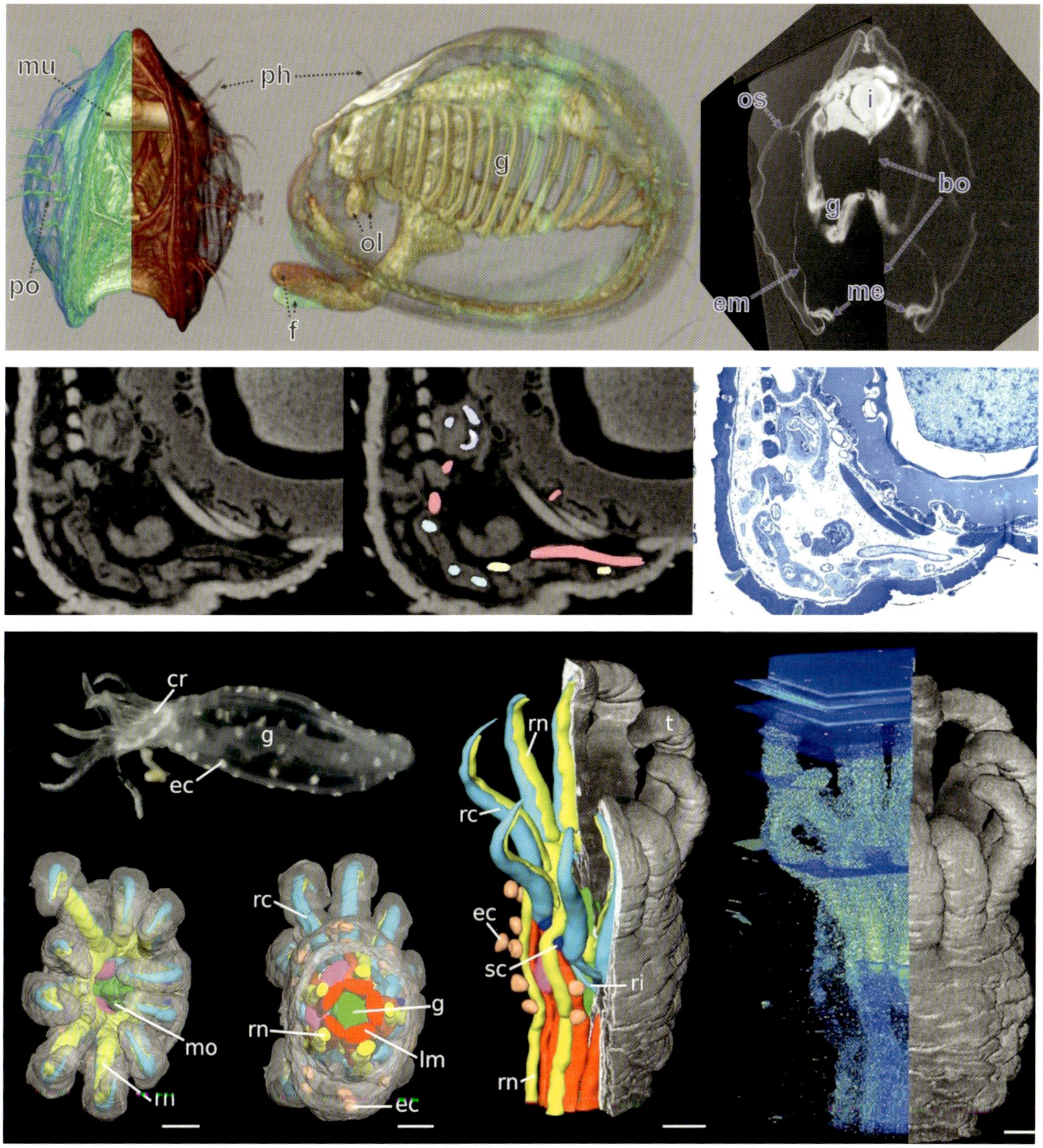

FIG. 4_*Correlative imaging approaches*

A_*Application of serial sectioning techniques and micro-CT to the same specimen of the mollusk* MYTILUS GALLOPROVINCIALIS.[89] *See also videos.*

For Creative Commons Attribution see the footnotes.

B_*Correlative CT slice of seal heartworm (*ACANTHOCHEILONEMA SPIROCAUDA*) larvae in the seal louse (Echinophthirius horridus). The manual segmentation result is shown in the center slice, and the corresponding semithin section on the right is used to validate micro-CT data interpretation*[90] *(see also fig. 2B).*

C_*Correlative approach using nano-CT and serial sections to reconstruct the morphology of the sea cucumber* LEPTOSYNAPTA CF. MINUTA *(live specimen in the upper left corner). Volume rendering of the nano-CT data set and the reconstructed organs shown as polygonal surfaces. Comparison of the serial section stack (left) with the nano-CT stack is shown on the right.*[91] *See also video.*

osmium-postfixed, and resin-embedded for optimal structural TEM preservation. Such samples can be readily imaged using micro-CT in high contrast. Subsequently, physical semithin and ultrathin sections are cut from the same specimen. This paper was one of the first of its kind, written in a period of renewed interest in correlative multimodal microscopic imaging. In several later studies, we used sections to validate interpretations of micro-CT findings_FIG. 4B. In a more recent study, Thomas Schwaha was involved in a similar correlative approach employing a nano-CT setup combined with histological serial sectioning techniques_FIG. 4C.[59]

Stephan Handschuh has also worked on several other method development projects related to the micro-CT imaging of biological samples. These have included, on the one hand, the establishment of microscopic spectral X-ray imaging protocols primarily focusing on microscopic dual energy CT (microDECT) together with Christian Beisser and Brian Metscher from the University of Vienna, and Bernhard Ruthensteiner from the ZSM Munich.[60] MicroDECT allows for the spectral separation of two elements or materials in a biological sample, thus generating color X-ray images from biological specimens in spatial resolutions below ten micrometers. To name but one example, this makes it possible to separate two different X-ray-dense contrast agents in one tissue sample_FIG. 3B. More recently, Stephan Handschuh has also collaborated with a team from the University of Manchester headed by Phil Withers in a project that aims to establish routines for the hyperspectral X-ray imaging of biological samples.[61] Stephan Handschuh's method development work has also included developing new tissue preparation routines. Together with a team from the Medical University of Innsbruck in a project headed by Anneliese Schrott-Fischer, Stephan Handschuh helped to optimize sample preparation for the visualization of nerve fiber pathways in the inner ear.[62] Together with Simone Gabner, Peter Böck, Dieter Fink, and Martin Glösmann (all Vetmeduni Vienna), Stephan Handschuh worked on developing a protocol for staining cartilage using the X-ray-dense contrast agent ruthenium red_FIG. 3C. For the first time, this has allowed for the 3D imaging and visualization of the entire developing skeleton in vertebrate embryos.[63] Together with Carolina Okada, Ingrid Walter, Christine Aurich, and Martin Glösmann (all Vetmeduni Vienna), Stephan Handschuh developed a workflow for the micro-CT imaging of unstained paraffin-embedded tissue.[64] This workflow, similar to the 2013 correlative workflow paper, also integrates micro-CT into an existing sample preparation and imaging pipeline. Micro-CT can be used to screen paraffin-embedded tissue biopsies, thus yielding 3D anatomy data that can be used to target regions of interest for subsequent histological and immunohistochemical analysis.

VIDEO

Teaching

New 3D imaging and visualization tools have also had a big impact on academic teaching. For more than a decade, both of us have been carrying out a lot of teaching on microscopic imaging techniques as well as image processing, visualization, and analysis. This teaching has taken place in different ways, including in university classes at the University of Vienna and Vetmeduni Vienna, in national and international workshops, and in one-on-one training for undergraduate students and PhD candidates. Together, we have taught more than two hundred students 3D image visualization and analysis.

In addition, we have aimed to increasingly integrate virtual 3D image data into our teaching_FIG. 5A. One example of this is a traditional comparative vertebrate anatomy class at the University of Vienna. Thomas Schwaha is in charge of this class, while Stephan Handschuh has been contributing since 2013 as an external lecturer. Traditionally, this class introduced students to vertebrate anatomy based on dissections and mounted skeletons of the main vertebrate groups (lampreys, cartilaginous fish, bony fish, amphibians, turtles, lizards and snakes, birds, mammals). Since 2013, we have increasingly included 3D images and models as teaching materials. These models are usually based on 3D tomographic images, which are annotated by the teacher and made available to students. This virtual content aims to supplement real biological specimens with additional information instead of replacing physical dissections with virtual dissections. Moreover, we always place a particular focus on telling students that a virtual dataset can never fully replace the physical specimen, focusing on the advantages of 3D data as an add-on to the real object. For example, students can study the heart and the main arteries of an anole lizard in a 3D model as these vessels are hard to comprehend solely by dissection due to the animal's small size_FIG. 5C. Similarly, students can study digital annotated models of different vertebrate skulls alongside real skulls. One advantage here is the possibility of virtually cutting into the skulls, thus making it possible to see interior parts of skull anatomy that are usually hidden_FIG. 5D. Or students can explore the cartilaginous head skeleton of a lamprey including the branchial basket, which is highly complex and almost impossible to comprehend by means of dissection _FIG. 5B. Over the years, we have created a number of digital datasets for teaching, but we equally focus on pointing to the large amount of data available from other scientific institutions, including the online repositories of both universities and museums. These repositories are now a rich resource both in terms of tomography image data[65] and 3D models.[66] The latter platform in particular is interesting for undergraduate classes because it includes a large number of annotated anatomy models.[67]

Over a Decade of Collaboration with Alfred Vendl and the Science Visualization Lab: Science Communication, TV Documentaries, and Art and Science Projects

We are scientists, and as such, we mainly communicate with other scientists. However, we should all be aware that communicating science to a wider audience is critical for raising the public's awareness, especially when considering the global ecological crisis and the dramatic increase in species extinctions. Thus, collaborating with the University of Applied Arts Vienna has given us a unique chance to introduce unfamiliar creatures to such a wider audience, both by means of TV documentaries and in art and science projects.

Our collaboration with Alfred Vendl's Science Visualization Lab at the University of Applied Arts Vienna started in 2010. By that time, Manfred Walzl from the University of Vienna had already been working with the Science Visualization Lab on many projects as a scientific consultant, and had been delivering and preparing biological specimens for electron microscopy imaging, including for the 2008 Emmy-Award-winning documentary *Nature Tech*. Our personal involvement in the projects of the Science Visualization Lab came from an idea to increasingly supplement real microscopic imagery such as stereoscopy and scanning electron microscopy with computer-generated imagery (CGI) in order to make portraying microscopic organisms and their behavior in their natural environment more flexible.

Stephan Handschuh joined the team that went on a virtual camera flight through the alimentary canal of a termite for the *Universum* documentary *Die geheime Welt der Termiten* (The Secret World of Termites).[68] This sequence was conceived of and directed by Alfred Vendl and was quite a demanding task because the termite species used in the animation had a total body size of approx. fifteen millimeters, and its alimentary canal is quite complex. These termites eat wood, and the sequence shows the complete processing of wood particles, from ingestion to the grinding of particles in the gizzard (a stomach that accomplishes mechanical chewing) and final digestion with the help of flagellate symbionts in the voluminous hindgut. On the imaging side, we chose to combine three methods to capture all of the anatomical features relevant to the model. We used micro-CT to image an osmium-stained specimen to get an accurate model of the whole termite gut. Furthermore, we used serial semithin sections for parts of the gut to obtain models in improved resolution, e.g., from the gizzard. Then, we used scanning electron microscopy images from the inner surface of the alimentary tract as well as from the flagellate symbionts to get nanoscopic surface information. Manfred Walzl performed the sample preparation; Stephan Handschuh did the micro-CT imaging, reconstructed serial microscopy sections, imaged segmentation, and created 3D models; and Rudi Erlach from the University of Applied Arts Vienna obtained the SEM images. Finally, these data were merged so that Reinhold Fragner and his Industrial Motion Art team could create an animated CGI sequence, for, at that time, they had already established a workflow to map scanning electron microscopy images onto digital polygon mesh models in order to create highly realistic animations. The full animation sequence can be viewed on *YouTube* as *The Impossible Journey Through a Little Wood Termite*.[69]

The second project we worked on together was the 2012 *Terra Mater* episode *Planet You*.[70] All of the people from the termite project team were still involved. Alfred Vendl directed the sequences, Manfred Walzl performed sample preparation, Stephan Handschuh carried out micro-CT and 3D modeling, Rudi Erlach obtained SEM images, and Reinhold Fragner and his team created CGI sequences. In addition, Thomas Schwaha joined the team as an expert in confocal laser scanning microscopy (CLSM). The episode included a lot of CGI, and the two of us were involved in two shots: one showing clothes lice on human hairs and skin (based on micro-CT images) and the other showing a tiny hair follicle mite dwelling in an eyebrow (based on CLSM images). Parts of the two shots can be viewed in the episode trailer on *YouTube*,[71] and the full episode can be viewed on the ServusTV homepage.[72]

After finishing the *Planet You* project, the Science Visualization Lab came up with a concept for a whole TV documentary on microscopic life in soil. In preparation for this project, a number of digital models from microscopic soil organisms were turned into »digital actors,« mainly featuring different sorts of mites[73] but also pseudoscorpions,[74] tiny centipedes,[75] and moss-dwelling tardigrades.[76] As part of this project, we also published a paper describing the full workflow for creating

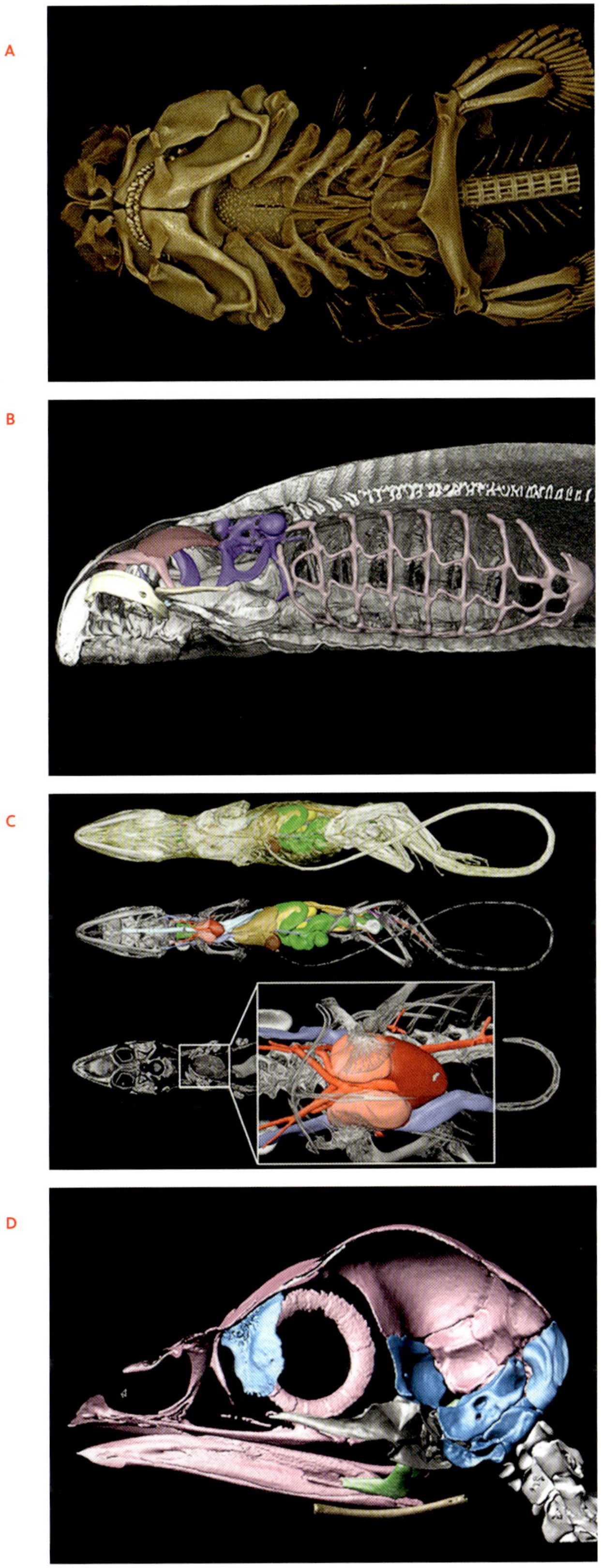

FIG. 5_ *3D visualizations in the teaching of animal morphology*
A_ *Volume rendering of a shark depicting the mineralized cranial cartilage skeleton, ventral view*
B_ *A surface model shows the cranial skeleton of a lamprey including the complex branchial basket. See also video.*
C_ *Digital dissection of an anole lizard. Some anatomical structures such as the vasculature are very hard to comprehend on the basis of real dissection due to the small size of the animal.*
D_ *3D PDF model of the skull of a juvenile chicken. Virtual cutting allows students to see interior parts of the skull that are not accessible in real and intact skull specimens. In this kind of file, every bone is annotated, and PDFs can be easily distributed among students because they can be viewed with a conventional Acrobat Reader.*

Images by Stephan Handschuh, and Thomas Schwaha, 2020

VIDEO

A

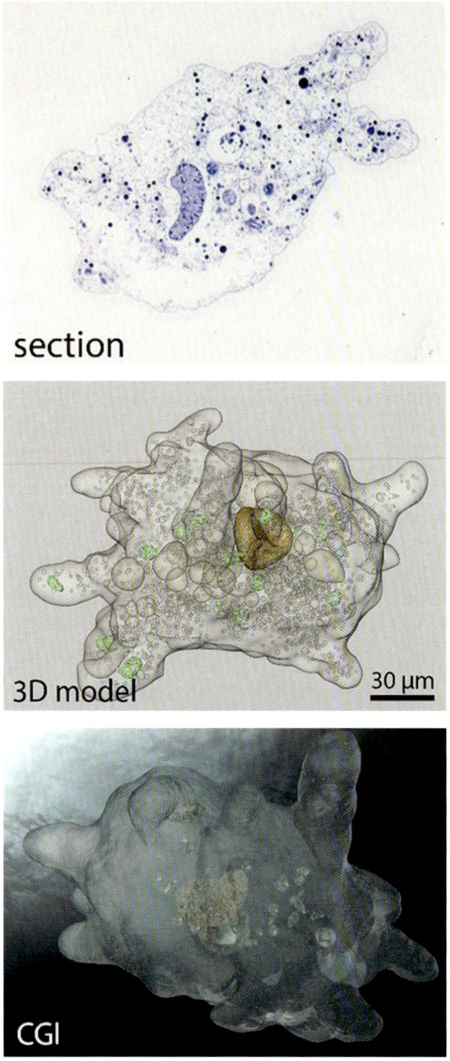

B

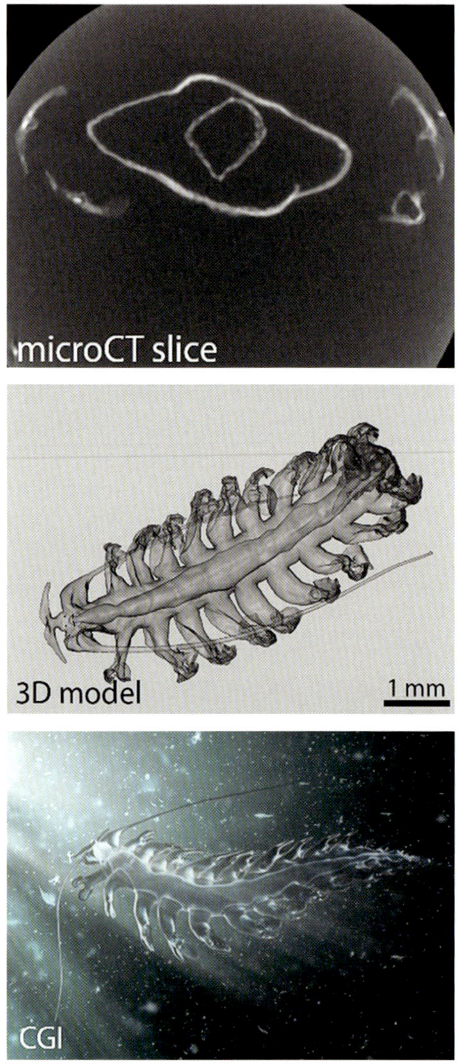

FIG. 6_ *Two »digital actors« from the art and science project* Noise Aquarium
A_ *Modeling of small planktonic species, demonstrated on a unicellular organism* Amoeba sp. *Serial semithin section micrographs (top) are segmented in order to create a 3D model (middle), which is then further processed and used in CGI (bottom).*

Images by Stephan Handschuh, 2015
Image (bottom) by Martina R. Fröschl, Science Visualization Lab, University of Applied Arts Vienna, 2016

B_ *Modeling of medium-size and larger planktonic species, demonstrated on the marine polychaete worm* Tomopteris *sp. Micro-CT slices (top) are segmented in order to create a 3D model (middle), which is then further processed and used in CGI (bottom).*

Images by Stephan Handschuh, 2016
Image (bottom) by Martina R. Fröschl, Science Visualization Lab, University of Applied Arts Vienna, 2016

VIDEO

A

B

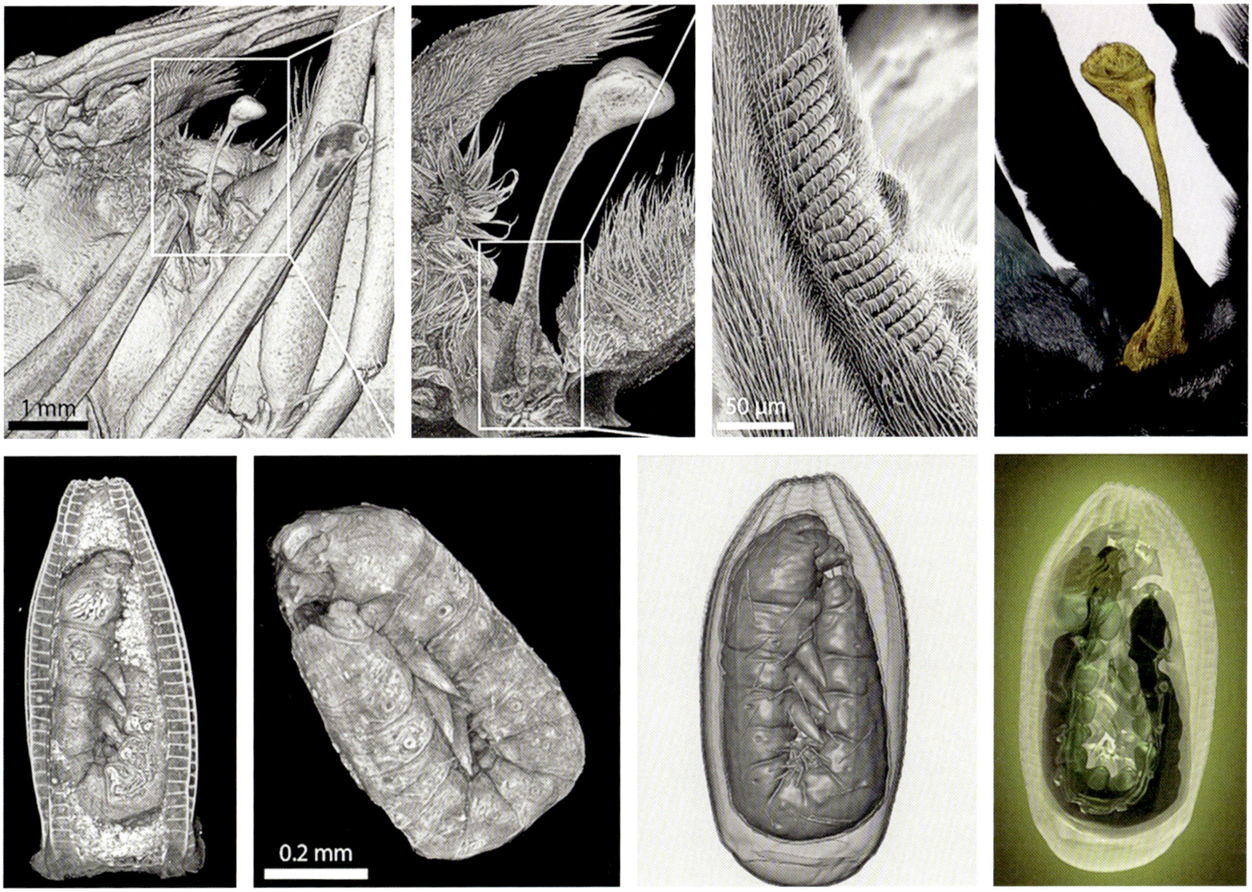

FIG. 7_ *Modeling of insect specimens for the Curiosity Stream documentary* PLANET INSECT

A_ *A fly haltere, a modified wing that servesas a sensory organ providing information about body movements during fly flight. We used a multiscale micro-CT approach with an overview scan providing information on the fly thorax (left) and higher-resolution tomographies providing more structural detail on the haltere (second from the left), complemented by SEM images from sensory fields close to the base of the haltere (second from the right). Finally, all these data were merged to create the CGI animating the movements of the haltere during flight (right).*

Images by Stephan Handschuh, 2022
Image (right) by Martina R. Fröschl, Science Visualization Lab, University of Applied Arts Vienna, 2022
Image (second from the right) by Rudolf Erlach, Science Visualization Lab, University of Applied Arts Vienna, 2022

B_ *Development of* PIERIS BRASSICAE *butterfly embryos inside the egg. Micro-CT image volumes (left, second from the left) were segmented in order to create 3D models (second from the right) of three developmental stages (day one, day four, day six, see also videos 13–15), which were then further processed and used in CGI (right).*

Images by Stephan Handschuh, 2022
Image (right) by Martina R. Fröschl, Science Visualization Lab, University of Applied Arts Vienna, 2022

VIDEOS

these models.[77] The project was never realized as a full TV episode, but many of the characters were later used to create the short animation film *The Incredible Water Bear*, which was directed by Reinhold Fragner, produced by Alfred Vendl, and released in 2014.[78] This project was also our first close collaboration with Martina R. Fröschl, who was in charge of editing, modeling, and texturing these sequences, and who focused in particular on studying the locomotion of the animals as part of her PhD project at the University of Applied Arts Vienna.

After finishing the project on soil organisms, Manfred Walzl left the team in order to retire, so the two of us were the only biologists left in the team. In 2015, Alfred Vendl asked us to create 3D models of three unicellular organisms (a cyanobacterium, a paramecium, and an amoeba_FIG. 6A) for the IMAX documentary *Voyage of Time* by US director Terrence Malick. Ultimately, these models were not used in the film, but Alfred Vendl decided to continue down the track of visualizing microscopic aquatic organisms and, in 2016, began a collaboration with Victoria Vesna from the Art|Sci Center UCLA that led to the art and science project *Noise Aquarium*.[79] This interdisciplinary project headed by Alfred Vendl and Victoria Vesna used computer animations by Martina R. Fröschl[80] to visualize the impact of noise pollution on microscopic planktonic animals.[81] Visual impressions were supported by sound by Paul Geluso and programming by Glenn Bristol that, ultimately, after starting as a linear video version, turned *Noise Aquarium* into an interactive virtual reality experience. As a thought-provoking installation, *Noise Aquarium* aims to raise awareness of the ecological crisis and the impact of noise pollution on microscopic life in the oceans, in this specific case focusing of tiny marine creatures unknown to most of us and not visible to the naked eye.[82] Between 2015 and 2020, we modeled numerous aquatic organisms including unicellular bioluminescent dinoflagellates *Noctiluca*, star fish larvae, phoronid worm larvae, the marine polychaete worm *Tomopteris*_FIG. 6B, the appendicularian *Oikopleura*, the chaetognath worm *Parasagitta*, the tiny copepod crustacean *Tigriopus*, and the larger malacostracan crustacean *Meganyctiphanes* (krill). Over the last few years, *Noise Aquarium* has been shown in many countries including Austria, Croatia, Switzerland, Poland, the UK, the US, Australia, and Singapore. In 2022, the story of *Noise Aquarium* continues. In 2020, the creatures of *Noise Aquarium* were used to create the short movie *LIFE*, produced and directed by Alfred Vendl and Martina R. Fröschl.[83]

VIDEOS

Our most recent project (2021–2022) was delivering 3D models for the Curiosity Stream documentary *Planet Insect*, produced by UK director Steve Nicholls. Steve Nicholls and Alfred Vendl conceptualized three CGI sequences for this documentary, including a fly haltere in motion, a camera flight around a moth antenna, and the development of a butterfly embryo inside the egg. While Martina R. Fröschl from the Science Visualization Lab did the CGI animation once more, the team of biologists grew to four members, also including Valentin Blüml and Christina Kaurin, who helped us with image segmentation and modeling. For the fly haltere, we captured multiple tomography scans and zoomed in from the whole fly thorax onto the haltere and its various sections. By using phase-contrast micro-CT imaging of semi-wet specimens in a humid air environment, we were able to create accurate models of the haltere and its neighboring structures. In addition to these 3D models, details of the surface texture were added based on SEM images by Rudi

Erlach. The final CGI sequence shows the haltere in motion while the fly is in flight _FIG. 7A. For the moth antenna, we also used a multi-scale tomography approach. After scanning a whole dry antenna, we focused on different regions of interest in order to provide more surface details. Scans of the dry antenna provided a high level of detail; however, upon drying, the side branches of the antenna tended to deform. We therefore also embedded a fixed, iodine-stained antenna in agarose and used this scan as a reference for the real shape of the antenna. For the CGI sequence, hairs were added to the antenna branches based on the CT reference, the antenna was colored, and a virtual camera was moved around the antenna. Finally, we modeled three different developmental stages of a butterfly embryo inside the egg. The CGI sequence used the morphing between the three modeled stages in order to illustrate embryo growth inside the egg_FIG. 7B.

Summary and Outlook

Since the beginning of our academic careers, technical developments in microscopic imaging, and 3D image visualization and analysis have had a huge impact on many biological and biomedical research fields. This chapter has reviewed the impact of these tools on our personal work, including research, method development, teaching, and art and science projects. Our technical training and fascination with microscopic techniques and digital 3D image data have given us the opportunity to collaborate with many scientists and artists. We believe that both microscopy techniques and image visualization and analysis tools will continue to develop at an overwhelming speed, continuously opening new doors to the hidden microscopic world.

1 See, e.g., T. Schlichthaerle, C. Lindner, and R. Jungmann, »Super-Resolved Visualization of Single DNA-Based Tension Sensors in Cell Adhesion,« *Nature Communications* 12 (2021): 2510; R. A. de Melo Reis, H. R. Freitas, and F. G. de Mello, »Cell Calcium Imaging as a Reliable Method to Study Neuron-Glial Circuits,« *Frontiers in Neuroscience* 14 (October 2, 2020).

2 J. Jonkman, C. M. Brown, G. D. Wright, K. I. Anderson, and A. J. North, »Tutorial: Guidance for Quantitative Confocal Microscopy,« *Nature Protocols* 15 (2020): 1585–611.

3 P. J. Keller, A. D. Schmidt, J. Wittbrodt, and E. H. K. Stelzer, »Reconstruction of Zebrafish Early Embryonic Development by Scanned Light Sheet Microscopy,« *Science* 322, no. 5904 (2008): 1065–9; J. Huisken and D. Y. R. Stainier, »Selective Plane Illumination Microscopy Techniques in Developmental Biology,« *Development* 136, no. 12 (2009): 1963–75; T. Chakraborty, M. K. Driscoll, E. Jeffery, M. M. Murphy, P. Roudot, B.-J. Chang, et al., »Light-Sheet Microscopy of Cleared Tissues with Isotropic, Subcellular Resolution,« *Nature Methods* 16 (2019): 1109–13.

4 M. Belle, D. Godefroy, C. Dominici, C. Heitz-Marchaland, P. Zelina, F. Hellal, et al., »A Simple Method for 3D Analysis of Immunolabeled Axonal Tracts in a Transparent Nervous System,« *Cell Reports* 9, no. 4 (2014): 1191–201; M. D. Rocha, D. N. Duering, P. Bethge, F. F. Voigt, S. Hildebrand, F. Helmchen, et al., »Tissue Clearing and Light Sheet Microscopy: Imaging the Unsectioned Adult Zebra Finch Brain at Cellular Resolution,« *Frontiers in Neuroanatomy* 13 (2019).

5 M. G. Gustafsson, »Surpassing the Lateral Resolution Limit by a Factor of Two Using Structured Illumination Microscopy,« *Journal of Microscopy* 198, no. 2 (2000): 82–7.

6 S. W. Hell and J. Wichmann, »Breaking the Diffraction Resolution Limit by Stimulated Emission: Stimulated-Emission-Depletion Fluorescence Microscopy,« *Optics Letters* 19, no. 11 (1994): 780–2; T. A. Klar, S. Jakobs, M. Dyba, A. Egner, and S. W. Hell, »Fluorescence Microscopy with Diffraction Resolution Barrier Broken by Stimulated Emission,« *PNAS* 97, no. 15 (2000): 8206–10.

7 E. Betzig, G. H. Patterson, R. Sougrat, O. W. Lindwasser, S. Olenych, J. S. Bonifacino, et al., »Imaging Intracellular Fluorescent Proteins at Nanometer Resolution,« *Science* 313, no. 5793 (2006): 1642–5; M. J. Rust, M. Bates, X. Zhuang, »Sub-Diffraction-Limit Imaging by Stochastic Optical Reconstruction Microscopy (STORM),« *Nature Methods* 3 (2006): 793–5; R. M. Dickson, A. B. Cubitt, R. Y. Tsien, W. E. Moerner, »On/Off Blinking and Switching Behaviour of Single Molecules of Green Fluorescent Protein,« *Nature* 388 (1997): 355–8.

8 J. Huff, »The Airyscan Detector from ZEISS: Confocal Imaging with Improved Signal-to-Noise Ratio and Super-Resolution,« *Nature Methods* 12 (2015): i–ii.

9 L. Moeckl, D. C. Lamb, and C. Braeuchle, »Super-Resolved Fluorescence Microscopy: Nobel Prize in Chemistry 2014 for E. Betzig, S. Hell, and W. E. Moerner,« *Angewandte Chemie International English Edition* 53, no. 51 (2014): 13972–7.

10 M. Weber, M. Leutenegger, S. Stoldt, S. Jakobs, T. S. Mihaila, A. N. Butkevich, and S. W. Hell, »MINSTED Fluorescence Localization and Nanoscopy,« *Nature Photonics* 15 (2021): 361–6; K. C. Gwosch, J. K. Pape, F. Balzarotti, P. Hoess, J. Ellenberg, J. Ries, and S. W. Hell, »MINFLUX Nanoscopy Delivers 3D Multicolor Nanometer Resolution in Cells,« *Nature Methods* 17, no. 2 (2020): 217–24.

11 A. T. Wassie, Y. Zhao, and E. S. Boyden, »Expansion Microscopy: Principles and Uses in Biological Research,« *Nature Methods* 16, no. 1 (2019): 33–41; O. M'Saad and J. Bewersdorf, »Light Microscopy of Proteins in their Ultrastructural Context,« *Nature Communications* 11 (2020): 3850.

12 W. Denk and H. Horstmann, »Serial Block-Face Scanning Electron Microscopy to Reconstruct Three-Dimensional Tissue Nanostructure,« *PLOS Biology* 2, no. 11 (2004): e329.

13 B. Inkson, M. Mulvihill, and G. Möbus, »3D Determination of Grain Shape in a FeAl-Based Nanocomposite by 3D FIB Tomography,« *Scripta Materialia* 45, no. 7 (2001): 753–8; L. Holzer, F. Indutnyi, P. H. Gasser, B. Münch, and M. Wegmann, »Three-Dimensional Analysis of Porous BaTiO3 Ceramics Using FIB Nanotomography,« *Journal of Microscopy* 216, no. 1 (2004): 84–95.

14 K. J. Hayworth, N. Kasthuri, R. Schalek, and J. W. Lichtman, »Automating the Collection of Ultrathin Serial Sections for Large Volume TEM Reconstructions,« *Microscopy and Microanalysis* 12, no. S02 (2006): 86–7; A. L. Eberle, S. Mikula, R. Schalek, J. Lichtman, M. L. K. Tate, and D. Zeidler, »High-Resolution, High-Throughput Imaging with a Multibeam Scanning Electron Microscope,« *Journal of Microscopy* 259, no. 2 (2015): 114–20.

15 See, e.g., S. Loomba, J. Straehle, V. Gangadharan, N. Heike, A. Khalifa, A. Motta, et al., »Connectomic Comparison of Mouse and Human Cortex,« *Science* 377, no. 6602 (2022): eabo0924.

16 J. Dubochet, B. Zuber, M. Eltsov, C. Bouchet-Marquis, A. Al-Amoudi, and F. Livolant, »How to ›Read‹ a Vitreous Section,« *Methods in Cell Biology* 79 (2007): 385–406; H.-M. Han, C. Bouchet-Marquis, J. Huebinger, and M. Grabenbauer, »Golgi Apparatus Analyzed by Cryo-Electron Microscopy,« *Histochemistry and Cell Biology* 140 (2013): 369–81.

17 T. Nakane, A. Kotecha, A. Sente, G. McMullan, S. Masiulis, P. M. G. E. Brown, et al., »Single-Particle Cryo-EM at Atomic Resolution,« *Nature* 587 (2020): 152–6; K. M. Yip, N. Fischer, E. Paknia, A. Chari, and H. Stark, »Atomic-Resolution Protein Structure Determination by Cryo-EM,« *Nature* 587 (2020): 157–61.

18 D. Cressey, E. Callaway, »Cryo-Electron Microscopy Wins Chemistry Nobel,« *Nature* 550 (2017): 167.

19 W. Kukulski, M. Schorb, S. Welsch, A. Picco, M. Kaksonen, and J. A. G. Briggs, »Correlated Fluorescence and 3D Electron Microscopy with High Sensitivity and Spatial Precision,« *Journal of Cell Biology* 192, no. 1 (2011): 111–19; C. J. Peddie, M.-C. Domart, X. Snetkov, P. O'Toole, B. Larijani, M. Way, et al., »Correlative Super-Resolution Fluorescence and Electron Microscopy Using Conventional Fluorescent Proteins in Vacuo,« *Journal of Structural Biology* 199, no. 2 (2017): 120–31.

20 K. D. Micheva and S. J. Smith, »Array Tomography: A New Tool for Imaging the Molecular Architecture and Ultrastructure of Neural Circuits,« *Neuron* 55, no. 1 (2007): 25–36; S. J. Smith, »Q&A: Array Tomography,« *BMC Biology* 16 (2018): 98.

21 S. Klein, B. H. Wimmer, S. L. Winter, A. Kolovou, V. Laketa, and P. Chlanda, »Post-Correlation On-Lamella Cryo-CLEM Reveals the Membrane Architecture of Lamellar Bodies,« *Communications Biology* 4 (2021): 137; D. L. Sexton, S. Burgold, A. Schertel, and E. I. Tocheva, »Super-Resolution Confocal Cryo-CLEM with Cryo-FIB Milling for In Situ Imaging of Deinococcus radiodurans,« *Current Research in Structural Biology* 4 (2022): 1–9.

22 M. L. Bouxsein, S. K. Boyd, B. A. Christiansen, R. E. Guldberg, K. J. Jepsen, and R. Müller, »Guidelines For Assessment of Bone Microstructure in Rodents Using Micro-Computed Tomography,« *Journal of Bone and Mineral Research* 25, no. 7 (2010): 1468–86; H. R. Buie, G. M. Campbell, R. J. Klinck, J. A. MacNeil, and S. K. Boyd, »Automatic Segmentation of Cortical and Trabecular Compartments Based on a Dual Threshold Technique for In Vivo Micro-CT Bone Analysis,« *Bone* 41, no. 4 (2007): 505–15.

23 J. T. Johnson, M. S. Hansen, I. Wu, L. J. Healy, C. R. Johnson, G. M. Jones, et al., »Virtual Histology of Transgenic Mouse Embryos for High-Throughput Phenotyping,« *PLOS Genetics* 2, no. 4 (2006): e61; B. D. Metscher, »Micro-CT for Comparative Morphology: Simple Staining Methods Allow High-Contrast 3D Imaging of Diverse Non-Mineralized Animal Tissues,« *BMC Physiology* 9 (2009): 11; B. D. Metscher, »Micro-CT for Developmental Biology: A Versatile Tool for High-Contrast 3D Imaging at Histological Resolutions,« *Developmental Dynamics* 238 (2009): 632–40.

24 P. M. Gignac, N. J. Kley, J. A. Clarke, M. W. Colbert, A. C. Morhardt, D. Cerio, et al., »Diffusible Iodine-Based Contrast-Enhanced Computed Tomography (Dicect): An Emerging Tool for Rapid, High-Resolution, 3-D Imaging of Metazoan Soft Tissues,« *Journal of Anatomy* 228 (2016): 889–909.

25 A. Sombke, E. Lipke, P. Michalik, G. Uhl, and S. Harzsch, »Potential and Limitations of X-Ray Micro-Computed Tomography in Arthropod Neuroanatomy: A Methodological and Comparative Survey,« *Journal of Computational Neuroscience* 523, no. 8 (2015): 1281–95.

26 R. L. Abel, C. R. Laurini, and M. Richter, »A Palaeobiologist's Guide to ›Virtual‹ Micro-CT Preparation,« *Palaeontologia Electronica* 15, no. 2 (2012); M. Dierick, V. Cnudde, B. Masschaele, J. Vlassenbroeck, L. van Hoorebeke, and P. Jacobs, »Micro-CT of Fossils Preserved in Amber,« *Nuclear Instruments & Methods in Physics, Research Section A: Accelerators Spectrometers Detectors and Associated Equipment* 580, no. 1 (2007): 641–3.

27 M. E. Dickinson, A. M. Flenniken, X. Ji, L. Teboul, M. D. Wong, J. K. White, et al. »High-Throughput Discovery of Novel Developmental Phenotypes,« *Nature* 537 (2016): 508–14.

28 M. A. Karreman, L. Mercier, N. L. Schieber, G. Solecki, G. Allio, F. Winkler, et al., »Fast and Precise Targeting of Single Tumor Cells In Vivo by Multimodal Correlative Microscopy,« *Journal of Cell Science* 129, no. 2 (2016): 444–56.

29 J. Schindelin, I. Arganda-Carreras, E. Frise, V. Kaynig, M. Longair, T. Pietzsch, et al., »Fiji: An Open-Source Platform for Biological-Image Analysis,« *Nature Methods* 9 (2012): 676–82.

30 A. Limaye, »Drishti: A Volume Exploration and Presentation Tool.« *Proceedings SPIE 8506, Developments in X-Ray Tomography VIII*, 85060X (October 17, 2012).

31 R. Kikinis, S. D. Pieper, and K. G. Vosburgh, »3D Slicer: A Platform for Subject-Specific Image Analysis, Visualization, and Clinical Support,« in *Intraoperative Imaging and Image-Guided Therapy*, ed. F. A. Jolesz (New York: Springer, 2014), 277–89.

32 H. Pfister, J. Hardenbergh, J. Knittel, H. Lauer, and L. Seiler, »The Volumepro Real-Time Ray-Casting System,« in *SIGGRAPH '99: Proceedings of the 26th Annual Conference On Computer Graphics and Interactive Techniques*, ed. W. Waggenspack (New York: ACM Press, 1999), 251–60.

33 R. Haase, L. A. Royer, P. Steinbach, D. Schmidt, A. Dibrov, U. Schmidt, et al., »CLIJ: GPU-Accelerated Image Processing for Everyone,« *Nature Methods* 17 (2020): 5–6.

34 B. Ruthensteiner, »Soft Part 3D Visualization by Serial Sectioning and Computer Reconstruction,« *Zoosymposia* 1 (2008): 63–100.

35 H. R. Buie, G. M. Campbell, R. J. Klinck, J. A. MacNeil, and S. K. Boyd, »Automatic Segmentation of Cortical and Trabecular Compartments.«

36 R. A. Russell, N. M. Adams, D. A. Stephens, E. Batty, I. Jensen, and P. S. Freemont, »Segmentation of Fluorescence Microscopy Images for Quantitative Analysis of Cell Nuclear Architecture,« *Biophysics Journal* 96, no. 8 (2009): 3379–89.

37 P. D. Lösel, T. van de Kamp, A. Jayme, A. Ershov, T. Faragó, O. Pichler, et al., »Introducing Biomedisa as an Open-Source Online Platform for Biomedical Image Segmentation,« *Nature Communications* 11 (2020): 5577.

38 S. Berg, D. Kutra, T. Kroeger, C. N. Straehle, B. X. Kausler, C. Haubold, et al., »ilastik: Interactive Machine Learning for (Bio)Image Analysis,« *Nature Methods* 16 (2019): 1226–32.

39 S. Minaee, Y. Boykov, F. Porikli, A. Plaza, N. Kehtarnavaz, and D. Terzopoulos, »Image Segmentation Using Deep Learning: A Survey,« *IEEE Transactions on Pattern Analysis and Machine Intelligence* 44, no. 7 (2022): 3523–42; P. Malhotra, S. Gupta, D. Koundal, A. Zaguia, and W. Enbeyle, »Deep Neural Networks for Medical Image Segmentation,« *Journal of Healthcare Engineering* 95 (2022): 9580991.

40 D. Ciresan, A. Giusti, L. Gambardella, and J. Schmidhuber, »Deep Neural Networks Segment Neuronal Membranes in Electron Microscopy Images,« in *NIPS '12: Proceedings of the 25th International Conference on Neural Information Processing Systems*, vol. 2, ed. F. Pereira, C. J. C. Burges, L. Bottou, and K. Q. Weinberger (New York: Curran Associates, Inc., 2012), 2843–51.

41 W. Ouyang, F. Beuttenmueller, E. Gómez-de-Mariscal, C. Pape, T. Burke, C. Garcia-López-de-Haro, et al., »BioImage Model Zoo: A Community-Driven Resource for Accessible Deep Learning in BioImage Analysis,« *bioRxiv* (2022), https://www.doi.org/10.1101/2022.06.07.495102.

42 F. Maes, A. Collignon, D. Vandermeulen, G. Marchal, P. Suetens, »Multimodality Image Registration by Maximization of Mutual Information,« *IEEE Transactions On Medical Imaging* 16, no. 2 (1997): 187–98.

43 M. A. Karreman, L. Mercier, N. L. Schieber, G. Solecki, G. Allio, F. Winkler, et al., »Fast and Precise Targeting«; S. Handschuh, N. Baeumler, T. Schwaha, and B. Ruthensteiner, »A Correlative Approach for Combining Micro-CT, Light and Transmission Electron Microscopy in a Single 3D Scenario,« *Frontiers of Zoology* 10 (2013): 44.

44 T. F. Schwaha, S. Handschuh, A. N. Ostrovsky, and A. Wanninger, »Morphology of the Bryozoan *Cinctipora elegans* (Cyclostomata, Cinctiporidae) with First Data on its Sexual Reproduction and the Cyclostome Neuro-Muscular System,« *BMC Evolutionary Biology* 18 (2018): 92.

45 T. Schwaha, J. M. Bernhard, V. P. Edgcomb, M. A. Todaro, »*Aethozooides uraniae*, a New Deep-Sea Genus and Species of Solitary Bryozoan from the Mediterranean Sea, with a Revision of the Aethozoidae,« *Marine Biodiversity* 49 (2019): 1843–56; S. H. Decker, D. P. Gordon, M. E. Spencer Jones, and T. Schwaha, »A Revision of the Ctenostome Bryozoan Family Pherusellidae, with Description of Two New Species,« *Journal of Zoological Systematics and Evolutionary Research* 59 (2021): 963–80.

46 E.g., A. Wurzinger-Mayer, J. R. Shipway, A. Kristof, T. Schwaha, S. M. Cragg, and A. Wanninger, »Developmental Dynamics of Myogenesis in the Shipworm *Lyrodus pedicellatus* (Mollusca: Bivalvia),« *Frontiers in Zoology* 11 (2014): 90; M. Seiter, L. Strobl, T. Schwaha, L. Prendini, F. D. Schramm, »Morphometry of the Pedipalp Patella Provides New Characters for Species-Level Taxonomy in Whip Spiders (Arachnida, Amblypygi): A Test Case with Description of a New Species of *Phrynus*,« *Zoologischer Anzeiger* 298 (2022): 10–28.

47 L. Johnson Chacko, D. Wertjanz, C. Sergi, J. Dudas, N. Fischer, T. Eberharter, et al., »Growth and Cellular Patterning During Fetal Human Inner Ear Development Studied by a Correlative Imaging Approach,« *BMC Developmental Biolology* 19 (2019): 11; D. Scarlet, S. Handschuh, U. Reichart, F. Podico, R. E. Ellerbrock, S. Demyda-Peyrás, et al., »Sexual Differentiation and Primordial Germ Cell Distribution in the Early Horse Fetus,« *Animals (Basel)* 11, no. 8 (2021): 2422.

48 L. Johnson Chacko, D. T. Schmidbauer, S. Handschuh, A. Reka, K. D. Fritscher, P. Raudaschl, et al., »Analysis of Vestibular Labyrinthine Geometry and Variation in the Human Temporal Bone,« *Frontiers in Neuroscience* 12 (2018): 107.

49 D. Fink, T. Yau, A. Nabbi, B. Wagner, C. Wagner, S. M. Hu, et al., »Loss of Ing3 Expression Results in Growth Retardation and Embryonic Death,« *Cancers (Basel)* 12, no. 1 (2019): 80; S. Rao, L. Tortola, T. Perlot, G. Wirnsberger, M. Novatchkova, R. Nitsch, et al., »A Dual Role For Autophagy in a Murine Model of Lung Cancer,« *Nature Communications* 5 (2014): 3056.

50 M. Vaidya, D. Lehner, S. Handschuh, F. F. Jay, R. G. Erben, and M. R. Schneider, »Osteoblast-Specific Overexpression of Amphiregulin Leads to Transient Increase in Femoral Cancellous Bone Mass in Mice,« *Bone* 81 (2015): 36–46.

51 D. Ebmer, S. Handschuh, T. Schwaha, A. Rubio-García, U. Gärtner, M. Glösmann, et al., »Novel 3D In Situ Visualization of Seal Heartworm (*Acanthocheilonema spirocauda*) Larvae in the Seal Louse (*Echinophthirius horridus*) by X-Ray Micro-C,« *Scientific Reports* 12 (2022): 14078.

52 S. Kunisch, V. Blüml, T. Schwaha, C. J. Beisser, S. Handschuh, and P. Lemell, »Digital Dissection of the Head of the Frogs Calyptocephalella Gayi and Leptodactylus Pentadactylus with Emphasis On the Feeding Apparatus,« *Journal of Anatomy* 239, no. 2 (2021): 391–404; S. Handschuh, N. Natchev, S. Kummer, C. J. Beisser, P. Lemell, A. Herrel, and V. Vergilov, »Cranial Kinesis in the Miniaturised Lizard Ablepharus kitaibelii (Squamata: Scincidae),« *Journal of Experimental Biology* 222, no. 9 (2019): jeb198291.

53 C. Zittra, S. Vitecek, T. Schwaha, S. Handschuh, J. Martini, A. Vieira, et al., »Comparing Head Muscles Among Drusinae Clades (Insecta: Trichoptera) Reveals High Congruence Despite Strong Contrasts in Head Shape,« *Scientific Reports* 12 (2022): 1047.

54 Y. M. Staedler, T. Kreisberger, S. Manafzadeh, M. Chartier, S. Handschuh, S. Pamperl, et al., »Novel Computed Tomography-Based Tools Reliably Quantify Plant Reproductive Investment,« *Journal of Experimental Botany* 69, no. 3 (2018): 525–35.

55 L. Chitimia-Dobler , B. J. Mans, S. Handschuh, and J. A. Dunlop, »A Remarkable Assemblage of Ticks From Mid-Cretaceous Burmese Amber,« *Parasitology* (2022): 1–36.

56 A. Walter, P. Paul-Gilloteaux, B. Plochberger, L. Sefc, P. Verkade, J. G. Mannheim, et al., »Correlated Multimodal Imaging in Life Sciences: Expanding the Biomedical Horizon,« *Frontiers in Physics* 8 (2020).

57 S. Handschuh, T. Schwaha, and B. D. Metscher, »Showing Their True Colors: A Practical Approach to Volume Rendering From Serial Sections,« *BMC Developmental Biology* 10 (2010): 41.

58 S. Handschuh, N. Baeumler, T. Schwaha, and B. Ruthensteiner, »A Correlative Approach.«

59 S. Ferstl, T. Schwaha, B. Ruthensteiner, L. Hehn, S. Allner, M. Müller, et al., »Nanoscopic X-Ray Tomography For Correlative Microscopy of a Small Meiofaunal Sea-Cucumber,« *Scientific Reports* 10 (2020): 3960.

60 S. Handschuh, C. J. Beisser, B. Ruthensteiner, and B. D. Metscher, »Microscopic Dual-Energy CT (Microdect): A Flexible Tool For Multichannel Ex Vivo 3D Imaging of Biological Specimens,« *Journal of Microscopy* 267, no. 1 (2017): 3–26.

61 R. Warr, E. Ametova, R. J. Cernik, G. Fardell, S. Handschuh, J. S. Jørgensen, et al., »Enhanced Hyperspectral Tomography For Bioimaging by Spatiospectral Reconstruction,« *Scientific Reports* 11 (2021): 20818.

62 R. Glueckert, L. Johnson Chacko, D. Schmidbauer, T. Potrusil, E. J. Pechriggl, R Hoermann, et al., »Visualization of the Membranous Labyrinth and Nerve Fiber Pathways in Human and Animal Inner Ears Using Micro-CT Imaging,« *Frontiers in Neuroscience* 12 (2018): 501.

63 S. Gabner, P. Böck, D. Fink, M. Glösmann, and S. Handschuh, »The Visible Skeleton 2.0: Phenotyping of Cartilage and Bone in Fixed Vertebrate Embryos and Foetuses Based On X-Ray Micro-CT,« *Development* 147, no. 11 (2020): dev187633.

64 S. Handschuh, C. T. C. Okada, I. Walter, C. Aurich, and M. Glösmann, »An Optimized Workflow For Micro-CT Imaging of Formalin-Fixed and Paraffin-Embedded (FFPE) Early Equine Embryos,« *Anatomia, Histologia, Embryologia* 51, no. 5 (2022): 611–23.

65 See, e.g., *MorphoSource*, https://www.morphosource.org.

66 See, e.g., *Sketchfab*, https://www.sketchfab.com.

67 See, e.g., »Blackburn Lab,« *Sketchfab*, https://sketchfab.com/ufherps.

68 https://www.imdb.com/title/tt2108385/.

69 »The Impossible Journey Through a Little Wood Termite,« *YouTube*, https://www.youtube.com/watch?v=7a34qKoaAvQ.

70 W. Thaler, *Die geheime Welt der Termiten*, November 8, 2011, https://www.imdb.com/title/tt2197082.

71 »Wir sind Planeten« (trailer), July 18, 2014, *YouTube*, https://www.youtube.com/watch?v=XXkQ8yTbc4g.

72 *Wir sind Planeten*, September 1, 2021, https://www.servustv.com/natur/v/1335957362999-1996707315/.

73 »Turtle Mite—Turntable,« May 16, 2013, *YouTube*, https://www.youtube.com/watch?v=wkqil1gVWo4; »Predatory Mite—Turntable,« May 16, 2013, *YouTube*, https://www.youtube.com/watch?v=IfQ3X--FkwM.

74 »Pseudo Scorpio—Turntable,« May 16, 2013, *YouTube*, https://www.youtube.com/watch?v=jUjClCax72s.

75 »Centipede (Lithobius)—Turntable,« May 16, 2013, *YouTube*, https://www.youtube.com/watch?v=-ZN7_ywo9OA.

76 »Water Bear (Tardigrade) Turntable,« May 16, 2013 *YouTube*, https://www.youtube.com/watch?v=e-XOLKJ9Ch8.

77 M. R. Fröschl, S. Handschuh, R. Erlach, T. Schwaha, H. Goldammer, R. Fragner, and M. G. Walzl, »Computer-Generated Images of Microscopic Soil Organisms For Documentary Films,« *Soil Organisms* 86, no. 2 (2014): 95–102.

78 »The Incredible Water Bear,« January 27, 2014, *YouTube*, https://www.youtube.com/watch?v=cp1WwNE6Lms.

79 *Noise Aquarium*, https://noiseaquarium.com; V. Vesna, »NOISE AQUARIUM: Iterations, Variations, and Responsive Ecotistical Work,« in *Plastic Ocean: Art and Science Responses to Marine Pollution*, ed. Reichle I. (Berlin: De Gruyter, 2021), 157–78; V. Vesna, A. Vendl, M. R. Fröschl, G. Bristol, P. Geluso, S. Handschuh, and T. Schwaha, »NOISE AQUARIUM,« *Leonardo* 52, no. 4 (2019): 408–9.

80 M. R. Fröschl and A. Vendl, »Computer-Animated Fluidity for Stiff Datasets and the Visualization of Underwater Noise,« in *Plastic Ocean*, 179–196.

81 R. D. McCauley, R. D. Day, K. M. Swadling, Q. P. Fitzgibbon, R. A. Watson, and J. M. Semmens, »Widely Used Marine Seismic Survey Air Gun Operations Negatively Impact Zooplankton,« *Nature Ecology & Evolution* 1 (2017): 195.

82 T. Schwaha and S. Handschuh, »From Live Imaging to 3D Modeling: A Guide to Documentation and Processing of Planktonic Organisms,« in *Plastic Ocean*, 197–212.

83 »›LIFE‹ 2D,« *YouTube*, https://www.youtube.com/watch?v=5V5NBV9heAc.

84 Modified from T. Schwaha, S. Handschuh, A. N. Ostrovsky, and A. Wanninger, »Morphology of the Bryozoan *Cinctipora elegans* (Cyclostomata, Cinctiporidae).«

85 Modified from T. Schwaha and M. Hirose, »Morphology of *Stephanella hina* (Bryozoa, Phylactolaemata): Common Phylactolaemate and Unexpected, Unique Characters,« *Zoological Letters* 6 (2020): 11.

86 Modified from T. Schwaha and A. Wanninger, »Unity in Diversity: A Survey of Muscular Systems of Ctenostome Gymnolaemata (Lophotrochozoa, Bryozoa),« *Frontiers in Zoology* 15 (2018): 24.

87 From D. Fink, T. Yau, A. Nabbi, B. Wagner, C. Wagner, S. M. Hu, et al., »Loss of Ing3 Expression Results.«

88 From D. Ebmer, S. Handschuh, T. Schwaha, A. Rubio-García, U. Gärtner, M. Glösmann, et al., »Novel 3D In Situ Visualization.«

89 Modified from S. Handschuh, N. Baeumler, T. Schwaha, and B. Ruthensteiner, »A Correlative Approach.«

90 Modified from D. Ebmer, S. Handschuh, T. Schwaha, A. Rubio-García, U. Gärtner, M. Glösmann, et al., »Novel 3D In Situ Visualization.«

91 Modified from S. Ferstl, T. Schwaha, B. Ruthensteiner, L. Hehn, S. Allner, M. Müller, et al., »Nanoscopic X-Ray Tomography For Correlative Microscopy of a Small Meiofaunal Sea-Cucumber,« *Scientific Reports* 10 (2020): 3960.

Traversing Invisible Walls: Facilitating Collaborative Experiences in Mixed Reality Environments

Jürgen Hagler, Jeremiah Diephuis

The *Virtual House of Medusa* is an interactive, co-located VR installation for a museum context that was developed in collaboration with the Federal Monuments Authority Austria and the Kunsthistorisches Museum Vienna to provide information about several fragments of Roman wall paintings. These archaeological artifacts were found at Lorch near Enns in Upper Austria and are one of the most important finds made of provincial Roman wall paintings in Austria. The mixed reality participants take the roles of archaeologists and experience the feeling that the past is being brought back to life piece by piece from a pile of shards.

Archaeologists have long faced the challenge of finding, analyzing, and piecing together the remnants of past civilizations to provide us with a door to our past. In most cases, only fragments of artifacts are found at any given site, and painstaking efforts are made to collect as much data about the surroundings as possible to better establish the context they were found in. The results of such work, unfortunately, are only partially visible, as there are often massive amounts of material collected, and much of it is simply too fragile and therefore not accessible to the general public. Moreover, making these artifacts accessible and comprehensible to a general audience, as in the case of a public exhibition, requires substantial additional effort, and these opportunities are limited in terms of the physical space available and the timeframe of an exhibition.

Although continuing efforts to digitize such content are definitely helpful, and the number of publicly accessible digital archives is increasing accordingly, digitalization alone is not enough to solve this problem. As with exhibitions, content needs to be processed and presented in such a structured way that it can be sufficiently understood. Archives of digital photographs, X-rays, and 3D scans may be of incredible value to experts, but they are simply not readable for a general audience. Virtual exhibitions, both as an asynchronous and a synchronous online activity, help to fill this gap, but their level of engagement tends to be limited to the user-based selection of which content should be shown. Mixed reality (MR) technologies have come a long way in the last decade, and they offer users more opportunities to interact with content by utilizing tracked head-mounted displays (HMDs) and controllers. What makes these approaches particularly interesting, however, is their immersive qualities, allowing users to interactively experience content much like they could in an actual museum or real-world archive. Additionally, another hidden door can be potentially opened by taking such an MR-based approach: a door that reveals the actual work processes that archaeologists and restorers have performed in their efforts to collect and assemble artifacts. This chapter addresses the conceptual and technical design of a mixed reality application entitled the *Virtual House of Medusa* (*VHM*), which was developed for use both as a virtual exhibition and an augmented exhibition activity about Roman frescos that were discovered near Enns, Austria. We will address the design challenges, the paths that were chosen, and the application's general reception in a series of exhibitions in museums and festivals.

Uncovering Walls

In 2014, we established the Playful Interactive Environments (PIE)[1] research group in the Department of Digital Media at the University of Applied Sciences Upper Austria, School of Informatics, Communication, and Media in Hagenberg. Originally, the goal was to combine the various research, artistic, and industry-related activities geared toward games and animation within the department. Since its founding, the research group has aimed to explore interactive spaces for different areas of application and investigate new and natural playful forms of interaction. PIE is primarily active in various interdisciplinary projects, cooperating with experts from multiple disciplines, such as theater, psychology, and computer games. In collaboration with research partners, we conduct perception research on computer-generated worlds and the design, implementation, and evaluation of interactive environments.

Since its inception, one of the guiding principles of PIE has been its focus on *co-located play*, a concept involving the sharing of playful experiences with others in the same physical and/or virtual space. The overall goal of this approach is to foster socialization, breaking down the walls between participants who may or may not know each other and increasing their engagement and creativity. In a typical installation developed by PIE, participants are brought together by the game mechanics implemented within the installation, sharing ideas and communicating with each other to do things like repair a virtual spaceship, build a tower, or defend planet Earth.

With the emergence of new immersive technologies in the last few years, collaborative experiences have become more viable in virtual environments. Mixed reality technologies utilizing mobile devices and head-mounted displays (HMDs), such as the HoloLens, Vive, and Meta Quest, provide users with fairly intuitive ways to interact with digital content in both virtual and hybrid scenarios, potentially with multiple people. However, virtual reality (VR) and augmented reality (AR) applications have largely remained single-user experiences. This is not too surprising as facilitating collaboration within VR environments can be quite challenging. The immersive qualities of VR can distract participants and inhibit their awareness and communication with others.

Within the framework of the research project Co-Located Virtual & Augmented Reality (CoVAR), funded by a grant from the University of Applied Sciences Upper Austria in 2016, PIE commenced research activities in the fields of AR and VR in a series of guided explorations. The focus was directed at two specific aspects: the aforementioned co-located play and AR/VR-supported collaborative processes. CoVAR was PIE's first extended endeavor to address co-located interaction and interface design within VR without any links to a specific use case. Thus, the grant

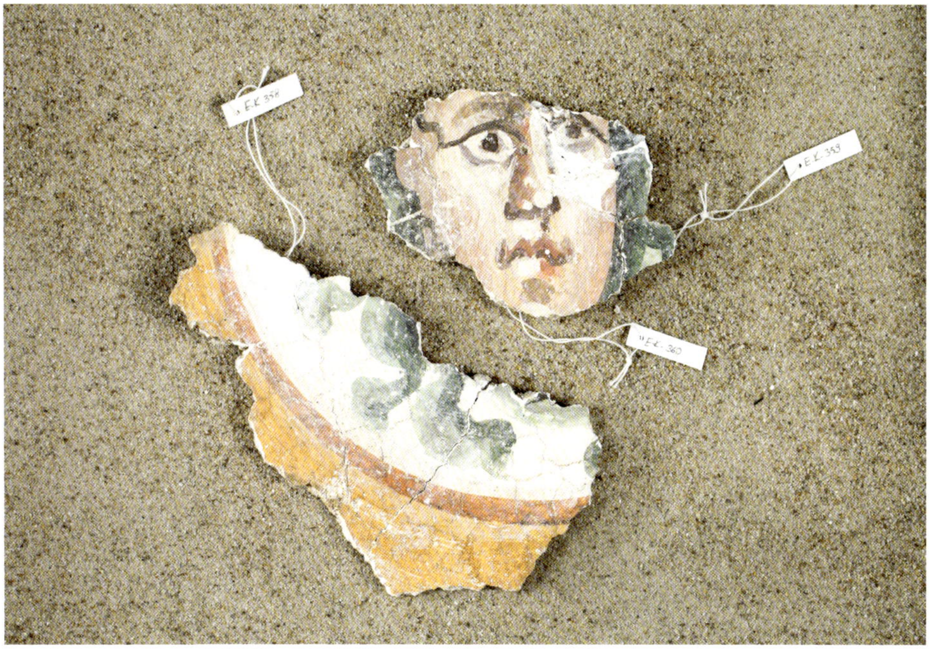

FIG. 1_*Two preserved and restored fragments from the Roman wall painting found in 2000 at the Danube Limes in Upper Austria. These fragments date from the third century CE and once decorated the walls of the House of Medusa.*

Photo by Irene Dworak, 2018

VIDEO_THE VIRTUAL HOUSE OF MEDUSA

allowed us a greater degree of flexibility and the potential to truly explore new subject areas and collaborate with new partners. And it was precisely at this time that the opportunity for *VHM* presented itself, and we began our initial research activities in the field of virtual heritage and archaeology.

Reconstructing Walls

Back in 2000, construction work had begun at Loch near Enns in Upper Austria when the remains of a previously undiscovered Roman villa were found, including several fresco fragments.[2] This area had historically been part of the Danube Limes, the border that separated the Roman Empire from other territories. The resulting excavation uncovered several hundred remnants and delivered the largest and most significant find of provincial Roman wall paintings ever discovered in Austria.[3] All in all, the archaeologists were able to assign the artifacts found to two specific rooms that had been part of a third-century Roman villa. Their work was made more difficult by the fact that the wall paintings consisted of up to four layers of plaster with decorative elements and magnificent figurative elements. Although these pieces were generally in exceptionally good condition, they were in disarray and needed to be reassembled with painstaking attention to detail.

Initially, the fragments were simply gathered and archived for several years. In 2011, the departments of Conservation-Restoration and Archaeology of the Federal Monuments Authority Austria together initiated a project to conserve, restore, and analyze them. Finding the appropriate fragments from the abundance of remnants proved to be a particular challenge. Essentially, these pieces were of various shapes and sizes, and some elements were either missing or not clearly visible, or it was simply too difficult to determine where they originally belonged. After several years of restoration work, the paintings were first displayed at the Kunsthistorisches Museum Vienna (KHM) from November 2017 to April 2018. Afterward, the paintings were shown as part of the exhibition *The Return of the Legion: Roman Heritage* for several months. Since then, the wall paintings have been part of the permanent exhibition at the Museum Lauriacum in Enns.

At the beginning of 2017, shortly before the completion of the work, we were contacted by Markus Santner, who was in charge of the restoration at the Federal Monuments Office. He invited us to visit the restoration studios in Vienna and discuss how to better display the restored material using a suitable media-technology-based approach. He gave us valuable insights, not only into the significance of the findings but also into the current work processes of archaeologists and restorers. He showed us various fragments and complex parts of the wall paintings, which were composed of many pieces. At this time, deliberations were already underway about how to exhibit the findings together with additional material such as visualizations of the Roman villa based on archaeological research. Furthermore, the Federal Monuments Office produced a video documentary about the archaeological excavation and restoration.

We discussed various approaches for augmenting the existing content, including interactive and playful forms that would go beyond traditional visualizations, 3D renderings, and video documentation. The idea was to develop an interactive VR application, a playful installation that would allow visitors to explore a 3D model of the House of Medusa. Fascinated by the studio tour at the Federal Monuments Office, we were inspired to not only show the house and its wall paintings but also to open the door to the hidden world of the restoration process.

Interfacing the Walls

The first prototype was designed to serve as a complementary installation and addition to the first KHM exhibition of a selection of the wall paintings in late 2017. The VR application aimed to make aspects of the archaeological artifacts visible that the originals simply could not easily show, although the exhibits on display generally provided a good overview. In addition, visitors were able to explore the *VHM* on their own. The installation made it possible for visitors to feel like they could open a door to the past and explore a structure that had long been claimed by the ages. To facilitate this experience, the *VHM* featured four virtual installations _FIG. 2:

1. At the first station, the participant received background information about the archaeological find. Here the participant learned that the walls consisted of multiple layers and that the villa had been completely repainted four times, indicating that it had most likely had four different owners.
2. At the second station, VR users were able to take the role of a conservator-restorer and assemble various fresco pieces together much like with a puzzle. In total, there were six fresco paintings to choose from. The main task was to examine the fresco fragments, which gave the players a sense of the complexity of the work performed by archaeologists and allowed them to identify the smallest details in the paintings.

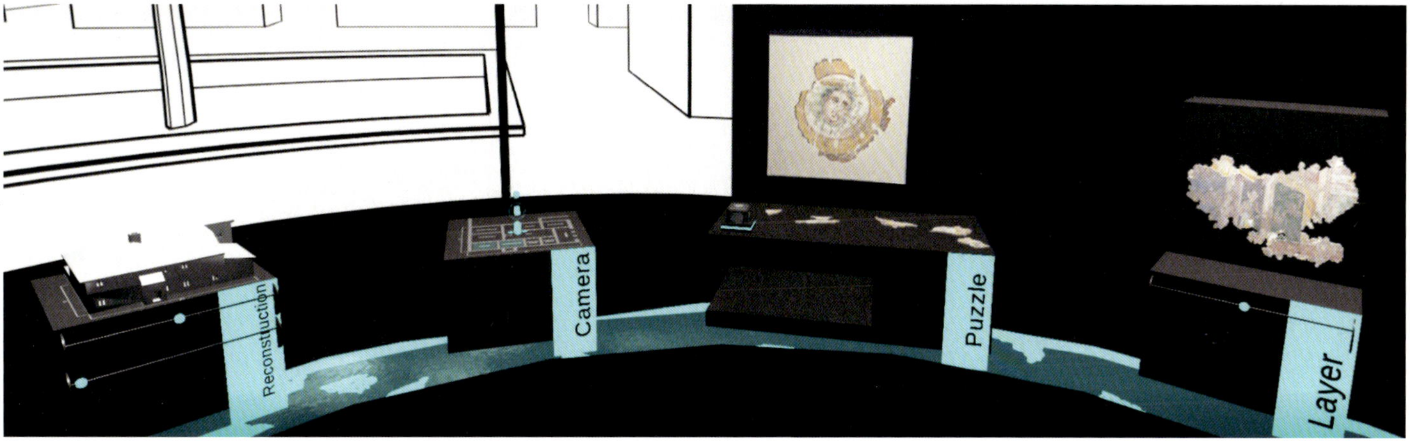

FIG. 2_*Four stations: participants could choose between four interactive stations, including an interactive display of the individual layers, a fragment puzzle, a virtual 3D camera, and an interactive model.*

3. The third station provided a virtual tour through the reconstructed building and gave an overview of all the wall paintings. VR users were also able to navigate through the house by moving a virtual camera over the floor plan. Rotating the VR controller allowed VR users to travel through time and explore the different decor phases.
4. The fourth station allowed VR users to explore a miniature model. Participants were able to rotate the virtual building and examine its structure from a bird's eye view.

The *VHM* installation consisted of a VR headset (HTC Vive) with one single-handed game controller (HTC Vive controller). Although the *VHM* can be utilized as a single-user installation, with one HMD and one VR controller, other participants could also get directly involved. In addition to simply viewing the content from the four workstations with an independent camera in the VR environment, visitors were able to directly collaborate with the VR user via touch input on a tracked tablet, which is referred to as an *MR guidance tool*. This device enabled up to four users to passively or actively participate along with the primary VR user. The co-player could control the animation for each participant's virtual avatar using a touch interface and create particle animation to draw attention to elements in the scene. Co-players could effectively point to and highlight virtual objects through the particles. This interaction was particularly useful at the second station, allowing the co-located players to assemble the fresco fragments as a collaborative puzzle activity.

Each guidance tool was effectively a modified tablet device, equipped with a HTC Vive tracker on the back of it. _FIG. 3 shows the *VHM*'s technical setup with the two separate devices (VR player and co-player), including the rendered screen view of each device. The players are visualized as very simple, abstract avatars. Co-players were able to control their own view of the scene and interact with the MR guidance tool. By swiping from the top edge of the tablet, co-players were also able to access a menu that facilitated multiple functions, such as resetting the installation; switching between different visualizations of the museum guide; switching between the four workstations; switching between different cameras (camera player, camera audience, etc.); controlling various avatar animations to improve social

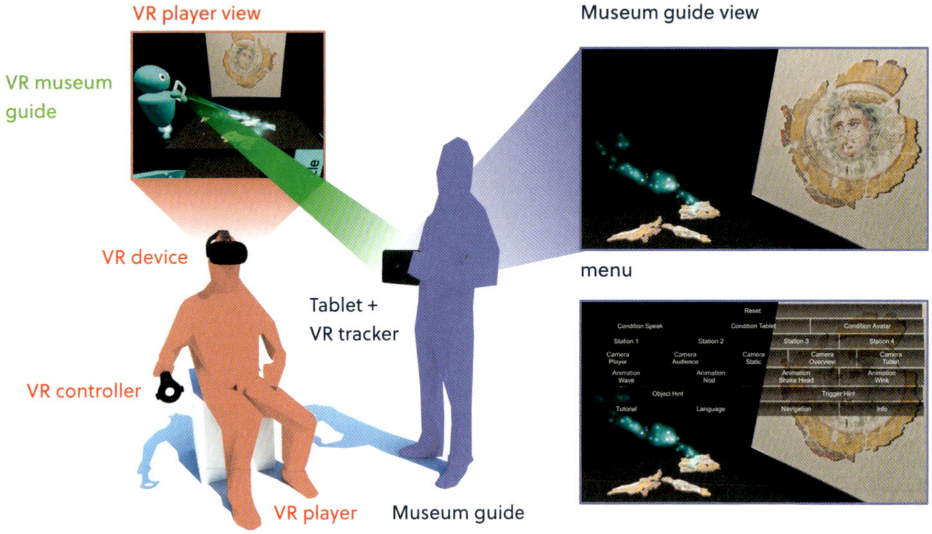

FIG. 3_*Technical setup: the VR player (red) uses an HMD (HTC Vive) with one controller, and the museum guide (blue) uses the MR guidance tool (a tablet equipped with a VR tracker). The guide is visualized in VR as an abstract avatar (green) and can interact with a VR player. The MR guidance tool offers multiple functions that are accessible via the menu.*

communication (wave, nod or shake head, wink); and showing hints or additional property functions (tutorial, language, etc.).

The *VHM* aimed to facilitate a shared virtual experience between the single VR user, additional museum visitors, and a museum guide as well. The museum guide introduced the VR installation to the HMD user and the spectators using the MR-based guidance tool. He or she could explain relevant events and mechanisms, in addition to helping participants navigate through the virtual experience.

Traversing the Fifth Wall

Interactive VR installations offer a great deal of potential for conveying additional perspectives on cultural and historical content. VR provides the possibility to more meaningfully interact with content to acquire a deeper understanding of it, for instance, with the ability to virtually touch, assemble, and/or disassemble and explore archaeological artifacts. In addition, VR users can switch between different roles — for instance, they can assume the role of an archaeologist or walk through historical landscapes. But VR museum installations using HMDs tend to neglect interactive possibilities for spectators, limiting their experience to simply following the virtual journey on an additional screen. In most cases, interactive VR installations must also be introduced and explained in detail by the museum staff. Additional opportunities for interaction and guidance are therefore of particular interest to museum guides looking to improve the introduction phase and the VR experience for the VR user and viewers.

In the fields of theater, film, comics, and even games, the concept of the »fourth wall« is frequently used to describe an imaginary boundary between the audience and the performance. Occasionally, this illusionary wall is »broken« when a performer directly speaks to or attempts to interact with the audience. In some of our previous research at PIE, we proposed the concept of the theoretical »fifth wall«

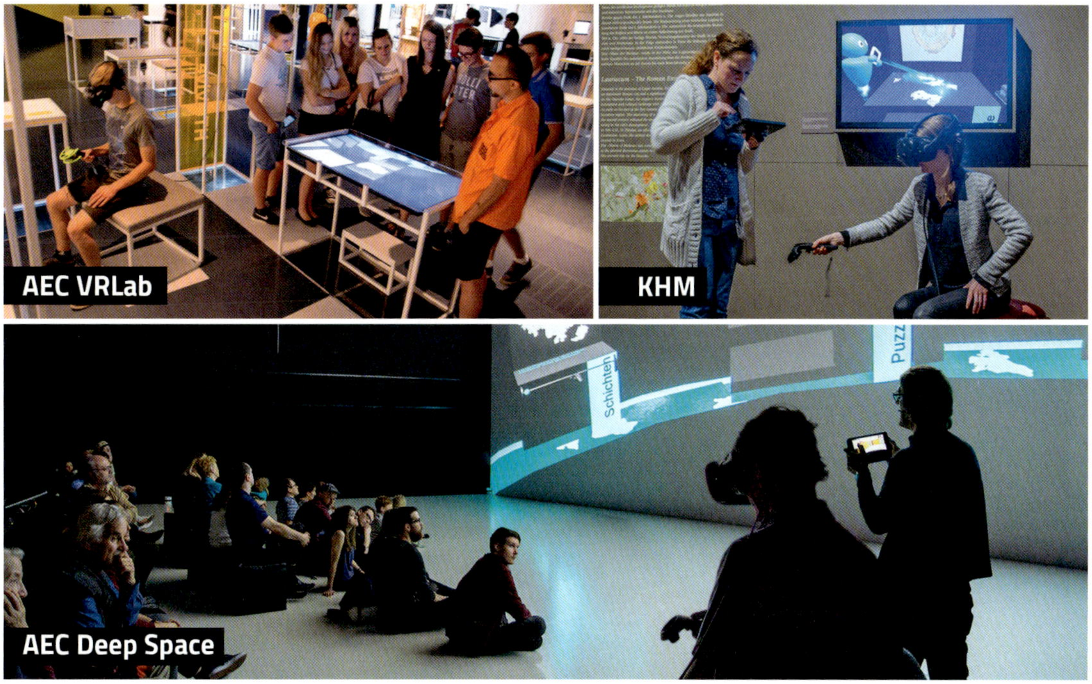

FIG. 4_ *Three different museum settings of the* VHM*: at the AEC's VRLab, a museum guide introduces the installation to the VR players using the MR guidance tool. Spectators observe the VR tour on the screen. A similar setting was exhibited at the KHM. At AEC's Deep Space 8K, a museum guide navigates the VR player through the* VHM*. A large audience can watch the journey through time on an extensive overview projection (16 × 9 meters).*

that typically separates individual participants in a collaborative experience, particularly with large displays or HMDs.[4] This fifth wall could be viewed as a sort of blinder that inhibits social presence and communication, effectively turning any collaborative experience into a parallel, single-user activity. VR experiences are often hindered by a sense of isolation for the benefit of immersion. Anyone who has tried to introduce someone to VR content for the first time understands the challenges of dealing with this communication barrier resulting from the lack of direct visual communication and the ambiguity of a dislocated voice. The *VHM* was developed to traverse this fifth wall, providing a flexible set of interaction paradigms to facilitate a collaborative yet immersive experience in a variety of settings.

The *VHM* was not only a research exploration but also an active part of eight different exhibitions in 2018. Its first iteration was shown at the KHM in January 2018 for just a few days as an additional installation in the first exhibition of the original artifacts of the House of Medusa. We then developed a more elaborate version for an exhibition for the GameZone of the Stuttgart International Festival of Animated Film 2018. The final version was then exhibited at the 2018 Oberösterreichische Landesausstellung for an extended period. We also presented the prototype at a number of other events, such as the Ars Electronica Center (AEC) and the 2018 Congress Visual Heritage Vienna, with the goal of evaluating how the VR player, the co-players, and the audience are integrated into the VR experience_FIG. 4. The research findings on the design potential and implications of the *VHM* have been published and presented at various conferences.[5]

After evaluating the *VHM* with a dedicated guide in multiple locations, the installation was then additionally exhibited unguided over six months at the Ars Electronica Center. The unguided version became part of the *VR Lab*,[6] a special

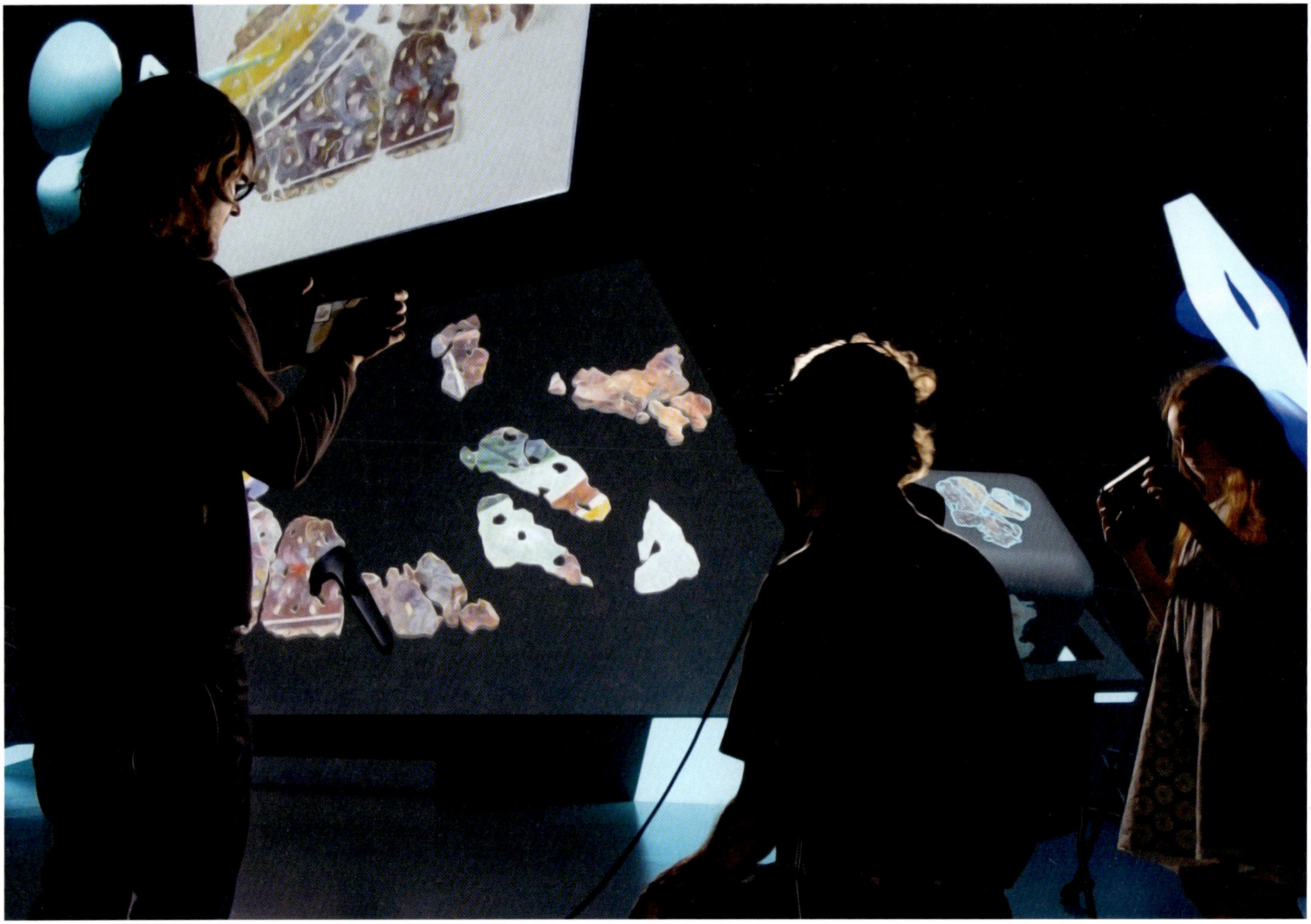

exhibition focusing on VR and AR. In addition, we developed a special version for the Ars Electronica's Deep Space 8K.[7] The big difference to previous settings was the challenge of integrating a much larger audience. The Deep Space 8K at the AEC is one of the few environments that can facilitate a VR experience for a large audience, featuring two 8K stereo 3D projections of 16 × 9 meters for the wall and floor. A Deep Space 8K guide introduces the *VHM*, and up to 150 spectators can take part in the experience.

Overall, the approach demonstrated in the *VHM* has proven its merit, facilitating a collaborative experience for small and large groups alike. Although the full technical setup with multiple tracked tablets requires some effort to maintain, the simplified setup with one MR guidance tool provides a flexible way for experts to more effectively integrate and even captivate audiences in co-located settings. Although the walls of the House of Medusa may no longer stand, approaches like those found in the *VHM* not only make them visible again but also serve to break down the unintentional walls erected between participants due to the nature of immersive technologies.

FIG. 5_*Traversing the fifth wall in the AEC's Deep Space 8K: a museum guide navigates the VR player and another co-player through the* VHM *while a large audience watches the journey on an overview projection.*

1 PIE, *Playful Interactive Environments*, 2022, accessed November 11, 2022, https://pie-lab.at/.

2 Markus Santner (ed.) *Das Haus der Medusa: Römische Wandmalereien in Enns* (Horn: Verlag Berger, 2017).

3 »Focus Monuments: The House of Medusa,« *Kunsthistorisches Museum Wien*, 2017, accessed November 11, 2022, https://www.khm.at/en/visit/exhibitions/2017/the-house-of-medusa/.

4 Jeremiah Diephuis, Michael Lankes, and Wolfgang Hochleitner. »Another brick in the (fifth) wall: reflections on creating a co-located multiplayer game for a large display.« Context Matters (2013): 162–167.

5 Juergen Hagler, Michael Lankes, and Andrea Aschauer, »The Virtual House of Medusa: Guiding Museum Visitors Through a Co-Located Mixed Reality Installation,« in *Joint International Conference on Serious Games* (Cham: Springer, 2018), 182–8; Juergen Hagler, Michael Lankes, and Jeremiah Diephuis, »Animating Participants in Co-Located Playful Mixed-Reality Installations,« in *2018 IEEE 1st Workshop on Animation in Virtual and Augmented Environments* (ANIVAE) (Piscataway, NJ: IEEE, 2018), 1–4; Michael Lankes, Andrea Aschauer, and Juergen Hagler, »The Virtual House of Medusa: Playful Co-located Virtual Archaeology,« in *Tagungsband FFH 2018* (Salzburg: 2018).

6 »VRLAB,« *Ars Electronica Center*, 2017, accessed November 11, 2022, https://ars.electronica.art/center/en/exhibitions/vrlab/; »The House of Medusa,« *Ars Electronica Center*, accessed November 11, 2022, https://ars.electronica.art/center/en/the-house-of-medusa/.

7 »Deep Space 8K,« *Ars Electronica Center*, 2015, accessed November 11, 2022, https://ars.electronica.art/center/de/deep-space-8k/.

Realisms and Realities: Constructions of Reality in Digital Art

Ruth Schnell
with Patricia Köstring

Constructions of reality constitute the world. These realisms depend on the development of technological image formulations, on modes of perception, and on time.

As a media artist, it is against this backdrop that I investigate, among other things, perception itself and visual representation. »Among other things,« because my work is shaped by my investigations into the political and social present, which are each interwoven with the anchoring and staging of technical media in space. Studying modes of perception is thus not to be understood as a technical supporting leg that gives stability to a free, artistic content leg.

My work deals with bodily and spatial experiences, in which the moving digital image and the viewer in motion are constitutive factors. It is about representations of space, time, and the body conveyed through media, as well as the dynamization of the intersections generated by collisions between real space and virtual space, or even a dynamic semiotic space. It is about staging modes of seeing, about juxtaposing and contrasting different ideas of space and, simultaneously, about changes that have taken place in the concept of the image as a consequence and product of the technologies that I myself use — sometimes contrary to their intended purpose.

This *mis-using* as part of an artistic research process opens up eye slits, as it were, on various realities. My media settings involve viewers in processes in which they experience the constructivity of orders of the gaze and patterns of perception as participants. The machine-manipulated introspection into how we see things creates the prerequisites for a different way of seeing and understanding the complexity of reality (and realities).

Artistic investigations into the codes of visual representation and modes of perception in the field of the electronic moving image are ultimately the result of changes that have taken place in the concept of the image resulting from and produced by technological developments.

New media have existed for about two hundred years if we date their origin to the advent of photography as the art of machine-assisted image production.

Telegraphy — splitting an image into a temporal sequence of points in order to (electrically) transmit an image over great distances — allowed the image to be redefined as a temporal form. Technological developments like film, electronic image recording, image transmission, and image manipulation did not just enable movement and, as a result, realities to be simulated or to move the viewer from outside the image to inside the image, they also compelled new models of time and space to develop.

Virtual and mixed reality technologies, on the other hand, have suspended the already blurred lines between reality and simulation by means of computable and completely manipulable digital images, thereby reinforcing an understanding of reality as a construction. The digital image is, so to speak, a possibility; it emerges in an open process of reciprocal dynamization. As one component of the immersive image space, the viewer loses the distance practiced in traditional modes of image viewing. The *trompe-l'œil* dreams of the image space as a portal that have been part of art since antiquity, as art historian Oliver Grau describes,[1] are now gaining previously unforeseen efficacy, although this power must be critically questioned as a potential strategy of overwhelming.

What is reality? What is actuality? How real is reality? What is even real? Are ideas and/or sensory experiences real? Back in 1995, technology expert and MIT professor Nicholas Negroponte anticipated bubbles and echo chambers, saying that the »daily me« would allow every media user to have their own individual »architecture of control,« self-curated perception.[2]

By talking about reality, we cultivate everyday myths that give us stability and relieve us of the strain of having to perceive a number of differently constructed realities at the same time. Chilean biologist and systems theoretician Humberto R. Maturana formulated one possibility of coping with this strain as follows: »We see with our legs.«[3] Maturana, who coined the term *autopoesis* together with Francisco Varela, puts the observer at the center of his considerations. Perception and interpretation — which constitute the world — take place in motion. In this sense, in the following I would like to look at selected works and groups of works in order to describe aspects of my artistic work with models of interaction and various immersive image modes, which, in various ways, ultimately require the viewer to move.

First-Person View: Subjectively Experiencing Layers of Reality

The work that I have been creating since the beginning of the 1990s using what I refer to as »light sticks« is one example of my investigations into visual perception, which take place in a crossover between the real architectural space and the virtual semiotic space. Through bodily movement, the virtual space of signs becomes an experience of space, a dynamic, boundless, virtual semiotic space, in a process of interplay between different parameters of time and movement. The interface is the eye of the viewer in motion. In the interplay between movement, perception, language, and thinking, a spatio-temporal mode of seeing constitutes itself. This also requires overlapping orders of the gaze.

On a flickering stick on which LEDs have been vertically arranged, translucent words or icons seem to push themselves into the viewer's field of vision. They remain in the space briefly, superimposing themselves over the architecture being

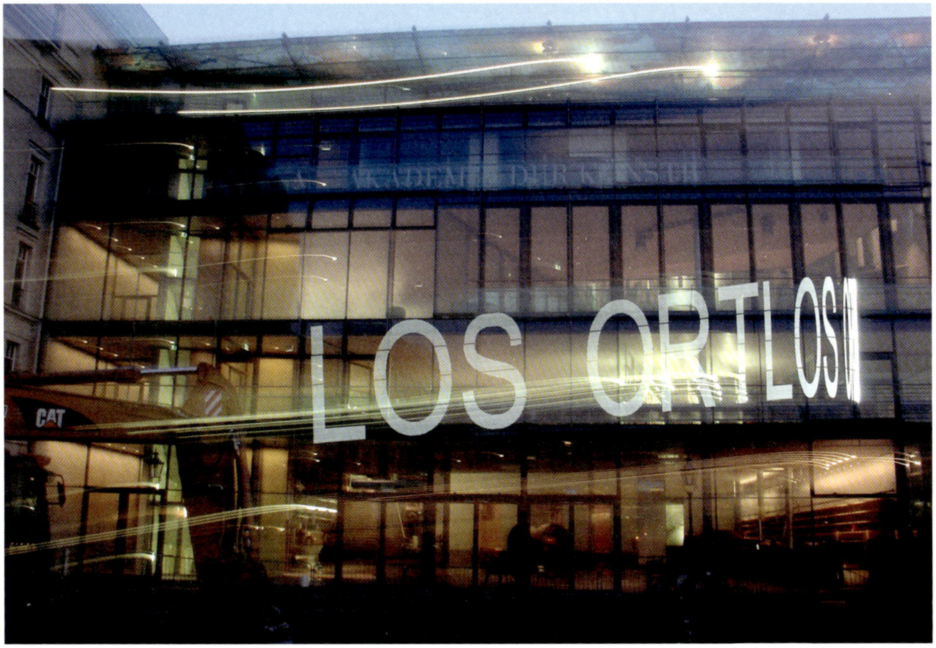

FIG. 1_ *Site-specific light installation* MISSION OF ART *(2005), new building, Akademie der Künste, Pariser Platz / Brandenburg Gate, Berlin*

Photo by Alexander Pausch, 2005

perceived. Here I make technical use of the persistence of human perception. Outside the frequency at which the human eye can still differentiate between individual images, the pulsating LEDs generate words, icons, or monochrome images that have been divided into dots and lines. They are rendered sequentially on the stick's vertical columns. As afterimages that are produced in the retina and interpreted in the brain, the signs generated through the light sticks appear to the viewer in motion as holographs in the space. The eye »scans« in horizontal movement and thereby becomes the image carrier. The viewer sees a number of situations in the present that are actually already in the past. The performative act of moving engenders this »present« and thereby the work, too, as a collaborative process. Ultimately, the work only realizes itself for the individual viewer; there is no collective mode of viewing.

Two fundamentally different orders of the gaze collide: while studying an art work requires deliberate seeing, the opposite mode of viewing is required in order to perceive the signs flickering across the light stick. The letters and images that have been transformed into a temporal sequence of high-frequency impulses only become legible when the viewer gives up the patterns of reception learned from reading. What is required is a freely floating gaze, a mode of viewing past the object, so to speak. Decoding the typefaces and icons requires unintentionality.

I have created light stick installations for different spatial settings in different dimensions. The temporary intervention *mission of art* (2005), for example, played on the facade of the Akademie der Künste in Berlin to mark the opening of the new academy building at Pariser Platz (at the Brandenburg Gate). Members of the academy each provided a different term. Like an all-over, the words seemed to extend from the glass facade of the academy to the surroundings at Pariser Platz by way of the interface of a four-meter-high light stick with super bright, white LEDs attached to it.

I have frequently made dialogical use of smaller light sticks, embedded within metal objects, while conceptualizing work series like *All Targets Defined* (2006) as diptychs. This is where the dilemma inherent to the modes of focused viewing clearly come to light. The thematic parentheses of the *All Targets* works are an examination of technologies and the reception of modern warfare. The diptychs are kept in a translucent green that is reminiscent of looking through night-vision goggles. While the panels on the left show roughly scanned motifs from images of war conveyed by the media from various camera standpoints, the second panels contain an LED stick with terms, figures, data, and facts from the world of military strategy, war technology, and the global politics shaped by energy interests. There is no mode of viewing that allows the viewer to take in the entire work: »Paradoxically, we can get the words only in passing, i.e., by glancing away. Isn't it representative of our times that learning the truth requires an effort not to make an effort to see what we are told to look at?«[4]

A work series begun in 2018 takes up these questions in relation to the layering of reality and the viewer striding through the various layers: *COMBATscience Augmented* and *COMBATscience Augmented II* (2018 and 2022) are mixed reality installations that use a pair of HoloLens data glasses as their interface. The biographies of the scientist and chief designer of Germany's poison gas war, Fritz Haber (1868–1934), and his wife, chemist and pacifist Clara Immerwahr (1870–1915), form the point of departure for these installations, which are critical reflections on science

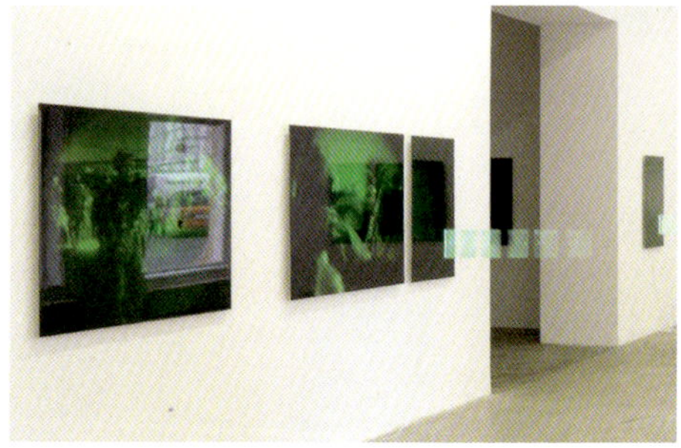

FIG. 2_ALL TARGETS DEFINED *(2006), light objects, exhibition view, Galerie Grita Insam, Vienna, 2006*

Photo by Peter Kainz, 2006

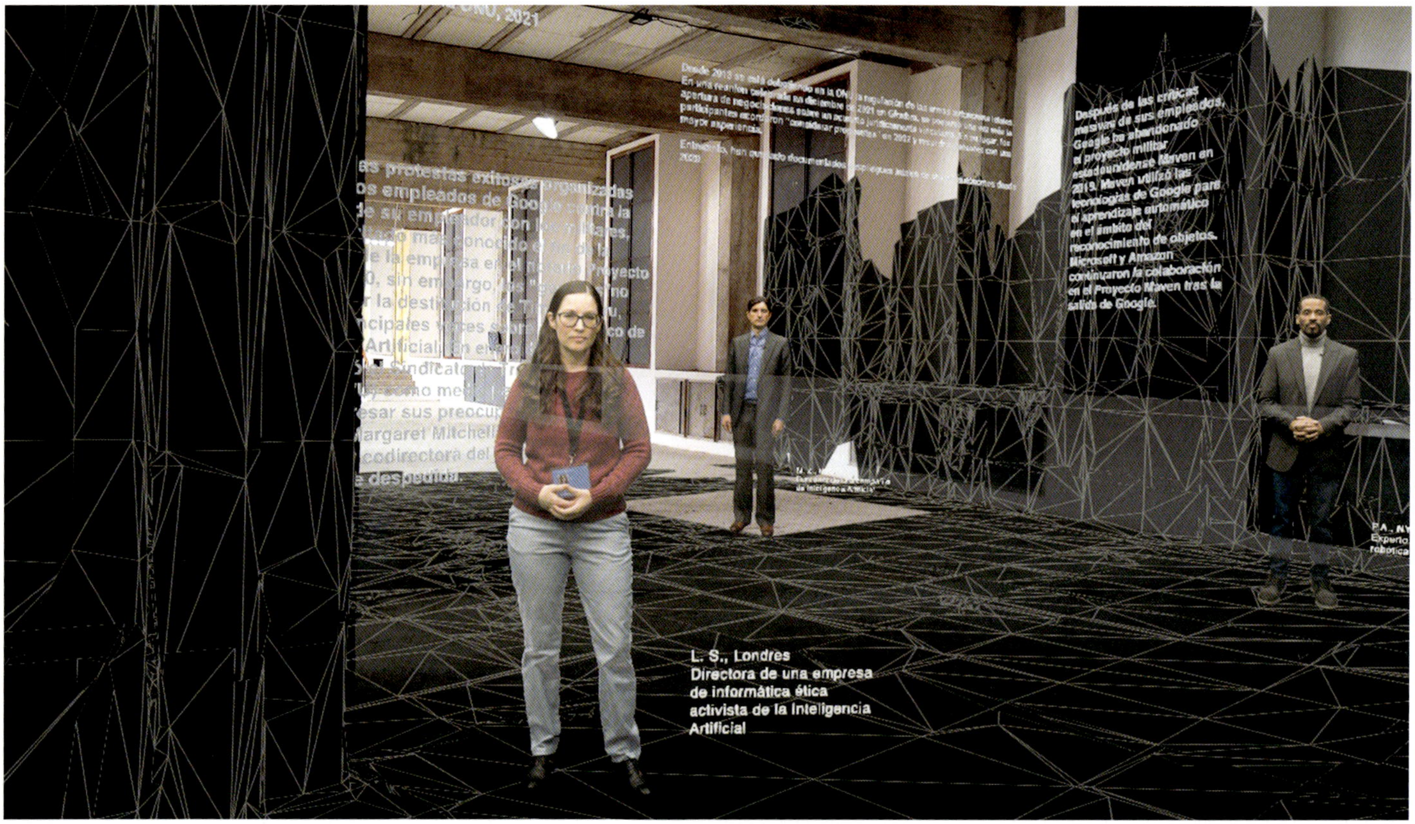

and its ethical implications, starting with the gas attacks in World War I and continuing on to the research being conducted into autonomous weapons systems in the present. Viewers move freely through the exhibition space using a HoloLens. Within the virtual architecture, which is designed specifically for the respective exhibition space, digital holographic scenarios unfurl for the users within the real surroundings. The main elements are a series of seven monologues and one dialogue performed by actors. Viewers activate the respective video sequences with a hand gesture, and the protagonists—like life-size moving holograms—seem to turn toward the viewer while reciting their text. 3D text elements and sound spots anchored in the digital scan of the real space complement the virtual environment.

The Illusion of Space and Excentric Viewer Standpoints

At the beginning of my artistic career, aspects of the juxtaposition of realities and performative viewing were already important, which, in 1989, for example, I staged in a complex interplay between different conceptions of reality. The three doors of the media environment *Tür für Huxley* (Door for Huxley), which was shown at sites like the 1989 Ars Electronica, simulates entry points to various realities. This installation hybridizes different representations of space. Certain media-specific orders of the gaze slide into each other, interact—and involve the viewer.

The viewer has the choice between passively viewing a conserved image reality, entering into a simulated space through a projected door controlled by sensors, or taking the standpoint of a voyeur. The door, ajar, projected on the left reveals an action space with mounted and artificially distorted scenes of stairs and doors from suspense films by Hitchcock and Siodmak. When the viewer steps into the sensor area, the door projected in the middle opens and shows the viewer themselves from the back, an imaginary »Other« from the standpoint of the viewer. If they try to approach their own image, it disappears in the simulated space. On the floor in front of the third, real door is a reproduction of a surrealist painting by Dorothea Tanning, the subject of which is also steps and doors. The image of the image is transmitted through a black-and-white camera onto a monitor behind the keyhole. »Original« and copy, obvious presence and secret, face each other.

FIG. 3A_COMBATscience Augmented II *(2022), mixed reality installation, screenshot of a first-person exhibition view, exhibition view 14, Bienal de la Habana, Havana, Cuba, 2022*

Photo by Ruth Schnell, 2022

VIDEO

FIGS. 3B, 3C_COMBATscience Augmented *(2018), mixed reality installation, screenshots of first-person exhibition view, exhibition view, Criatech Festival, Aveiro, Portugal, 2021*

Photos by Ruth Schnell, 2021

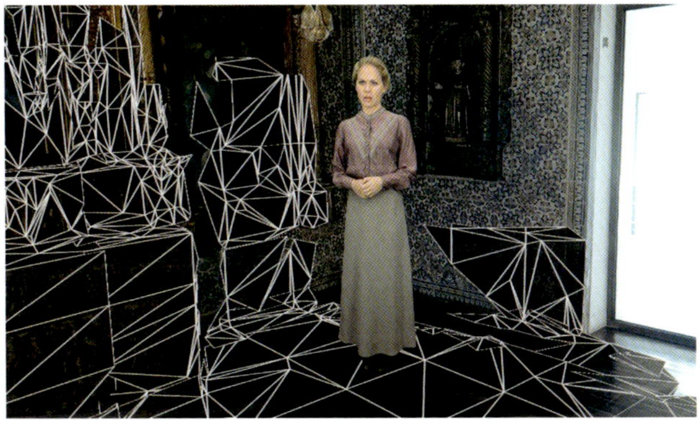

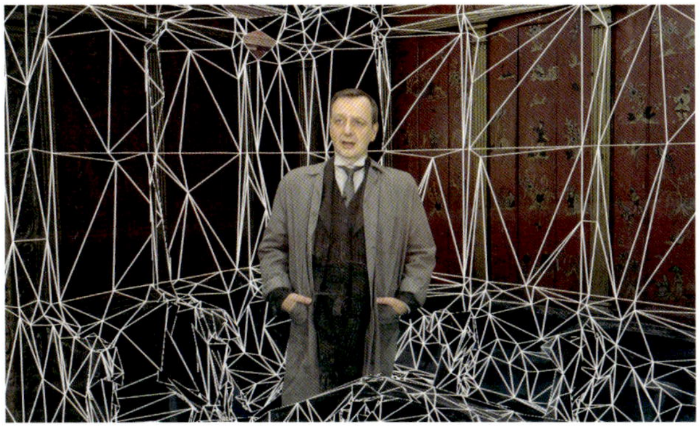

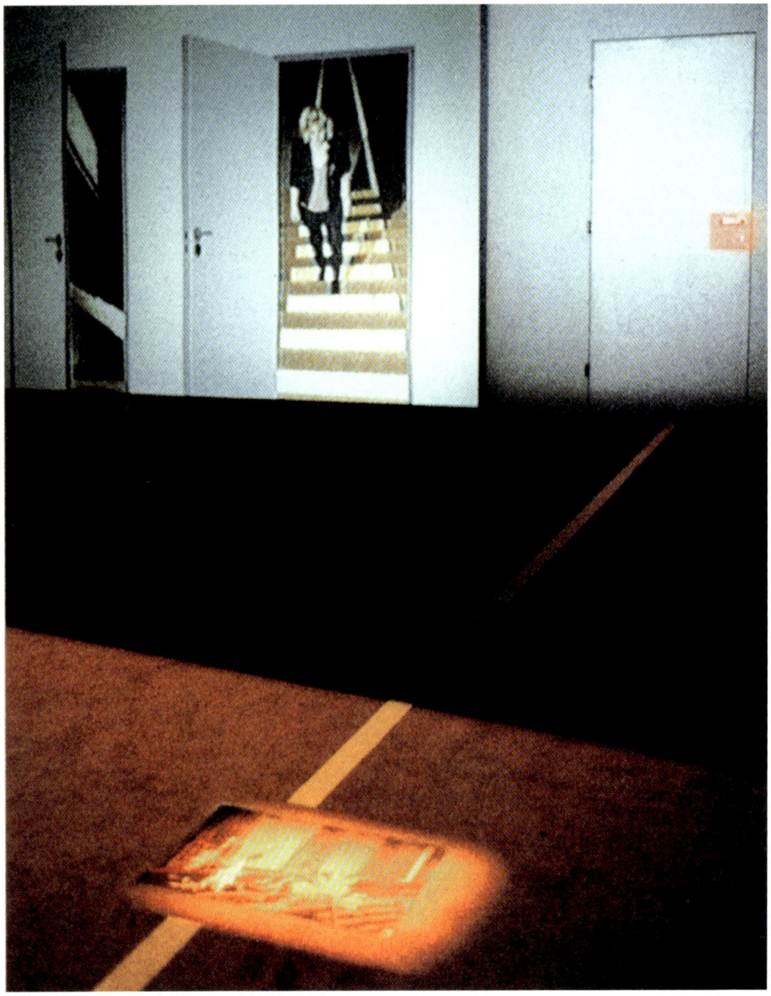

FIG. 4_ TÜR FÜR HUXLEY *(Door for Huxley), interactive computer-video installation, exhibitions views, Ars Electronica 1989*

Photos by Ruth Schnell, 1989

VIDEO

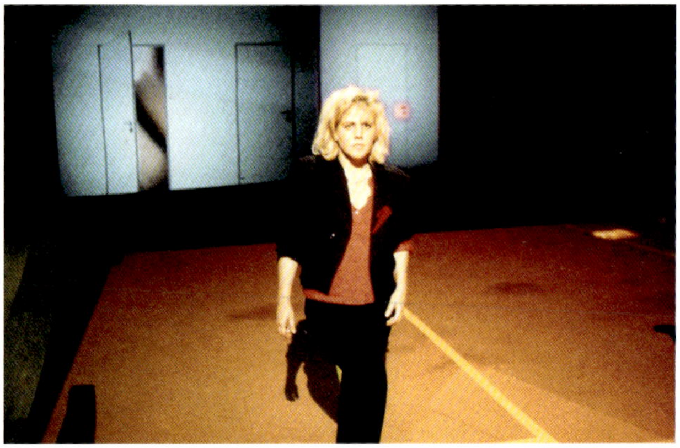

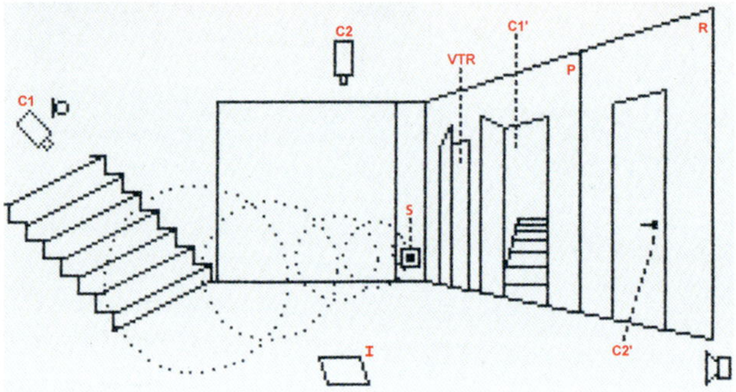

The viewer sees themselves from behind, locating themselves in the part of the mirror space that they would normally not see but simultaneously finding themselves in another space. This spatial impression is intensified when the viewer moves away from themselves as they try to approach the inverse mirror image in the other space. The simultaneity of these different, somewhat entangled spaces evokes uncertainty. To achieve this, I partially repurposed various media *dispositifs*: this involved, for instance, *trompe-l'œil* techniques and illusionistic architecture; however, instead of the texture and color of painting, a moving video picture is used and, instead of the static depiction of architecture, a moving projection corresponding to the real architecture of the space.

The recipient's gaze doubles as a camera eye, as the visitor is both actor and agent at the same time, and becomes part of the game being played with representations of space in the suspense film. The surveillance monitor suggests that someone has control over the space and the events taking place »there«; the viewer, in the position of the keyhole voyeur, is surprised by the »here.«

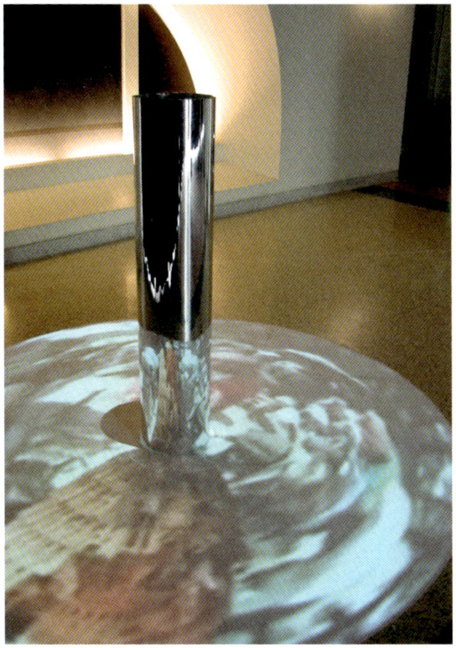

FIG. 5_ BABEL *(1993/1996), anamorphic installation, exhibition view, California Science Center, Los Angeles, US, 1994*

Photo by Ruth Schnell, 1994

VIDEO

The viewer's perspective is also crucial in a series of works that I made in the 1990s. The installation *Babel* (1993/1996) projects anamorphically distorted video sequences onto the floor. It uses archival material from television reports about refugees in Rwanda and a section from Brueghel's painting *The Tower of Babel*. The viewers stand and walk around in the distorted image area and see the straightened-out image in the mirror cylinder overlapping with their own distorted mirror image. The bent infinity mirror forms its own space. This creates the impression that the images are circling around the cylinder. A continuous stream of people passes the viewer, leaving the tower on one side and entering it again on the other. The viewer is able to circumvent the mirror but cannot walk »behind« it. This installation stages a gaze that is unable to escape the mirror.

The principle of anamorphic distortion, which requires the viewer to take a certain standpoint to decode the content of an image, can also been found in the (unrealized) site-specific light installation *house* (2014), a twenty-meter-high, monochrome spatial drawing made from customized, bent red neon pipes. *house* was conceptualized for the multistoried atrium of an office building in Vaduz, which the work, fitted centrally into the air space, would have traversed from the bottom of the first floor to the fourth floor.

The work is at first reminiscent of a climbing graph (like a stock-exchange chart) extending into the spatial dimension. Viewed from a certain position at the foot of the foyer steps, the spatial convolution foreshortens into the image of a graphic figure: what the viewer sees is the front of a house, the »Haus vom Nikolaus,« a children's puzzle and mathematical problem.

In the lobby of a Viennese luxury hotel, *Dynamic Sphere* (2012), a ceiling piece made from gold leaf, plays with the idea and appearance of illusionistic painted ceilings. A gold leaf design was applied to one hundred square meters of the 250-square-meter ceiling. From a certain perspective, the intense distortion in the design, caused by the low height of the space—the degree of distortion was calculated using a simulated model—simulates a dome with a frieze of climbing plants.

The Dynamic Interweaving of Constructed Spaces

In my large-format, dynamic video installations in particular, the proprioceptive body—understood here as the represented body in its relationship to the viewer in motion, who takes a stance toward this representation—plays a big role. The way that the content of the image is scaled and distorted by architectural components is also constitutive of these installations.

Gegen die Zeit (Against Time) is a dynamic video installation originally conceptualized for the secular Johanneskirche in Feldkirch and later shown during the Linz 2009 project »Tiefenrausch.« The projection of a woman's arm holding either a brush or a cloth moves over the surfaces of the slightly darkened architecture. Viewers hear a scrubbing noise that is synchronous with the movements. The confrontation between architecture and projection distorts the image, thereby remodeling the real space. The projection itself is conceived not as a frame but as the movement of cut-out images. This creates the impression that the movement of the image is being caused by the movements of the projected section of the body. The projection into the space takes place by means of a computer-controlled mirror.

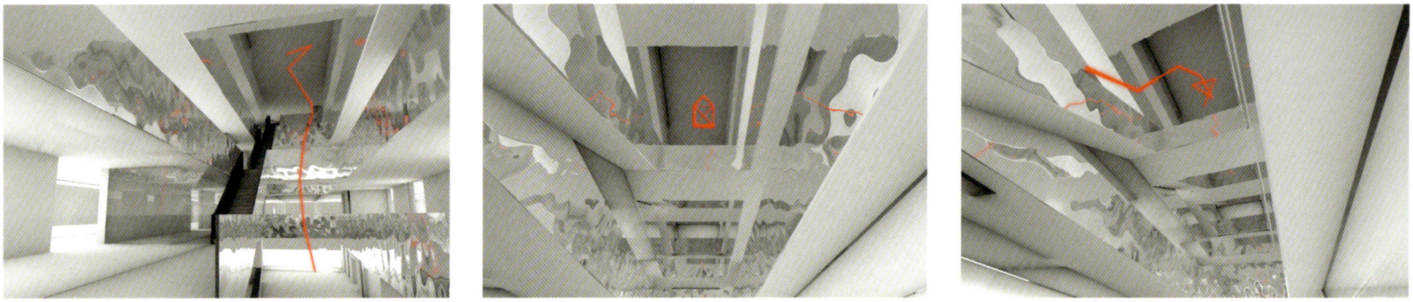

FIG. 6_ *Site-specific light installation* HOUSE *(2014), Vaduz (not realized), renderings*

Photos by Ruth Schnell, 2014

FIG. 7_ DYNAMIC SPHERE *(2012), site-specific ceiling drawing, lounge at the Ritz-Carlton, Vienna*

Photo by Mischa Erben, 2012

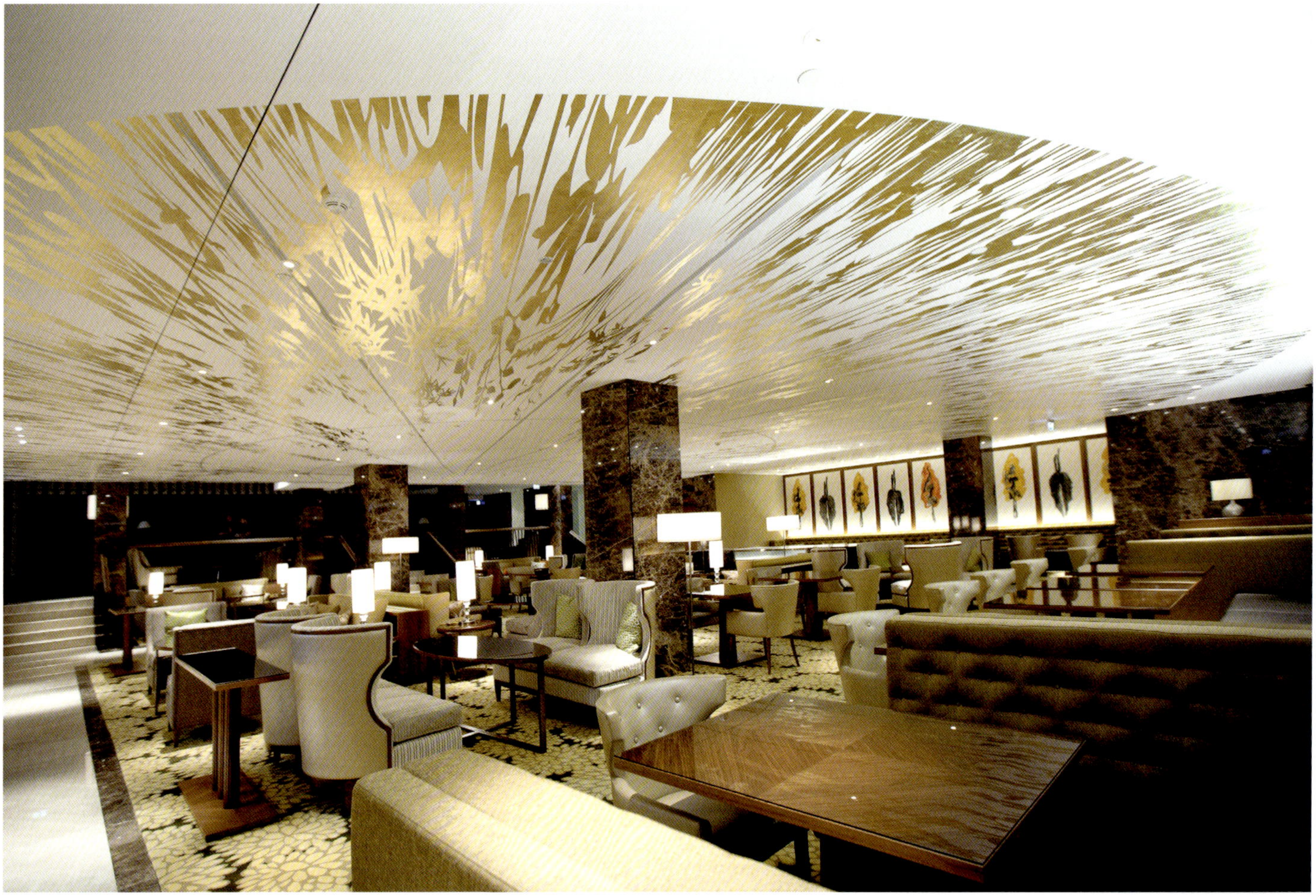

Because the length of the video composition and the intervals of the mirror's movements do not match, the projected sequence and movement in the space come together again and again in ever-new constellations.

The dynamic image projection *Topography of Movement* (2016) displays two oversized moving hands. They seem to explore their surroundings and move or swipe over a surface that they are lying on. The two cropped projections cover the walls and parts of the ground and ceiling of the exhibition space. A projection mirror further dynamizes them, depending on how the movement patterns of hand and mirror collide. The projection of a left and a right hand point to an (absent) body. Even though it is impossible to take a subjective viewer's perspective due to the intense magnification of the image content, the action is spatialized by the plausibility of the movements and by the simultaneously audible noises made during the recordings. It becomes possible to experience the space sensuously.

Territorism (2002), a multipart video projection developed for the Kunsthaus Bregenz (KUB), was played on the entrance and side facades of the KUB and on part of the facade of the neighboring state theater in the evenings of summer 2002. The moving and distorted image components created a simulated spatial structure that intertwined the solitary building structure of the KUB with its surroundings. However, it was only possible for viewers to experience the performance as a whole when they moved around.

On the glass surface of the art museum, a hand was projected that ran a model tank over the museum's facades. A voice coming from loudspeakers attached to the building simulated shots and explosions as if in a war game. A second projected hand seemed to lean on the buildings. The movement sequences of both hands alluded to someone playing on the ground; the view of the facade tipped into a view from above.

A Gap in Acceptance / Scaling in 2D

The oscillation between the desire to recognize and classify events in relation to one's own body on the one hand and the gap in acceptance (the rejection of the similar) through scaling and distortion on the other are also at work in a series of two-dimensional works. Here, images from medicine and medical research reveal themselves to be hypotheses about the visible, and it is only the ability of the viewer to face what they are seeing with their own knowledge, their own sensations, that makes the images legible.

The piece *Mirrors of the Unseen* (2011), which I made for the Hohenems State Hospital, presents an investigation into technologically generated images, including questions of translatability and the interpretation of information. The images that are spread across five stories on the ceilings of the corridors and patients' rooms make reference to visualization techniques used in the field of medicine. On display are »structures of life«—microscopic images of cells, blood platelets, the body's own bacteria, abstracted through scaling, image editing, and rearrangement. The ornamental formations that meander over the ceilings of the rooms, suggesting a drawing, seem to transcend the borders between the patients' rooms and corridors.

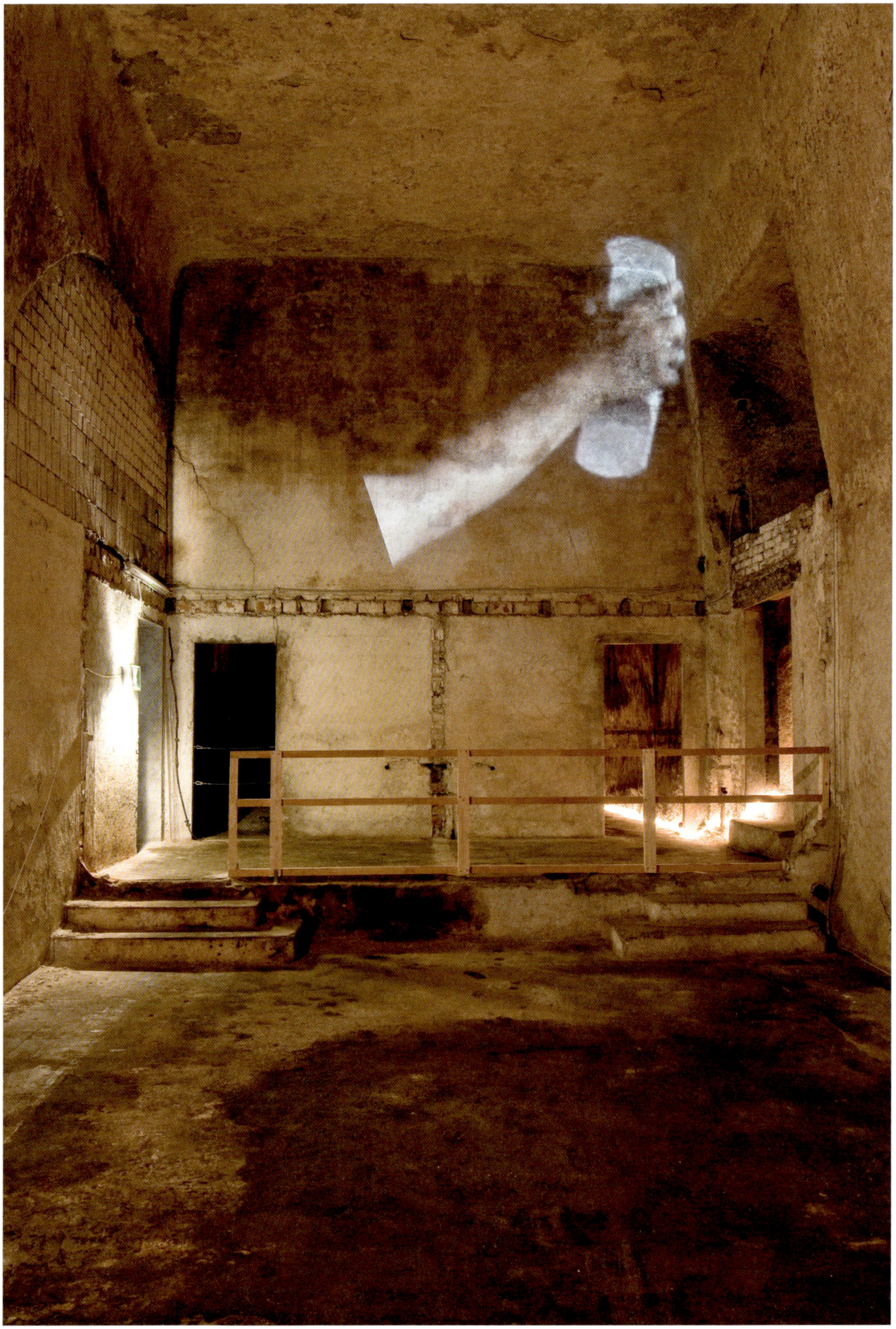

FIG. 8_GEGEN DIE ZEIT *(Against Time), dynamic video installation, exhibition view, Linz, 2008*

Photo by Otto Saxinger, 2008

VIDEO

When the viewer's gaze follows the lines and networks, the ornaments become form, become a view of a »cosmos of the invisible.« Patients can thus view themselves as observers or as part of the image. This piece also very deliberately makes reference to the visualization techniques of Renaissance perspective painting, to the views and room openings presented as *trompe-l'œil*. Whereas the fake architectural views in *trompe-l'œil* painting usually open up to the sky or Arcadian landscapes, viewers here take a look inside, so to speak, into a microscopic albeit abstract world.

The aluminum tableaux entitled *Mirrors of the Unseen* (2010/11) from the same period also present a glimpse inside the body, as it were. The technical images on which the motifs attached to the plates by UV printing are based are microscopic images used in the field of medicine, which, in turn, are the products of various imaging techniques. Whereas the work at the hospital has an impact on the viewer's

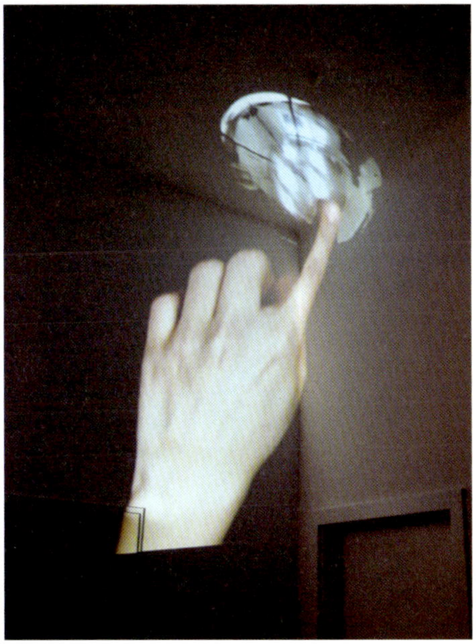

FIG. 9A_ Topography of Movement *(2016), dynamic projection, detail*

Photo by Ruth Schnell, 2016

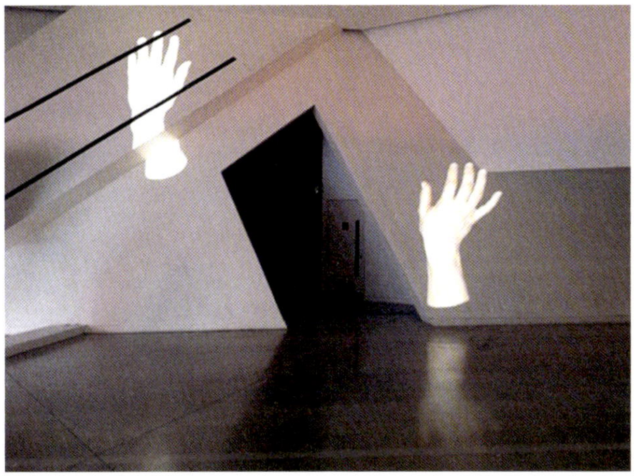

FIG. 9B_ Topography of Movement *(2016), dynamic projection, installation view, ISEA, Run Run Shaw Center for Creative Media, Hong Kong*

Photo by Ruth Schnell, 2016

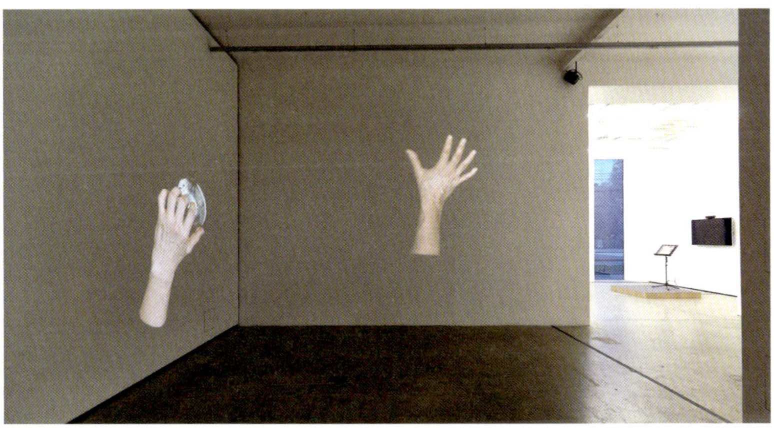

FIG. 9C_ Topography of Movement *(2016), dynamic projection, installation view,* Digital Synesthesia, *AIL, Vienna*

Photo by Peter Kainz, 2016

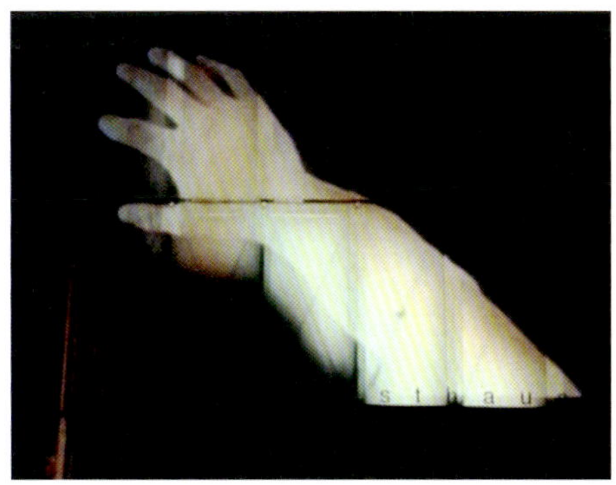

FIG. 10_ Territorism *(2002), dynamic video projection, Kunsthaus Bregenz*

Photo by Ruth Schnell, 2002

VIDEO

relationship with the illusionistic space it constructs, this series opens up new aesthetic fields of interpretation in relation to the panel painting. The dimensions of the tableaux, at 2 × 1 meters, are about the same as those of a mirror, encompassing the figure of the viewer and pointing back into the reality of the visible.

Immersion / Permeability

I would like to conclude this essay with a field of artistic research that has occupied me and my colleagues at the University of Applied Arts Vienna (above all Martin Kusch as the Head of the Fulldome / VR & AR Lab) for years. With colleagues from other art institutions, but also in the research environment of the Fulldome / VR & AR Lab, we are working on developing new artistic grammars for immersive fulldome, VR, and AR environments. Martin Kusch describes fulldome environments as follows:

> *The word »fulldome« refers to immersive dome-based video projection environments. The dome, horizontal or tilted, is filled with real-time (interactive) or pre-rendered (linear) computer animation, live captured images, or composited environments. Although the current technology emerged in the early-to-mid 1990s, fulldome environments have evolved from numerous influences, including immersive art and storytelling, with technological roots in domed architecture, planetariums, multi-projector film environments, flight simulation, and virtual reality. […] The flat enriched image, being a data model or specimen, an entity, or an epistemic instrument for the creation and structuring of cognition, is being transformed to a three dimensional space — a space with the potential of an* immersive experience, *a shared virtual reality. These immersive processes have broader implications for the resuscitated and emerging field of Virtual Reality.*[5]

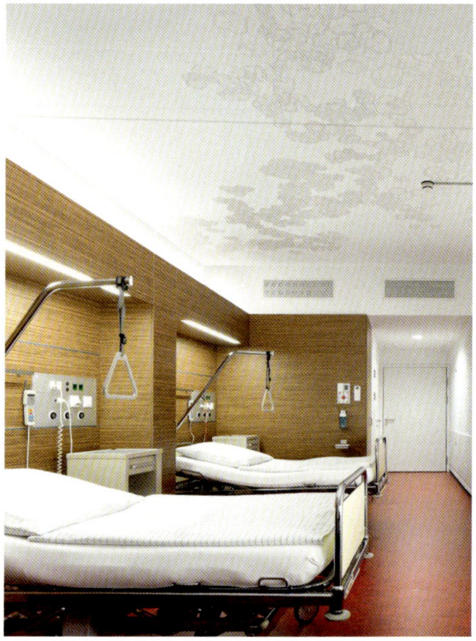

FIG. 11_ MIRRORS OF THE UNSEEN *(2010/11), site-specific ceiling drawing, Landeskrankenhaus Hohenems*

Photo by Adolf Bereuter, 2010

What I personally find especially interesting here is once again the layering of reality levels, the construction of permeability, and the work done with spatial sound.

facades (2015) is an audiovisual animation to be played in a fulldome. This piece once more questions traditional concepts of space, reality, and perception, and turns the immersive experience into an experience of instability. Constructed from arrangements of point clouds, a virtual urban landscape takes shape. The facades and streets seem elastic; decentralized streetscapes overlap; panoramas flip up. The movement of the projected content simulates both spatial dynamics and a first-person perspective. Sound accompanies the movement of the images and is based on the audio recording of a text published by media theoretician Friedrich Kittler about the NSA back in 1986, »No Such Agency«: whispered fragments of the text are acoustically situated within the projection and seem to drift by and converge in the virtual space.

Outlook From the Artist's Perspective

Working with images in the field of digital media requires a fundamentally experimental attitude — not just when it comes to hardware and software, but also in the realm of ideas and imagination, and when coming up with concepts. There are

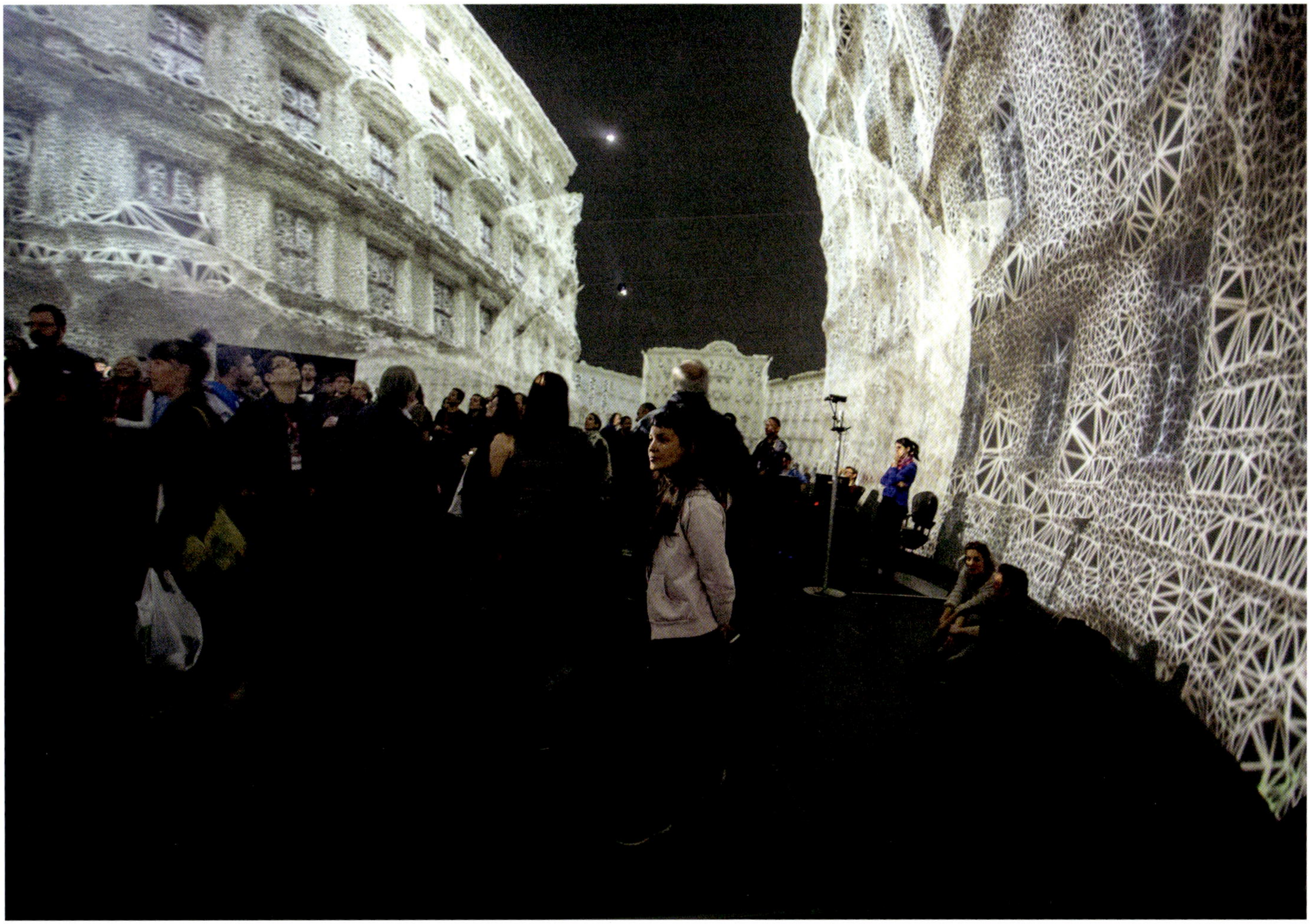

FIG. 12_ FACADES *(2015), immersive audio-visual animation, installation view, Satosphère Montréal*

Photo by Sebastien Roy, 2015

numerous conflicts that arise during the conceptualization, design, and programming of, e.g., interactive environments. Existing patterns of perception and notions of space collide with demands on visual concepts like the ones used when working with dynamic spaces — especially in the field of the simultaneously active variables of time, space, and movement. I am curious about working on new models of imagination and implementing them in artistic environments, always in line with current technological developments and in close connection with findings from scientific research.

1 Oliver Grau, »Immersion and Interaction: From Circular Frescoes to Interactive Image Space,« *medienkunstnetz.de*, accessed February 25, 2023, http://www.medienkunstnetz.de/themes/overview_of_media_art/immersion/8/.

2 Quoted in Cass Sunstein, »1. THE DAILY ME,« in *#Republic: Divided Democracy in the Age of Social Media* (Princeton: Princeton University Press, 2018), 1–30, here 1, https://doi.org/10.1515/9781400890521-002.

3 Quoted in Heinz von Foerster, »To Know and Let Know: An Applied Theory of Knowledge,« *Cybernetic* 1 (1985), 47–55, here 51.

4 Diplomat and cultural manager Christoph Thun-Hohenstein in his text accompanying the *21 Positions* exhibition held at the Austrian Culture Forum in New York 2007.

5 Martin Kusch, »Virtual Together—The Full-dome Medium as an Artistic Research Field,« in *Art Research Envelope* 2, ed. Alexander Damianisch and Margarete Jahrmann (Vienna: University of Applied Arts Vienna, 2019).

Vision Impairments in Extended Reality

Katharina Krösl

In this chapter, we look at vision impairments, their impact on visual perception, and how extended reality (XR) can help us to increase our understanding of such impairments and our empathy for challenges that people with vision impairments face in their everyday lives. I present the state of the art in vision impairment simulations and current research challenges in the field. Looking at the example of my research project XREye, I explain how XR simulations of various eye diseases can be built, tailored to hardware limitations and users' vision capabilities, and evaluated with the help of affected people. This chapter will highlight how we can leverage emerging technologies such as virtual and augmented reality to help make our society more accessible and inclusive for everyone.

Introduction

»Seeing, contrary to popular wisdom, isn't believing. It's where belief stops, because it isn't needed anymore,«[1] is a famous quote by Terry Pratchett and a fitting description of the motivation behind scientific research and visualization in the field of visual computing. For most people, sight is the most important sense through which they experience the world and connect with their fellow humans. »A picture says more than a thousand words,« is just one common saying that emphasizes the important part vision plays in human communication. But is this saying true at all, since it is well known by now that every individual looks at the world through his or her own unique pair of eyes? We live in a world where more than 2.2 billion people are affected by vision impairments, according to the World Health Organization (WHO)[2]—this is more than one-third of the current world population. These impairments can range from minor and common refractive errors (e.g., shortsightedness or farsightedness) to eye diseases or even blindness. In the future, even more people will be affected by vision impairments due to an aging population, increasing urbanization, and the behavioral and lifestyle changes taking place in our society. Are we ready for these growing numbers of visually challenged people? Are we even ready right now to offer safe and inclusive living environments, especially in highly complex areas such as cities? How are non-disabled developers, urban planners, or lighting designers supposed to know what life with a vision impairment is like, and what accommodations need to be made in order for a person to be able to live a self-determined life? To move safely through the busy inner city or navigate public buildings? How are medical and caregiving personnel supposed to fully understand the diverse effects various visual impairments can have? After all, two people with the same eye disease can suffer totally different effects and will therefore need different assistance. But what if we were able to truly understand and even experience vision impairments firsthand in order to better comprehend them and therefore develop our world according to the needs of a growing number of people with vision impairments? Visual computing and, in particular, extended reality (XR)[3] offer answers to this challenge.

In Austria, the Viennese research center VRVis Zentrum für Virtual Reality und Visualisierung Forschungs-GmbH (VRVis Center for Virtual Reality and Visualization)[4], where the research project XREye was developed (which you will learn more about on the following pages), is at the forefront of bringing XR research and inclusive digitalization together. To do so, VRVis draws on more than two decades of experience in the research and development of solutions based on the cross-sectional technology of visual computing for a wide range of applications and industries, providing a strong foundation for creating innovative solutions with the benefit of digitalization for people at their very core. Helping doctors, scientists, architects, designers, and—quite frankly—everyone better understand human vision in all its different dimensions is one of the most important tasks we face as a society. We have identified virtual reality (VR) and augmented reality (AR) as unique »eyeopeners« here, literally allowing us to see through the eyes of people with vision impairments. Doing so helps us to better understand the challenges of people with vision impairments and the hurdles they face, which in turn can be used as a basis for making the world around us a more accessible and inclusive place.

The Impact of Vision Impairments

The aging of the world population is having a particularly grave impact in relation to the occurrence of vision impairments, as many conditions, like presbyopia, cataract, glaucoma, and age-related macular degeneration (AMD), show a higher prevalence in age groups of forty years and older. According to data from the US National Eye Institute (NEI), we can expect twice as many people to be living with vision impairments in 2050 than were recorded in 2010.[5] Since the portion of the population that is affected by vision impairments is constantly increasing, accessibility is becoming more important in different areas of our society. This includes not just architectural design but also lighting design, or the design of signage and visual guiding systems that enable everybody to find their way safely and efficiently. Unfortunately, people with vision impairments are hardly ever taken into account in city planning or architectural and lighting design. There are simply no proper tools or guidelines available to architects and designers to evaluate the accessibility of their design concepts for people with vision impairments. Designers often have to rely on their own experience and understanding of what is accessible for someone with poor eyesight. However, inclusive and barrier-free urban design is now becoming more and more important as the number of people affected by vision impairments increases annually. The biggest challenges here are removing and mitigating barriers to empathy and creating a common understanding of the effects of vision impairments on affected people and their environment.

In medical eye exams, ophthalmologists use standardized tests to assess the severity of different symptoms of an eye disease or vision impairment. Knowledge about vision impairments has been gained in medical experiments or by treating patients, and we can learn about the effects of vision impairments in books, medical papers, and from medical professionals, or by talking to affected people. The problem is that textual descriptions can't give us a true understanding of the impact of vision impairments on the visual perception of affected people. Images or videos can help us to get an idea of what the world looks like with a certain eye disease, but looking at an image or video on a 2D screen is still vastly different from experiencing firsthand a vision impairment that affects one's whole field of view. Conversations with affected people can increase our empathy and comprehension of the challenges people with vision impairments face in their everyday lives, but reports by affected people are not necessarily accurate. Some eye diseases progress rather slowly, and affected people sometimes only notice symptoms once they get quite severe, especially if both eyes are affected in such a way that there is no noticeable difference between the vision in the left and right eyes.

Current tools, descriptions provided by text or in 2D images, and even state-of-the-art 3D simulations often cannot provide enough in-depth insights to fully grasp the effect of a specific vision impairment and how it affects somebody's visual perception. With the help of live camera imagery, we are able to simulate vision impairments graphically, but most current methods are insufficient for realistically depicting the impaired vision as they simplify the simulated effects. Considering all these obstacles, we must inevitably ask: How can a person with healthy eyesight truly understand the full effects of impaired vision and begin to understand the difficulties of, for example, reading escape route signs and navigating safely out of a building? Also, truly everyone—be they relatives of people with vision impairment, or caregiving or medical personnel—can benefit from experiencing the world through another person's eyes; immersion in a realistic VR or AR simulation of a visual impairment can help us to understand and feel empathy.

Reliable and detailed vision impairment simulations are crucial to kick-start inclusivity by design. We need inclusive design in all areas of our lives, like in seemingly mundane everyday objects, urban planning, or even the development of standards to make public spaces accessible to all people, especially with regard to emergency and escape route signage. Architects and lighting designers who develop escape route signposting and lighting systems are important creators in the planning phase of buildings, infrastructure, and public areas. International standards and norms provide guidelines for design processes, but there currently just aren't enough guidelines that accommodate vision impairments. Therefore, many designers rely on their own expertise to take vision impairments into consideration when planning emergency signage—if they even have room to accommodate this at all. All of this reveals the immense lack of data we face when we want to consider people with vision impairments in the design process as well as the lack of tools available to evaluate designs for accessibility. User studies could help us to obtain the necessary data to increase our understanding of the challenges and difficulties that can be caused by vision impairments, essentially reducing the barriers to empathy, and we have identified vision impairment simulations as a key tool on the path to gaining more insight into vision impairments.

Vision Impairment Simulations

To state the obvious first: we need to understand how accessibility works for different people with different challenges, and only then will we be able to define clear directives and guidelines to make accessibility permanent. To facilitate inclusive design, architects and urban planners need helpful rules such as norms and standards to guide them in their work and tools to evaluate the inclusiveness and accessibility of their designs with regard to vision impairments.

We need user studies in order to obtain data, which is the foundation for analyzing the impact of vision impairment on a person's perception of the world. However, such studies are challenging. To generate sufficient, reliable statistical data in order to revise standards and norms in regard to the evaluation of architectural accessibility or lighting design, large studies are needed. Additionally, it is anywhere from difficult to impossible to precisely assess all the characteristics, the extent, and the subjective perception of the visually degrading effects caused by diseases such as cataracts, diabetic retinopathy, glaucoma, and AMD. Different study participants might have an eye disease with a similar characteristic and severity (e.g., a similarly clouded lens) when measured in eye exams, but since they experience the symptoms in an individual and subjective manner, it is hard to determine whether two study participants are impacted by their vision impairment in the exact same way. This makes it very challenging to conduct a user study because finding a large enough user group with the same form and severity of vision impairment in order to obtain enough data for reliable statistical analysis is sometimes impossible. A completely different option is to fall back on participants with normal eyesight and simulate the vision impairment graphically. If we can get a large group of normal-sighted users to experience the exact same form of vision impairment in a computer-generated simulation, running a study to investigate the impact of such degraded vision suddenly becomes a possibility. But to do so, we need very realistic, immersive vision impairment simulations, which pose a number of challenges that I address in my research.

Research Challenges

Simulations of vision impairments have already been done in the past using modified goggles, 2D images, 3D simulations, or VR or AR simulations but have been limited in their realism, immersion, and/or adjustability. As we see the population with vision impairments and eye diseases growing, intensive research in this area is required. At VRVis, we are contributing to this field with our research. In our research project »XREye,« we are collaborating with ophthalmology experts and working on developing plausible simulations of various vision impairments. To achieve this ambitious goal, we have to overcome different challenges:

1. There is no one-fits-all solution. Vision capabilities differ from person to person. Thus, if we want to create a simulation that looks the same for everyone, we need to come up with a way to adjust it to the actual vision capabilities of users.
2. Most eye diseases cannot be simulated with one visual effect alone. They cause different symptoms of varying severity. Therefore, we need an approach that enables us to combine multiple symptoms and adjust them individually.

3. There is no ground truth. Since many eye diseases are experienced differently from person to person, and because some symptoms cannot even be assessed precisely, we don't have a ground truth with which we can compare our simulations. This means we need to come up with another way to evaluate the quality or faithfulness of our simulations.

The Journey So Far: Extended Reality as a New Way to Experience Vision Impairments

There are different ways to create simulations with different levels of immersion and using different display modalities, such as goggles, 2D images, 3D simulations, or XR (VR/AR) simulations. Some researchers have already followed the idea of simulating vision impairments in VR or AR. I am also convinced that these immersive settings are best suited to helping us understand the effects of impairments on visual perception because they enable us not just to look at visual representations of vision impairments but to experience them ourselves. My own work is based upon and influenced by the insights gained from previous research on simulating vision impairments in VR and AR.

VIRTUAL REALITY SIMULATIONS_In recent years, we have seen significant advancements in VR technology, which have opened the door to advanced immersive simulations. However, even before modern VR displays sparked public attention, research was being conducted on vision impairment simulations. Some early simulations of visual field loss caused by glaucoma, double vision (diplopia), and AMD had already been developed in 2000 by Ai et al. in order to teach medical professionals and increase awareness among relatives and friends of patients.[6] Since VR head-worn displays (HWDs) were not that commonly available and did not provide sufficient visual quality back then, researchers like Ai et al. often used projection-based VR display devices with shutter glasses (e.g. CAVEs, ImmersaDesk). Though such systems could not provide anything like photorealistic graphics or truly immersive environments, they paved the way for more research in this area.

In 2016, when more modern VR hardware like the Oculus Rift became available, Väyrynen et al. created an immersive system to educate architectural designers about the impact of vision impairments on people and their ability to navigate through a virtual 3D city model.[7] Even though their simulations were simplified depictions of the respective eye diseases, their work marked an important step toward more sophisticated vision impairment simulations and inspired much of the research that followed. That same year, Maruyama et al. developed a virtual accessibility evaluation system in VR.[8] Their system does not enable users to move freely through the virtual environment, but it does allow them to follow digital humans that move around autonomously and to see the environment through their eyes with simulated impairments. The purpose of the system was to help to find areas in digital building models that required improvements due to missing or unclear signage that resulted in reduced accessibility for people with vision impairments.

Choo et al. introduced their system Empath-D to enable Empathetic User Interface Design.[9] They envisioned the use of VR/AR displays to allow designers to test the usability and accessibility of their websites or applications. A Samsung Gear VR with a Galaxy Note 4 was used as an AR device. Their application was aimed at designing for motor problems, using impairment profiles and representing the interactions of people with certain impairments, including visual impairments. The Empath-D system was later adapted by Kim et al. to work in a purely virtual environment for the accessibility-aware design of smartphone apps, using a commodity VR HWD.[10] A physical smartphone and the hand holding it are tracked (including finger tracking), and the app whose accessibility is to be evaluated is mapped onto the physical smartphone. The vision impairment simulations included in the Empath-D system were still simplified versions of the respective eye diseases, and the system did not include eye tracking, which is crucial to realistically simulate vision impairments that affect different areas of the visual field differently, such as glaucoma, cortical cataracts, or posterior subcapsular cataracts.

In 2020, Alexander et al. presented a research demo that simulated AMD in a virtual home environment. It was designed for a university physical diagnosis course to help medical students develop empathy for geriatric patients.[11] Similar to previous work on AMD simulations, images of their application show a dark shadow in the central field of view. Most AMD simulations simplify the disease pattern in their depictions of AMD in such a way. They focus primarily on simulating the typical vision loss in the center of the field of view, without including any other effects to simulate additional symptoms caused by this eye disease.

AR SIMULATIONS_There has been some research on using AR technology to simulate vision impairments. Flatla and Gutwin,[12] and McAlpine and Flatla[13] worked on assessing the type and severity of a person's color vision deficiency (CVD), then used the results to simulate that person's reduced color vision. They modified the live video feed from the camera of a mobile device to show on the display what the world looks like for a person with a CVD. The goal of their work was to allow people with normal vision to explore different severities of CVD.

Focusing on applications for accessibility inspections, Ates et al. implemented a simulation of vision impairments as a stereoscopic video shown in a VR media player when wearing an HWD.[14] Their solution is based on photos from the NEI,[15] and simulated impairments can be adjusted. However, like the photos from the NEI, their implemented impairments are simplified approximations of the respective vision impairments and do not recreate the impaired vision of specific persons. The cataract simulation, for example, is realized by simply blurring the image without including any simulations of other symptoms.

To increase empathy for people with audiovisual sense impairments, Werfel et al. developed an AR and VR system using real-time audio as well as visual filters.[16] They implemented simulations of AMD, diabetic retinopathy, and retinitis pigmentosa based on illustrations from the German Association for the Blind and Visually Handicapped (DBSV). They did combine different symptoms for their simulated eye diseases but did not consider gaze-dependent effects or the actual vision capabilities of the users experiencing their simulations.

Jones and Ometto worked on solutions targeted at developing teaching or empathy aids, as well as tools for accessibility evaluations.[17] Besides providing options to adjust simulated impairments, they also integrated eye tracking in their VR/AR simulation; however, they only achieved near real-time performance. Later, Jones et al. included simulations of eye diseases such as glaucoma in their software Open-

VisSim, using gaze-dependent regions of variable blur.[18] They conducted studies in VR and AR, measuring participants' task completion times under two types of simulated glaucoma, and also ran some experiments under photopic and mesopic lighting conditions, which revealed that vision impairments can affect visual perception differently under different lighting conditions. Their glaucoma simulation was still a simplification of this vision impairment but could be improved by combining it with other effects in their OpenVisSim software.

Aniruddha et al. proposed a parameterized model to simulate the AMD of specific users.[19] Their model includes luminance degradation, the perceptual deficit region, rotational distortion, and spatial distortion, which are reported symptoms for AMD. The authors further proposed applying the inverse of the adjusted parametric model to compensate for the vision impairment of affected people. They conducted a small preliminary study with healthy participants, giving them simulated AMD in one eye.[20] Participants then had to adjust the parameters of the model in order to replicate the existing simulated AMD for the second eye. Even though this approach is promising, it needs to be tested with people who actually have AMD in order to evaluate whether the simulation is able to replicate a user's vision with AMD sufficiently for the inverse model to correctly compensate for the vision impairment instead of distorting the user's vision even more.

Challenge 1: We Do Not All See the Same

We do not all see the same; even people with »normal eyesight« don't necessarily see equally well. What we consider normal eyesight or normal vision is not one score that a person has to reach on an eye exam but a whole range of eyesight levels that we classify as »normal.« There are different ways to measure eyesight and different units to describe vision. To demonstrate the complexity of this topic, we will now look at it in a bit more detail. One unit often used in research on perception is visual acuity (VA). VA is used to quantify a person's ability to see small details. It is measured by showing the test person different optotypes (standardized symbols used in medical eyesight tests, such as the Landolt C) at different sizes at a given distance in order to find out which size the test person can recognize and which they cannot. Usually, VA is specified relative to 6/6, the Snellen fraction for the test distance of six meters,[21] or 20/20 in feet, or the decimal value of those fractions (»decimal acuity«). As a general rule, a person has normal vision if he or she can recognize a detail that spans one arc minute (one-sixtieth of a degree), which corresponds to a size of ~1.75 millimeters at six meters' distance. Theoretically, this can be tested at any viewing distance, provided that the detail in question is scaled appropriately in relation to the distance. For short-sighted people, it is quite possible to see objects very close, whereas VA is limited beyond a certain distance. This means that the test distance should not be too short (e.g., not under one meter). A standard test distance is six meters or twenty feet.

For the classification of vision impairments, normal vision is defined as a range from 0.8 to 1.6 decimal acuity (dA) or better.[22] However, in the context of VA calculations, we refer to normal vision as a reference value of 1.0dA, or 6/6 vision (in the metric system) or 20/20 vision (using feet as units). When we perform an eyesight test, the Snellen fraction is used to compare the result of this test to a standard person with normal eyesight in recognition distances in meters or feet. It is written as the test distance at which the tested person was able to recognize a test symbol of a certain size, divided by the maximum distance at which a normal-sighted person would still be able to correctly recognize the same test symbol.

The World Health Organization (WHO) differentiates between three stages of visual impairment and one for blindness, using the reference value set of 6/6 (in meters) or 20/20 (in feet).[23] As shown in_TABLE 1, VA can vary widely, and the severity of each stage, even blindness, is given as a range of VA values. People with a VA worse than 0.05 decimal (20/400 feet or 3/60 meters) are considered legally blind. That said, many legally blind people are still able to see shapes or shadows. Some are even able to read, providing that the text is large enough for them to recognize.

stage	Snellen fraction		decimal acuity
mild	< 20/40 ft	<6/12 m	<0.5
moderate	< 20/60 ft	<6/18 m	<0.3
servere	< 20/200 ft	<6/60 m	<0.1
blind	< 20/400 ft	<3/60 m	<0.05

TABLE 1_*The stages of visual impairment defined by the WHO are depicted in terms of Snellen fractions (in feet and meters) and decimal acuity.*[24] *Smaller VA values indicate more severe impairments.*

Our Eyesight in Extended Reality

A multitude of factors affect our perception in XR (VR/AR) and the real world. During the development of our simulations, we identified the factors that we believe have the most impact on the realism of VR or AR simulations:

— vision capabilities of participants (with normal or corrected sight)
— fixed focal distance of the HWD, leading to a vergence-accommodation conflict
— possible misplacement of the HWD
— resolution of the HWD
— latency, refresh rate, and flicker of the display
— dynamic range and color correctness of the display

We need to consider these factors if we want to create a visual experience of a simulated vision impairment that is consistent for all users. We can target some of these factors to make the visual perception in VR as similar to real life as possible under the limitations of current hardware and software. We can also exercise caution in how we select the users we recruit for our studies. We need to recruit normal-sighted participants because otherwise we risk degrading a user's vision more than intended by adding our simulation on top of a preexisting vision impairment. That said, even normal-sighted people have varying vision capabilities that need to be accounted for. Users of HWDs sometimes experience a mismatch between the vergence and accommodation of their eyes,[25] which is caused by the HWD's fixed focal distance relative to the eyes of the user. This vergence-accommodation conflict can result in eye strain and, as a consequence, reduce the VA of a user while in XR. Experiences in XR should therefore be limited to short time periods. Furthermore, we have to make sure the headset is worn correctly because any misplacement of the HWD could also cause visual fatigue and reduce VA. The resolution of an HWD is usually lower than what the human eye can perceive, which is also a factor that needs to be taken into account in the simulation since this low resolution reduces the VA of a user in VR to a level that is already considered a mild vision impairment. An HWD might also have a noticeable latency, especially when used for video see-through AR. Additionally, the dynamic range of most displays is significantly smaller than what a human eye can perceive,[26] and colors are not generally represented as they are experienced in the real world and are dependent on the display calibration. The limited color bit depth can even lead to »banding artifacts« or »contouring artifacts.«[27] Lastly, a low refresh rate could cause the brightness of a display to drop for a very short but noticeable time, resulting in visual flicker. Latency, dynamic range, and color representations, as well as the refresh rate of a display, are hardware limitations that we can only mitigate by using modern HWDs. However, making sure our software is fast enough to render at a maximum of eleven milliseconds per frame is crucial to avoid adding any additional latency. Furthermore, with the limited color range that most HWDs are able to display, it is important to perform calculations in the correct color space when creating the simulation in order to avoid limiting color perception further.

How to Calibrate Simulated Symptoms in XR

In order to use vision impairment simulations in VR or AR to evaluate accessibility and measure recognition distances or the readability of signage, we should recruit only participants with normal sight for user studies. Otherwise we would risk adding our simulation to a participant's preexisting vision condition and unintentionally creating a simulation of degraded vision that is worse than intended. However, even for normal-sighted participants, we need to take individual vision capabilities, as well as the deficiencies of the HWD, into account if we want to develop realistic vision impairment simulations. The goal is to create the same experience for every user, no matter the hardware limitations or varying vision capabilities of users. In our research project XREye, we decided to tackle this challenge by applying a methodology that allows us to calibrate simulated symptoms of vision impairments to the same level of severity for different users. We used this methodology to calibrate two common symptoms caused by different eye diseases: reduced VA and reduced contrast.

SIMULATING REDUCED VA_A reduction in VA is one of the most common symptoms caused by vision impairments. We can simulate this symptom by blurring the image that is shown inside a VR HWD. There are different ways to do this. Commonly, an image filter like a Gaussian blur is used. Hogervorst et al. found that there is a relation between the parameters of a Gaussian blur applied to the image and the VA value a person can achieve under this blurred vision.[28] This means that, in order to simulate a certain level of reduced VA, we just need to find the parameters for the

Gaussian blur that corresponds to this level of VA, which we can do for each user in a calibration step. Considering previously discussed hardware limitations (such as the low resolution of the VR display), a user's vision is already less sharp compared with the real world when they put on a VR HWD, even before any blur filter is applied to the image. This means that, for reduced VA, we can only simulate a level of severity that is worse than the reduction of VA already caused by the hardware or by the vision capabilities of a user (e.g., if someone is a bit shortsighted).

CALIBRATING REDUCED VA_In order to calibrate this effect in a way that enables every user to experience the same level of reduced VA regardless of their own VA, we implemented an eyesight test in VR. During normal eye exams, VA is measured in a standardized eyesight test. Optotypes, such as letters on a Snellen chart, the Lea Symbols Chart,[29] or Landolt Cs are shown to a test subject who has to correctly recognize them in different sizes. The smallest correctly recognized size then determines a person's VA.

The Landolt C or Landolt ring is a ring with a gap at one of eight possible positions: horizontally left or right, vertically up or down, or diagonally in between. We can adapt a standardized eyesight test, in which a person has to correctly recognize the direction of the gap in the ring, and use it to calibrate our effect for reduced VA. We know which size of the Landolt C a person with a certain VA can recognize from a certain distance in the real world. This means that, for our test in VR, we first select a suitable test distance, such as six meters. We then choose the level of VA we want to calibrate our simulation to, for example, 0.25 decimal acuity. The smallest Landolt C that a person with 0.25 decimal acuity is able to recognize during a normal (non-VR) eyesight test is four times larger than for a person with normal sight. Hence, we simply increase the size of the optotype in our VR simulation to four times its original size. Of course, a person with normal sight has no trouble identifying the correct position of the gap in the Landolt ring at this size, even inside our VR simulation. But now, we apply our blur effect and run the standardized vision test with a slightly blurred image. After each correctly identified row of Landolt Cs, we increase the blur effect until the user is no longer able to correctly identify the gaps in the ring. We have then calibrated the simulated reduced VA for this user to the predefined level and know the corresponding parameters of the Gaussian blur that we can then use to simulate 0.25 decimal acuity for the current user. _FIG. 1 illustrates this procedure. This enables us to calibrate the effect for users with different vision capabilities, but the calibration has to be performed for each individual user. For a person with very good eyesight, the calibration will apply more blur than for a person who already has a mild refractive error, but in the end, both users will experience the same level of degraded vision through our simulation. This methodology can also be applied to the calibration of other effects, such as our simulation of reduced contrast vision, where we incrementally decrease the contrast during the eyesight test until we reach the desired level of reduced contrast.

Calibrating the vision of many different people to the same level of VA allows us to run experiments targeted at investigating the effects of VA on accessibility, for example, on the recognizability of signage. During our research project XREye, we conducted a quantitative study entirely in VR to investigate the maximum recognition distance (MRD) of escape route signs in buildings. The MRD is the maximum

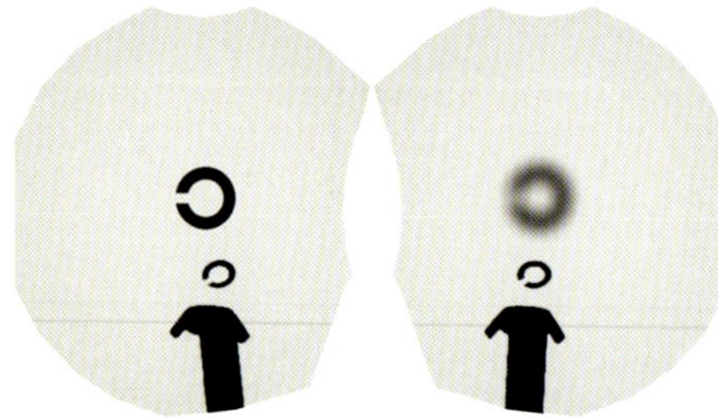

FIG. 1_*Illustration of VA calibration in VR, with unmodified vision at the beginning of the calibration (shown for the left eye) and blurred vision at a later stage during the calibration procedure (shown for the right eye in this image). The optotype that has to be recognized is a Landolt ring, shown at a fixed distance (enlarged in this illustration). Another Landolt ring is displayed directly above the HTC Vive controller. The user has to align the gaps of both Landolt rings by moving their thumb along the trackpad on the controller to turn the Landolt ring that is displayed directly above it and then confirm their choice by pressing the trigger. As long as at least three out of a set of five Landolt rings are aligned correctly, the set is counted as correctly recognized and the blur applied to the image is increased.*

VIDEO

distance at which a person can still correctly recognize a sign. This is important because these maximum values are included in standards and norms on signage as the minimum suggested distance between two signs. Hence, abiding by these norms should ensure that, at any given time, a person has one escape-route sign in sight that they can read.

We evaluated the MRD of escape route signs under reduced VA in our VR study with people with normal or corrected sight (wearing glasses or lenses) and with graphically simulated impaired vision.[30] We calibrated the vision of each user in the HWD to specific levels of reduced VA by applying our symptom calibration methodology and validated it in eyesight tests in VR. We measured the MRDs for different sizes of escape route signs, different viewing angles, and different levels of vision impairment, and compared our measurements to the values prescribed by current norms and standards for the installation of signage. This study opened a door to new insights into how reduced VA can affect the readability of signage. Our results showed that the distances between signs prescribed by norms and standards at the time did not sufficiently consider people with vision impairments.

Challenge 2: Eye Diseases Cause Multiple Symptoms

The WHO estimates that about 2.2 billion people worldwide are affected by vision impairments. These impairments include presbyopia (1.8 billion) and other refractive errors such as myopia and hypermetropia (123.7 million), cataracts (65.2 million), AMD (10.4 million), glaucoma (6.9 million), corneal opacities (4.2 million), diabetic retinopathy (3 million), trachoma (2 million), and other eye diseases.[31] Some impairments, such as refractive errors, are very common and, in most cases, are easy to correct with glasses or contact lenses. Other eye diseases, such as AMD, have a sustained impact on visual function and can lead to central vision loss.[32] While highly treatable, cataracts are one of the leading causes of vision impairment (33 percent) after refractive errors (43 percent) and are, at 51 percent, the leading cause of blindness.[33] All these vision impairments and eye diseases have one thing in common: they have a negative impact on a person's vision — but in very different ways.

Refractive Errors

Refractive errors create blurry vision for affected people and are the major cause of vision impairments worldwide.[34] The most common reason for this blurry vision is an increase or decrease in the axial length of the eye. This condition is called *myopia* (also known as nearsightedness or shortsightedness) if the eye grows too long or *hyperopia* (farsightedness or longsightedness) if the eye grows too short from front to back.[35] Myopia causes blurry vision due to images being focused at a point in front of the retina, which creates a blurred image on the retina. People with myopia have good near vision, whereas their vision becomes more blurred with distance.[36] Hyperopia, where the eye focuses images on a point behind the retina due to the eyeball being too short or a deformation of the cornea, causes blurred vision of near objects, while objects faraway can appear clear.[37] Presbyopia also causes blurred vision of near objects due to aging and a reduction in the accommodation abilities of the eyes.[38]

Cornea Disease

The cornea is the transparent front layer of the eye. Different conditions can affect the cornea and therefore a person's vision, such as injuries, allergies, inflammation (keratitis), dry eye, or corneal dystrophies.[39] Pink eye, also called allergic conjunctivitis, can cause swelling and redness, itchy burning, and watery eyes. Affected people can experience blurry vision and sensitivity to bright light.[40] Keratitis is an eye infection linked to wearing contact lenses. This condition can occur when contact lens wearers do not take care of their lenses as instructed, increasing the possibility of germs invading the cornea. Viruses, bacteria, fungi, or parasites can then cause inflammation of the cornea, leading to irritated, red eyes, pain, sensitivity to light, blurry vision, and watery eyes.[41] Dry eye is a very common condition where the eyes don't produce enough tears, which leads to red eyes and scratchy, stinging, or burning feelings in the eyes. It can also cause vision problems, such as blurry vision or sensitivity to light.[42] Corneal dystrophies, such as keratoconus, Fuchs's dystrophy, lattice dystrophy, and map-dot fingerprint dystrophy can cause fogging or a swelling of the cornea, or material build-ups in the cornea. Besides itchy feelings and pain, symptoms can include blurry vision, clouded vision, and sensitivity to light.[43]

Age-Related Macular Degeneration

The macula is the central area of the retina, containing the most cones, the highest density of which is at the fovea, which is responsible for central vision. AMD is a leading cause of vision loss, primarily affecting people over fifty years of age.[44] AMD is not responsible for full blindness, but due to the decrease or even loss of central vision, it has a significant impact on a person's perception, which makes reading, getting around the house, or recognizing faces a challenge.[45]

There are two main types of AMD: dry AMD is characterized by deposits that build up beneath the retina and can lead to atrophy, the destruction of photoreceptive cells, while wet AMD causes abnormal blood vessels to grow under the retina, which leak into the retina and ultimately create scar tissue. All stages of AMD can affect vision differently.[46] Symptoms primarily affect the center of the field of view and include blurry vision, reduced brightness, loss of central vision, and distorted vision, often described as »straight lines looking wavy.«

Cataracts

If the lens of the eye is affected by opacities, the phenomenon is called cataracts. Cataracts impair the visual field but can also lead to full vision loss when not treated in time. Cataracts can be classified in three categories, depending on the region of the lens that is affected and the specific characteristics: nuclear, cortical, or posterior subcapsular cataracts.[47]

Nuclear cataracts manifest as ubiquitous yellow tinting (or clouding) in one's vision, with increased straylight (light that is scattered by parts of the eye due to optical imperfections such as particles in a clouded lens[48]). The reason for this is an accumulating cluster of yellow-brown pigment or protein in the nucleus area of the lens, which is responsible for the homogeneous clouding of the lens, often with a yellowish/brownish tint, leading to increased light scattering.[49]

Cortical cataracts appear as dot-like opacities, peripheral radial opacities (called »shades«), or spoked opacities near the periphery. They are caused by an opacity forming at the lens cortex due to protein aggregation or damage to the fibers in this area. The most frequently occurring form of cataracts are cortical cataracts with spoked opacities near the periphery.[50]

Posterior subcapsular cataracts are caused by defective fiber production in the lens and result in opacities forming at the posterior pole of the lens, perceived as dark shadows in the center of one's vision. It is the least common of the three forms of cataracts mentioned here. Because of its location near the visual axis, its effect on vision is very severe, even with a low density of defective fibers.[51] All three forms of this eye disease can develop to various extents within the same eye. The severity can be graded in a slit-lamp examination in mydriasis (dilation of the pupil).[52]

Based on patient reports and insights from ophthalmologists, and according to the NEI,[53] the following symptoms can be directly linked to cataracts:

- cloudy or blurry vision (reduced VA)
- faded colors (reduced contrast)
- tinted vision (color shift)
- troubles seeing at night (bloom/glare)
- increased sensitivity to light (bloom/glare)
- halos around lights
- double vision

Lens opacities influence vision depending on their location and on pupil size. Under bright lighting conditions (e.g., in daylight), the pupil's diameter is small, and only opacities within the pupillary zone are likely to affect vision. The pupil diameter becomes larger when light is reduced, which affects vision due to the increasing amount of straylight falling on the retina.[54]

Simulating Eye Disease Patterns Using an Effects Pipeline

All the eye diseases discussed above cause a number of different symptoms. Additionally, each symptom can be present at a different level of severity, which creates numerous different eye disease patterns that can affect people. Therefore, a single visual effect is, in most cases, insufficient to create a realistic simulation of a vision impairment. We need to be able to create complex eye disease patterns as a combination of different effects, each simulating a symptom, and we need to be able to adjust each effect individually.

In our research project XREye, we have developed methods to simulate complex eye disease patterns like the conditions described above in XR (VR, AR, or when viewing 360-degree images). This allows people with normal sight to experience the respective vision impairments firsthand using a VR or AR device. We have utilized the newest generation of VR and AR devices, which now provide opportunities to create more immersive and realistic simulations than ever before. However, developing a vision impairment simulation remains a challenging task because different factors influence the perception of users in VR. As I have already discussed, we have to consider users' vision capabilities, the impacts of immersive hardware on a user's vision, as well as the impacts of the respective vision impairment or eye disease on different aspects of the human visual system. We can account for these factors by calibrating our simulation to a predefined level of severity. However, since eye diseases can cause multiple symptoms, we also need to be able to calibrate each symptom individually and then combine them into a simulation of a particular eye disease pattern. This is possible when using an effects pipeline, which enables us to apply one effect (simulating one symptom) after the other. VR devices such as the HTC Vive Pro are an inexpensive option for testing such simulations in VR as well as video see-through AR. For the AR mode, the Vive's front cameras can be used to record the surrounding real-world environment of the user. This video stream is then displayed inside the HWD. Each image frame is modified by applying several different effects in sequence before it is displayed in the HWD. A Pupil Labs[55] binocular eye tracker add-on can be used inside the HTC Vive Pro HWD to track the gaze of the user and, along with with the user's gaze, »move« simulated effects that are supposed to only affect one part of the visual field. Depending on the vision impairment we want to simulate, we can combine different simulated symptoms in our effects pipeline.

SIMULATING CORNEA DISEASE_ Cornea disease can have different causes, which result in different symptoms for affected people, such as blurry vision, sensitivity to light (e.g., caused by pink eye, keratitis, dry eye), and foggy or clouded vision (e.g., caused by corneal dystrophies). Some patients describe this vision condition as »looking through opal glass.« To recreate such a visual impression, we can apply the following effects to build our effects pipeline:

1. reduce VA
2. reduce contrast
3. apply color shift
4. add high-frequency noise

In a first step, we reduce the VA by applying a Gaussian blur filter to the image. Then, the contrast of the image is reduced. There are different approaches toward reducing the contrast in a color image. One simple method is to blend the image with the color gray, but this also results in a desaturated image. A more advanced approach is to reduce the differences in luminance values in the image by applying a histogram compression to the lightness values of all pixels. To achieve an effect that matches the perception of the human eye, these calculations have to be done in the appropriate color space, such as the CIELAB space.

Next, we apply a color shift to our low-VA and contrast-reduced image. A color shift can be realized as an interpolation between image colors and a target color, but this also reduces the contrast in the image. There are also more sophisticated ways to simulate a color shift. If a person has opacities in the lens of their eye or in the cornea, these opacities absorb or reflect parts of incoming light. The color of the visible opacities is the part of the light that does not reach the retina. This means that the resulting color shift perceived by the affected person can be simulated by subtracting the reflected or absorbed part of the light from the image. Consequently, blueish opacities would allow vision to appear more red or brown since the blue part of the light is subtracted.

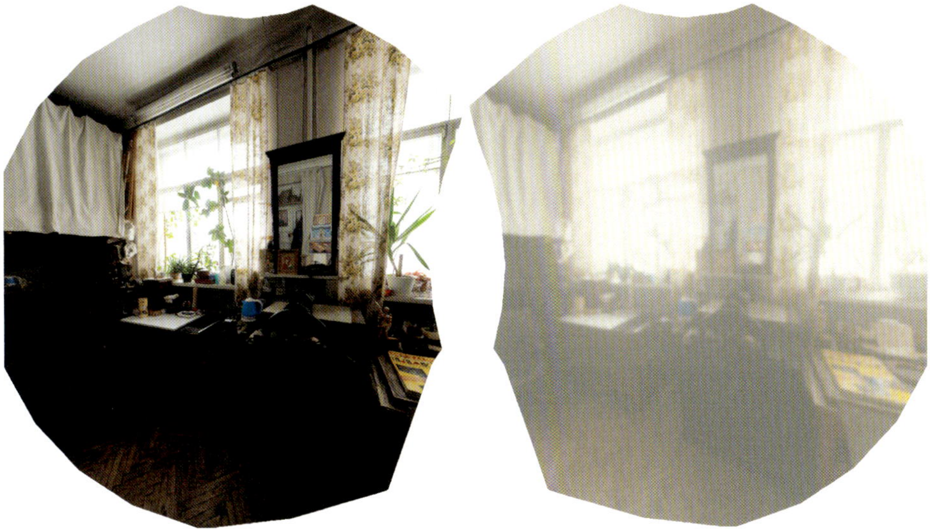

FIG. 2_ *Simulation of cornea disease when looking at a 360-degree image inside an HWD, with unmodified vision for the left eye and our simulation applied to the vision of the right eye*

VIDEO

To create the impression of opal glass, we can add a high-frequency noise texture, which can be softened by means of interpolation with a white color image, and blend this material with our low-VA, contrast-reduced, and color-shifted image in the last step of our effects pipeline. The simulation of this type of cornea disease can be seen in_FIG. 2. This simulation was developed based on reports from patients and in consultation with experts in ophthalmology. Other characteristics of this eye disease can be realized by calibrating or adjusting the parameters of each effect or by adding additional stages to the effects pipeline.

SIMULATING AGE-RELATED MACULAR DEGENERATION_AMD mainly affects central vision, which can become blurry or distorted and is often darkened, desaturated, or low in contrast. To simulate this eye disease, we can use the following effects pipeline:

1. distort
2. desaturate
3. reduce contrast
4. darken central vision

In the first step, we apply distortions to the image. A simple way to do this is to randomly use a water texture to shift the image pixels in different directions. Blending multiple radial distortions together creates multiple inward or outward bulges in the image.

In the second step, we apply radial desaturation to create a washed-out image in the center of a user's field of view. This can be done in the same manner as the desaturation step performed in the cornea disease effects pipeline above. To additionally achieve the radial falloff of this effect, a radial gradient exponential can be used as a weighting function.

After reducing the contrast of the image (as described for the cornea disease simulation), we can simulate the loss of central vision by blending a dark radial

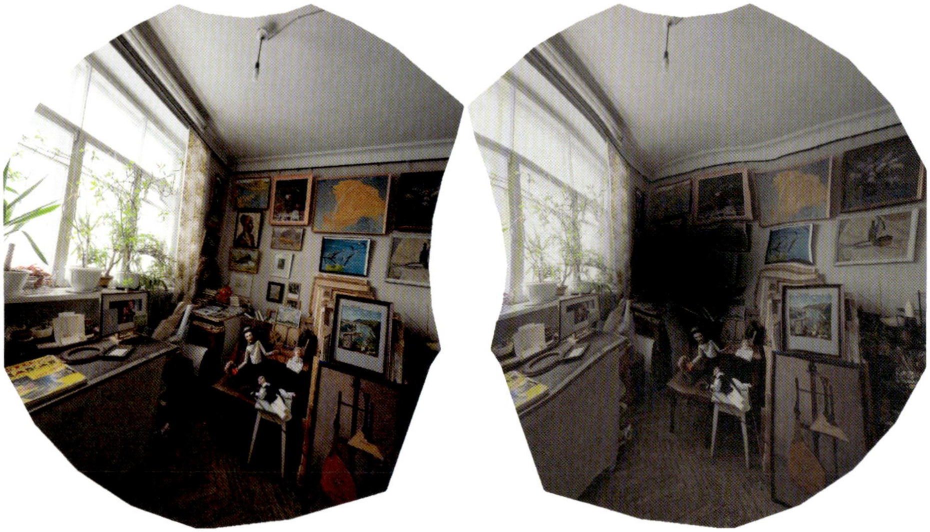

FIG. 3_*Simulation of wet AMD when looking at a 360-degree image inside an HWD, with unmodified vision for the left eye and our simulation applied to the vision of the right eye*

VIDEO

image above the central field of view or by reducing the luminance of the pixels in this area. _FIG. 3 shows an example of our simulation of wet AMD in AR. We developed this simulation based on related work and expert knowledge from ophthalmologists.

SIMULATING CATARACTS_Common symptoms of cataracts, such as blurred vision, contrast loss, tinted vision, dark shadows, and increased sensitivity to light, can also be simulated individually and then combined to create a simulation of a specific cataracts disease pattern, by using the following effects pipeline:

1. reduce VA
2. reduce contrast
3. apply color shift
4. simulate dark shadows
5. simulate sensitivity to light

The first three effects can be implemented as previously described. To simulate dark shadows in the periphery (caused by cortical cataracts) or in the center of the field of view (caused by posterior subcapsular cataracts), a shadow texture can be blended above the image. The size of this shadow texture is then scaled according to the brightness of the pixels around the gaze point of the user in order to simulate the reaction of the pupil, widening when looking at dark or contracting when looking at very bright areas.

In the final step of our effects pipeline, a bloom or glare effect is rendered around bright areas or light sources to simulate the increased sensitivity to light that often affects people with cataracts. The intensity of this effect can also be scaled depending on whether the user is directly looking at a light source or not.

According to the ophthalmologists we have worked with, symptoms such as halos around lights or double vision are less common for cataract vision but could also be added as separate stages of the effects pipeline. We based all of our vision impairment simulations on medical expert knowledge by working closely with ophthalmologists. However, evaluating the realism of such simulations remains a challenge, since many vision impairment symptoms are highly subjective and cannot be assessed precisely in eye exams.

Challenge 3: Evaluating Simulations

If we conduct research, we need to evaluate our results. In the field of visual computing or computer graphics and rendering, this is often done by comparing the results to a ground truth. This ground truth could be a rendered image or some measurements, depending on the application. When it comes to vision impairment simulations, we are faced with a dilemma: there is no ground truth that can tell us exactly what the vision of a person with a complex eye disease looks like. We can obtain measurements by conducting eye exams to measure certain vision capabilities or measure visible opacities in the eyes of an affected person, but it is very difficult, if not impossible, to precisely determine how a particular eye disease affects a person's vision. The same symptoms can be experienced very differently between people. Ophthalmology experts usually also have only secondhand knowledge from medical publications or treating patients but have not experienced these vision impairments themselves. The affected people are the real experts when it comes to judging the realism and faithfulness of a vision impairment simulation.

By creating an interactively adjustable vision impairment simulation system that allows us to modify each symptom at runtime, we are able to conduct user studies with people with vision impairments. In doing so, we are able to not just use their feedback to evaluate our simulations but also to tweak our simulations live during our experiments with affected people, based on their feedback, and to try to match the vision impairment they experience with our simulations as closely as possible in order to, ultimately, obtain even more realistic depictions of those impairments.

Symptom Matching

To validate and improve the realism of the VR or AR simulations of vision impairments created during our research project XREye, we developed a novel approach to evaluate and adjust the simulated symptoms together with patients affected by the same symptoms that we included in our simulations. To this end, we created software that enables us to show two different images to the left and right eye of a user. One image is the unmodified video stream of the user's surroundings, captured in real time by the front-facing cameras of the HWD (e.g., from an HTC Vive Pro). The other image is the same video stream but modified with our simulation of a vision impairment. The key point is to conduct these experiments with users with only one eye affected by the respective vision impairment and a second healthy eye. This could, for example, be a cataract patient between surgeries, when one of their eyes is already corrected, allowing them to see clearly, but the other eye is still affected by the eye disease. These patients can look at our cataract simulation with their corrected eye and compare it to the vision they have in their remaining eye affected by cataracts when looking at the unmodified video see-through AR stream. During such an experiment, we can adjust our simulation at runtime to match a participant's impaired vision and consequently obtain parameter values

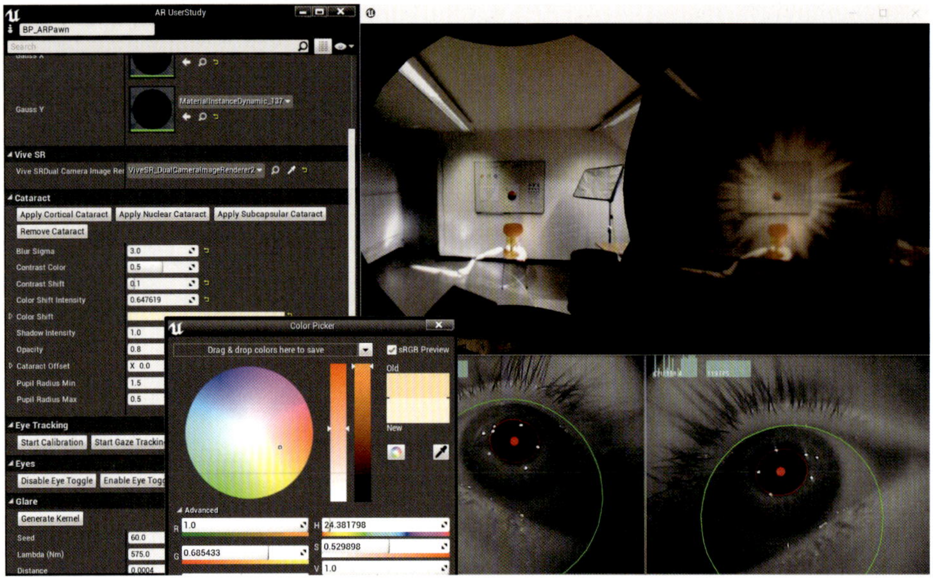

FIG. 4_*Symptom matching via the interactive adjustment of cataract simulation (including eye tracking for gaze-dependent effects) at runtime, with unmodified video stream shown to the left eye and our simulation applied to the live AR video stream shown to the right eye*

© VRVis, 2022

VIDEO

for simulated symptoms that can be used to realistically simulate such vision impairments. This symptom-matching methodology can be used to adjust the simulation of any vision impairment where the patient has clear vision in one eye when the experiment is conducted. Vision impairments such as cataracts, glaucoma, cornea disease, AMD, diabetic retinopathy, and others can be adjusted using this methodology if only one eye of a participant is affected and if the available simulations provide the functionality to adjust all the included effects at runtime. _FIG. 4 shows how we tested this symptom-matching methodology and the flexibility of our system by interactively adjusting our cataract simulation at runtime together with participants to match their own cataract vision.

Opening the Door to New Insights

In this chapter, I have shown that it is now possible to develop immersive and realistic vision impairment simulations by leveraging the possibilities of emerging technologies such as VR and AR, and by conducting quantitative and qualitative user studies with people with healthy eyesight and graphically simulated impairments. With our approach, we can calibrate individual symptoms, combine them to simulate eye disease patterns, and adjust them with patients using our symptom-matching method. In our research project XREye, we have complemented this methodology with a system that builds on state-of-the-art VR, AR, and eye-tracking technology, integrating an effects pipeline to combine simulated symptoms as well as different viewing modes for VR, AR, and 360-degree images.

To evaluate our methodology and system for different applications, we have conducted three user studies, each one building upon the results from the previous study: in our first study,[56] we evaluated norms and standards for signage and compared our results to the values prescribed by these norms. This study revealed that norms and standards that were in effect in Austria at the time did not sufficiently take people with vision impairments into account. As a result of publishing our findings, we were asked to contribute our expertise by participating in an Austrian Standards committee. This led to a revised norm on visual guiding systems that also considers the limited visual acuity of people with vision impairments.[57] Following the guidelines in this new norm can reduce barriers for people with vision impairments and make public spaces more inclusive and accessible for everyone.

We also tested our approach toward accessibility considerations in a VR user study[58] and investigated the influence of cataracts on visual perception under different lighting scenarios. We measured MRDs under simulated cataract vision (for people with healthy eyes) as well as the influence of different illumination scenarios on perception under these conditions. We calibrated the VA as well as the contrast vision of participants and simulated eye disease patterns for three types of cataracts by combining the reduced VA and reduced contrast with other symptoms using our effects pipeline. The quantitative study confirmed our initial assumption that cataracts affect vision significantly and degrade a person's ability to read signage. We also found that indirect lighting can be beneficial for people with cataracts as it can reduce discomfort from blinding effects and should therefore take precedence in the homes of affected people (and possibly also homes for the elderly, since this is a common eye disease in the population over sixty years of age). Furthermore, participants reported that taking part in this study helped them to better understand the difficulties people with cataracts face every day. These findings highlight the need for proper simulation and evaluation tools in this area.

Finally, we conducted a study[59] to evaluate and adjust our cataract simulation with cataract patients between surgeries. This study allowed us to demonstrate the feasibility of our methodology and the system we had developed to create realistic depictions of eye diseases together with patients. In this primarily qualitative study, participants looked at our cataract simulation blended over a video see-through AR stream with their healthy eye that had already been operated on and compared it to their own cataract vision when looking at the unmodified AR stream with the other eye that still had cataracts. We tested the symptom-matching methodology and flexibility of our system by interactively adjusting the simulation at runtime together with participants to match their own cataract vision and asked them to rate each symptom in terms of realism. This pilot study showed that AR solutions can help to create vision impairment simulations with symptom matching and is the first step toward creating more realistic simulations.

Our research project XREye has enabled us to demonstrate how to create various immersive vision impairment simulations that can help to increase understanding of vision impairments. Our approach can also be used to help standardization committees to revise standards and norms. Based on our research, we can provide architects and designers with tools to create more accessible designs. Furthermore, such applications can be used to train medical personnel or serve as a basis to develop novel diagnostic methods for vision impairments in the future, and to generally help people with normal sight to open a door to a world that looks different from the one that they are used to seeing.

1 Terry Pratchett. *Pyramids*, vol. 7 (London: Corgi, 1990), 216.

2 World Health Organization, *World Report on Vision* (Geneva: World Health Organization, 2019).

3 Refers to any combination of real and virtual environments and objects, including virtual reality (VR), augmented reality (AR), and mixed reality (MR).

4 VRVis is funded by the BMK, the BMAW, Styria, the SFG, Tyrol, and the Vienna Business Agency within the scope of COMET—Competence Centers for Excellent Technologies (879730), which is managed by the FFG.

5 NEI Office of Science Communications, Public Liaison, and Education, »Prevalence of Adult Vision Impairment and Age-Related Eye Diseases in America,« accessed November 9, 2018, https://nei.nih.gov/eyedata/adultvision_usa.

6 Zhuming Ai, Balaji K. Gupta, Mary Rasmussen, Ya J. Lin, Fred Dech, Walter Panko, and Jonathan C. Silverstein, »Simulation of Eye Diseases in a Virtual Environment,« in *Proceedings of the 33rd Annual Hawaii International Conference on System Sciences* (Maui, 2000), https://www.doi.org/10.1109/HICSS.2000.926803.

7 Jani Väyrynen, Ashley Colley, and Jonna Häkkilä, »Head Mounted Display Design Tool for Simulating Visual Disabilities,« in *Proceedings of the 15th International Conference on Mobile and Ubiquitous Multimedia* (New York: Association for Computing Machinery, 2016), 69–73.

8 Tsubasa Maruyama, Satoshi Kanai, and Hiroaki Date, »Vision-Based Wayfinding Simulation of Digital Human Model in Three Dimensional As-Is Environment Models and its Application to Accessibility Evaluation,« in *International Design Engineering Technical Conferences and Computers and Information in Engineering Conference*, vol. 50077 (New York: American Society of Mechanical Engineers, 2016), V01AT02A064, https://doi.org/10.1115/detc2016-59170.

9 Kenny T. W. Choo, Rajesh K. Balan, Tan Kiat Wee, Jagmohan Chauhan, Archan Misra, and Youngki Lee, »Empath-D: Empathetic Design for Accessibility,« in *Proceedings of the 18th International Workshop on Mobile Computing Systems and Applications* (New York: Association for Computing Machinery 2017), 55–60, https://doi.org/10.1145/3032970.3032981.

10 Wonjung Kim, Kenny T. W. Choo, Youngki Lee, Archan Misra, and Rajesh K. Balan, »Empath-D: VR-Based Empathetic App Design for Accessibility,« in *Proceedings of the 16th Annual International Conference on Mobile Systems, Applications, and Services* (New York: Association for Computing Machinery, 2018), 123–35, https://doi.org/10.1145/3210240.3210331.

11 Drew Alexander, Thuy Nguyen, Patrick Keller, Jason Orlosky, Shilpa Brown, Elena Wood, Onyeka Ezenwoye, and Wanda Jirau-Rosaly, »Design of Visual Deficit Simulation for Integration into a Geriatric Physical Diagnosis Course,« in *2020 IEEE Conference on Virtual Reality and 3D User Interfaces Abstracts and Workshops (VRW)* (Atlanta: IEEE, 2020), 838–9, https://www.doi.org/10.1109/VRW50115.2020.00270.

12 David R. Flatla and Carl Gutwin. »›So That's What You See‹: Building Understanding with Personalized Simulations of Colour Vision Deficiency,« in *Proceedings of the 14th International ACM SIGACCESS Conference on Computers and Accessibility* (New York: Association for Computing Machinery, 2012), 167–74, https://doi.org/10.1145/2384916.2384946.

13 Rhouri MacAlpine and David R. Flatla, »Real-Time Mobile Personalized Simulations of Impaired Colour Vision,« in *Proceedings of the 18th International ACM SIGACCESS Conference on Computers and Accessibility* (New York: Association for Computing Machinery, 2016), 2015, 225–228, https://doi.org/10.1145/2677199.2680551.

14 Halim C. Ates, Alexander Fiannaca, and Eelke Folmer, »Immersive Simulation of Visual Impairments Using a Wearable See-through Display,« in *Proceedings of the Ninth International Conference on Tangible, Embedded, and Embodied Interaction* (New York: Association for Computing Machinery, 2014), 301–2, https://doi.org/10.1145/2661334.2661346.

15 NEI Office of Science Communications, Public Liaison, and Education, »Prevalence of Adult Vision Impairment.«

16 Fabian Werfel, Roman Wiche, Jochen Feitsch, and Christian Geiger, »Empathizing Audiovisual Sense Impairments: Interactive Real-time Illustration of Diminished Sense Perception,« in *Proceedings of the 7th Augmented Human International Conference 2016* (New York: Association for Computing Machinery), article 15, 1–8, https://doi.org/10.1145/2875194.2875226.

17 Pete R. Jones, and Giovanni Ometto., »Degraded Reality: Using VR/AR to Simulate Visual Impairments,« in *2018 IEEE Workshop on Augmented and Virtual Realities for Good (VAR4Good)* (Reutlingen, 2018), 1–4, https://www.doi.org/10.1109/VAR4GOOD.2018.8576885.

18 Pete R. Jones, Tamás Somoskeöy, Hugo Chow-Wing-Bom, and David P. Crabb, »Seeing other Perspectives: Evaluating the Use of Virtual and Augmented Reality to Simulate Visual Impairments (openvissim),« *NPJ Digital Medicine* 3, no. 1 (2020): 1–9.

19 Prithul Aniruddha, Nasif Zaman, Alireza Tavakkoli, and Stewart Zuckerbrod, »A Parametric Perceptual Deficit Modeling and Diagnostics Framework for Retina Damage Using Mixed Reality,« in *International Symposium on Visual Computing* (Cham: Springer, 2019), 258–69.

20 Nasif Zaman, Alireza Tavakkoli, and Stewart Zuckerbrod. »A Mixed Reality System for Modeling Perceptual Deficit to Correct Neural Errors and Recover Functional Vision,« in *2020 IEEE Conference on Virtual Reality and 3D User Interfaces Abstracts and Workshops (VRW)* (Atlanta: IEEE, 2020), 269–74, https://www.doi.org/10.1109/VRW50115.2020.00055.

21 Herman Snellen, *Optotypi ad visum determinandum (letterproeven tot bepaling der gezichtsscherpte; probebuchstaben zur bestimmung der sehschaerfe)* (Utrecht: Weyers, 1862).

22 August Colenbrander, »Visual Standards—Aspects and Ranges of Vision Loss,« April 2002, accessed September 9, 2017, http://www.icoph.org/downloads/visualstandards-report.pdf.

23 World Health Organization, *World Report on Vision*.

24 World Health Organization, *World Report on Vision*.

25 Gregory Kramida, »Resolving the Vergence-Accommodation Conflict in Head-Mounted Displays,« *IEEE Transactions on Visualization and Computer Graphics* 22, no. 7 (2015): 1912–31; Ernst Kruijff, J Edward Swan, and Steven Feiner. »Perceptual Issues in Augmented Reality Revisited,« in *2010 IEEE International Symposium on Mixed and Augmented Reality* (Seoul: IEEE, 2010), 3–12, https://www.doi.org/10.1109/ISMAR.2010.5643530.

26 Laura R. Luidolt, Michael Wimmer, and Katharina Krösl, »Gaze-Dependent Simulation of Light Perception in Virtual Reality,« *IEEE Transactions on Visualization and Computer Graphics* 26, no. 12 (2020): 3557–67.

27 George A. Koulieris, Kaan Akşit, Michael Stengel, Rafał K. Mantiuk, Katerina Mania, and Christian Richardt, »Near-Eye Display and Tracking Technologies for Virtual and Augmented Reality,« *Computer Graphics Forum* 38, no. 2 (May 2019): 493–519, https://doi.org/10.1111/cgf.13654.

28 Maarten A. Hogervorst, and Wim J. M. van Damme, »Visualizing Visual Impairments,« *Gerontechnology* 5, no. 4 (2006): 208–21.

29 Lea Hyvärinen, Risto Näsänen, and Pentti Laurinen, »New Visual Acuity Test for Pre-school Children,« *Acta Ophthalmologica* 58, no. 4 (1980): 507–11.

30 Katharina Krösl, Dominik Bauer, Michael Schwärzler, Henry Fuchs, Georg Suter, and Michael Wimmer, »A VR-Based User Study on the Effects of Vision Impairments on Recognition Distances of Escape-Route Signs in Building,« *The Visual Computer* 34, no. 6–8 (2018): 911–23.

31 World Health Organization, *World Report on Vision*.

32 National Eye Institute, »Eye Conditions and Diseases,« accessed December 18, 2019, https://www.nei.nih.gov/learn-about-eye-health/eye-conditions-and-diseases.

33 Donatella Pascolini and Silvio P. Mariotti, »Global Estimates of Visual Impairment,« *British Journal of Ophthalmology* 96, no. 5 (2012): 614–18.

34 Donatella Pascolini and Silvio P. Mariotti, »Global Estimates of Visual Impairment.«

35 National Eye Institute, »Refractive Errors,« accessed December 18, 2019, https://www.nei.nih.gov/learn-about-eye-health/eye-conditions-and-diseases/refractive-errors.

36 National Eye Institute, »Refractive Errors.«

37 National Eye Institute, »Refractive Errors.«

38 National Eye Institute, »Refractive Errors.«

39 National Eye Institute, »Corneal Conditions,« accessed December 18, 2019, https://www.nei.nih.gov/learn-about-eye-health/eye-conditions-and-diseases/corneal-conditions.

40 Centers for Disease Control and Prevention, »Germs & Infections,« accessed May 10, 2020, https://www.cdc.gov/contactlenses/germs-infections.html.

41 National Eye Institute, »Corneal Conditions.«

42 National Eye Institute, »Corneal Conditions.«

43 National Eye Institute, »Corneal Conditions.«

44 National Eye Institute, »Eye Conditions and Diseases.«

45 National Eye Institute, »Eye Conditions and Diseases.«

46 John A. Shoemaker, *Vision Problems in the US: Prevalence of Adult Vision Impairment and Age-related Eye Disease in America* (Bethesda; National Eye Institute, 2002).

47 Ralph Michael and Anthony J. Bron, »The Ageing Lens and Cataract: A Model of Normal and Pathological Ageing,« *Philosophical Transactions of the Royal Society of London B: Biological Sciences* 366, no. 1568 (2011): 1278–92.

48 Thomas J. T. P. Van Den Berg, »Importance of Pathological Intraocular Light Scatter for Visual Disability,« *Documenta Ophthalmologica* 61, no. 3–4 (1986): 327–33.

49 Michael and Bron, »The Ageing Lens and Cataract.«

50 Michael and Bron, »The Ageing Lens and Cataract.«

51 Michael and Bron, »The Ageing Lens and Cataract.«

52 Leo T. Chylack, John K. Wolfe, David M. Singer, Cristina M. Leske, Mark A. Bullimore, Ian L. Bailey, Judith Friend, Daniel McCarthy, and Suh-Yuh Wu, »The Lens Opacities Classification System III,« *Archives of Ophthalmology* 111, no. 6 (1993): 831–6.

53 NEI Office of Science Communications, Public Liaison, and Education, »At a Glance: Cataracts,« accessed October 24, 2019, https://www.nei.nih.gov/learn-about-eye-health/eye-conditions-and-diseases/cataracts.

54 Ralph Michael, Laurentius J. Van Rijn, Thomas J. T. P. Van Den Berg, Rafael I. Barraquer, Günther Grabner, Helmut Wilhelm, Tanja Coeckelbergh, Martin Emesz, Patrik Marvan, and Christian Nischler, »Association of Lens Opacities, Intraocular Straylight, Contrast Sensitivity and Visual Acuity in European Drivers,« *Acta Ophthalmologica* 87, no. 6 (2009): 666–71.

55 Pupil Labs GmbH, »VR/AR,« accessed October 24, 2019, https://pupil-labs.com/products/vr-ar/.

56 Krösl, Bauer, Schwärzler, Fuchs, Suter, and Wimmer, »A VR-Based User Study on the Effects of Vision Impairments on Recognition Distances of Escape-Route Signs in Building.«

57 Austrian Standards, »ÖNORM A 3012: »Visuelle Leitsysteme für die Öffentlichkeitsinformation,« 15 March, 2021 (Vienna: Austrian Standards, 2021).

58 Katharina Krösl, Carmine Elvezio, Matthias Hürbe, Sonja Karst, Michael Wimmer, and Steven Feiner, »ICthroughVR: Illuminating Cataracts through Virtual Reality,« *2019 IEEE Conference on Virtual Reality and 3D User Interfaces (VR)* (Piscataway: Institute of Electrical and Electronics Engineers, 2019), 655–63.

59 Katharina Krösl, Carmine Elvezio, Laura R. Luidolt, Matthias Hürbe, Sonja Karst, Steven Feiner, and Michael Wimmer, »CatARact: Simulating Cataracts in Augmented Reality,« *2020 IEEE International Symposium on Mixed and Augmented Reality (ISMAR)*, (Piscataway: Institute of Electrical and Electronics Engineers, 2020), 682–93.

Life and Biology

Peter Mindek

What does life look like? Think of the most fascinating thing you can imagine. It looks something like that.

I love being alive. It's one of my favorite things. All the interesting stuff that has ever happened to me only happened to me because I was alive at the time. I can recommend it. And it's not just me—looking at pretty much everything else that is alive, it all more or less seems to feel the same way. Everything eats and procreates and dodges cars just to keep that living thing going.

Most people have asked, at some point during their lives, what *is* life? The answer is, *It's complicated*, and so they let it go. So did I. That's usually the point where life becomes too abstract to invest any mental energy into it. Life just is, we have it, now get over it and think about sports and shopping and houses instead. Living can be awesome, but questioning life itself is dull.

Or is it? There's an entire scientific field devoted to studying life. And it's not just some academic niche that no one except a few nerds cares about—*biology* actively participates in making living much more comfortable for all of us. It helps us to get to know our bodies and how they work, to understand the diseases that attack us and find remedies and cures for them, and to understand our environment and how to take care of it. All of this makes our lives longer and more enjoyable. And who wouldn't want to know everything about that, right?

Well, as it turns out, most people. Our first encounter with biology comes at a very delicate age, when we are elementary school subjects. And as such, biology is presented in a pretty boring way. Rather than focusing on all the awesome stuff and understanding how life works, most schools focus on memorization (at least that was the case when I was growing up; luckily, new education concepts are now becoming more and more commonplace, and that is a very good thing). And to engage children, memorization is a really suboptimal strategy. There are two types of kids—those with bad memories and those with good ones. Those with bad memories quickly get frustrated, while those with good ones won't see the point of training them and will get bored.

Indeed, in school, biology is reduced to the most boring part of any scientific field—taxonomy. Of course, as an adult, I understand the importance of categorizing and classifying knowledge, but as a twelve-year-old kid, I really didn't care about memorizing the names of species and their body parts. The logic behind presenting biology to small kids in this way seems sound—kids care about animals! *For sure* they will want to learn more about them. *For sure* they will want to know what other animals exist and how they are related—and most importantly, how we, the glorious adults, have made sense of it all by putting all of them into a number of boxes.

Well, not really. More than animals, children care about *stories*. Kids are empathetic; they care about what the animals are up to, not what class they belong to. Kids care about animals because they know them as the main characters in their favorite stories. They want to know if the hedgehog managed to get the apple into his tiny house after all; they couldn't care less that a hedgehog is a mammal from the Erinaceidae family (by the way, I had to look that up right now as well). Dinosaurs illustrate this quite nicely. Since they are all dead, we adults don't think it's that important to teach kids how we classify them. And so, all we tell kids about dinosaurs are stories—how they roamed the Earth and ruled it with their mighty size and sharp claws until the universe decided to end their empire once and for all with a single blow (as teenagers, they learn that dinosaurs were basically chickens, but by then that's a very cool story too). Kids love that kind of thing. As a result, dinosaurs rank among the top animals for prepubescent kids worldwide.

But the adults are kind of right on this one: knowing about dinosaurs is not as useful as knowing some other things. For instance, knowing how our own bodies work can have a direct impact on our quality of life. Understanding the biology of the human body lets us treat it right and give it all it needs. It's kind of like with a car. Sure, when you don't need to understand how it works, somebody else will fix it for you once it breaks down. Here and there, you might need to exchange a part, but eventually, the car will break down beyond repair, and you will have to buy a new one. Knowing a bit about the engine lets you treat it right, and the car will last longer until that happens. Same with your body. Only you can't buy a new one.

The human body is a never-ending source of stories to be told. For example, the one where we are made of cells. Cells are where all the magic of life happens. We can't see them, so explaining how they work takes additional effort. Seeing something and then being told what it was is quite different from hearing about something we have never seen before. After all, why should you care if you can't even see it?

Albert Barillé did a particularly brilliant job in his animated series *Once Upon a Time... Life*,[1] which explains how the human body works on the cellular level. It breaks this incredibly complex, invisible world down into stories in which the cells of our body are the main characters. They run around, solve problems, and enjoy their day-to-day lives. Now we're talking! Tell me more!

Speaking about life. Do you know what else is complicated? Fractals. Fractals are geometrical shapes that maintain their complexity no matter how far you zoom in on them. Imagine an island that is shaped like a potato. Easy enough. Now zoom in. The coastline becomes a pretty complex curve, revealing the island to be a much more complicated shape. Keep going. Zoom in to see individual stones forming the coastline, giving it a completely new level of complexity. Keep zooming. The surface of each stone turns out to be pretty rough when you look at it up close, meaning that the island isn't a basic potato at all. If you follow its coastline really closely, it might be one of the most complex shapes you have ever come across. We're not done yet. Zoom in. See the individual atoms the stones are made of. Crazy! What's the opposite of basic again?

Life is just like that. When looking at Earth from space, we see a colorful marble of green and blue. Zooming in, we see oceans, forests, mountains, and deserts. Mountains sound pretty good, so we take a look there, see a goat flexing its climbing abilities on what seems to be a perfectly vertical wall. Look closer and you see a tick wandering Planet Goat, searching for the perfect spot to take a bite and suck on that sweet, sweet protein cocktail. Keep going closer and you'll see the cells the tick is made of — mindless biological computers that fulfill the tasks they have been programmed to carry out like clockwork. Get even closer and you'll see that a cell is actually an enormous city of protein molecules interacting with each other, doing things that make you question what the nature of free will is and all sorts of other things. Going even further, you'll see atoms bonding with each other, driving those molecules, leaving you with a simple message: life is just like a fractal. It's everywhere, on every scale, every order of magnitude, and its complexity doesn't change, no matter how much we zoom in.

And that's exactly what the problem is with explaining biology through stories. It's not like nobody has thought of that before. It's just very hard to build a story around something that has so many layers. Where would you even begin? Sure, it's possible to explain any of those layers individually, but putting them into perspective without it being overwhelmingly confusing is no easy task.

Fractals fascinate me. So does life. In school, I had the best biology teacher you can imagine. She was smart and caring, and she was able to talk in an extraordinarily engaging way. Still, her job was to teach us taxonomy and so, despite the perfect setting, I didn't care about biology at all. Why couldn't her job be to teach us about dinosaurs or robots or computers?

Funny thing is, that's exactly what she was doing. I just didn't get it. It wasn't part of the syllabus for her to tell me. I had to learn the names of individual organelles instead of learning that the cell is the most complex *computer*, which works thanks to its own army of *nanorobots*, orchestrated by an ancient *program code* that was built over generations, containing within itself everything from our present to our past. If only she had been allowed to tell me that...

Fast-forward ten years: I studied how to use computers to produce images of very complex things. Medical scans and such. Thanks to this expertise, I was hired for a project where I was supposed to work on displaying a model or a virus. It was a really complex model that contained all the atoms the virus is made of. Looking at such a model for the first time is kind of like looking at Google Earth for the first time — at first glance, you go, *Neat, the Earth*! But then you realize you can zoom in to see your city, your street, your house — and your mind is blown. I looked at the virus model and I thought to myself, *Neat, a virus*. And then I zoomed in...

Life is complicated. It's definitely worth telling stories about. But saying that life is complicated is pretty abstract. One person imagines taxes, one person imagines relationships, money, society, or politics. Few people imagine molecular machines driving the processes inside our cells that ultimately make us experience this wonderful thing we call life. And that's a pity because few things are more mind-blowing than that.

> *Life is complicated because we cannot see it. It is definitely worth showing.*
>
> *Seeing that virus model, I knew I wanted to tell stories about life from the perspective of the nanoworld. I want to tell people* what it looks like, what's going on in it, *and* what the driving force is behind it. *I want other people to experience that moment I had when I got a glimpse into this invisible world we are all made of. And to truly experience it, you have to see it.*[2]

That's why I like to tell my stories visually. Human imagination is a wonderful thing, but if you want to let it wonder about something real, you better show it to it. By *showing* what we want to tell, we are setting the bounds of reality, within which the mind can navigate. That puts the mind at ease, it can safely float and enjoy the ride without the risk of getting lost, all the way until it gets successfully blown.

Virus—Part 1

We've talked a lot about life, which is funny because we are going to start with something that isn't even alive—not really anyway. A virus is a parasite that needs a living host cell to replicate itself. Outside the cell, a virus exists in the form of separate particles called virions. Virions are nothing but genetic material wrapped in a coat of proteins (and sometimes lipids coming from the membrane of the last cell the virus infected). They have no metabolism and, on their own, they cannot replicate. That means we don't consider them alive.

Virions are not aware of their surroundings; they don't know where they came from or where they're going. Well, they don't know anything. It's when they randomly happen to bump into a compatible living cell that they start to act out. When they touch a certain protein on the surface of a cell (which consists of a fatty membrane with a lot of protein molecules in it), the cell proactively takes them inside (cells, despite being alive, also don't know anything; you can imagine them as biological computers that simply do as they're told—and if something attaches to their membrane, they are instructed to take it inside, and that's that).

Normally, a cell continuously produces the parts it needs to sustain itself and eventually multiply. The cell is a piece of incredibly complicated machinery that follows the programming written in the genetic code stored in its nucleus. The cell produces all the necessary protein molecules and transports them to where they are needed. Once a virus gets inside the cell, the machine is compromised. The virus carries its own genetic code (in the form of nucleic acid—either DNA or RNA) that it uses to reprogram the cell's machinery. It forces the cell to start producing parts of the virus, which later assemble into new virions, until the cell's resources are completely depleted, and the cell dies. Its membrane gets ripped up, and the new virions are spat out with an opportunity to bump into a new cell and repeat the whole cycle again.

Since their business model is hijacking cells to do their living, viruses are much simpler than any unicellular organism that has to do its metabolizing and replicating on its own. Because they are relatively simple, it is possible to use current technology to build a detailed model of an entire virion (a virus particle before it infects a cell), right down to its individual atoms. Such models are very useful — you want to know your enemy down to the last atom.

Our scientific collaborators built this kind of precise model of an HIV virion (human immunodeficiency virus; the virus that causes acquired immunodeficiency syndrome, or AIDS) according to the latest biological knowledge. Their problem was that the model was huge. Can you imagine how many atoms a single virus particle is made of? The answer is *a lot*. It takes a long time for the computer to display such a huge model on its screen. They wanted to do all sorts of things with the model, and not being able to display it quickly would have been pretty annoying, or even prohibitive to their research. That's where the research unit I was working in at the time came into play.

We knew how to instruct computers to display huge datasets very quickly, and so we developed software that could show this virus model in real time. That allowed the scientists to not only look at it but also interact with it. And they could *tell stories* with it. Because the more people know how the virus works, the better the chance that somebody will figure out how to end it.

Our software was a success story, but we didn't stop there, and we kept on improving it. We kept making it faster and its output clearer and prettier. It was able to generate stunning illustrations of what we call a *mesoscale* — things that are smaller than cells (the microscale), but larger than protein molecules, which are the basic building blocks of everything that is alive (nanoscale). An example from the mesoscale is a virion — a tiny particle way smaller than a cell, consisting of thousands of protein molecules. Illustrating it by hand in all its molecular detail is extremely difficult and time-consuming, so letting a computer do it was a good idea.

And then the COVID-19 pandemic came. In the beginning, it was very confusing, and it was important to find efficient means of fighting SARS-CoV-2, the virus that caused it. A lot of scientists immediately started studying the virus in order to learn how exactly it works, so that they could figure out a way to stop it.

AN INTERMISSION_ By now, you probably already know that I love telling stories and, at this point, I believe it's important to tell you one about how life works on the molecular scale. I am a programmer, not a biologist, so it's something you should probably really ask your local biologist about, but until then, I'll give it my best shot.

The basic building blocks of life are protein molecules. There are many types of proteins, and they all have different shapes. You are probably familiar with molecules like water — H_2O — which is made from two hydrogen atoms and one oxygen atom. Proteins are much larger molecules than that. They are made of hundreds, sometimes thousands of atoms. They are long chains that, upon creation inside a cell, fold into complicated, three-dimensional shapes. These shapes are crucial. Everything that a cell does happens because the proteins inside of it are interacting with each other. Two proteins with compatible shapes bind together, changing their shape to become one protein that matches another protein, which then binds to them, changing the shape again… These cascades of protein interactions, called *pathways*, are what make the cell alive. And cells being alive make us alive. All of this is possible because certain protein shapes fit together like pieces of a jigsaw puzzle. On the molecular scale, the *function* is the function of *shape*. In the molecular world, knowing the shape is the first step to knowing how things work.

Back to the SARS-CoV-2 virus: now we know that, on the molecular level, shape matters a lot; that's why scientists made a molecular-scale model of the virus to study the shape of its proteins, and the shape of the whole virus particle. To study this model, we used our software to create very detailed renderings of this new virus to help scientists study their theories about its 3D structure.

Theories, you ask? Well, yes. You cannot see a virus because it's too small. Even with the most powerful microscope, we can only make out its rough shape, but we cannot see enough details to resolve its internal 3D structure. And it is crucial to know what that looks like in order to understand exactly how the virus works. Scientists have to figure out the structure indirectly during various experiments. Afterward, they put together all the data they have collected and come up with theories of how all of this could make sense and what the internal structure of the virus might look like by fitting together all the knowledge they have gathered. The result is a 3D model that we help to visualize. By studying the model, biologists can determine whether what they can see or can't see makes sense. The measured data is solid so, if the 3D model doesn't seem to make sense, they change their theories, make a new model, and look at it again. Now you see why tediously illustrating every iteration of the virus model would not be such a great idea. Just let the computer do it!

The only issue is that computers are dumb. You can't say, *Draw this model for me!* to a computer and expect it to figure out what it is that you want. You can't tell a computer to *Draw a bunch of atoms!* You have to tell the computer how *exactly* it has to do it. And you have to say it like you're telling someone really, really dumb.

ANOTHER INTERMISSION_ How do you draw an atom? Let's discuss this in a bit more detail. We know what things look like. When we encounter something new, we can examine it with our senses and figure out its shape. After that, we create a mental model of the thing and remember it. Once someone mentions it or shows you a picture of it, you immediately recognize the object and can imagine its 3D form, even though you can't thoroughly examine it this time.

For some things. though, our mental model is based solely on observations made using instruments. Some things we will never be able to examine directly. We will never be able to walk around enough to understand that the Earth is round. We have had to build instruments and conduct experiments that have shown us that the Earth is indeed round.

Or that thing with your face: you will never be able to see your own face directly. You might have heard that a lot of people don't like themselves in photographs. That's because their mental model of their face is created from the reversed image that they see every day in the mirror. And since nobody is perfectly symmetrical, the photographs, which are not reversed, seem a little bit off to them. In fact, most front cameras on phones mirror the image they capture so that people will like their selfies better.

With all that in mind, what do atoms look like? Well, first of all, we will never be able to see atoms directly. They are smaller than the wavelength of visible light and this essentially means that our eyes, which use the light that reflects off things to see them, will under no circumstances be able to resolve individual atoms. Special electron microscopes can be used to *visualize* (but not really see) individual atoms, and in these images, atoms look like little blurred spheres. This checks out, doesn't it? Because from what we have learned in school, atoms are supposed to be little spheres consisting of a nucleus and a bunch of electrons orbiting it, right?

Well, in reality, it's not that simple. Atoms are tricky, and the world on the atomic scale doesn't quite work like the world around us that we are used to. Atoms indeed have a nucleus that is orbited by electrons, but in a weird, *quantum* way. Electrons form strange probability clouds around the nucleus that look more like balloons and doughnuts than spheres. All of this is pretty complicated, and so, in school, when they first teach us about atoms, they just tell us that they are little spheres. They need us to build *some kind of* mental model, and in this case, reality would just be a bit too confusing.

On top of that, we don't have any means to visualize the electron orbitals the way they actually look; we can only calculate their shapes mathematically. What we can visualize using electron microscopes and other techniques is where the atoms are, and in those visualizations, atoms look like blurred spheres. In the end, for most people, atoms are little spheres.

And so, the answer to the question *How do you draw atoms?* is *It depends* — when you try to explain what an atom looks like, a picture capturing the real shape of the electron orbitals might be appropriate. But when you try to explain how life works on the atomic scale, the best way to show atoms is to follow the mental model that your audience has of an atom and portray the atoms as little spheres. Otherwise, the risk is that your audience will get that strange feeling they get from looking at non-reversed photographs of their faces. The weird-looking atoms will distract them from what you actually want to tell them. They will ask, *What on Earth are those weird balloons and doughnuts?* instead of focusing on the fact that all those atoms form the molecules that ultimately make you alive. This is why, when we want to draw atoms in stories about biology, we draw them as spheres, even though they are not.

Mesoscale biological models are essentially lists of atoms. The model has a record for every atom that tells us exactly where that atom is, what chemical element it is, and what biological structure it belongs to (for instance, a protein molecule). To visualize such a model, we start by telling the computer that we need it to draw a sphere at the 3D position of each atom.

Computers have a long history of drawing things. For them, drawing is nothing more than calculating, or computing if you will. That's all they can do anyway. It all started with 2D pictures that were plotted via printers and later with various displays. But soon after that, people wanted to draw 3D objects from the real world. That involves calculating how light would bounce off various surfaces, what parts of them would be in shadow, and what parts would be lit. After all, everything we have ever seen in real life has only been thanks to the light bouncing off it, going through objects, and ending up in our eyes. And our brains have adapted to recognize the 3D shapes of things based on how light interacts with their surfaces. So, if you calculate the amount of light reflected from each point in your scene and tell the computer to plot those points on the screen, you get something that people will recognize as a 3D world.[3]

The next thing you need to take into account when calculating what a 3D world looks like is that visible light is made of different wavelengths. And each material reflects different wavelengths of light differently. We perceive these different wavelengths that make it into our eyes as different colors. So, when we plot a 3D model, we need to tell the computer what hue each surface should be. Together with the lighting information, the hue is used to compute the resulting color in each point in the 3D world that we are drawing.

Which bring us to the question: what colors should those spheres representing the atoms be? Well, let's start by saying that atoms don't really have colors as we know them from looking around us (and they are not really spheres either). Since they are smaller than the shortest wavelengths of visible light, it doesn't make sense to talk about realistic colors. Instead, we can do one of two things: either we can mimic the instruments that allow us to »see« the atoms and make them black and white (that's how electron microscopes usually plot their scans made using electrons instead of photons of visible light), or we can invent artificial colors that have nothing to do with reality but that communicate some additional information to the audience. A good strategy is to use different colors to tell what type of atoms they are. When we look at those spheres representing atoms from up close, we can color them according to the chemical elements they belong to. And when we look at them from further away, we can simply assign different colors to each protein type and color the atoms accordingly. In this way, when looking at the resulting visualization, the audience will be able to distinguish individual molecules from each other much easier since they will all have different colors.

Besides light and shadow, there is one more thing that allows us to understand the three-dimensional world around us. Thanks to the way the lenses in our eyes bend incoming light, objects that are further away appear smaller to us. And not just objects — distances between objects appear smaller as well. So, if we look at two objects, and one of them is further away from us than the other, their relative position to each other will appear to change as we change the viewpoint from which we are looking at them. This is called parallax, and it helps our brain figure out how far away things are. (This is particularly handy since we have two eyes with overlapping fields of view. If we look at something, we can easily estimate how far it is relative to the other things around it, as each eye has a slightly different viewpoint, and our brains can compare images from both eyes and do all the math. Pigeons have eyes on the sides of their heads, which gives them a greater overall field of view, but each eye sees a different part of the world. So, in order for them to use the parallax effect, they have to bob their heads back and forth to look at things from slightly different viewpoints to be able to estimate how far away things are.)

We did all of this for the SARS-CoV-2 model.[4] We placed it in a virtual mesoscale world that modeled the environment a virus would commonly exist in — near the surface of an epithelial cell inside a human's respiratory system. And it looks like_FIG. 1.

To summarize: in order to draw a picture with a recognizable 3D mesoscale model, in the place of each atom you draw a sphere (which in reality they are not), calculate which parts are in shadow and which parts are illuminated (which in reality they cannot be), assign it a color (which in reality it doesn't have), and project everything onscreen in such a way that atoms that are further from us are smaller and closer together, just like we would see them with our own eyes (which in reality we cannot).

Does that sound like it doesn't make any sense? Perhaps. Does the mesoscale actually look like this? Absolutely not. The mesoscale doesn't *look* like anything. Remember, everything we've ever seen in our lives has been light interacting with the objects and materials around us. All of the mental models that we use to recognize shapes are based on light coming into our eyes. Protein molecules cannot interact with light the same way as the objects on our scale do; they cannot reflect it to produce different colors, shadows, and highlights. But we want to show their shapes, because, as I said before, shape is everything. And so, we instruct the computer to treat atoms like objects from our world, at our scale, and draw them as such. We invent their colors and make them easily distinguishable. And before our eyes, molecules appear, with their shapes deeply rooted in reality and visualized in ways that make the scene understandable for anyone. We cannot capture a photograph of the nanoworld, but we can *show* it.

Virus—Part 2

The previous sentence might sound a bit disappointing. We can't take a photograph of the nanoworld? Why not? Well, once again, it has all to do with light. A photograph is a recording of light reflecting off objects and being focused by a lens on a projection plane, where it creates an image of what the lens is looking at. But light can only resolve things that are roughly half the size of its wavelength. The shortest wavelength that we can see is around 380 nanometers, but a virus particle is only roughly one hundred nanometers across. So, it sounds like photographs are a bit impossible.

Except, we don't have to use visible light to capture a virus. We don't have to look at it directly. We can use photons of much shorter wavelengths and detect them using special detectors that will reconstruct the image. The computer will then assign colors to individual measurements and plot an image of things that we could never see with our own eyes. And we can do one more thing: when capturing the image, we can place the specimen on a tiltable table and capture a series of images with the specimen tilted at different angles. We can then use this series to reconstruct a 3D scan of the specimen.

This sounds pretty straightforward, but actually, it isn't. First of all, if we did this, the electrons would be scattered by the air molecules surrounding the specimen, and we wouldn't get a clear picture. We can fix that by placing the specimen in vacuum. But biological specimens contain a lot of water that would boil off in vacuum, destroying the specimen. The solution is to freeze the specimen very quickly to a very low temperature. If this is done fast enough, the water molecules don't have time to form the crystals that would again destroy the specimen.

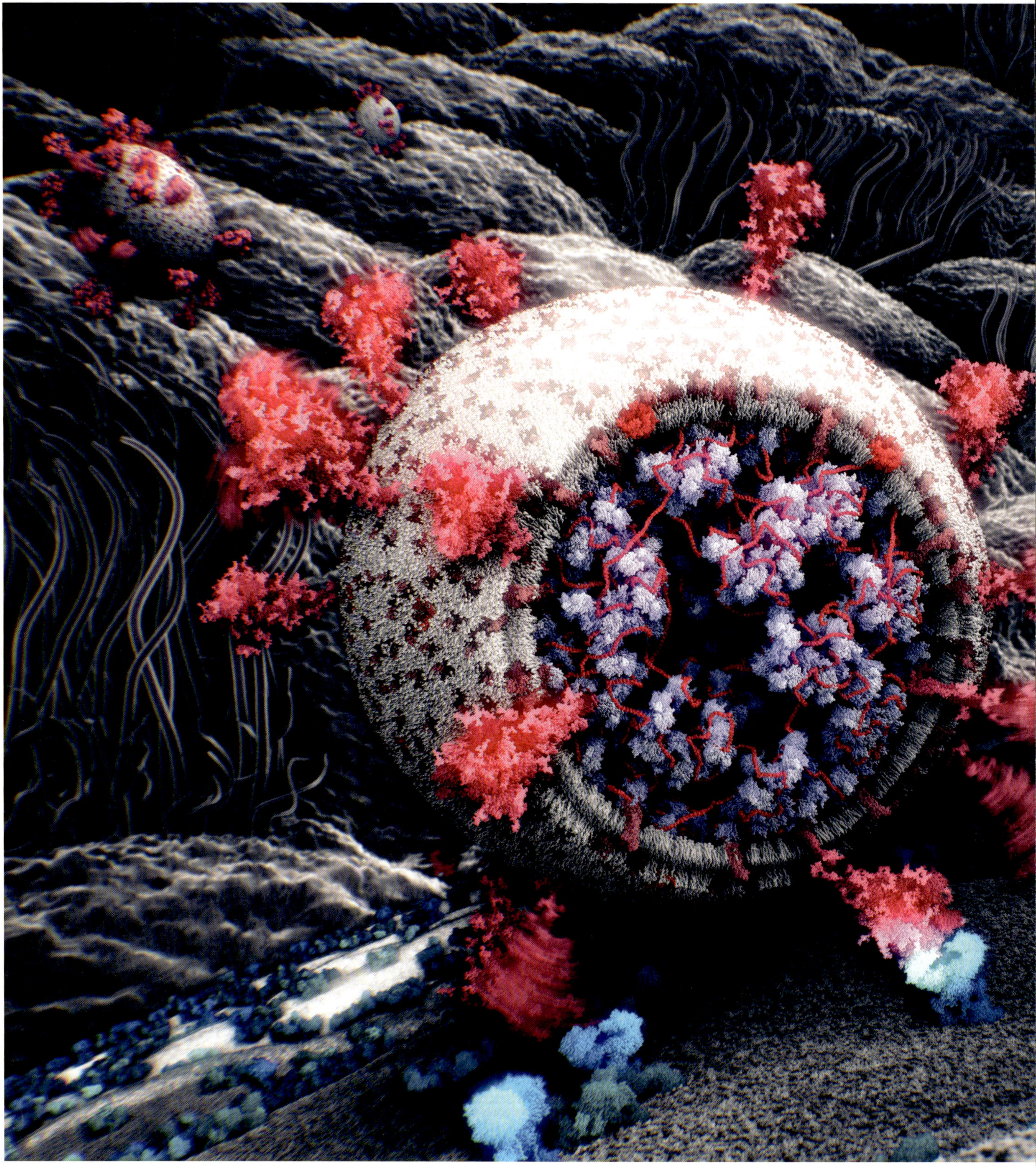

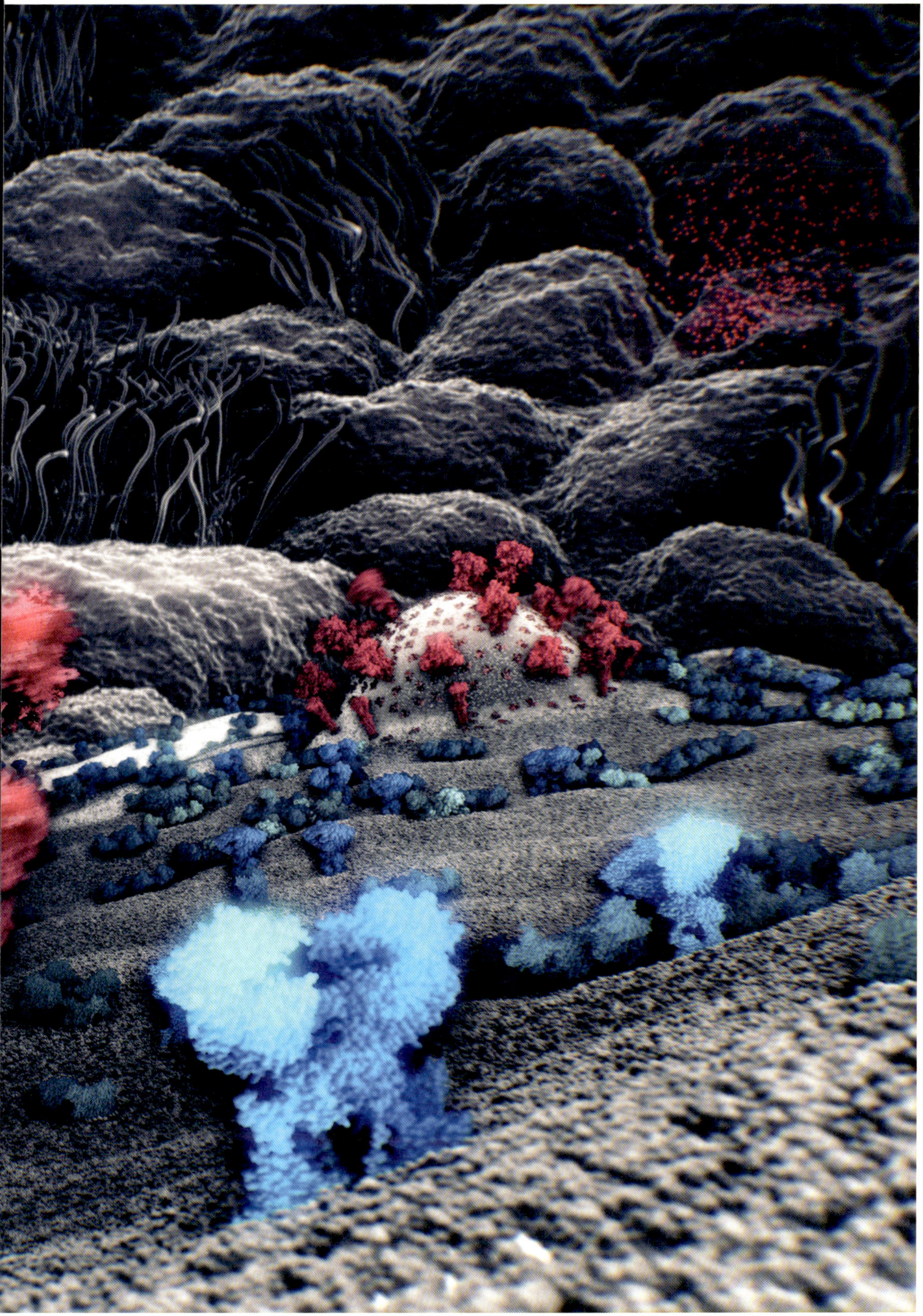

FIG. 1_*SARS-CoV-2 virion attacking an epithelial cell: a visualization in atomic resolution of a SARS-CoV-2 virion entering a cell by fusing with its membrane. The spike proteins of the virus (marked red) attach to the membrane proteins of the cell (marked blue) to initiate this process.*

VIDEO_*Making of*

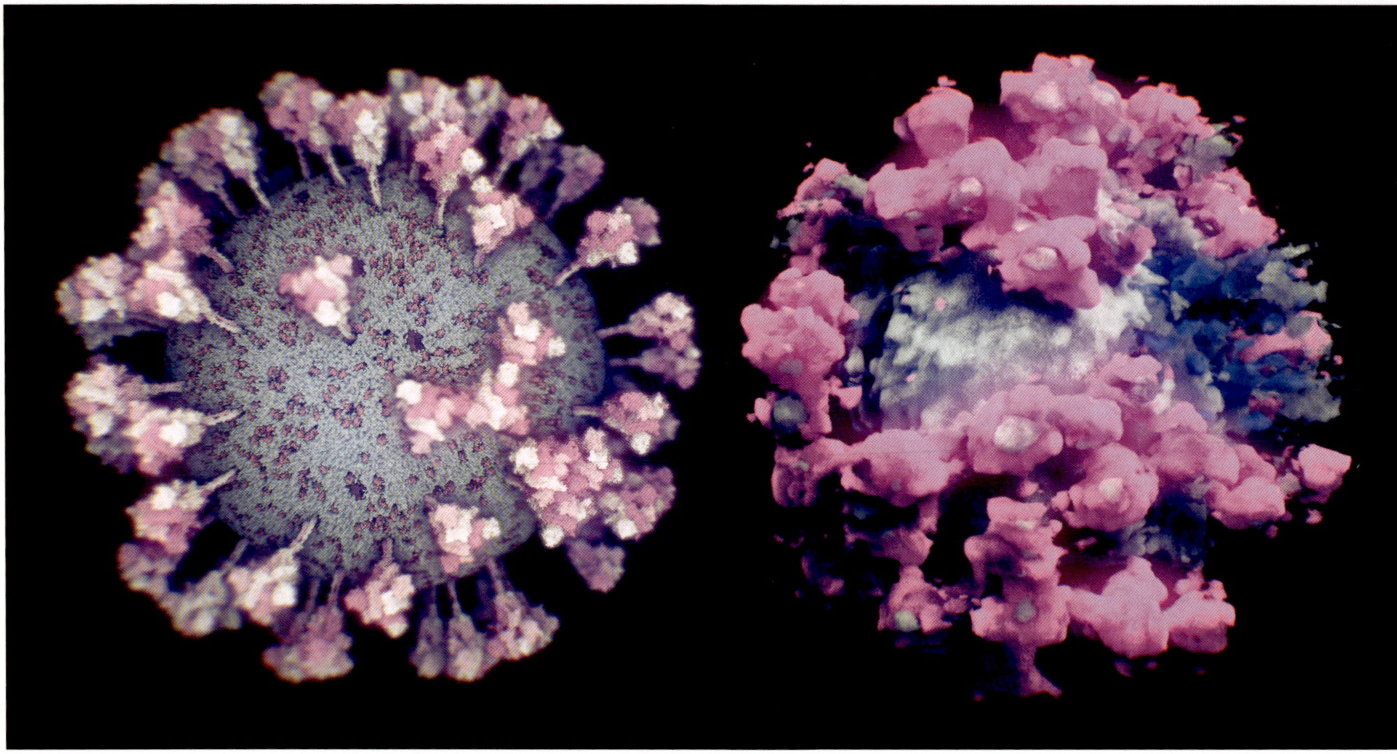

We worked with scientists who did all of this to capture a 3D scan of a real SARS-CoV-2 virion. They gave us the scan, and we tried to display it the same way as we display medical CT or MRI scans. We could see the virus, but the results were pretty underwhelming. We had previously created those beautiful visualizations of the atomistic models of viruses, and now, when we wanted to show a real scan or an actual virus particle, we were only able to see a noisy, black-and-white image that only very roughly revealed the shape of the virus.

Learning more about the biology of the virus isn't the only reason we make images of it. When the global COVID-19 pandemic started, people quickly understood that it had been caused by a virus, and they understood that this virus was something very, very small. Small enough that they could not see it. But seeing it helps us to realize this is a real disease caused by a real pathogen. It exists in our bodies, and it moves from person to person with a single purpose — to multiply.

This is why we, and many other people, built 3D models showing what the virus looks like. We put a face to the invisible enemy, and we showed people what it is that we are up against. But artificial models weren't enough. Without studying all the scientific papers that inform the creation of such a model, it is very abstract for people to see that this is indeed what the virus looks like. After all, it is only a man-made model of the virus, and a model can be wrong. To confirm that the model correctly represents reality, we wanted to put it side by side with a scanned image of a real virus particle cleared of all the noise and other artifacts introduced by the scanning process. In this way, it would be possible to directly see if our model has the same shape and if it exhibits the same features as the real thing.

FIG. 2_ *The first 3D visualization of a real SARS-CoV-2 virion: a comparison of a SARS-CoV-2 virion model in atomic resolution (left) with a rendering of an electron microscopy scan of a real virus particle (right). As colors as we know them do not exist on this scale, electron microscopy scans are only black and white. We augmented the scan with artificial colors to distinguish individual parts of the virus particle.*

VIDEO_ *SARS-CoV-2 virion*

My part in all this was to develop a computer algorithm that simulates the way light travels through semi-transparent media, such as gelatin. That might sound pretty arbitrary, but I did it for the same reason that we wanted to display the virus model. The data captured by an electron microscope can be interpreted in many different ways. When scientists look at it, they want to see a representation as close to what was actually captured as possible—they want to see raw data. Raw data from electron microscopes, as I mentioned before, are noisy and lack any color. While these data are useful for scientists, laypeople would not be impressed. Most of them would not care for such an image much because they wouldn't see a lot in it.

That's why we introduced the same things we had used with our virus model to help people understand it—light, shadows, and colors. These things do not exist in the nanoworld, because light as we know it doesn't exist there. But by applying them to real microscopy data, we can generate images that precisely describe reality in a way people can easily comprehend. Here's a comparison of our virus model and the scan of a real virus particle displayed with the algorithm that I helped to develop.

We applied the same artificial colors as in our virus model to highlight the relationship between the model and the real thing. Virus particles, like everything else in the nanoworld, don't have any colors. But coloring different parts of it differently makes it much easier to look at_FIG. 2.

And that's how I showed what the nanoworld looks like.

Microtubules

Do you remember how we had to freeze the specimen in order to capture an image of it? Indeed, to this day, we don't have the means to directly observe how things in the nanoworld move. That's a pity because the movement of molecules has direct consequences for all the biological processes that keep us alive. Life starts with the shapes of molecules, but the way these molecules interact with each other is where the magic happens.

We can learn a lot about the movement of molecules from various experiments and observations. For instance, scientists sometimes take an electron microscopy image of a protein molecule that they think moves in a certain way. To do so, they have to freeze it, effectively disabling all its movement. Only they don't take a picture of just one molecule—they take a picture or many molecules of the same type. All of them perform the same movement, but when flash-frozen, they are all in a different phase of the movement. Scientists then isolate the images of the molecules in different phases and manually sort them to create an animation of the movement. Every frame of that animation will, in fact, be a different molecule, so it's far from perfect. But since the molecules of the same protein are pretty similar, this technique gives us a good idea of how the molecules move.

One interesting example of moving things from inside our cells is *microtubules*. They are long, hollow tubes that form scaffolding within a cell, helping to maintain its shape. They also act as telegraph lines that cells send signals through, allowing different parts of the cell to communicate with each other. Moreover, microtubules act as railroads that cells move materials along to where they are needed. When I said cells are like cities, I had good reason. And microtubules play a big part in that.

What makes microtubules such a flexible system is their ability to quickly assemble and disassemble according to the cell's needs. They are made of small molecules of a protein called tubulin. Tubulin molecules float around in the cell's cytoplasm — the liquid full of protein molecules that fill the empty space inside the cell. The tubulin molecules tend to link up into long chains called protofilaments. A microtubule is formed as several protofilaments connect together into a flat sheet that bends until it forms a hollow tube that closes up continuously like a zipper. When the microtubule is no longer needed, its protofilaments split up, and the microtubule falls apart.

FIG. 3_*Microtubule catastrophe*

VIDEO_*A procedurally generated model of a microtubule inside the densely crowded environment of the cell's cytoplasm*

No clear recordings of this process exist. There have been imaging studies that suggest that this is how microtubules grow and shrink, but nobody has ever seen them in action. When I first heard that, I knew our visualization technology would be perfect for showing this process in a realistic way. All we had to do was to model a microtubule and model the movement of the tubulin molecules. That might sound easy, but it really isn't.

To model something, we first need to understand it. So how do these molecules know how to form such complicated structures? Well, they are molecules, so they don't know anything. They don't have a goal or will, they just are. Their movement is random, and it is caused by other molecules bumping into them. And it happens very, very frequently — several billion times a second. Let that sink in. This is because the inside of a cell is very busy. Not only is it crowded with molecules, but the molecules are also moving very fast. If there was nothing in its way, a small molecule would move from one end of a cell to the other in one millionth of a second. But since it's so crowded, a moving molecule immediately bumps into something, and so its movement is more like a random walk. But despite that, it can still move across the entire cell in one-fifth of a second — while constantly colliding with other molecules, which change its direction. So, as you can imagine, a molecule visits many places throughout the cell very quickly. Sooner or later, by pure chance, it will be pushed to exactly the spot where it is needed, where it will chemically bond with another molecule to fulfill its destiny.

Computers are fast too, but not *that* fast. We could simulate all this, and eventually we would end up with a virtual microtubule slowly forming in our simulated environment. But this would take years. The fastest supercomputers today can only simulate a few nanoseconds of molecular interactions within a reasonable amount of computing time. And so, if we want to create a computer animation of a process as complicated as the formation of a microtubule, we have to simplify it a lot.

Still, we want the animation to be realistic, otherwise, what would be the point? And realistically, a human could not watch individual molecules move around as they react together and form larger structures — it would be one large blur. What I ended up doing was modeling the results of all that random molecular movement — the closing sheet of protofilaments forming a tube. I wrote down a mathematical formula that describes the positions of the molecules that form this structure — quite a nerdy way of creating an animation, but it has its advantages. For instance, after I showed it to biologists, they were able to tell me that certain aspects of it should be modified according to the latest research, and all I needed to do to update the animation was to change a parameter or two in my equation instead of manually redoing the animation from scratch. Sometimes being a nerd pays off.

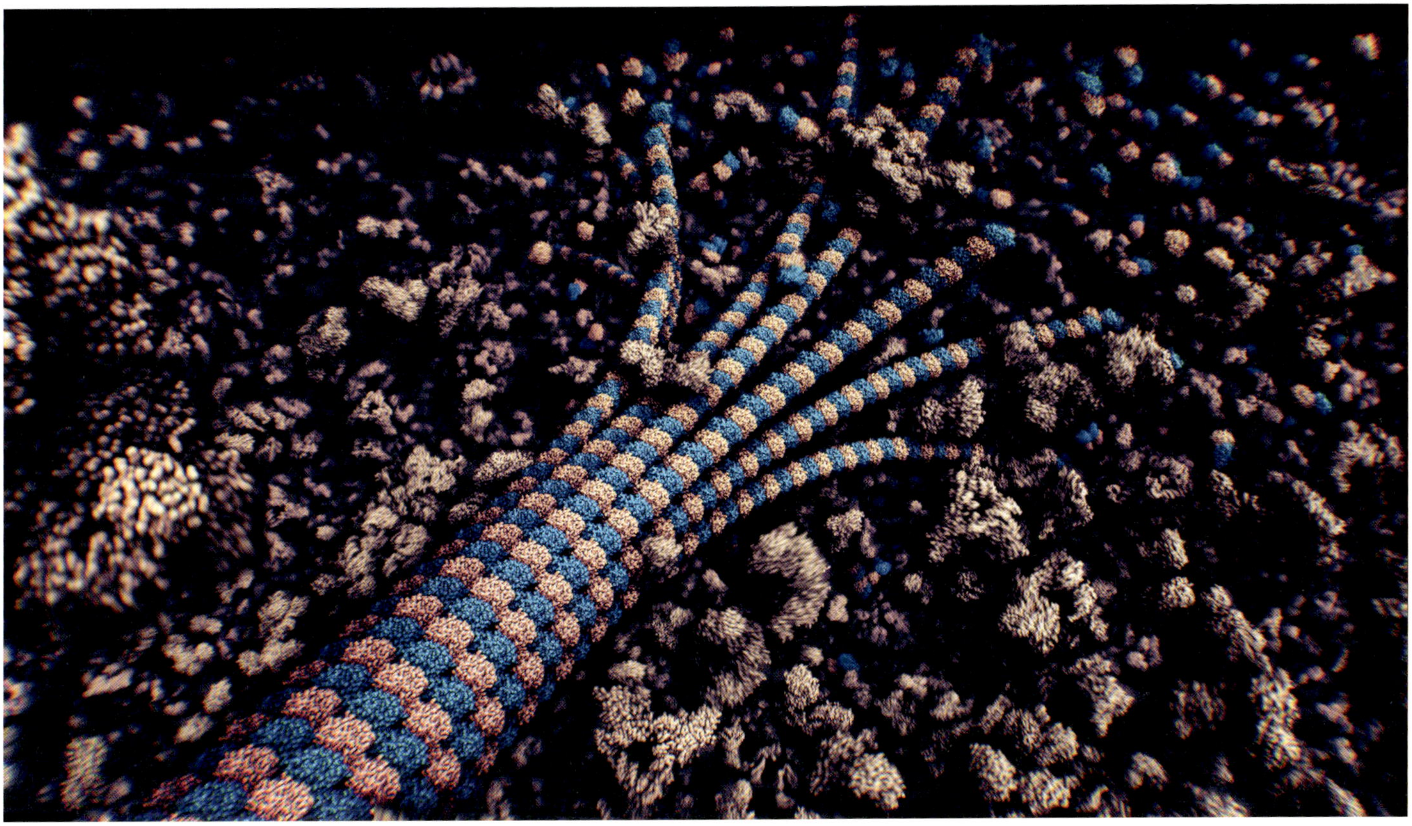

I also modeled the random movement of the tubulin molecules. Instead of using a realistic simulation, I modeled the movement using a simple noise function that makes the molecules move randomly. People looking at this animation, which explains a biological process, will not be able to look at individual molecules and keep track if the molecules are realistically bouncing off of each other (a billion times per second). All they need to see is the chaotic movement that results in an organized process, like the growth of a microtubule. So, what I did was let the molecules randomly wander the 3D computer-generated world and, one by one, transition into the structure described by my mathematical formula. The result was a realistic-looking visualization of what scientists currently believe is how microtubules—the skeletons of our cells—assemble themselves_FIG. 3.

And that's how I showed what's going on in the nanoworld.

DNA

Life is like a fractal, and its complexity never decreases, no matter how much we zoom in. We run around having fun and experiencing drama, doing chores, working and eating and sleeping. If we take a closer look, we see our organs, pumping blood, processing food, blinking, and turning songs into earworms. Those organs are made of cells that feed on the stuff we feed on and turn it into energy; they multiply, synchronize to make our organs work, and they kill themselves when they are no longer needed (which might sound sad but, if they refuse — and they sometimes do — we get cancer, and that is even sadder). Each cell is a living city containing many organelles — large structures that are essential to the cell carrying out all its tasks. They are like the institutions that make a city work. Organelles are made of millions of proteins and other molecules that interact with each other in complicated cascades of chemical reactions.

Proteins are made of amino acids, and those are made of atoms. Atoms like to bond with each other because they are held together by forces that force them to. And as the atoms bond, they form molecules, and molecules bond together too, and that's where life begins.

We showed the molecular processes essential for life and how proteins interact to keep cells alive. But it would be interesting to show what drives these processes — where it all begins.

When searching for the origins of life, one molecule stands out. Its name is deoxyribonucleic acid, or DNA for short. It is found in every living organism we have examined so far (some viruses have RNA instead of DNA but, as you remember, they are not really *living* anyway). DNA might be the most essential molecule for all life on Earth. And, arguably, it is the molecule at the beginning of it all.

I mentioned the importance of proteins for life. But where do they come from? Cells make them using complex molecular machines called ribosomes. Ribosomes chain amino acids together in a specific order. This order is crucial because it dictates which shape this chain will fold into when the protein is formed and, as I said before, shape is everything. And the order in which the amino acids are chained together is written in the DNA. DNA encodes the chemical composition of all our proteins and therefore their shape, their function, and the entire mechanism of us being alive. Indeed, DNA is pretty damn essential.

DNA is a long molecule made of a sequence of small subunits called nucleotides. There are four different nucleotides in DNA — guanine, adenine, cytosine, and thymine. The order in which they are chained up encodes the order of amino acids that will be used to create proteins.

When cells divide to create new cells, their entire DNA has to be replicated so that the new cell can get its own copy to make its own proteins. DNA replication is (like most things in life) an extremely complicated process, and illustrating it illustrates the atomic forces driving the molecules that drive life. I was lucky enough to be asked to make an animation of this process (well, a small part of it, but still).

I used the same approach as with the microtubules — I wrote down a mathematical formula that describes what DNA looks like while being copied and what all the proteins involved in this particular process are doing, all at their actual speeds. But unlike with the microtubule, I didn't want to just show the outcome — I wanted to dig deeper and show what exactly drives this process. That means I had to *zoom in* — and more importantly, to *slow it down*.

In humans, the molecular machinery that replicates DNA does so at a speed of roughly one hundred nucleotides per second. That is too fast to see in real time, so I modeled the process of nucleotides being copied at a much slower pace.

So, now I had a model of the same process, but one of them was zoomed in and slowed down. Transitioning between them is pretty straight forward: we can use the virtual camera to bring the viewers closer, and as we do, we cross-fade between

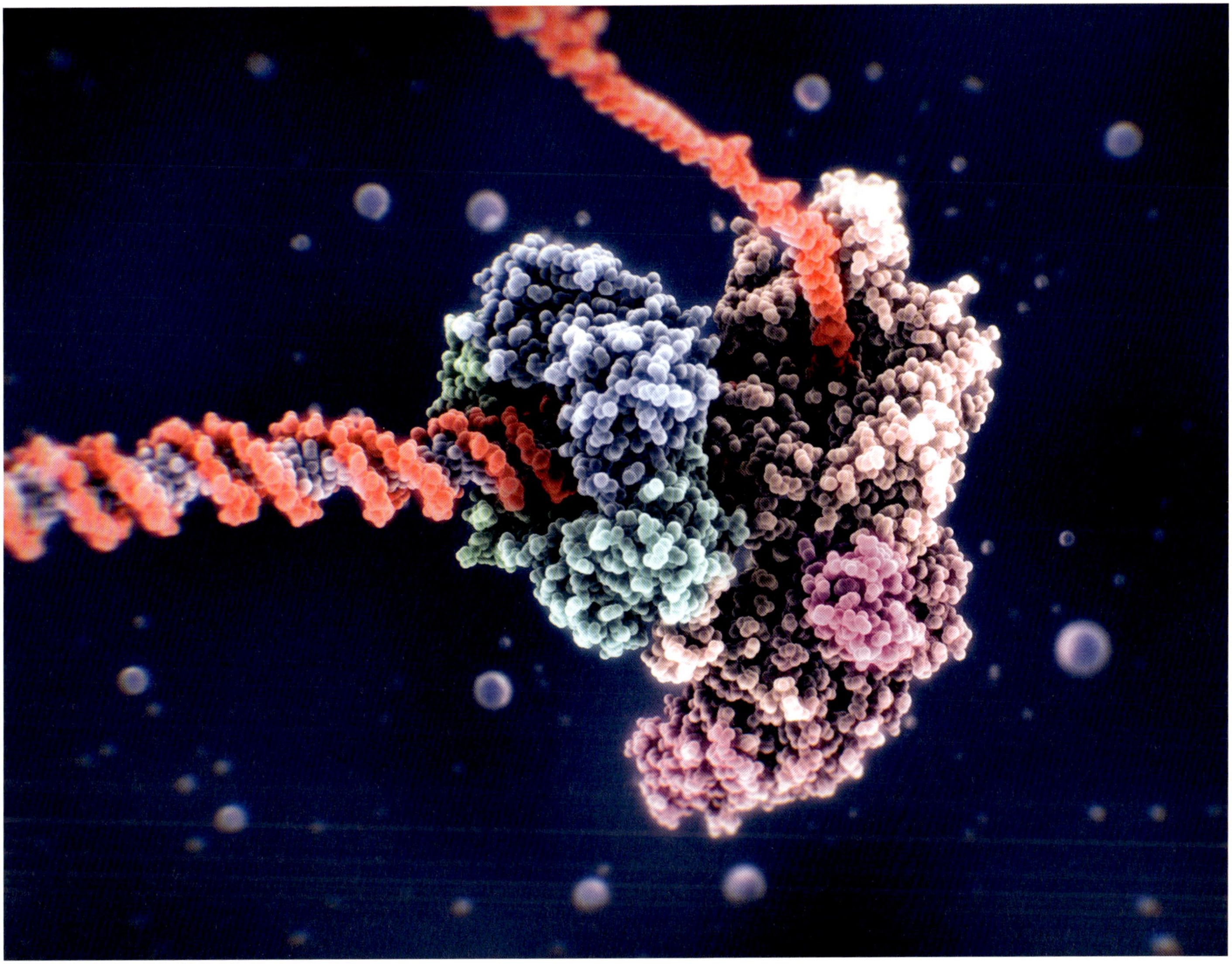

FIG. 4_*DNA replication of the lagging strand: a procedurally generated model of a DNA molecule being copied*

VIDEO_*DNA replication of the lagging strand*

the two models. We can also continuously slow down time as we switch between the real-time model and the slowed-down model. The issue is: people generally don't know how fast things are in the nanoworld. So, if we just show the same thing at two different speeds, they have no idea which one is the real speed. Just because something is obvious to me because someone told me about it before doesn't mean it will be obvious to my audience as well.

To communicate which part of the animation is being shown at the real speed and which one has been artificially slowed down, I used sound. The animation has an ambient soundtrack made of synthetic sounds and an underwater recording played in a loop. This sound is pretty recognizable to people and conceptually fits the animation, since there is a lot of water inside cells. Of course, it is complete fiction, as we cannot record sound at such a small scale. For viewers, made-up ambient sound feels much more natural than absolute silence, because absolute silence is something we never really experience. Just like we use light effects from our scale to illustrate the shapes of the nanoworld, we can use sound effects from our scale to illustrate the *space* of the nanoworld.

And we can slow the sound down. People recognize this sound effect from movies and music. If they are not familiar with the math behind transforming the frequencies of the recorded sound waves to change the playback speed, people will hear that the sound has been artificially slowed down. So, when we do this while molecules in the animation suddenly start moving slower, the sound tells them that time has now slowed down and that what they are seeing right now isn't happening at the real speed.

Using several channels to tell a story is a technique as old as storytelling itself. Sound is not important to animate molecules, or to animate atoms bonding with each other. And so, nothing gets disrupted when we use this channel to get the idea across to our audience that we are switching between two different scales.[5]

I used this technique because my goal wasn't just to show how DNA gets replicated. I wanted to bring my audience from the mesoscale into the nanoworld of atoms and show what happens inside the protein complex that plays a role in the process of DNA replication. I wanted to show that, rather than *magic*, life is *clockwork*. The proteins are cogwheels. And the atoms are the cogs_FIG. 4.

That's how I showed chemistry—the driving force of life.

Conclusion

When I was young, I fantasized about creating a computer program that would generate any possible picture, one after the other. I was trying to be realistic, so I thought I would limit the picture size to 20 × 20 pixels. And let's say the pixels can only be either black or white, so that there aren't *too* many combinations. It would be just a few lines of code, and if we were patient enough, the program would eventually generate *everything* (well, everything that fits into a 20 × 20, black-and-white image, but still). Imagine that: every pattern, every animal, every cheese manufacturer's logo. A face of every person that has ever lived.

I never made such a program. Turns out that, if we ran it, the program would create more pictures than there are atoms in the observable universe before it was finished. That might take a bit too long, so I ended up pursuing a career in scientific visualization instead. Here, my job is to instruct computers to draw very specific pictures — those that tell a story.

There's an advantage to teaching a computer how to create storytelling images and animations instead of making them manually: the more stories you tell, the better the computer gets at creating them. Many aspects are shared between visual stories about different topics, and computers are great at generalizing those aspects. This means that new visual stories created with the help of computers will always be easier to make than the previous ones.

We will never be able to create *all* the possible stories. But hopefully, one day, we will be able to show everything that we know. Being able to tell enough stories to understand life on every scale, from the tiniest atom up to entire ecosystems, is the dream of many educators. With the help of computer-generated visualizations, we are slowly progressing toward this goal. Right now, we are at the beginning of this journey. We are building the necessary technology that will take us there one day. Ask me how it's going in ten years' time. Until then, I will keep telling stories about the very origins of life, stories about the nanoworld. And I will keep making computers better at helping me to do so.

Talking about life can take us to many different places. All of them are interesting, all of them are worth visiting. But I am a programmer: I'm interested in how things work and what's at the bottom of it all. That's why I'm so interested in molecules. I believe that trying to understand how life emerges from basic chemistry is something that will keep me occupied for a very long time. And whenever I understand a tiny drop in this vast ocean of reactions, interactions, and their consequences, I will tell a story about it. And I will present it in a way that allows everyone to see how beautifully complex life is. And that the dinosaurs and us are not all that different.

1 Albert Barillé, *Il était une fois … la vie*, 1987–1988, https://www.imdb.com/title/tt0284735.

2 Ngan Nguyen et al., »Modeling in the Time of COVID-19: Statistical and Rule-Based Mesoscale Models,« *IEEE Transactions on Visualization and Computer Graphics* 27, no. 2 (2021): 722–32, https://doi.org/10.1109/TVCG.2020.3030415; May 6, 2020, https://nanovis.kaust.edu.sa/sars-cov-2-virus-model.

3 Peter Mindek et al., »Visualization Multi-Pipeline for Communicating Biology,« *IEEE Transactions on Visualization and Computer Graphics* 24, no. 1 (2018): 883–92, https://doi.org/10.1109/TVCG.2017.2744518.

4 Ngan Nguyen et al., »Modeling in the Time of COVID-19.«

5 https://www.rcsb.org/structure/6TNY.

Geometry as a Key to Hidden Doors?

Georg Glaeser

»The laws of nature are written in the language of mathematics.« This quote is attributed to Galileo Galilei. But how can nature know mathematics? Geometers and mathematicians have probably always favored the branches of mathematics that have turned out to be great tools for comprehending nature. It seems that mathematics has undergone an »evolution« comparable to the one that we know from nature.

Spatial Imagination

Ever since I was young, I have been fascinated by geometric forms — be they abstract, forms in arts, or forms that occur in nature or technology. Despite having always been a »half-biologist,« I decided to study mathematics and geometry because I realized that they are mighty tools to describe nature. At the age of seventeen (that is, in 1972), I had — amid numerous books about animals — several books about geometry in my bookshelf. I was fascinated by perspective drawings of bridges and accurate constructions of snail shells, as well as by the geometry of motion. Back in those days, there were no computers on the market, and everything had to be drawn by hand, which required not only patience but also extensive knowledge (for example_FIG. 1).

In the course of the many years that followed, I have come to realize that this has been of great advantage to me: since learning to program a computer is not hard for mathematicians — you only need to apply the same kind of structured thinking that determines your whole approach to tackling problems — computers have become a new and very efficient tool to help us solve tasks — no more, no less (for a

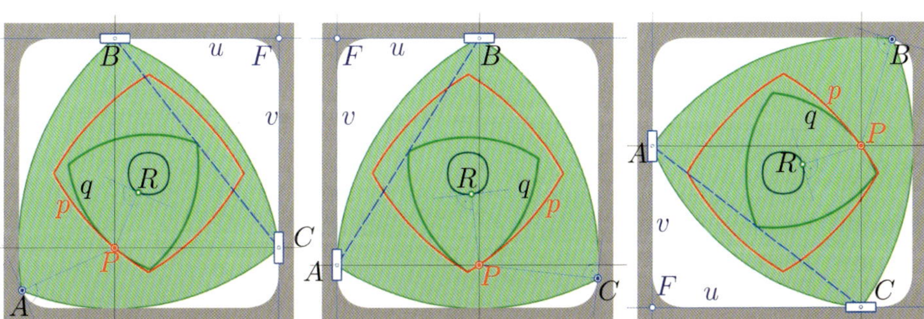

FIG. 1_*How to drill an almost quadratic hole in theory*

Images by Georg Glaeser, 2022

corresponding example, see_FIG. 2, which is a screenshot of an accurate computer simulation). The underlying brainwork, paired with spatial imagination, still remains the most important component.

Geometry is usually considered to be one branch of mathematics. Indeed, the mental processes required for geometry are closely related to mathematical logic. Furthermore, since geometry is a very comprehensive field, it is often thought of as an independent science. In any case, geometry and mathematics complement each other quite nicely. Geometry can be conducted in several ways, but the kind of geometry that we use here seeks to utilize the reader's three-dimensional imagination as often as possible.

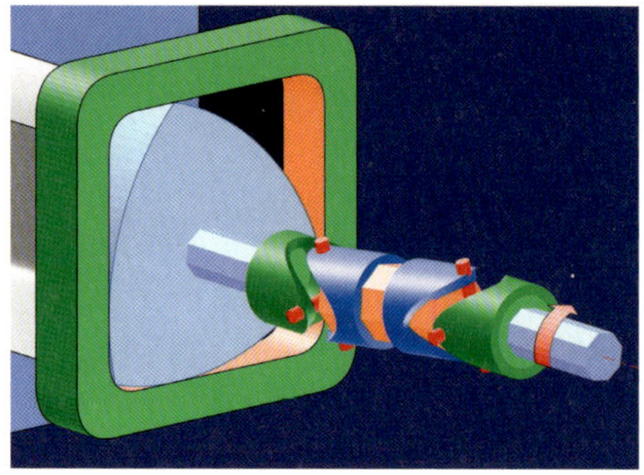

FIG. 2_*Computer simulation: how to drill an almost quadratic hole in practice*

Image and video by Georg Glaeser and Franz Gruber, 2019

VIDEO

Since geometry not only utilizes but also improves this »sixth sense« of spatial imagination, younger readers often find it easier to progress in this discipline compared to other branches of mathematics. The latter require very solid foundational knowledge as errors that may be overlooked by inexperienced eyes may lurk in every line of calculation. The kind of geometry with which we are concerned, however, always yields results that can, to some extent, be verified by our imagination. It is for this reason that geometric (»synthetic«) proofs are usually preferred to strictly mathematical (»analytic«) ones. It is by no means uncommon for these proofs to contain such deep levels of understanding that the attentive mind will inevitably generate new or additional results.

Due to my unceasing enthusiasm for photography, my books about geometry, mathematics, and computer graphics evolved into something akin to »picture books of geometry.« For an example of this, see the series of photos of a fish below, in which I managed to capture the extreme dislocation of its jaws_FIG. 3: the sling-jaw wrasse (*Epibulus insidiator*) has developed a cunning hunting trick. While catching prey, it rapidly extends its mouth into a long tube with which it sucks up small fish. When I took these photos, I was observing *Epibulus* as it performed a »training exercise« for this mechanism without the presence of any prey. The fish appeared to be yawning, and these photos could be used to study this complex motion process without slow motion.

This is only one of the numerous examples of natural constructions, processes, and evolutionary principles that are yet to be studied and adapted in engineering. In_FIG. 3, there are three screenshots of an animation of the wrasse's jaw protrusion created by me and my coworker Franz Gruber: the underlying kinematic mechanism, which contains multiple joints, is a very sensitive process due to the length of the movable bone structures.

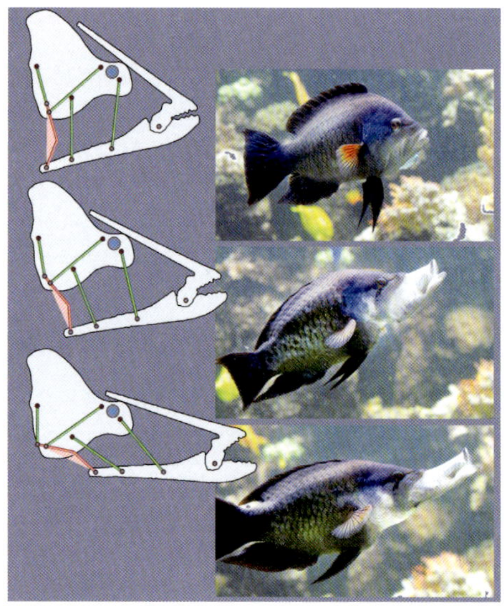

FIG. 3_ *The sling-jaw wrasse demonstrates its mechanism for catching prey.*

Image and video by Georg Glaeser and Franz Gruber, 2019

VIDEO

Galilei and the Moons of Jupiter

The night of January 7–8, 1610, was a stellar moment in the history of astronomy. With the help of the newly invented refracting telescope, Galileo Galilei discovered the four largest moons of Jupiter, which then came to be known as the Galilean moons. The names of the individual moons are derived from the mythological lovers of the ancient god Jupiter: Io, Europa, Ganymede, and Callisto. What made this discovery even more special in those days was that the four moons were seen to revolve around Jupiter, that is, around a planet other than the Earth. This discovery undermined the then-accepted geocentric model of the universe and clearly supported the heliocentric model, which had already been proposed by Nicolaus Copernicus seventy years prior (but had so far failed to gain widespread acceptance). Only a few months before Galileo's discovery, Johannes Kepler had formulated the first two of his laws, which also take the sun as the center of the planetary orbits.

_FIG. 4 shows what Galileo might have seen on different nights (the photos, though, were taken by me). The fact that Galileo was able to deduce a spatial picture like the one in _FIG. 5 was only due to his excellent spatial imagination: the almost

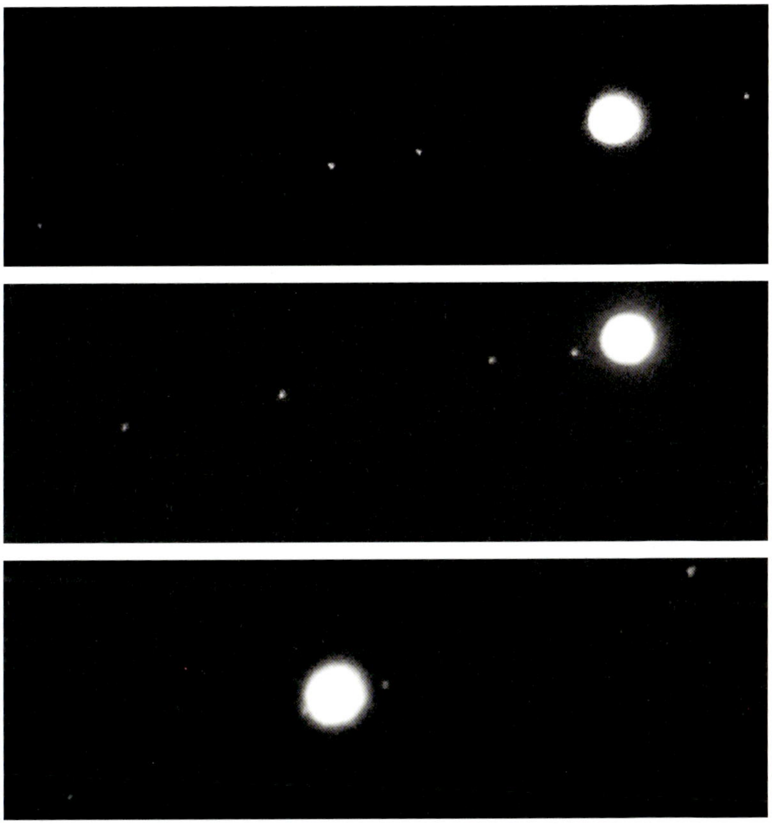

FIG. 4_ *This is how Galileo might have seen the moons of Jupiter on three consecutive nights...*

Images by Georg Glaeser, 2022

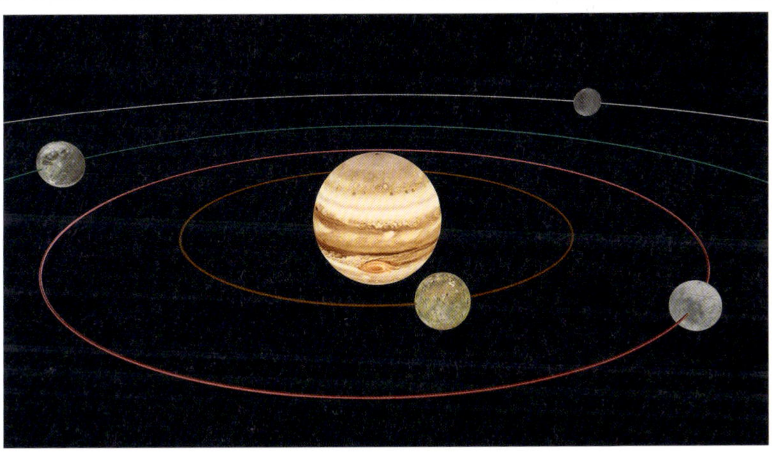

FIG. 5_ *...and this was his conclusion.*

Image by Georg Glaeser, 2022

VIDEO

circular paths of the individual moons do not appear as circles or ellipses but as straight lines. This is due to the fact that their orbital planes and the orbital plane of Jupiter are almost identical. By the way: Galileo Galilei is said to have coined the phrase, »The laws of Nature are written in the language of mathematics.«

Newton and Gauß and Their Geometric Contributions to Astronomy

When Isaac Newton proved Johannes Kepler's conjecture that the paths of the planets are ellipses, he utilized properties of ellipses that have almost been forgotten since then. Twenty-three-year-old Carl Friedrich Gauß gained his reputation by rediscovering the dwarf planet Ceres, which had only been observed on very few occasions. He utilized the fact that an ellipse is uniquely determined by a single focal point (the sun) and three additional points. Gauß himself considered that this type of problem »commended itself to mathematicians by its difficulty and elegance.« Two hundred and fifteen years later, with Ceres still on its elliptic path, NASA's Dawn spacecraft has orbited the dwarf planet, transmitting stunning images of this icy remnant of our early solar system.

The Nature of Geometry and the Geometry of Nature

The wordplay in the title is perhaps an indication of this chapter's intentions — namely, to bridge the gap between nature and geometry. It is well known that many of the surfaces or solids described by geometry can be found in nature, in both the macrocosm and the microcosm. This raises the following question: does geometry merely »plagiarize« nature, or has it developed independently, with coincidental parallels to the results of evolution? Geometry helps us understand, for example, why Platonic solids are »obvious« solutions to many problems in nature and that, if certain conditions are given (such as regularity or symmetry), there are actually no other solutions. Nature usually produces regularity or symmetry if an equilibrium of forces or an optimal packing surface is advantageous. Evolution finds optimal forms on its own. More than two thousand years ago, the ancient Greeks Plato and Archimedes could not have known that »their« solids are so frequently found on a molecular scale. Pythagoras, on the other hand, defined music by means of proportions. It is therefore unsurprising that musical rules can be beautifully visualized geometrically. Another valid question may be: is music, as defined by Pythagoras, rooted in »natural phenomena«? Stanislaw Ulam calls mathematics a self-contained microcosm that nevertheless has a strong capability to capture and model not only any kind of thought process but perhaps the entire scientific process. To some degree, this probably also applies to geometry, as it is a subdiscipline of mathematics.

Symmetry and Biology

Geometrically, two figures are symmetrical in the plane if they can be made to coincide by mirroring them on a straight line (the mirror axis) — and possibly also shifting them along the same straight line. An impression of the left foot looks essentially the same as that of the right foot, but we can still always tell whether it was the left or right foot — especially by the position of the big toe: naturally, the inherently non-symmetrical shape of a single foot derives evolutionarily from the prehensile

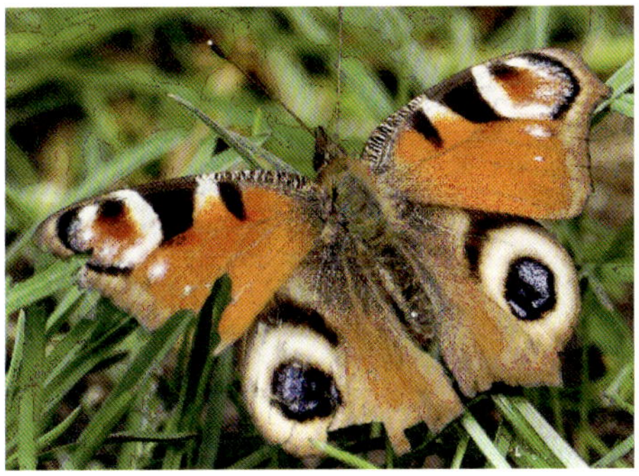 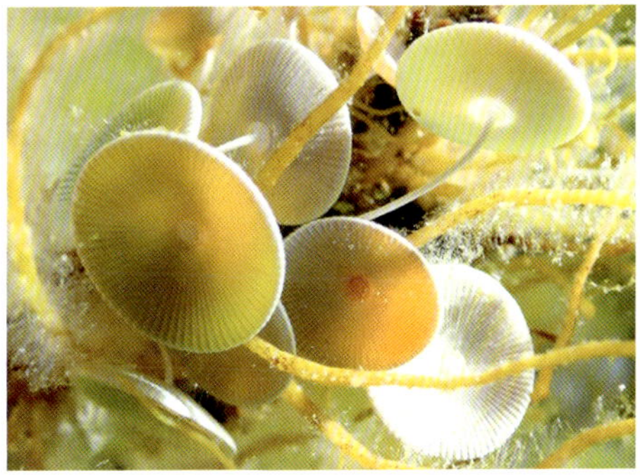

FIG. 6_*Left: symmetrical shapes; right: »infinite number« of symmetries*

Photos by Georg Glaeser, 2022

feet of our ancestors. Since we move on two legs, we are better able to keep our balance and generate ground pressure at the right place when walking and running.

The left-hand side of_FIG. 6 shows a peacock butterfly. Its symmetrical pairs of wings are already badly damaged. It is a specimen that has hibernated and will produce a new generation in spring. Why the wing pairs are symmetrical is immediately obvious for physical reasons: if they were not, the butterfly would have a hard time flying. In this respect, wings that are severely damaged on one side are a great disadvantage during flight. This can also be seen in birds, which renew their feathers in a process known as molting. But why are these *patterns* symmetrical?

Well, they serve, among other things, as a visual signal for species recognition. So, if a butterfly or bird had a pattern on its left wing that was different from the pattern on its right wing, it would be much more difficult for its own species to recognize it. For butterflies, the eyespots are also a significant predator deterrent, mimicking the eyes of larger animals, which will frighten the predator when the butterfly opens its wings.

Symmetrical plane figures can be transformed into each other by being mirrored on their »symmetry axis.« Regular figures like the square in particular can thus be transformed and even mirrored in several ways. The shape with maximum symmetry is the circle, which can be transformed into itself in an infinite number of ways by mirroring it onto itself on an arbitrary diameter. If we look at the umbrella algae on the right of_FIG. 6, the umbrellas lie *almost* in one plane. These algae, which have a single giant cell nucleus in the area of the root cells, reproduce when the nucleus divides into thousands of smaller nuclei, which migrate into the chambers of the umbrella and produce flagellated spores. Because of the circular arrangement, the freely moving spores can leave in all directions. This is reminiscent of the flying seeds of the dandelion_FIG. 4.

In space, a symmetry plane (mirror plane) takes the place of the symmetry axis. Our likeness in a mirror or water reflection is not identical to us: while the coordinates of the individual points in planes parallel to the mirror plane remain unchanged, the third coordinate, namely the distance to the mirror plane, is reversed_FIG. 7.

Mathematically speaking, there are also bodies or surfaces with a particularly high degree of symmetry in three-dimensional space. A cube can be mirrored in nine different ways along its symmetry planes. A sphere can withstand any reflection across any plane through its center (see the withered dandelion in_FIG. 8). Here,

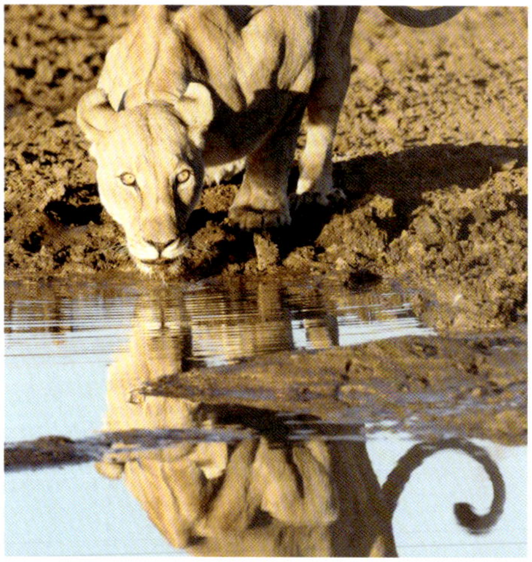

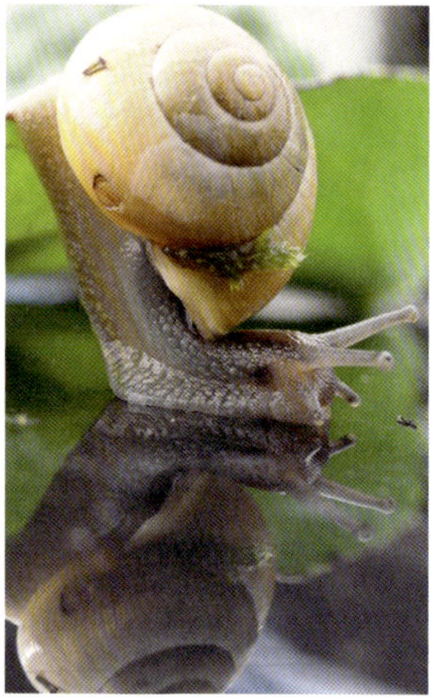

FIG. 7_*Reflections in nature*

Photos by Georg Glaeser, 2022

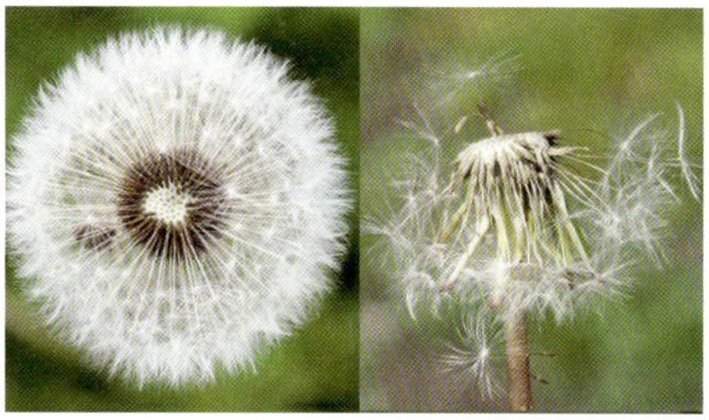

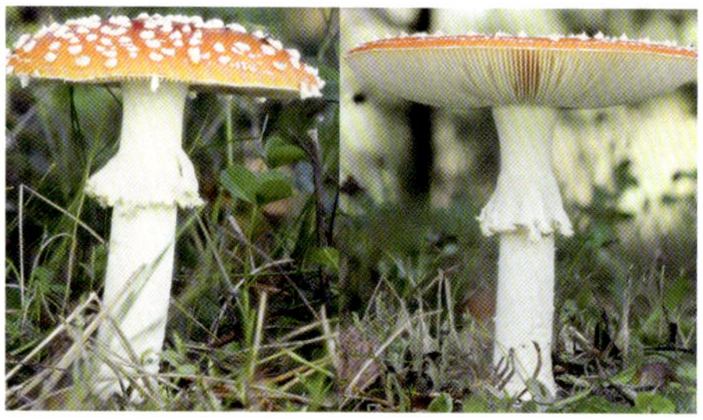

FIG. 8_*Withered dandelion with high degree of symmetry*

Photos by Georg Glaeser, 2022

the spherical shape is created by compressing as many equally sized, »parachuted« flying seeds as possible into a very small space. The sphere is the place where all points are at the same distance from a fixed center.

Rotationally symmetrical objects like cylinders or, more generally, surfaces of revolution still have an infinite number of planes of symmetry. These pass through the axis of rotation of the object. A nice example of this is the toadstool in_FIG. 5.

SYMMETRY IN THE ANIMAL KINGDOM_Symmetry obviously plays an enormously important role in nature, both in the plant kingdom and in the animal kingdom, and for a variety of reasons. The deer in_FIG. 6 would not be able to balance its powerful antlers so effortlessly and would have to exert much greater force to move its head smoothly if its antlers were not symmetrical. _FIG. 9 on the right shows a simulation of antler growth (changing with the age of the deer).

Let us first address symmetry in animals:

— Sponges (that is, underwater sponges, which have little to do with fungi and belong to the animal kingdom), as the simplest multicellular animals, have no symmetry at all.
— Jellyfish, anemones_FIG. 10, and other cnidarians are radially symmetrical. These animals also move in all directions. The anemone on the right takes advantage of its optimal circular shape to absorb a maximum amount of food in the smallest possible space.

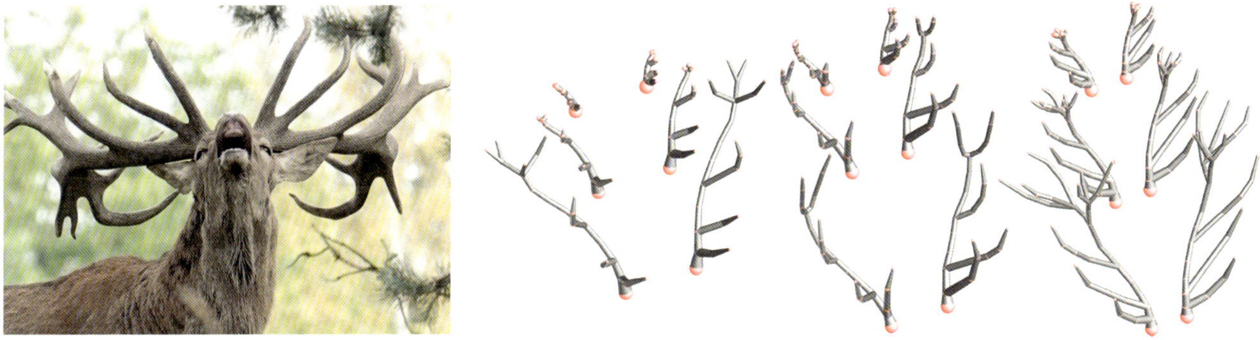

FIG. 9_ *Left: balancing powerful antlers; right: growth simulation*

Photos by Georg Glaeser, 2022

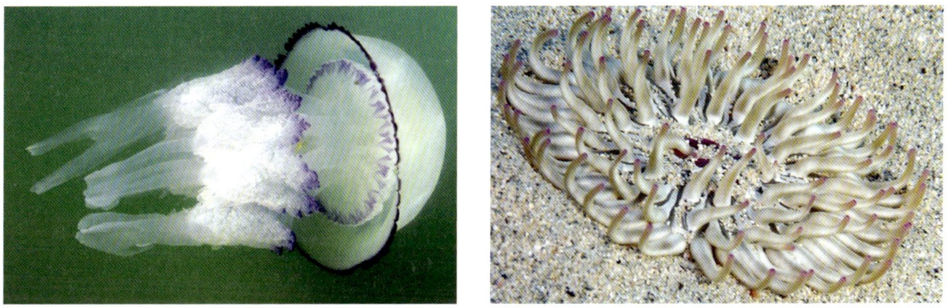

FIG. 10_ *Radial symmetry (jellyfish, anemone)*

Photos by Georg Glaeser, 2022

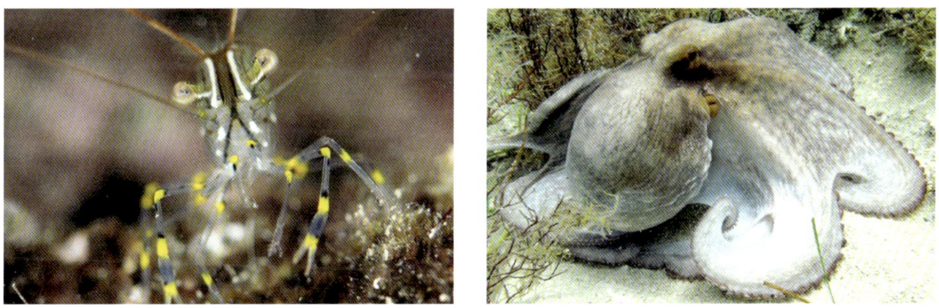

FIG. 11_ *Shrimp and octopuses are, like vertebrates, bilaterally symmetrical.*

Photos by Georg Glaeser, 2022

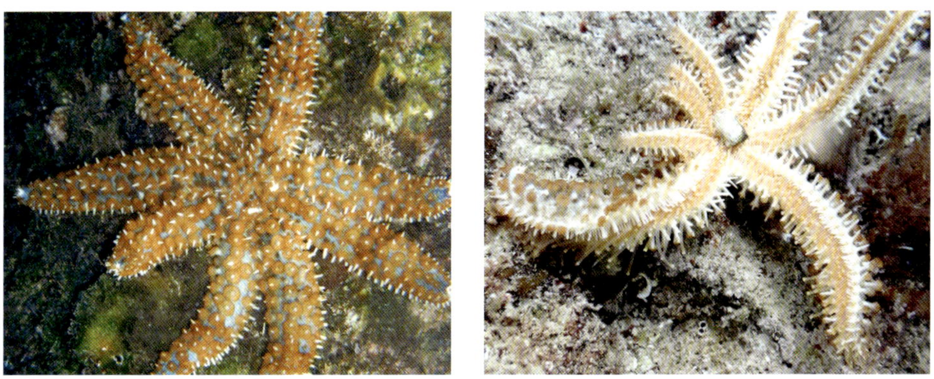

FIG. 12_ *Starfish usually display five-rayed symmetry.*

Photos by Georg Glaeser, 2022

— Vertebrates_FIG. 7, LEFT, but also most other animal phyla_FIG. 11, have a single plane of symmetry: they are bilaterally symmetrical and thus have a »front«—the direction of locomotion. Even though the octopus has eight arms, it is still only bilaterally symmetrical.
— Echinoderms (starfish, sea urchins), on the other hand, often display five-rayed symmetry (they then evolved into animals with six, seven, and many more arms)—astonishingly, this is because their larvae are bilaterally symmetrical and undergo metamorphosis, during which they develop different symmetry. Even several internal organ systems are five-armed. This development has no parallels in the evolutionary history of animals.

_FIG. 12 shows a starfish that normally has only five arms. Mutations or other circumstances (lost arms that have been replaced twice) have, exceptionally, produced an eight-armed representative here.

It has been shown that symmetry is a determining factor in the choice of a mate for animals, even though perfect symmetry is, in fact, not so easy to achieve biologically. When only females can choose mates, they opt for the more symmetrical individuals when in doubt, probably because the females are more likely to attribute good genetic characteristics to those more symmetrical candidates. This has also been found in humans: women generally prefer men who have symmetrical faces.

SYMMETRIES IN THE PLANT KINGDOM_In contrast to the animal kingdom, bilateral symmetries seem to be less common in the plant kingdom. For example, the horned clover on the left of_FIG. 13 and the donkey spurge on the right of_FIG. 13 display bilaterally symmetrical flowers, which in turn are arranged rotationally symmetrically.

Multiple symmetries, such as the quintuple symmetry in bluebells_FIG. 14, are more common. Biologists believe they know the reason: calyxes are there to attract insects, which then perform pollination. When a calyx looks almost identical from all sides, it can be used as a landing platform that insects can approach from all directions. This proves to be very advantageous when pollen also needs to be attached everywhere—and pollen production is costly. Flowers with bilateral symmetry resort to another trick: their shape forces pollinating insects to adopt a certain position. As a result, these plants manage to get fertilized with significantly less pollen, which is particularly important when little nitrogen is available.

As an aside: in the early 1980s, I felt inspired by bluebells and, as a result, created one of my first computer drawings_FIG. 14, RIGHT. The problem at that time was that only raster graphics were available. This forced programmers to develop several algorithms, e.g., a hidden surface algorithm and an algorithm for the detection of contours.[1]

When looking at sunflowers, daisies, and many other flowers (e.g., chamomile in_FIG. 15), the human eye recognizes spirals. Sometimes they wind to the left, sometimes to the right.[2] Discerning which individual disk flowers the spirals stem from is often not that easy or clear. The spirals are approximately logarithmic spirals and are also related to the Fibonacci numbers. Each number in the sequence 2, 3, 5, 8, 13, 21... is the sum of the two preceding ones (e.g., 3 + 5 = 8, 5 + 8 = 13).

FIG. 13_ Plants are more often rotationally symmetrical.

Photos by Georg Glaeser, 2022

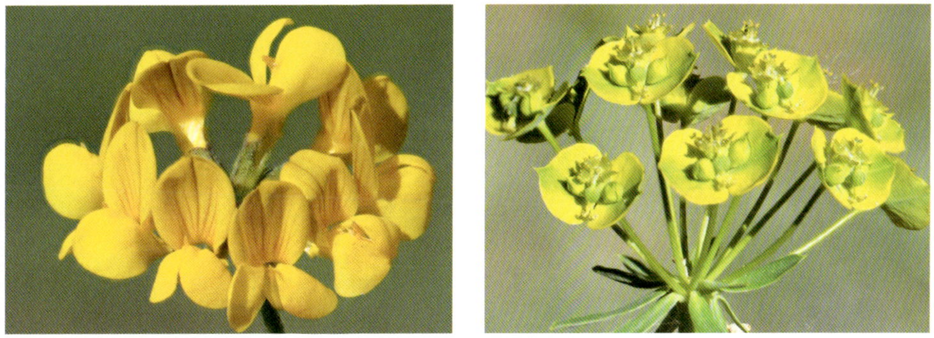

FIG. 14_ Quintuple symmetry in bluebells and computer simulations

Photos by Georg Glaeser, 2022

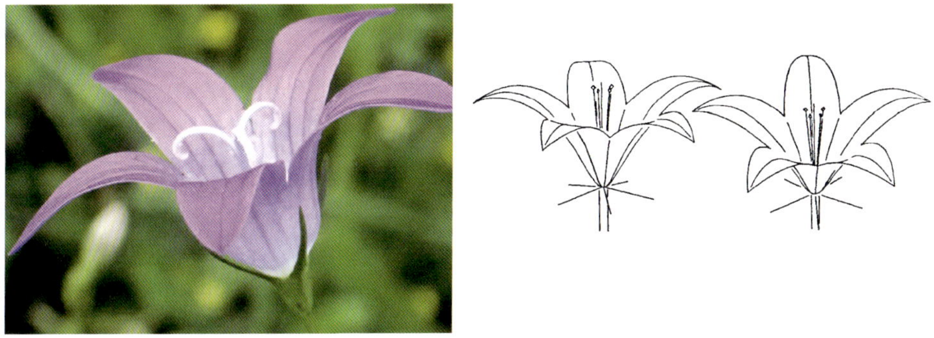

FIG. 15_ The human eye recognizes spirals.

Photos by Georg Glaeser, 2022

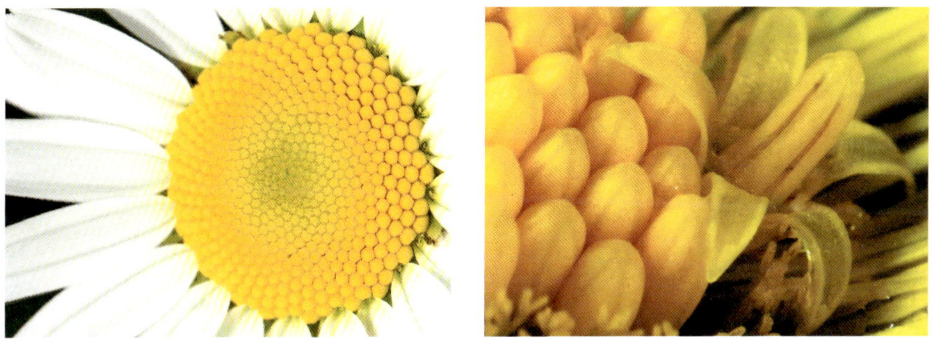

The reason for this arrangement is always the same: the plant wants to distribute as many seeds as possible in the smallest possible area. The most favorable way for the plant to grow is by placing the next disk flower in the golden angle while simultaneously exponentially increasing its distance from the center. Each plant will pass on its »chosen« angle to its offspring. Minor mutations will cause changes in the angle. The plant that has chosen the best angle can create more offspring. Accordingly, the optimal angle will prevail over and over again.

In detailed images_FIG. 15, RIGHT,[3] the spirals are no longer visible. Here you can see that the aim is to pack in as many seeds as possible. The burst single flower on the right is only one millimeter in size, but it is reminiscent of the calyxes of bellflowers, dozens of times larger.

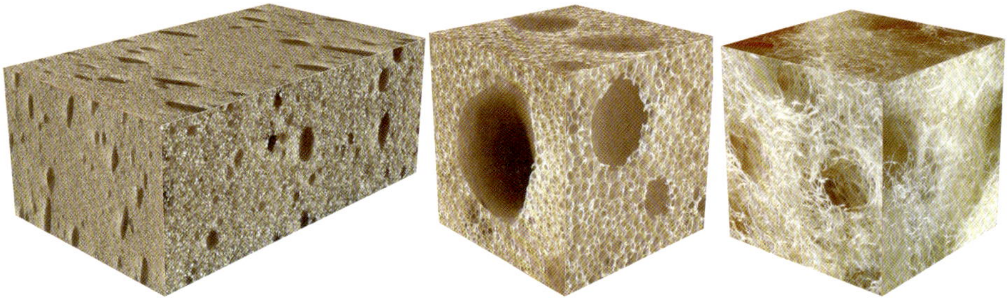

FIG. 16_ *Blackboard sponge, electronically processed at various magnification levels*

Photos by Georg Glaeser, 2022

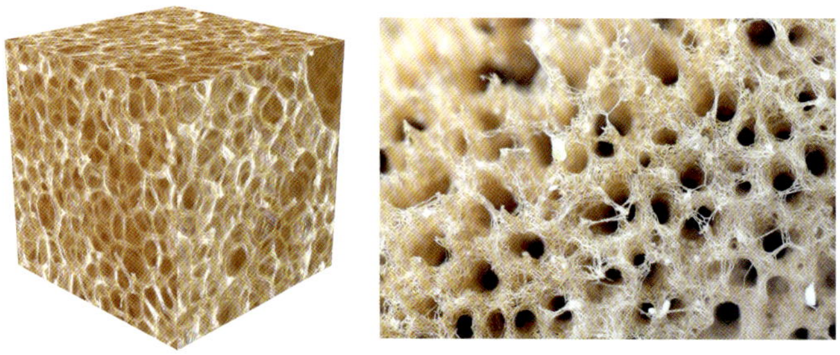

FIG. 17_ *Artificially produced table sponge vs. marine sponge*

Photos by Georg Glaeser, 2022

Of Sponge Structures, Random Patterns, and Voronoi Diagrams

ASSOCIATIONS_A blackboard sponge_FIG. 15, LEFT, triggered in my mind—that of a mathematician and geometry teacher with an affinity for nature—several reactions at once: first, there was the familiar sensation of using the sponge, which had been soaked and squeezed to remove excess water, to clean the chalkboard in the lecture hall efficiently and without creating dust. Secondly, the sponge reminded me of the sea sponges (not cuboidal, of course) that I had found washed up on the seashore and that could even be used in the bathtub to gently clean the skin. Thirdly, the structures of the artificially created sponge begged comparison with those known from recent computer-assisted research on cell arrangements.

Today, table sponges and bath sponges are artificially produced, which, in view of the dwindling stocks of marine sponges, is an absolute necessity._FIG. 17 shows a detailed comparison of the structures. The mode of operation is very similar in both cases, namely, absorbing as much water as possible in a short time. Accordingly, their dry weight is low.

BIOLOGICAL SPONGES_The structure of sponges living under water resembles those of mushrooms, the spongy fungi that we find in a forest. However, fungi are classified as belonging neither to the animal kingdom nor to the plant kingdom, instead forming a large, independent third kingdom of living beings. The sponges living in water, on the other hand, are classified as a phylum of multicellular animals.

STABLE FOAM STRUCTURES_If we perform a simple experiment and shake a small cube that has been filled with soap solution and then sealed, the result, as in_FIG. 18 on the left, is a foam mixture that looks aesthetically pleasing and is reminiscent of the structures in a sponge.

What is even more impressive about our experiment, however, is that the structures change quite rapidly at first and then more and more slowly._FIG. 19

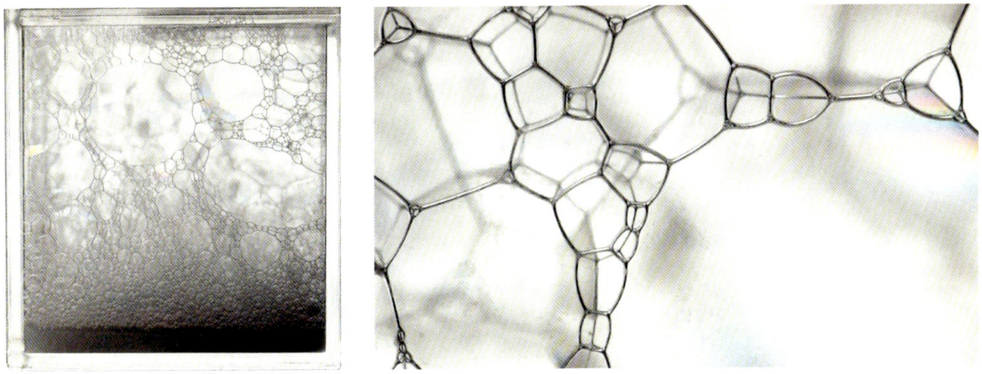

FIG. 18_ *Foam structures (at high magnification on the right)*

Photos by Georg Glaeser, 2022

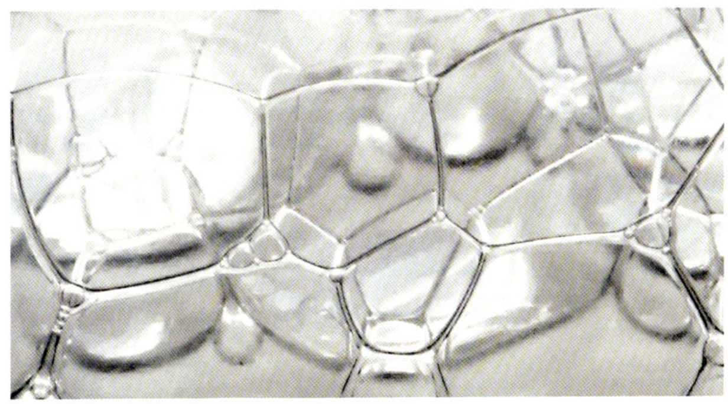

FIG. 19_ *Consolidation of the foam overnight*

Photo by Georg Glaeser, 2022

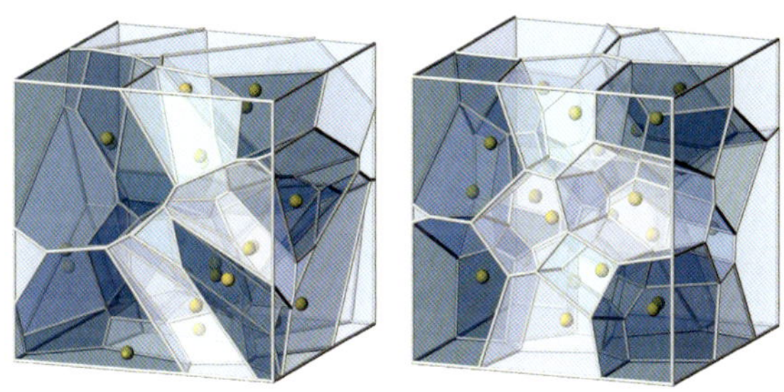

FIG. 20_ *3D Voronoi diagrams in a cube (left »unbalanced«; right »consolidated«)*

Images by Georg Glaeser, 2022

VIDEO

shows a detailed image from the same series of experiments. Overnight, the resulting cells have adapted their respective sizes well to each other, become stable, and formed a balanced end product, so to speak.

Especially when looking at these consolidated foam structures, a mathematician will be reminded of spatial Voronoi diagrams, a type of spatial structure named after the Russian mathematician Georgy Feodosevich Voronoy_FIG. 20.

VORONOI DIAGRAMS_To appreciate the Voronoi diagram's significance in natural processes, let me briefly explain how such structures are defined, and, to better understand them, let us start with the two-dimensional case. This is mathematically formulated as follows: we want to selectively divide a plane into convex regions by means of a number of given points (centers). All regions around the centers are then without indentations or constrictions, and all points of a region are closest to their center_FIG. 21.

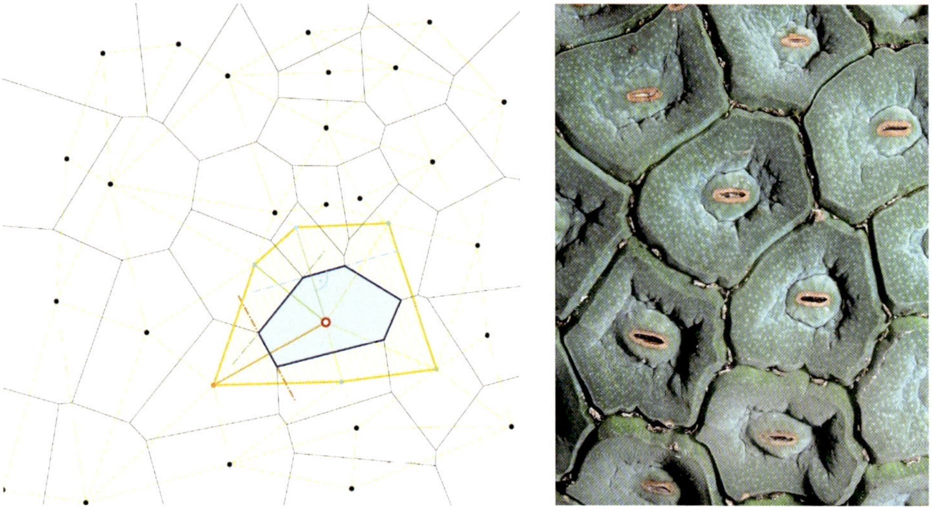

FIG. 21_ *2D Voronoi diagram (left) and surface structure on the fruit of a monstera plant*

Images by Georg Glaeser, 2022

FIG. 22_ *Dried mud*

Photo by Georg Glaeser, 2022

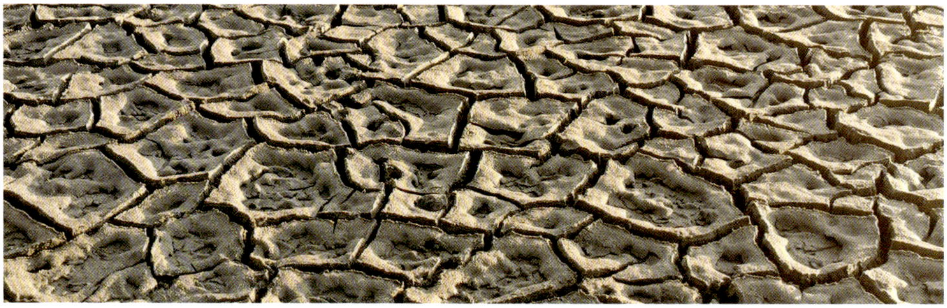

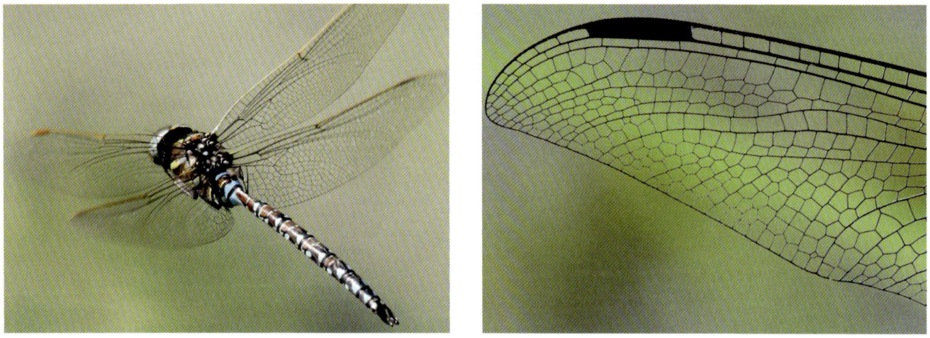

FIG. 23_ *Stabilizing effect of the veins in a dragonfly wing*

Photos by Georg Glaeser, 2022

FIG. 24_ *Leaf structures under the magnifying glass*

Photo by Georg Glaeser, 2022

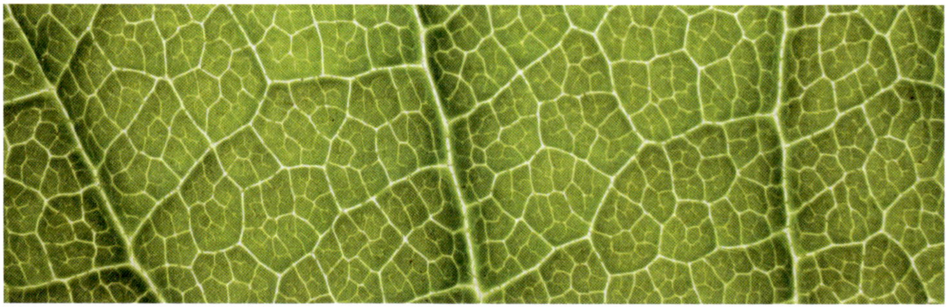

For each of the given points, we need to determine all perpendiculars (symmetrals of the line) with respect to the remaining points. For each point, these perpendiculars to the neighboring points form a convex region. All regions together form the Voronoi diagram_FIG. 21, LEFT. In the three-dimensional case_FIG. 20, the word »perpendiculars« should be replaced with »symmetry plane.«

A photo of the fruit of a *Monstera deliciosa*_FIG. 21, RIGHT, shows that nature produces very similar structures when it comes to dividing the surface of a plant fruit into regions. We must take into account that nature's solutions often depend on many additional parameters, so that any claims to mathematical exactness can only apply to a limited extent here. Thus, we will always find a slight indentation or constriction, and the actual result is only a good approximation of a mathematically defined Voronoi diagram.

_FIG. 22 shows a comparable image from the world of inanimate nature. After a muddy pond has dried out, the mud, in which stress has built up due to the evaporation of the water, will crack open at various points. Now imagine that within this shallow pond, due to slightly varying water depths, centers of tension with forces of somewhat equal magnitude emerge and »try« to bind as much material as possible around their respective center. The mud will crack approximately in the middle of two centers (the center symmetry). From this point of view, Voronoi diagrams also illustrate equilibrium states in stress fields.

In this respect, it is unsurprising that we can also recognize structures in insect wings that are often good approximations of Voronoi diagrams_FIG. 23. Likewise, leaf structures under the magnifying glass are strikingly reminiscent of such diagrams_FIG. 24.

OPTIMIZATION BY ITERATION_The scramble and tug-of-war in a field of tension—be it in dead or living matter—provides a stable equilibrium only after several repetitions. On the computer we can simulate such a procedure well by empirically striving for a slightly improved version of each intermediate layer. Specifically, we can replace the original cell nuclei with the centroids of the calculated regions of the diagram. We will then recognize that this optimization converges toward an ideal solution after several steps. In practice, however, it is sufficient to perform the process about a dozen times, which also imitates processes in nature._FIG. 25 illustrates how an unbalanced Voronoi diagram with randomly chosen nuclei becomes a stable, balanced diagram after several iterations. In_FIG. 20, we can see this optimization in the three-dimensional case: with each iteration, the current cell nuclei has been replaced by the volume centroids of the calculated cells.

GEOMETRICALLY COMPLEX RESULTS WITHOUT ANGLE MEASUREMENTS_

_FIG. 26 serves as a final example of Voronoi diagrams. Nature did not have to inject a solid understanding of geometry into the genome of bees and wasps to give them the ability to produce perfect honeycomb-shaped structures. Moistening their building material with saliva, these insects will stick equally sized cylindrical or slightly conical wax cells to each other and arrange them in a somewhat systematical manner, which almost inevitably results in the classic hexagonal honeycomb structure. At the edges of these honeycombs, you will sometimes find pentagonal or heptagonal cells.

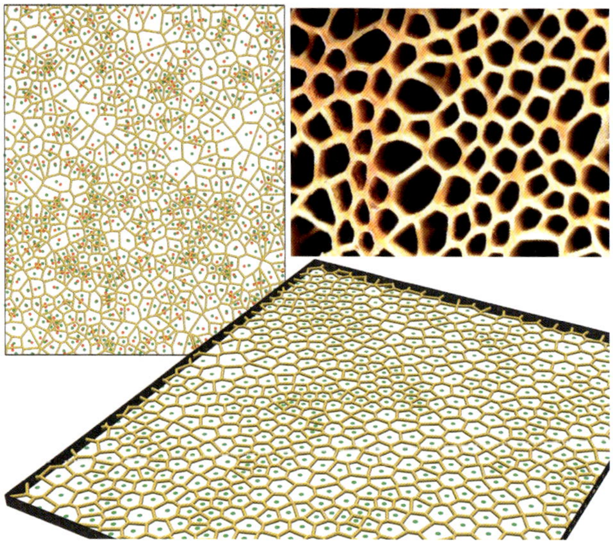

FIG. 25_ *If randomly selected cell nuclei (marked in red) in the Voronoi diagram at the top left are repeatedly replaced by slightly different, more suitable nuclei (marked in green), the diagram becomes more balanced with each iteration (bottom). The top right shows the underside of a porcini mushroom.*

Images by Georg Glaeser, 2022

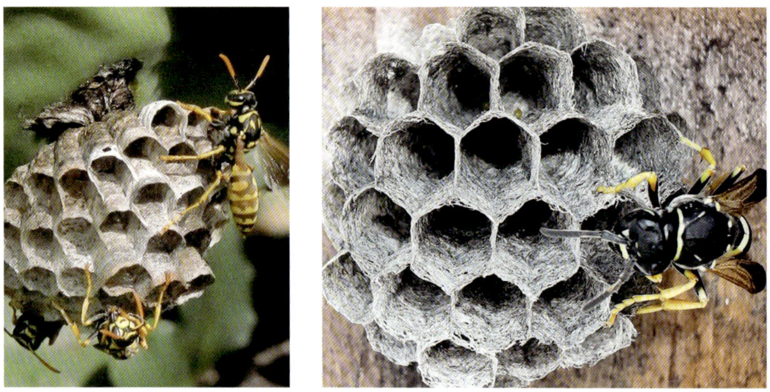

FIG. 26_ *Wasps create the hexagonal shapes of their honeycomb-shaped structures by lining up slightly conical cells and then letting their building material dry.*

Photos by Georg Glaeser, 2022

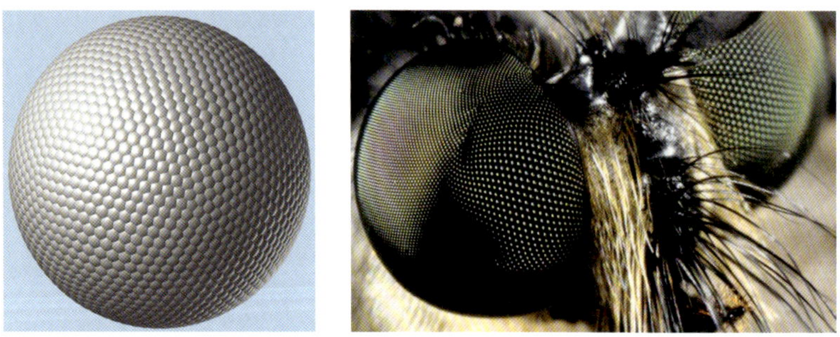

FIG. 27_ *Left: simulation of a compound eye, derived from a refined icosahedron as triangle mesh*

Photos by Georg Glaeser, 2022

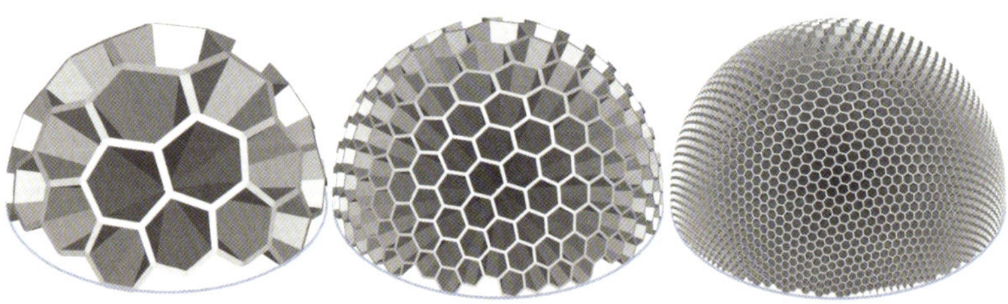

FIG. 28_ *This image is reminiscent of compound eyes or wasp honeycomb*

Images by Georg Glaeser, 2022

POLYGON MESH ON A SPHERE_Though spheres can be approximated more or less well using many quadrangles or triangles, not all the faces of these quadrangle or triangle compounds will be mutually congruent. Fairly good triangulation can be achieved by utilizing the »refinement« of an icosahedron. Spheres with a suitable radius are placed in the triangle's centroid. These spheres intersect each other and form polygon patterns that usually consist of hexagons and sometimes also of pentagons_FIG. 27. On the right of_FIG. 27, for comparison, we see the eye of a robber fly.

HOW TO TILE A SPHERE EXCLUSIVELY USING HEXAGONS_If we do not want to use pentagons, the task is more complicated. It is impossible to cover a sphere with hexagons of equal size, but what if we use different sizes? If this were possible, would the algorithm work not only on spheres but also on more general surfaces? Given the number of faces on a sphere (or hemisphere), we might take the following approach: we would find a set of more or less balanced distributions of the given points (this could be done using formulas or with the aforementioned algorithm) and then determine the spherical Voronoi diagram. We could draw our inspiration from a photo of a wasp's honeycomb-shaped cell_FIG. 28.

Geometric Simulations »as Proof«

Let us focus on some biological discoveries: small fish often exhibit very sophisticated swarming behavior, not unlike many bird species.

Interestingly enough, there is no alpha leader that determines the direction the swarm takes. This choreography is instead the result of many individual reactions. The goal of a single fish is merely to remain close to its neighbors (about a single body length). While foraging, individuals tend to be move leisurely, swimming in the direction of a food source.

When a predatory fish takes interest in the swarm, the individuals in its vicinity will naturally want to flee from the danger_FIG. 29. Their neighbors may not even have realized the cause of the disturbance, but they still reflexively follow their neighbors on their escape routes. This leads to a chain reaction (shock wave), which may even involve individuals that are far away from the predator.

Once the danger has passed, the fish that are farthest away will once again seek the close company of their relatives_FIG. 30. After a certain amount of time, and possibly until the next predatory attack, »order« will return to the swarm. This supposed chaos is actually governed by only a few parameters, which hardly restrict the variety of shapes that a swarm can take at any given moment (see also_FIG. 31).

Evolution and Geometry

How do birds »know« that non-spherical eggs are better for their offspring (assuming that spherical eggs roll away more often and thus break more frequently)? If a bird were to lay spherical eggs on an inclined surface or in otherwise unstable conditions, then it might discover that an elliptical form would be more practical, if only the animal had the mental capacity for this analysis. If it had that capacity and could somehow affect the shapes of its own eggs, it would certainly do so in the next attempt.

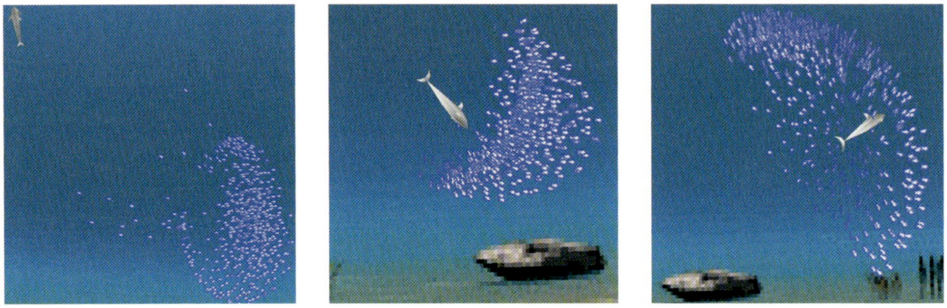

FIG. 29_ *Impending danger and escape*

Images by Georg Glaeser, 2022

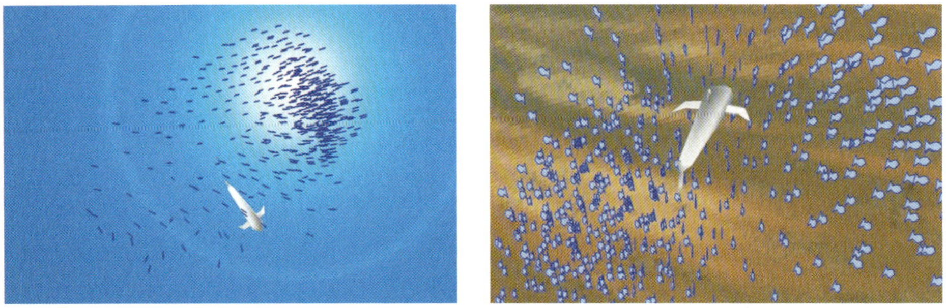

FIG. 30_ *Back to the swarm ... and immediately away from it!*

Images by Georg Glaeser, Claudia Carozzi, Stefan Felkel und Christoph Fessl, 2004

VIDEO

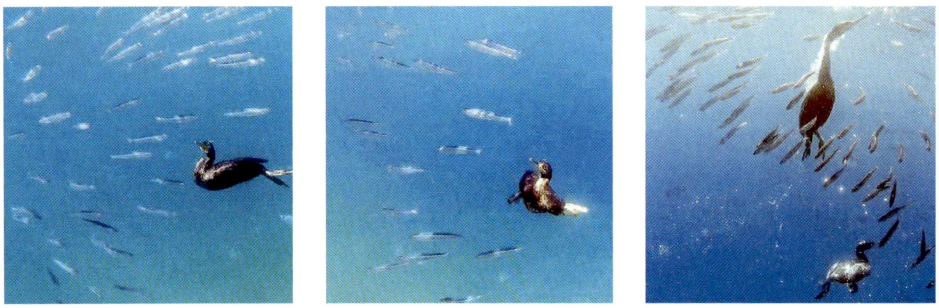

FIG. 31_ *An everyday occurrence under water: cormorants hunt small fish. They try to encircle the fish, but this proves to be quite a challenge amidst all the chaos...*

Images by Georg Glaeser, 2022

While the *individual* may be incapable of such a feat, the *species* as a whole is able to surmount this challenge given evolutionary timescales. Evolution is as ingenious as it is simple. We must now deviate slightly from the main topic — for good reason, as a solid understanding of evolutionary principles is essential for the section that follows. Biologists are thereby asked to indulge my generalizations, but since evolutionary thinking does not yet enjoy widespread awareness — although it has been 150 years since its landmark introduction — and is even the target of some hostility, there is no harm in clarifying some of the basic concepts.

MUTATION_ We shall first and foremost consider the principle of »inexact replication« (mutation): no animal in nature produces offspring that are completely identical to itself, and no lifeform produces identical offspring all the time. This may in part be influenced by the physical difficulty of »repeating« a complicated genetic code without any error. (As a point of comparison, you might try to memorize the first 250 decimal places of π.) The act of exact replication is further complicated by the varying states of the environment in which such copying takes place. However, it stands to reason that even perfectly identical laboratory conditions would, sooner or later, be host to imprecise copying procedures.

When the ancestors of modern-day reptiles and birds »initiated« the colonization of land, they might have laid eggs that were spherical, like those of fish. If that was the case, then many of those eggs must have rolled away and were probably destroyed in the process. This waste might have been compensated for with the more prolific production of eggs, which would have cost more energy. However, due to the principle of inexact copying, some early reptiles may have produced more spherical eggs, while those of other reptiles were less so. The non-spherical eggs, being more stable on the ground, are then more likely to have stayed intact and produced healthy offspring.

INHERITANCE_A second principle of nature now applies: offspring usually inherit most of their phenotypic properties from their parents. So, if more parents are born from *less symmetrical eggs*, then — depending on the strength of this selective pressure — the next generation should exhibit a higher proportion of individuals laying non-symmetrical eggs. If the selective pressure is strong, then it does not require many generations before the kinds of reptiles that lay *classically shaped eggs* — and not those that lay spherical eggs — become dominant.

IT IS THE SUM THAT MATTERS_We might now argue — quite wrongly — that this principle should apply to every trait of the organism, and that it should take only a few generations before a perfect solution is found for every problem (which in the case of the eggs would be purely geometrical). However, this runs contrary to a third principle of nature: the *best sum of adaptations* is usually the successful one, and this also depends on a constantly changing environment. If some reptiles, for instance, continue to lay spherical eggs but are »better at evading predators« due to another trait that is affected by the same genes, then this might turn out to be a more successful adaptive strategy than »not laying spherical eggs.«

It *may* be the case that the individuals who are better at evading predators also adapt to lay non-symmetrical eggs without any negative effects on their evasive strategies. For lack of a better term, this would kill two birds with one stone. It may simultaneously transpire that another group of »better evaders« acquires a new trick by way of mutations that somehow lessen the disadvantage of spherical eggs. Whatever the situation might have actually been, it seems that the advantage of more or less non-spherical eggs was so significant that this trait is now shared by virtually *all* oviparous land-based animals.

PHYLLOTAXIS, OR: HOW EASILY ARE WE DECEIVED?_What appears to be a simple explanation for the non-spherical shape of eggs may require more profound scrutiny in other cases. A typical example is the arrangement of sunflower seeds, which is also known as *phyllotaxis*. Much has been written and theorized about this topic — and despite all of that, the solution turns out to be as ingenious and simple as evolution itself.

A PREMATURE CONCLUSION_Let us briefly discuss the underlying theories: if we look at a sunflower long enough, we will inevitably find spirals in the pattern. In fact, it is possible to discover two opposing spirals! We may also notice that the number of spirals in sunflowers is always what is referred to as a Fibonacci number. The *golden mean* is closely linked to these famous numbers, as is exponential growth. (By repeatedly multiplying a number with itself, we get a geometrical sequence that — if displayed in a coordinate system — yields an exponential curve.) We might be quite impressed by these insights: it seems that sunflowers are »aware« of the golden mean and grow exponentially — while most other plants do not!

A COMPLETELY DIFFERENT EXPLANATION_Now on to the most likely explanation (claiming to know the *definitive solution* for such matters could be misguided): a sunflower multiplies by means of its seeds, which are usually eaten by birds and dispersed because the birds that pick them tend to lose seeds from time to time. One thing is clear: for this strategy to work, a sunflower has to produce as many seeds as possible. This increases the probability that the species will survive in the long run.

Let us imagine a plant species (a predecessor of the modern sunflower) which may have adopted the following successful growth strategy during the course of its evolution: »Grow your seeds by creating a principal seed at the approximate center and position each further seed such that the direction changes and the position is translated slightly away from the center.« In order for this procedure to be quantifiable, the rotation angle would have to be constant, as would the translation distance. Up to a certain point, this strategy would, indeed, allow the flower to produce new seeds. The plant would reach its natural limit when there was no more space for a new seed because it was occupied by an existing seed.

Among the infinite number of combinations of angles and distances, there is certainly an optimal one. Evolution would eventually find it over the course of many generations, as it did with the shape of eggs: mutation is constantly at work, changing angles and distances ever so slightly, and the flowers that are lucky to have hit upon a good combination will carry more seeds for their reproduction. »Their« advantageous combination will be passed on to their progeny, which have a chance of »improving« it further.

The result is quite impressive: a very »close« approximation of a modern sunflower_FIG. 33, LEFT. The center is a little tight — the seeds there would have to be smaller than those at the outer edges. But the fact still remains that this is a remarkably effective method of packing seeds into a blossom. We can also detect the aforementioned spirals. Indeed, this method creates them entirely without the need for exponential growth! The angle of rotation is quite peculiar, dividing the whole revolution of 360° at the golden mean.

The images in_FIG. 33 at the center and on the right show further possibilities if the translation distance is allowed to decrease by a certain percentage as the seeds move away from the center. However, this is mere »cosmetics.« If this procedure were to be »extrapolated« onto a sphere, as illustrated in_FIG. 33, then the seed density would increase even further.

Incidentally: from a daisy to a meter-high sunflower — an explanation for their specific proportions can often be found in the »golden mean,« which has fascinated many numerologists. This is not meant to be derogatory: even their deliber-

FIG. 32_ *Different translation distances*

Images by Georg Glaeser, 2022

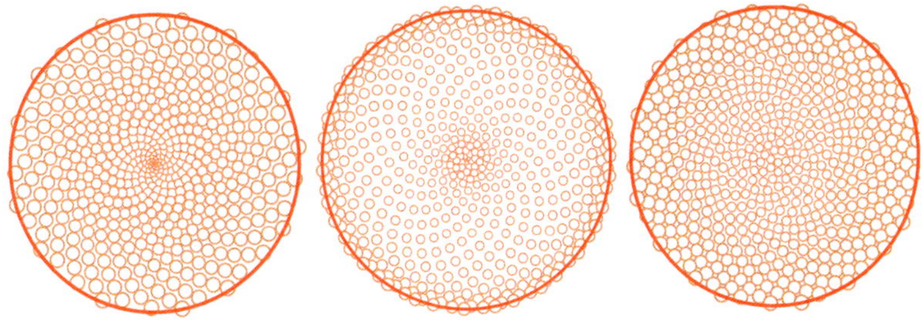

FIG. 33_ *Computer simulations; right: optimization by extrapolation onto the sphere or ellipsoid*

Images by Georg Glaeser, 2022

VIDEO

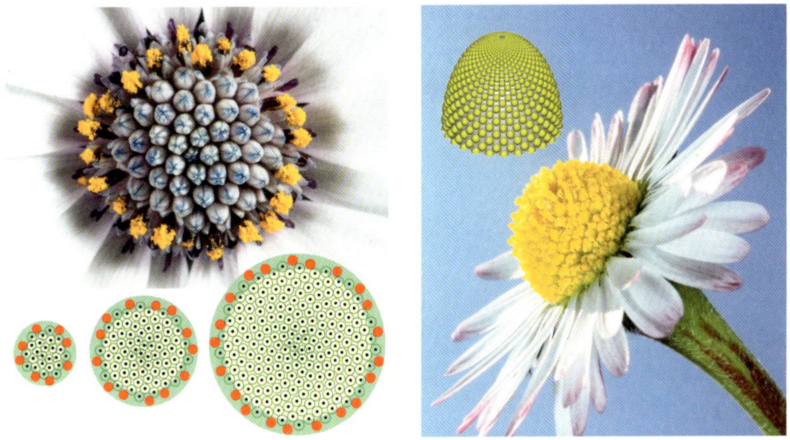

ations may be mathematically valid. It can be shown that the golden mean is the »most irrational of all numbers«:

$$\phi = 1 + \cfrac{1}{1 + \cfrac{1}{1 + \cfrac{1}{1 + \cfrac{1}{1 + \ldots}}}} \approx 1.6180339887$$

The flower »needs« this irrationality, which makes it possible to apply the rotation-translation step to the seeds as often as possible.

THE DEVELOPMENT IS NOT ALWAYS CLEARLY VISIBLE_By changing the golden angle ever so slightly, we get patterns as in_FIG. 34 on the right. Such shapes are familiar to us from nature: for instance, the blossom on the left of_FIG. 34 already closely resembles the pattern.

Let us leave it at that, without proposing any further hypotheses. In geometry, spirals are relatively easy to produce: even when they are defined quite differently, they may be very similar locally. The spiral pattern on the left of_FIG. 34 might owe its appearance to a completely different principle than the filling pattern of the blossom.

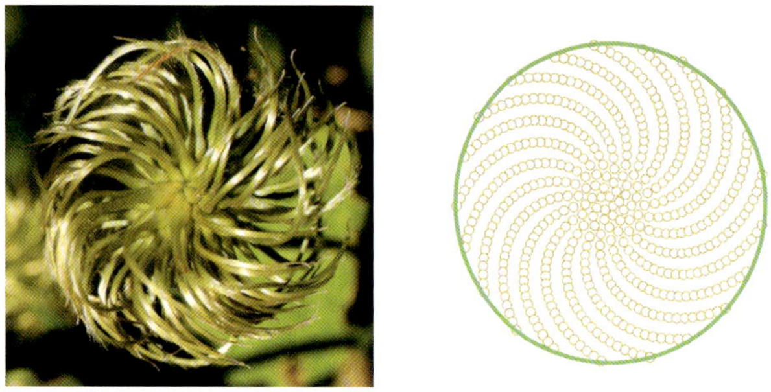

FIG. 34_ *Changing the »golden angle« by tenths of a degree*

Images by Georg Glaeser, 2022

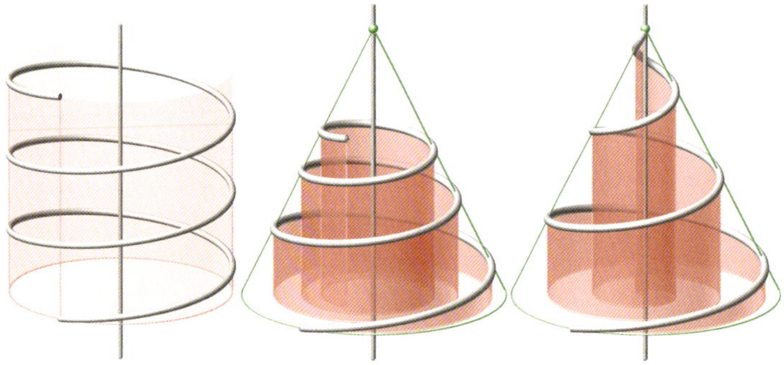

FIG. 35_ *Helix, helispiral, and the »classical« spiral*

Images by Georg Glaeser, 2022

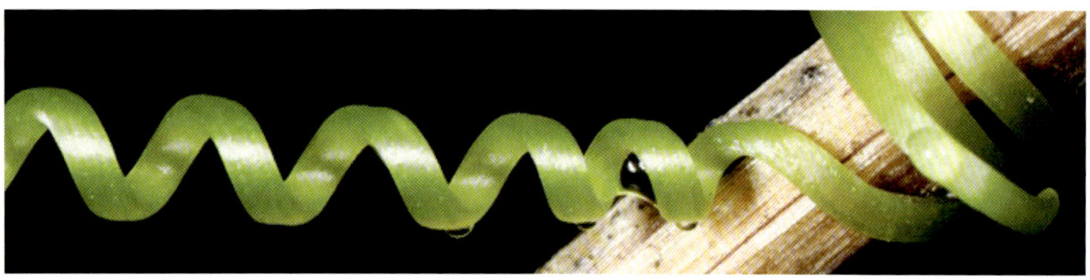

FIG. 36_ *A result of a helical motion in nature*

Photo by Georg Glaeser, 2022

Spiral Shapes

In three-dimensional space, in analogy to the rotation around a center, there is also screwing around an axis. The translation of a space point along the axis is proportional to the rotation angle ϕ about the axis, and the path curves are then helices_FIG. 35, LEFT. If, at the same time, the distance to the axis is increased/decreased linear-proportionally to ϕ, the path curves are called helispirals _FIG. 35, MIDDLE. If both the speed of the translation and the change in distance to the axis are altered exponential-proportionally to ϕ, the path curves are »classical« spirals_FIG. 35, RIGHT.

_FIG. 36 shows a rare example of a helix that can be found in nature. Remarkably, the helispiral motion and the classical spiral motion are frequently found in nature when it comes to the growth of horn-like material_FIG. 37 or calcareous shells_FIG. 38. They are among the very few complex structures in nature whose shape a computer is capable of reproducing with almost unbelievable precision.

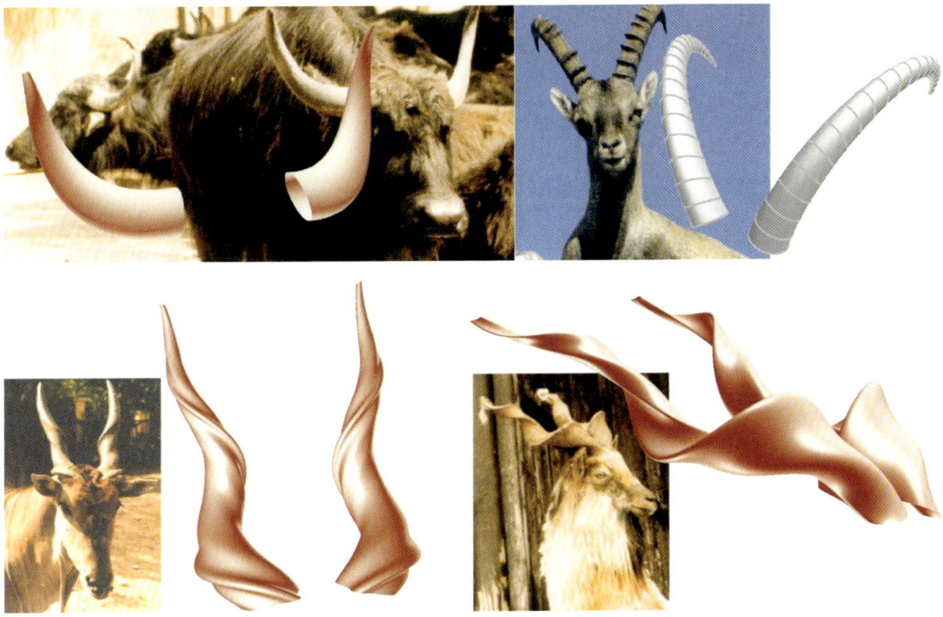

FIG. 37_ *Results of helispiral motion in nature*

Images by Georg Glaeser, 2022

VIDEO

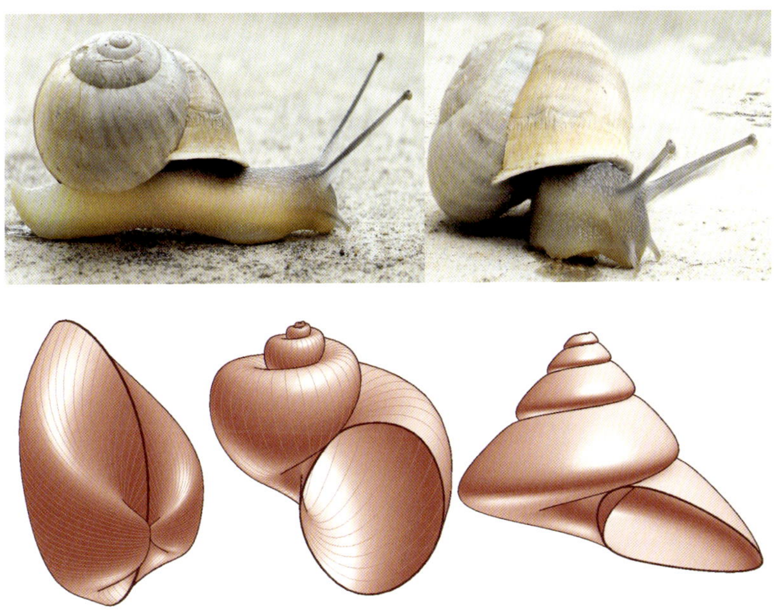

FIG. 38_ *Results of classical spiral motion in nature*

Images by Georg Glaeser, 2022

1 Georg Glaeser, *Fast Algorithms for 3D-Graphics* (New York: Springer Verlag, 1994).
2 Georg Glaeser, *Geometry and its Applications in Arts, Nature and Technology*, 2nd ed. (Vienna: Springer, 2020); Georg Glaeser, *Nature and Numbers* (Berlin: De Gruyter, 2013).
3 Glaeser, *Geometry and its Applications*.

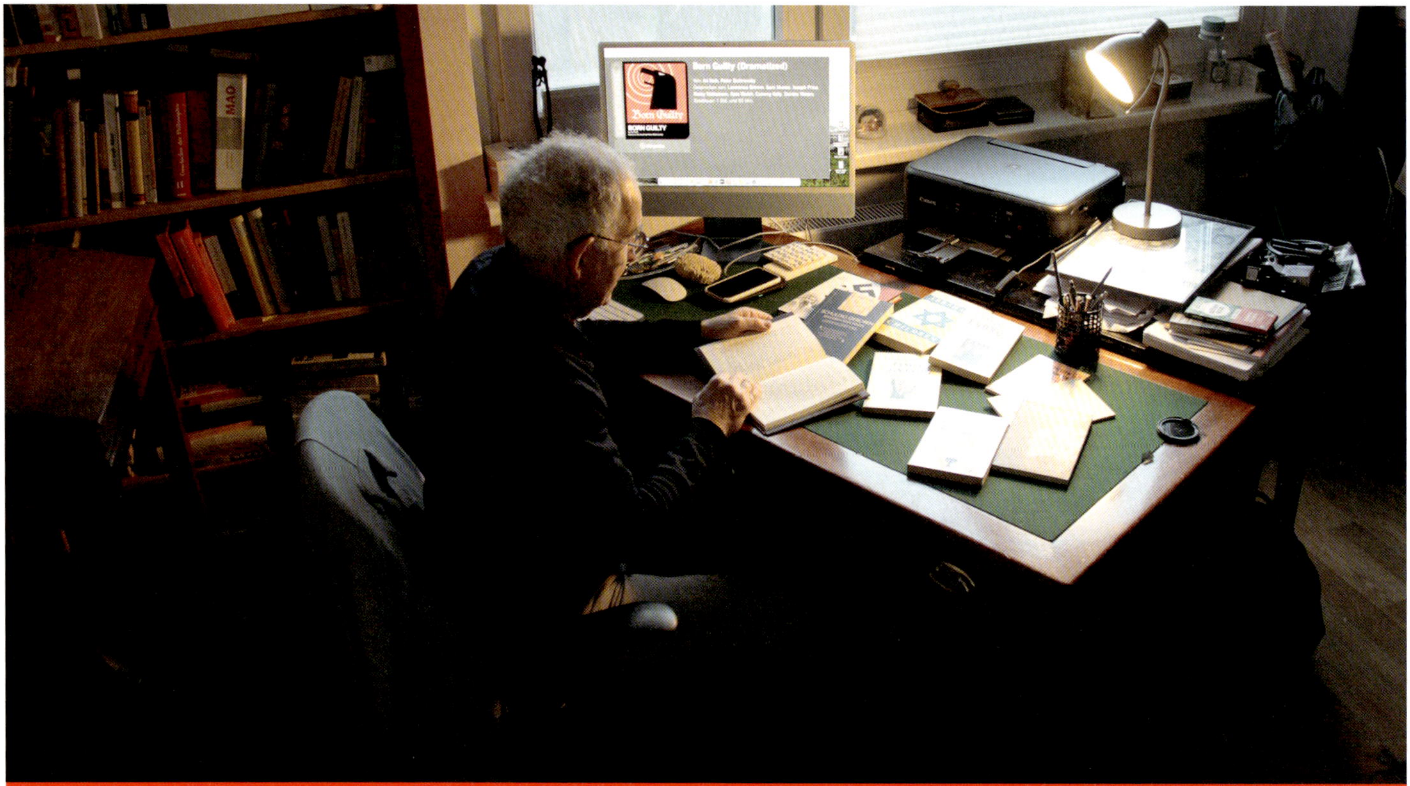

FIG. 1_ *The writer's kingdom — doors to perception*

© Alfred Vendl, 2023

Visualizing Words: The Diversity of Literary Adaptations

Peter Sichrovsky

There are no rules for or limits to adaptation. It is the task and responsibility of each artist to create a compelling adaptation in their own field while showing respect toward the artist who created the original work. However, there are also exceptions, such as when problematic works are criticized or even made fun of. This means that there is only one rule for the adaptation and visualization of texts: there are no rules_FIG. 1.

Interview Adaptations

One of the definitions of adaptation is making a literary work fit another text or art form while changing the content of the work as little as possible. The term is of Latin origin and means »adjustment« or »the ability to adjust to other circumstances.« In medicine, it means the ability of the retina of the eye to adjust to various lighting conditions. In literature, we speak of adaptation when a written text is translated into another art form, which can be music, a play, a film, an audioplay, or a ballet.

Right at its etymological root, the term »adaptation« contains a misunderstanding. In the beginning, it was predominantly used to describe physiological processes, later the adaptation of electronic systems (adapters). The specialist terminological use of its figurative meaning in the field of the arts has been shaped by these everyday semantics. When adapting a work of art to another artistic genre or another art form, there is always a danger that it will be misunderstood as »mere« adaptation, which implies both the valorization of the original and the depreciation of the adaptation.

Born Guilty

I would now like to address some of my personal experiences with the various stage and film adaptations of my books. In 1985, my book *Wir wissen nicht was morgen wird, wir wissen wohl, was gestern war* was published with the subtitle *Junge Juden in Deutschland und Österreich* (later published in English as *Strangers in their Own Land: Young Jews in Germany and Austria Today*). It was about the postwar generation of Jewish families, and I spoke with men and women who had been born after the war. It was not about survivors or refugees, but about people who, ultimately, had not experienced anything firsthand but had still grown up with the experiences of their parents, uncles and aunties, grandparents, and parents' friends.

War was part of their everyday lives, even when they had been born years later. They lived with stories that were not their own but that still had a crucial impact on their lives. On the basis of many pages of documented conversations, I wrote monologues in which the affected parties spoke about their lives in the first person. In both this book and the one that followed, there is also a dialogue between siblings. My questions do not appear.

In 1987, the second interview book was published, *Schuldig geboren*, with the subtitle *Kinder aus Nazifamilien* (English: *Born Guilty: Children of Nazi Families*). It was a collection of interviews with the daughters and sons of parents who had been active National Socialists in Germany and Austria.

Once again, the condition was that the interviewees had not lived through the war themselves but had indirectly experienced that time through their parents and other relations. It was not about important names, neither in the book about the descendants of the victims nor in the collection about the children of the perpetrators. I was interested in utterly insignificant families, the banalities of everyday life during this period, and how the parents' experiences had influenced their offspring.

A few weeks after the publication of both the first and second books, directors began contacting me requesting the theater rights. They wanted to adapt the books into plays. In the years that followed, several dozen stage adaptations were produced around the world, from Australia to Israel, from Austria to the US_FIG. 2. In 1990 and 1991, *Schuldig geboren* was one of the most frequently performed plays in the German-speaking world. A theater troupe traveled through small German towns performing it on stages that did not put on their own productions. An adaptation by an Israeli director was shown at the same theater in Tel Aviv for a whole year. At the Khan Theater in Jerusalem, there was a period of several months during which Israeli soldiers attended performances of *Born Guilty* as part of their training.

Some productions selected several characters from the two books and mixed them into a single play. A Spanish production got the actors to speak into the camera with their backs to the audience and projected an oversized image of their faces across the stage onto a screen. Some broke the monologues down and got the actors to recite them to each other.

American writer Ari Roth went in a completely different direction_FIG. 3. He wrote a new play and turned me, the interviewer, into the main character, also calling him Peter.

He spoke with me for several weeks and created a fictional character who, in the play, decides to write a book about the children of the Nazis. His idea turned very odd when he set out to find someone to play the role of »Peter« in Chicago and ended up hiring my cousin, who worked there as an actor and also looked like me. I thus saw myself on the stage working on my own book using anecdotes from my own life; my ex-wife, my children, and my parents also made appearances. His productions were awarded a number of prizes in the US and were performed in numerous theaters_FIGS. 4, 5, 6.

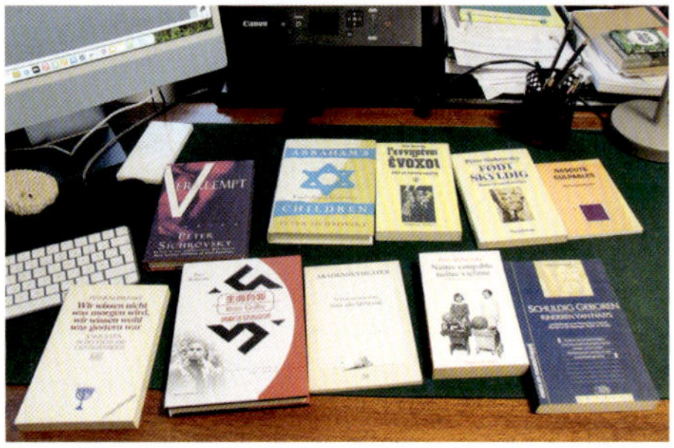

FIG. 2_ *Book publications by Peter Sichrovsky*

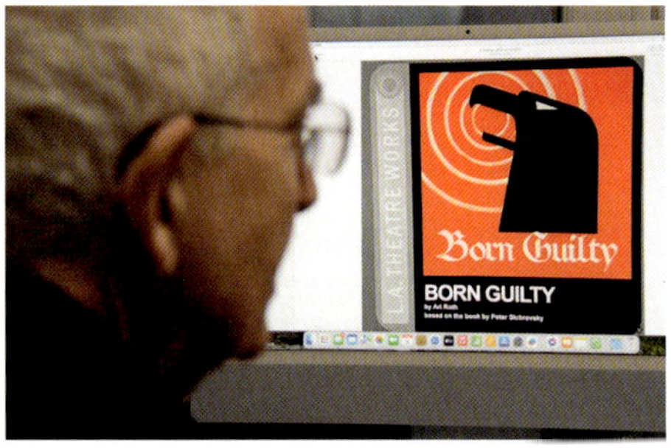

FIG. 3_ *Theater play of* BORN GUILTY, *memories of Ari Roth in the US*

George Tabori

One of the directors invited me to work on his production with him — George Tabori, who produced a staging of *Born Guilty* in Vienna at the Der Kreis theater.

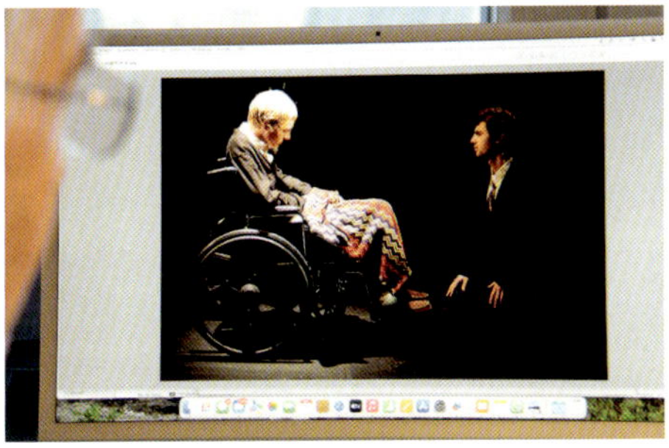

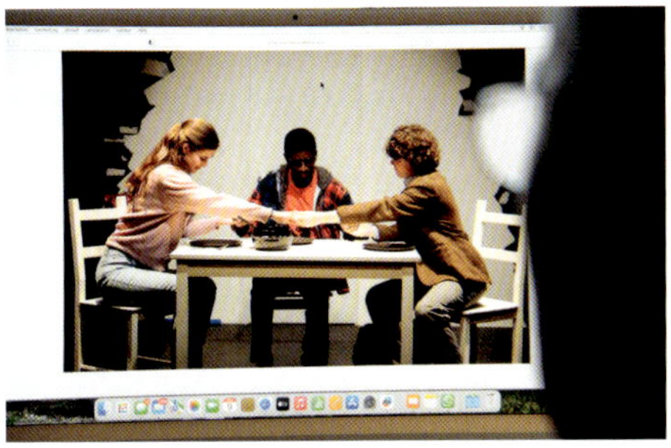

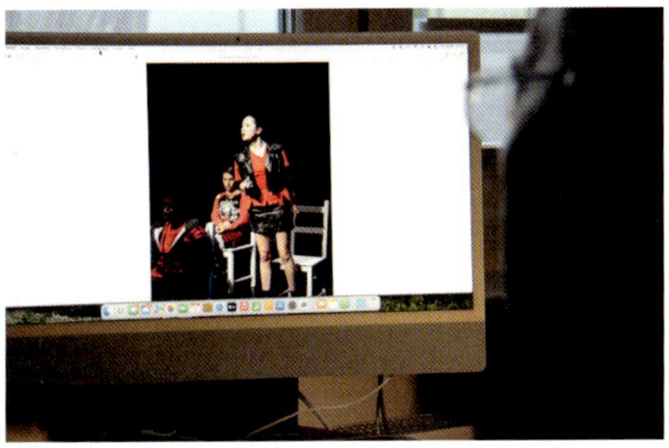

FIGS. 4, 5, 6_*Memories of the theater play* BORN GUILTY *in Washington, New York, and Chicago (US)*

Born in Budapest in 1914, Tabori was the bilingual son of a journalist. His father was a member of the Galileo Circle, which Georg Lukács also belonged to. Until he was seven years old, his parents raised him as a Catholic and only revealed to him later that he was a Jew. Tabori began a hotel apprenticeship at the Hotel Adlon in Berlin after finishing his *Matura* at high school. In 1933 he left Germany and emigrated to London, where his older brother Paul lived. Between 1941 and 1943, he worked as a foreign correspondent in Sofia and Istanbul, and traveled to Palestine and Egypt. In his biography he says that this was probably the best time of his life. During the war, he cooperated with the British secret service, faked his own suicide by writing a farewell letter to his parents from Istanbul, and worked as an officer of the British Army's intelligence agency under the name of George Turner. His father was murdered in Auschwitz. His mother escaped deportation by chance. He addressed his mother's fate in the story *My Mother's Courage*, which was also turned into a film. His brother Paul was the head of the British PEN Club's division of authors in exile.

In 1947, he moved to the US, where he would live and work for the next twenty years. His intellectual environment was a »who's who« of émigrés, with names like Bertolt Brecht, Thomas Mann, Lion Feuchtwanger, Arnold Schönberg, Theodor W. Adorno, and Heinrich Mann. He got to know the family of Charlie Chaplin and had a brief affair with Greta Garbo. He later described that time as follows: »Imagine you are living in a house and, every Sunday, Büchner, Kafka, Flaubert, Mahler, and others come to visit you…« He wrote screenplays for Alfred Hitchcock, Joseph Losey, and Anatole Litvak.

In 1947, in New York, during the rehearsals for a play by Brecht that he was attending as a journalist, he offered his translation services to Charles Laughton, who was despairing at the German text. After completing the job, he stopped working as a journalist and writing novels and screenplays, and dedicated himself to the theater. The artistic bridge he built from writing to directing and theater-making produced a unique combination of skills for visualizing literary texts as dramatized versions of themselves on stage.

One of his best friends was Elia Kazan, to whom he owed his directing method. About his methodology he wrote: »Most of the things that can be seen there come from the actors. All I did was react to their ideas.« Tabori developed a style—that I also had a chance to observe—where he only emphasized positive aspects about each of the actors' portrayals during preparations and rehearsals and simply ignored the negative.

In 1968, Tabori returned to Germany for the first time. Helene Weigel had invited him to an event celebrating Brecht's seventieth birthday. He spent the next few years working as a director in Germany until he took over the management of the Der Kreis theater in Vienna in 1987.

DER KREIS_Form 1983 to 1987, I lived in New York, where I met Oscar Bonner, who convinced me to come to Vienna to help him found the newspaper *Der Standard*. I was not enamored with the idea. I had left Vienna in 1980 for Berlin before moving to New York, and was happy that I was not living in Vienna anymore. The work on the two interview books about the children of the victims and perpetrators had successfully prevented me from thinking about my intentions to perhaps go back to Vienna one day, until Oscar Bronner came to me with his idea, at which point

I was also slowly running out of money in New York. During the preparations for the founding of the newspaper, I received a call from the Der Kreis theater. It was Tabori inviting me to have a coffee with him.

A tall, thin man with thick white hair and a mustache was waiting for me. He looked old, was already over seventy, seemed slow and tired, and I wasn't sure what he actually wanted from me. I wasn't really interested in the theater back then. Modern plays bored me, my German teachers had turned me off the classics, and I saw myself as more of a journalist, even though I had already written a number of books by that time.

Tabori said he would like to stage the book *Schuldig geboren* and wanted to know if I had anything against it, as it was also a matter of the rights to the book. I told him I didn't care, that it was a matter for the publishing house. He laughed and asked me if I really didn't care about what he wanted to do with my book. I nodded and answered: »No, I don't care at all.« My reaction seemed to surprise and amuse him. He said he had actually expected more problems from my side and had been prepared to win me over me if he had had to. Now it was I who had to laugh, and I assured him once more that the book was over and done with as far as I was concerned. I wouldn't stand in the way of anyone who wanted to work with it now. We got along well, spoke a similar language, and seemed to think in a similar way, even when we were both making fun of each other.

»However, I will only do it if you work on it with me,« he said suddenly, using the informal German word for »you«: *Du*. It was not the *Du* that agitated me, as he seemed to sit there in front of me like the grandfather I had never met; rather, it was his request that I work with him. »I don't know anything about the theater, and I'm not really interested in it. I'll make you a suggestion: I'll come to the premiere, and everything else is your responsibility,« I said. »Then I won't do it,« he said. »Then don't,« I replied. But he did not give up and convinced me to come to his rehearsals and talk to any actors who had questions. He said he wasn't asking me to work on the production, that it would be more of an offer from the author, which he said was quite common in stagings of modern plays. I no longer know how he managed to bring me around, but I agreed. I had underestimated the old man. With his quiet voice and slow sentences, his method was like that of a spider spinning its web, thread for thread. He simply didn't react to any of my disparaging, sometimes cynical comments; he ignored them, praised the book, spoke enthusiastically about some of the characters among the interviewees, and, after an hour and several coffees, I agreed to everything he was asking from me.

On the way home, I tried to understand what had happened. I was supporting a theater production of my book and had also said that I would be willing to work on the rehearsals—although theater, firstly, bores me and, secondly, I was unable to work in groups. I would have liked to go back to the cafe and explain to him that I had changed my mind, but I couldn't do it—I couldn't do it to him. I just liked him too much.

THEATER REHEARSALS_A week later I arrived at the theater. He introduced me to the actors and his colleagues, and I explained to everybody with a mischievous grin that I had just come to watch. I wasn't interested in the theater, but they could ask me questions whenever they liked. I sat down on one of the free seats in a corner of the rehearsal space. Tabori asked everybody to form a circle and then got into the circle himself with the actors. He waved me over: he wanted me to join the circle too. When I hesitated, he said: »These aren't the rehearsals yet, we haven't even read the text. This is just us getting to know each other.« Once again, he had convinced me. He asked everybody in the circle to whisper a secret to one of their colleagues, something that they were thinking about. But the listeners were not allowed to react. Later, everybody would have to try to guess how the silent listeners would have reacted. Tabori stood next to me. He whispered something in my ear. When it came to talking about how I would have dealt with his secret, Tabori said: »Peter would say: don't worry about it, it's all quite normal and not really that important.« He had actually captured my attitude quite precisely.

In the weeks that followed, Tabori interrupted the reading of the texts with role-plays and other activities that didn't seem to have anything to do with my book but that forged a sense of community among the actors. He also encouraged everybody to speak about their own parents, about memories, and experiences. The subject of the book didn't play a role; he was interested in every idea, every anecdote, every experience, regardless of whether it had anything to do with the book or the play that was being developed.

Then he had a large table set up on the stage and asked the artists to take a seat at it. He wanted me to take a seat too. There are fourteen people in my book. Each woman, each man tells the story of their fate in the first person. I had also included a telephone conversation with a potential interview candidate in the book, although he had declined to give an interview after that conversation. Tabori was fascinated by the telephone conversation. He said: »That puts you in the book too, which means that you have to get involved.« The actors chose a character from the book, and Tabori asked the people sitting at the table to speak to each other in the first person, just like in the book.

DIALOGUES—MONOLOGUES_ All of a sudden, dialogues were being produced using monologues. He thus pulled the printed text onto the stage for the first time, where it could be watched. This was the beginning of the visualization process. The dialogue between the author and the readers without an audience developed into a public dialogue on the stage that everybody was able to experience. The individual characters that had been written independently of each other for the book and lined up consecutively in the form of chapters entered into relationships. To illustrate this, Tabori decided to open the performance with the chapter »A Telephone Conversation,« that is, with an interview that did not take place. The telephone conversation begins with the harmless request for an interview and ends in an argument with the potential interviewee.

He then tried to extract more creativity from the group situation. The goal couldn't just be to get the actors to confront each other on the stage with the texts; the most crucial step to finish the play had not been taken yet—finding a form that could be used to reach the audience.

The experiment with the table on stage and the monologues among the actors did work as an interim step but not as a satisfying form of adaptation. In his next attempt, Tabori asked the technicians and all the other collaborators to sit in the auditorium and suggested that the monologues be read from the stage directly to the audience. It worked terrifyingly well; we stared spellbound at the speakers on stage, who told us the stories of their lives. Observing their facial expressions and movements helped us to listen. The text had left the sofa, as it were, and could now be watched on stage.

The next step was simple and logical. A number of actors suggested the following: »We won't speak to each other here, sitting at the table; we'll speak to the audience, one after the other; each person will stand alone on the stage.«

I later suspected that Tabori had planned this form of adaptation from the beginning but had elicited and permitted a process, a development, to take place that led the actors to the same result. By supporting certain ideas without criticizing other suggestions, he steered the process in a certain direction. It all looked very democratic, but it wasn't. It was just friendlier than intervening with negative and positive reactions to various suggestions. Tabori produced a creative atmosphere among the people involved by lending his support to what he thought was good and ignoring what he didn't like. He invited the actors to support the monologues with different ideas, with different clothing, objects that they held in their hands, or movements that would fit the language.

After many weeks with all kinds of solution models, the play was finished and so was the idea for adapting the book to the stage. Tabori and his actors agreed that the characters in the book would speak to the audience in monologues, just as the chapters had been delivered to the readers.

The foundation of the individual chapters, the characters' self-portrayals, were interviews and conversations between the interviewees and me. Back then, I used an interview technique that I had developed myself: I usually began by speaking about myself, about my childhood, my relationship with my parents, until the interviewees suddenly said: »That was the same for me,« or, »That was quite different for me.« This led to a conversation without me cornering them with questions from the outset. I did not accuse anybody of anything in any of the interviews, not even those who defended their parents. In a relaxed atmosphere, most of the interviewees were prepared to talk patiently and at length about their childhoods and memories of their parents and relatives. For some, it untied a knot of silence, and after that they would call me telling me they had remembered some thing or other, that they wanted to see me, or even maintain a friendly relationship.

THE SILENCE OF PARENTS_ In the beginning, many of the interviewees found it difficult to speak about their parents, as they had all been active and enthusiastic Nazis under National Socialism. Some parents never talked about that time; others defended their former decisions and were not prepared to acknowledge any mistakes. The parents' attitudes toward their own pasts were transmitted to the next generation, which, however, at school and through the media and other forms of information, had to indirectly live through that historic period as one of the most horrific in the history of Germany and Austria.

A historically unique rupture with the parental generation took place. The process of separation that usually takes place in puberty and adolescence overlapped with the transition that was taking place from dictatorship to democracy. As in every intergenerational conflict, everything that the parents thought and did was criticized, even without the addition of the Nazi period. This was now joined by condemnations of that time as a period of dictatorship and criminality. This might have been the reason why the book was discussed in other societies, like in Spain after the changes that took place in the wake of Franco, in China with respect to the Cultural Revolution, in Russia after the fall of the Soviet Union, and in South America. In societies where dictatorships had ended, the next generation was asking: What did my parents do back then? Were they responsible too? Were they involved in crimes? Was my father a murderer? Did my mother hide everything that my father did? And has all of that had an influence on me?

One of the monologues ends with a sentence spoken by a woman whose father was in the SS. »Can wolves become sheep from one generation to the next?« Tabori had the idea of getting the woman to play with a stick, alternately placing it on the ground, picking it up again, pointing at the audience, or simply holding it in her hand. In the last sentence about the wolves and the sheep, she broke it in half. It was one of the most memorable scenes of the entire performance.

The premiere was a huge success. The play ran for weeks to a full house. A dozen theaters invited the dramaturg from the Der Kreis theater to put on the play in the same form on their stages. What was the reason behind its success? Was it simply its content? My well-written book? That was just a small piece of the mosaic. What was decisive was Tabori's staging — the way he tricked the audience into playing the role of the interviewer was ingenious. The actors spoke directly to the spectators, stood at the edge of the stage without moving much, and kept the spectators glued to their seats with their monologues. This transferred the original concept of the journalist interviewing potential »victims« about their relationships with their parents onto the audience, which took on the role, creating a play in a dialogue between artists and spectators.

THE TELEPHONE CONVERSATION_ But in order to immortalize me as the author in the play, Tabori included the »Telephone Conversation« chapter in his staging. He recorded it and played it before the play began as the spectators arrived in the auditorium and looked for their seats. This allowed him to capture the audience's undivided attention as they took their seats in silence, listening.

With *Born Guilty*, Tabori produced one of the most successful stage adaptations in the history of German literature. The visualization of a written text for spectators worked better here than in many other adaptations, which are usually only able to stage a small excerpt of a work. Unlike many other literary works, this book was not a product of someone's imagination, of the thoughts of an author writing at a desk. It was based on interviews, dialogues, and therefore scenes from everyday life, and was not the solely creative, silent work of an author. The conversations were first edited and transferred to paper, which was then translated into a confrontation between actors and the audience.

Der Spiegel wrote the following about the premiere in Vienna:

> *The confessional prose kept sliding into a perverse role reversal — some of the perpetrators' children tried to self-pityingly present themselves as a special kind of primary sufferer. Tabori's troupe can take credit for shattering this attempt to »level the difference between perpetrators and victims« (Peter Körte in* Konkret*) both mercilessly and theatrically. This sometimes took place by means of denunciatory comedy and sometimes by means of powerful (stage) brutality. One Nazi daughter*

pulled the SS boots over her arm to make a grotesque Hitler salute before describing the most pitiful of all figures from the industry of people coming to terms with the past: a history teacher, »long hair, beard, Norwegian sweater,« who begins by rambling on for hours during lessons about the resistance mounted against the powerful but goes straight to the principal when his students rebel. Her denunciation of self-pity took place on double ground. For almost everybody involved was a member of the generation that had risen up in the 1960s with the goal of putting an end to the unbearable silence of the fathers' generation, demanding enlightenment, if not penitence, at the top of their lungs. For a good two weeks, Tabori, a quiet 73 year old with a slow drawl, had gotten the young actors to read, recite, play out the interview texts — until the point of unavoidable psychological crisis. That, too, came on stage.

In a piercing voice, Therese Affolter, the forty-three-year-old officer's daughter, exposed her brother as someone constantly bemoaning his »victim role.« It was a scene of roaring speechlessness. The brother, 38, abused his Nazi dad for not joining the resistance with insistent, lagging disobedience — while sonny boy smoked joints and even took part in Antifa protests, before fleeing into helplessness (quite the weary revolutionary from the movies).

This scene culminated in a separation and a chaotic brawl. But occasionally, like in a hot flirtation, the brother put the SS boots on his sister — a teasing scene like on the park bench (»no, don't do it«), when her knickers fall down to great snickering. Unrequited love becomes frivolity. The truth, says Tabori, is obscene and unbearably funny.

Adapting Sociopolitical Problems

Keeping it in the family, I would like to describe two more examples of successful adaptations. The Israeli director Noam Brusilovsky is considered an expert at adapting sociopolitical events and personalities to the stage. The German public broadcaster, the ARD, awarded him its audio play prize for *Adolf Eichmann: Ein Hörprozess* (Adolf Eichmann: An Audio Trial). In 2021, the Klagenfurt Municipal Theater commissioned him to adapt the horrific case of child abuse by Dr. Franz Wurst for the stage. This led to the project *Nicht Sehen* (Don't Look)_FIG. 7, in which my son, actor Axel Sichrovsky, played the main role.

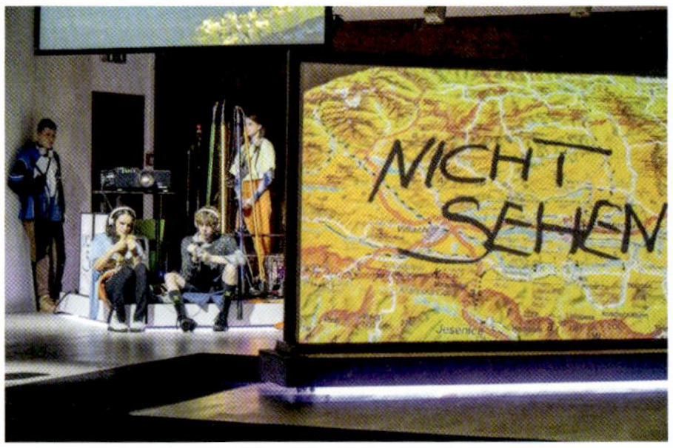

FIG. 7_*Stage scenery*
NICHT SEHEN *(Don't Look)*

Courtesy of Stadttheater Klagenfurt, 2022

The victims spoke on stage alongside professional actors, as did a psychoanalyst, a security guard, and experts who provided specialist commentary on the case and described case studies. The performance had nothing to do with traditional theater, but the two hours on the stage were captivating and terrifying, and left behind lasting memories.

A political scandal that had gone on for years was rendered in concentrated form without any long-winded analysis or extensive reports. After two hours, everybody had understood what had happened back then. The theater, which otherwise gives space and a place to theatrical interpretations of everyday situations made by authors, directors, and actors, was now offering itself up as a documentary stage, as if it was presenting a comprehensive news show on the theater stage. It takes the talent and the aptitude of directors to adapt political and societal events in such a way when they have not yet been addressed in art.

The Klagenfurt Municipal Theater published the following text about the production:

> *The visible and the invisible, as well as the possibility of telling their story, are at the heart of Noam Busilovsky's documentary staging, in which actors, town residents, and adolescents from the municipal theater's acting club are involved. In a staged »tour« through the town, the play takes a new look at the places where the events unfolded and their history. By examining orthopedagogical research literature, victims' reports, and voices from the town, the institutions' closed structures are broken apart on stage and the various stories woven together acoustically and visually. This creates an interplay between what can be seen and heard — between what we think we know and what we can only guess at.*

The Austrian magazine *Profil* wrote the following about the production:

> *Young director Noam Brusilosky has worked a small miracle with the play* Nicht sehen *and the collectively produced script. The theater has seldom had such an impact. From the files of the victim protection agency, with appearances by experts and a very young group of lay actors, he has brought this scandal into the present day. The thirteen- and fourteen-year-old students are extremely angry. Rightly so. They had already learned a lot from the internet, the dark net, violence, porno — but this is really »extreme,« says one boy. A girl sneers: »Saw nothing, heard nothing? Surprise, surprise.«*
>
> *The Görtschach youth home was closed in 2013. Today it is a home for underage refugees who have fled their countries on their own. As Syrian Muaz Abou Noumeh, whose journey across the Mediterranean ended in Görtschach, stood on the Klagenfurt stage and told his story of arriving in Kärnten, the helpers ordering pizza and placing chocolate on the beds — »the first nice moment in ages« — many of the audience members had tears in their eyes. A little bit of humanity after all that violence. Some victims gave accounts of being abused by groups of men: »Work limousines — log books — trendy types.« The experts cannot exclude the possibility that there was a pedophile ring that erected a wall of silence with social prestige and influence. There is no proof of it.*

The theater production of *Nicht sehen* was a form of visualization that had less to do with optics and more to do with providing information, a documentary with actors, laypeople, and experts from practice. But on a stage without these visual, dramaturgical interventions, it would probably have been too boring and would not have had the same impact on the spectators.

Adapting Children's Books

Die Schule der magischen Tiere (The School of Magical Animals) is a children's book series by German author Margit Auer with illustrations by German illustrator Nina Dulleck. The first volume was published in 2013. So far, twelve volumes have appeared in the main series and seven volumes in the spin-off series *Die Schule der magischen Tiere — Endlich Ferien* (The School of Magical Animals — Finally Vacation Time). The books have been translated into more than twenty languages and keep appearing on bestseller lists.

The series takes plays at Winterstein School in Germany. The name of the town is not specified, nor is the age of the school children. Unlike at normal German schools, there are magical animals at this one. These animals can speak, and each of them becomes friends with one of the children. Each child's best friend is a »magical animal.«

In 2020, the books were turned into a film based on a screenplay that has little to do with the books. The film opened in cinemas on October 14, 2021. In early November 2021, the film distributor announced that one million moviegoers had already seen the film. With 1.5 million viewers in 2021, *Die Schule der magischen Tiere* was the most successful German film of the year. This is why it was the first children's film to win the award for the film with the most viewers at the 2022 German Film Awards. In September 2022, the new film, *Die Schule der magischen Tiere, 2. Teil* (The School of Magical Animals, Part 2) premiered. My grandson Loris Sichrovsky, son of actor Axel Sichrovsky, plays one of the three main roles_FIG. 8.

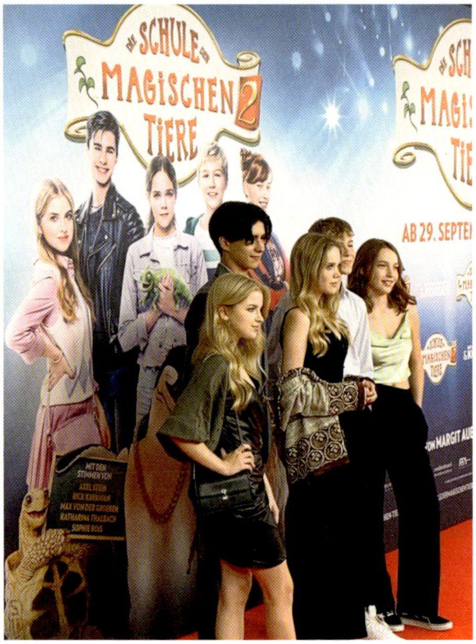

FIG. 8_ *Actors from the film* SCHOOL OF MAGICAL ANIMALS

The filming of this book series was not a simple task for the screenplay authors. They had to cram a whole series of books into the short plot of a ninety-minute-long film. But the most important characters had to stay identifiable for the viewers. The sequel turned out to be even more difficult. Unlike the famous and most successful children's book screen adaptation—the *Harry Potter* films—the children in the books by Margit Auer stay the same age; they do not get older from book to book. But because the film producers wanted to film part 2 with the same actors, who had not just gotten older but had also gone through puberty, they had to come up with an entirely new plot.

Peter Gutting from www.film-rezensionen.de writes:

> *Director Gregor Schnitzler and screenplay author Viola M. J. Schmidt have done everything they could to pull off a successful opening. By request, there are real actors and real settings, so it is not just animated. But the filmmakers did not want to work with real animals, so the fox, the turtle, and a blackbird were computer-generated and copied into the real film retrospectively. The producers did not shy at any costs to get this to look reasonably realistic. Some of the adults are charismatic magicians like the teacher and her brother, the owner of the magical pet shop, Mister Mortimer Morrison. On the other hand, some of them, to the joy of the children, are laughingstocks, like the school principal, who is overplayed as a neat freak, and his clumsy janitor.*
>
> *Author Margit Auer actually wanted to avoid imbuing her books with pedagogical lessons and instead turned the magical animals into thechildren's friends, who give them the odd useful tip, thereby indirectly pointing out bad behavior and damaging attitudes to their charges. However, in the film version, the messages about the value of friendship and the damaging effect of discord come across much more directly than in the books. The filmmakers were clearly more interested in carrying their young audience members away than getting them to use their imaginations. They might have achieved this, in particular as the effect of the songs and the casting of the children's roles had already been examined during test screenings.*

To visualize the story on the screen, the children, who had no previous acting experience, had to communicate with the speaking animals and sometimes even dance with them in the »empty space.« This is relatively easy in a literary text, where the action is simply described. It is much harder in a film, revealing a special form of visualization and adaptation involving the mixing of animation and real actors.

Summary

Adapting works that were originally narrative is nothing new for dramaturgs and directors. Since antiquity, people have been adapting material that was already available as literature to the stage and therefore turning it into something that can be experienced visually. Some works by great dramatists like Aeschylus, Sophocles, and Euripides were based on the works of Homer. Shakespeare also used epic sources for his drama productions. In the late nineteenth century, medieval epics were adapted for the stage within the scope of growing nationalist sentiment.

Since the 1920s, there has been a new emphasis on the practice of adaptation. Director and theater manager Erwin Piscator adapted a number of novels for his Proletarian Theater, including Tolstoy's *War and Peace* and Hasek's *The Good Soldier Švejk*. First and foremost, this had to do with content: the old classics did not reflect the political changes that were taking place in the Weimar Republic. Directors wanted to put current events on stage and show them to people who did not read books. The fantasy of better reaching the masses of workers by means of stage productions, visualizations, and topical material, however, failed. People who do not read books do not go to the theater either.

At the same time, Bertolt Brecht was developing his theory of epic theater. For both theater-makers, it was about opening up new material, narrative modes, and theatrical means that would reflect the experience of reality after the First World War and help to develop an »art of spectating« (Brecht) that was no longer based on illusion but on critical distance. The visuals on stage were supposed to reflect upon political conflicts and not just encourage the audience's fantasy with fairytales and historical material. Brecht, too, imagined he would be able to support the workers' revolution with his art, but the audience at his plays was the same one that had been watching the classics before.

The trend toward sociocritically epicizing the theater came to an end with the dawn of the Nazi period in Germany. Nor was the proletarian revolutionary tradition of the 1920s continued at the beginning of the postwar period. Most of the attention was being directed toward classical, realistic heritage, that is, toward the traditional dramatic forms.

It is only since the 1970s and 1980s that there has been another increase in in the number of adaptations of current material in theater programs. Their primary purpose is to expand upon theatrical and visual offerings and to open up interesting, complex material that does not exist in traditional drama or in a comparable quality. The synergy effect of material that has already been successful as a book or a film is another reason for adapting it. At the same time, the concept of drama has changed so significantly in recent decades that open structures and switching between situative portrayal and narrative report within one text are now usual practice. This has brought above all a younger generation to the theater. Modern visualizations of topical material using contemporary methods have modernized many dusty stages. The State Theater in Graz, for example, performed an experiment in the form of a stage project with rival football fans from the two Graz teams.

Not all literary originals are suitable for theater productions or filming in the same way. Some can be adapted almost true to the original, others require more interpretative transformation that can take them away from the original. To differentiate between the possibilities of adapting literature, the four types described by Helmut Kreuzer in 1981 are still frequently used today:

1. Adaptation as the appropriation of raw literary material: film and stage adaptations of this kind make use of selected plot elements or characters from the original; literature here supplies the material. This approach is a form of adaptation that has little to do with the original.
2. Adaptation as illustration: this kind of adaptation stays true to the events of the plot and the character constellations in the original as far as possible without taking into account the impact or rules of form in the new medium. These works, often referred to as illustrated literature, are generally disparaged as films or stage productions.
3. Interpretative transformation: taking into account the media-specific conditions, not just the content but also the relationship between form and content, the meaning and the impact of the original are translated into the image. This kind of analogous work can retain the spirit of the original, even if that spirit is radically changed.
4. Adaptation as documentary: recordings of theater productions, readings, interviews with authors, documentaries about places and artists, documentaries about the biographies of the protagonists or the localities of literary works.

The visualization of literary works is gaining increasing significance in art today. There are no limits to adaptation and therefore visualization. Theater, film, ballet, comics, opera, and other art forms reveal the spectrum of creativity and fantasy in the realization and adaptation of literary material: novels, poems, stories, newspaper reports, non-fiction books, documentaries, and even interviews.

Texts that convey a message and motivate artists from other fields to adapt them in their own field are an important, creative expansion and also disseminate the information in the original material among an audience that might not be familiar with the original texts or that might not have any interest in or access to them. Some of the most well-known and successful adaptations include the film adaptations of the *Harry Potter* novels by J. K. Rowling and the *Lord of the Rings* series by J. R. R. Tolkien.

There are no rules for or limits to adaptation. It is the task and responsibility of each artist to create a compelling adaptation in their own field while showing respect toward the artist of the original work. However, there are also exceptions, such as when problematic works are criticized or even made fun of. In summary this means that there is only one rule for the adaptation and visualization of texts: there are no rules.

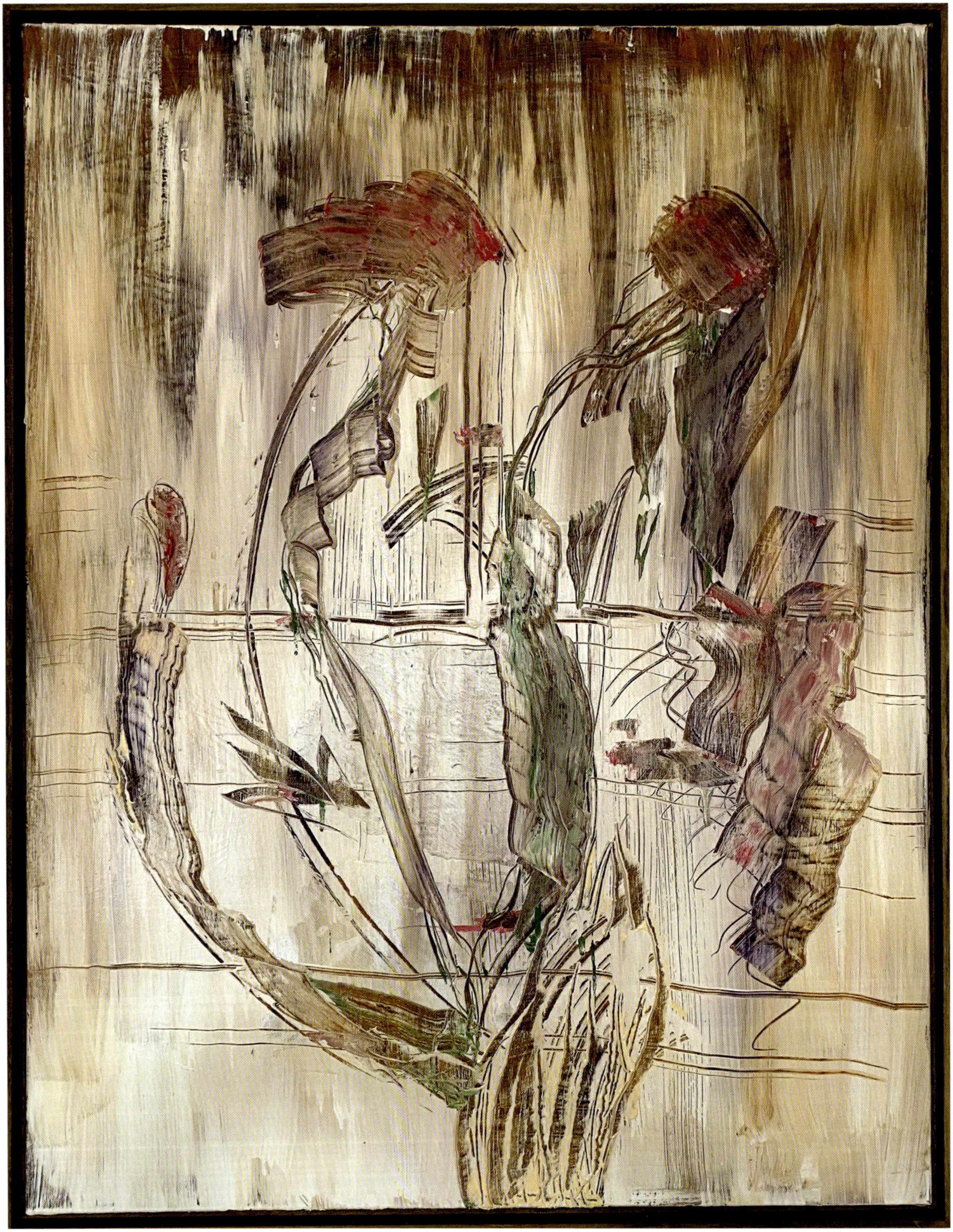

FIG. 1_
Rosenkrieg

© Peter Rumpler, 2020

»Break on through to the other side«*

Peter Rumpler

In his *Metaphysics*, Aristotle noted that human beings always prefer visual sensory perception. »For not only with a view to action, but even when we are not going to do anything, we prefer seeing (one might say) to everything else.«[1] This is joined by one general characteristic of perception, namely that we are more likely to perceive something that is not there than we are to not see something that is. This is the result of the logic necessary to survival that failing to see a danger could be punished with death and therefore with the extinction of the species. Human perception, and here above all visual perception, is therefore always heading toward a surplus of possibilities of world and is thus ideally predisposed to immersions of all kinds.

New Access to Deeper Layers

It is remarkable that the editor of a collected volume of essays on the scientific benefit of visualization has chosen an approach that has its origins in psychedelics. Alfred Vendl, one of the first on the quest to find the reality behind the mere surface, seems to have made a pact with the mystic here and suggests that we take the path less trodden together with Blake and Huxley. However, he chooses not the drug-induced path but the path that always manages, by means of extraordinary technical effort and creative performance, to penetrate deeper into objects and to depict the invisible so realistically that the viewer believes they are in a new world.

»If the doors of perception were cleansed everything would appear to man as it is, infinite,« wrote painter and poet William Blake in his 1793 book *The Marriage of Heaven and Hell*. According to Blake, truth only reveals itself to be infinite once the doors of perception have been cleansed. In 1932, Aldous Huxley published his famous *Brave New World*.[2] He then became familiar with the work of Blake during drug experiments. So, in 1953, he published *The Doors of Perception* and, in 1956, the book *Heaven and Hell*. He uses the same words as Blake but searches for the path to genuine knowledge by other means. In *Heaven and Hell*, he autobiographically describes his psychedelic path to truth. Until his death, he experimented with mescaline and wrote about it in his book *The Doors of Perception*. Without making any special mention of it, Huxley followed Blake on his search for pure knowledge. He achieved this by »cleansing« those »doors« with mescaline. It was only by »cleansing« the »doors to perception« of the pollution of the civilization process by means of drug-taking that it would become possible to access the authentic world. This was a project that found many imitators and produced a movement that reflected the zeitgeist. For Huxley, it was literally about achieving pure perception, a desire that every scientist shares with him, to genuinely penetrate reality: the process of acquiring pure knowledge ad infinitum. For both painter Blake and writer Aldous

* The Doors, Jim Morrison

Huxley, »heaven and hell« were the religious poles that such a simple reality spans. This is certainly the simplest Manichaeism of good and evil, light and dark; epistemology that simply argues once more within a sacred frame. Blake painted the fitting symbolist and natural-mystical images, which look Romantic and reveal elements of fantasy. The people he depicted are very corporeal and look like they are imprisoned within their muscles. The boundaries between things and people blur in pastel. They are reminiscent of the pompous, Romantic pictures on some Crowley tarot cards, or of the characters in the Wagnerian universe. The blurring of the boundaries into exaltation is the message!

Blake's work was rejected by his contemporaries. It was clearly ahead of its time, and its style and content would be taken up again two hundred years later. It was only then that it was picked up by the esotericism on the US West Coast, which emerged from the growing self-experience movement, a field that professed to be scientific but was just a melting pot for fragments of theories of varied provenience pretending to be scientific. For Huxley, this seems to have been the bridge that led from sacred visions to mystically looking within. In the worldviews of the similarly drug-infused love and peace movement, the world was supposed to be as colorful as possible but also hermetic. It was to achieve this state of perception through drug consumption. However, the new gaze was primarily directed within, meaning that the most objective knowledge was subjective knowledge. I feel it, therefore it is! Solipsism as the early bird of an identity politics that refused scientific »objectivity.« On the search for one's own ego, the psychedelic promised to illuminate what was inside in the brightest colors. Visualizing what was felt within was now the reality at stake. This new equality of a wide range of very different theorems, summarized by Paul Feyerabend as »anything goes,« and a multitude of new age theories (F. Capra, K. Wilber), breathed, together with marijuana smoke, a world of thought that dictated that only the whole is true. Psilocybin and LSD soon became new experimental platforms for mind-expanding hallucinogenic experiences. One might say that the kaleidoscope was the predecessor to augmented reality, which anticipated what the members of a community that expected its perceptions to be overwhelmed wanted to see. They accepted perceptual distortion, believing that they would understand it as the long-desired expansion of consciousness. »Wow!« was the reaction to finally crossing the boundary to higher truth.

God's proclaimed death left behind a vacuum for man, who now urgently required guidance. The mind, which had been weakened by intoxication in its insularity, was now particularly susceptible to the doctrines of salvation and meaningful self-discovery on offer. A generation that had nothing left for God, which meant that it was suffering from metaphysical homelessness, was especially open to mystic and sacred visions. Musicals like *Jesus Christ Superstar* and *Hair* held gods that could be touched at the ready together with a coming higher age of Aquarius. During these performances, the clouds of marijuana smoke were a worthy replacement for incense. They also made it easier to receive and integrate the salvific messages.

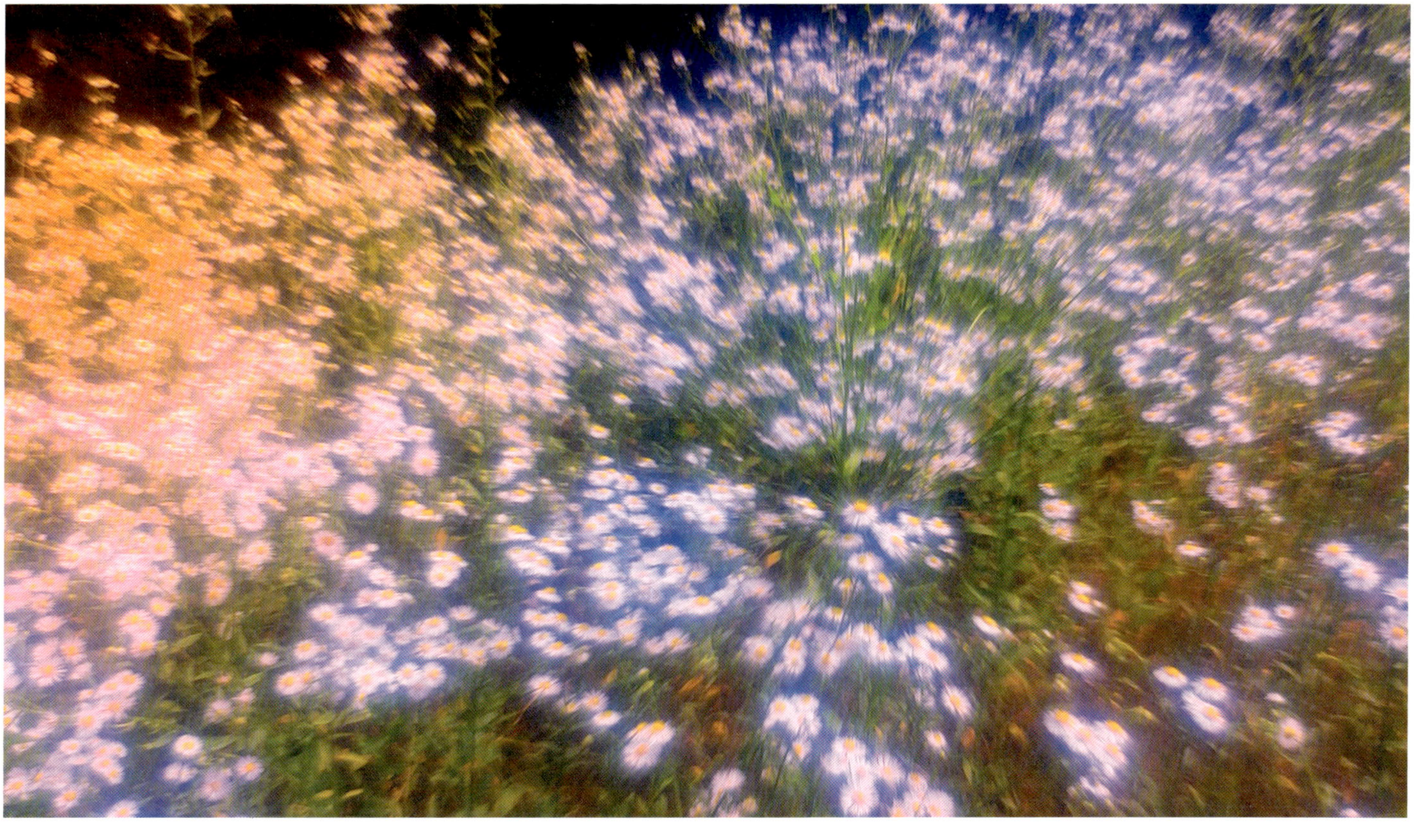

FIG. 2_ ERIGERON ANNUUS—ALTERED PERCEPTION

© Martina R. Fröschl, 2023

VIDEO

Good Morning Starshine

The ideological exaltation of drug consumption was welcomed due to its cleansing and mind-expanding effects. Many beatniks followed drug pope Timothy Leary's message of »turn on, tune in, and drop out« and set off into their own worlds. Unfortunately, this led also to psychotic, and even lethal, »drop outs.« Carlos Castaneda also helped to habilitate drug consumption with his literary works. Although they were banned, drugs were in. Talking about the drug experiences one had had was the thing to do.

The Doors of Perception

The name of the band The Doors was also a reference to Huxley and Blake. Together with Joplin, Hendrix, and Jones, Jim Morrison joined Club 27, the members of which died of overdoses at that age. But Carlos Castaneda, Timothy Leary, Allan Ginsberg, William S. Burroughs, and many others remained the preachers of this new special path to knowledge, which was paved with drugs.

The adage, »Those who can still remember it weren't really there,« accurately and sufficiently explains the epistemological potency and position of that age. It was all so wonderfully colorful—from the clothing to the living quarters and club rooms. Paisley patterns of all kinds conveyed psychedelic impressions. This led to the development of a very unique artistic trend that tried to remove the limits of and take up the colorful flower power feeling visually. The Beatles also sang about the drug-induced perception that allowed the boundaries to blur in their lyrics about »Lucy in the sky ... with kaleidoscope eyes.«

Picture yourself in a boat on a river
With tangerine trees and marmalade skies
Somebody calls you, you answer quite slowly
A girl with kaleidoscope eyes
Cellophane flowers of yellow and green
Towering over your head
Look for the girl with the sun in her eyes
And she's gone
(LENNON/MCCARTY)

Suddenly, everything was colorful, and floating patterns anticipated the fractal art of the chaos theory that would only achieve fame much later. (Incidentally, the major, short-term fascination with chaos theory, which also spilled over into popular culture, might have been responsible for the beauty of the colorful, strange Mandelbrot set images: visualizations of iterations of arithmetical shapes.)

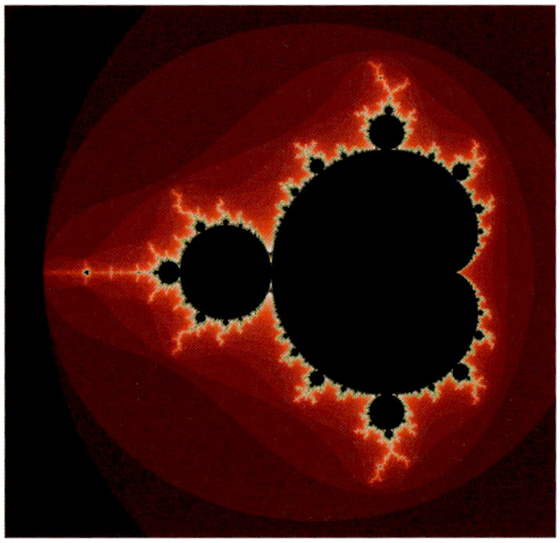

FIG. 3_MANDELBROT

VIDEO

Music also began to acoustically realize these drug-induced distortions and pseudo-spiritual intentions, with gentle Indian sitar music and spheric sound buildings accompanied by professional light shows. Concerts thus became total works of art created by drug-influenced musicians and their extremely amplified music for their audience, also massively under the influence of drugs. Here, too, the boundaries were successfully suspended and became fluid. Jimmi Hendrix's ecstasies of distortion and the spherical sounds of Pink Floyd exemplify this.

Of course, some also attempted to do psychotherapy under the effects of LSD or psilocybin, but it was not possible to achieve consistent results at that time. It is only recently that there have been encouraging results using psychedelic substances to treat depression, and investments are already being made in these substances. (P. Thiel is investing in the company Atai.)

Intoxication

So far, I have attempted to describe the attitude to life that had stepped up to cleanse the doors to perception in order to penetrate reality so well that it would become possible to know it in its essence as infinite. I remember my own mind-bending LSD experiments. They were spectacular and an expression of what had been previously unknown: flying curtains, consecutive images of movements in slow motion, and feelings that gained material qualities as they became visible. Sensory experiences were also overwhelming, and we repeated them. They were profound experiences that manifested themselves the next day in flash backs — as intensive, reoccurring experiences. These extraordinary experiences with others created something like another shared reality, which we had all experienced on the *trip* that we had gone on together.

But there were moments when we actually believed we had realized something ingenious. The next day it turned out to be some disappointing banality. It was

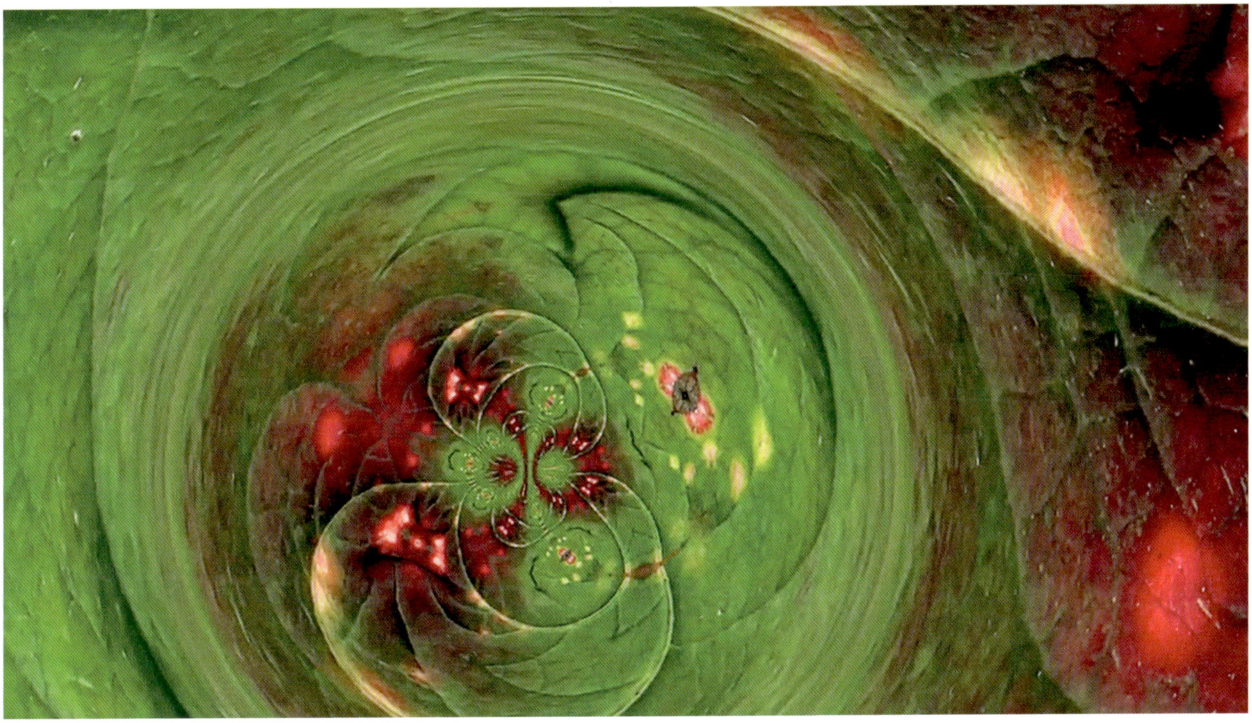

FIG. 4_ Colors of a Dying Landscape — *excerpt one*

© Martina R. Fröschl, 2014

VIDEO

FIG. 5_ Colors of a Dying Landscape — *excerpt two*

© Martina R. Fröschl, 2014

VIDEO

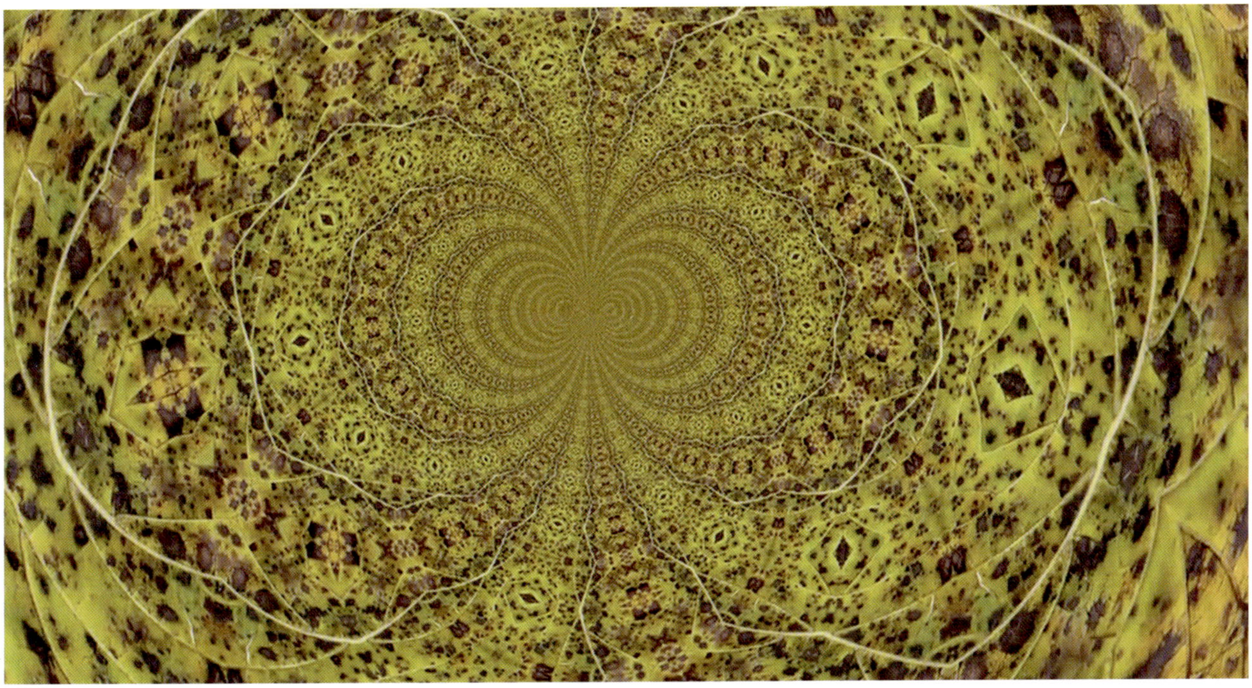

simply the moment that had seemed enlightening to us; it was not a cleansing but a wonderful deception by means of drug-induced highs. These visualizations were of no consequence epistemologically, but it was easy to tell friends about them because they were so extraordinary and wonderful. Intoxication!

But those adventures and experiences clearly had a far-reaching, motivating power. The euphoric feeling of upheaval during that psychedelic period was so contagious that even the CIA did not shy at any expense or effort to train decorated agents to stare at goats for as long as it took to make them fall over by means of telepathy and the power of thought. It was probably beneficial to everybody involved in the training program that this was carried out with the assistance of drugs. As we would expect today: the agents failed to meet their objective, which had been pursued in all seriousness and with large budgets. The men in training collapsed, and the lab goats probably stared at these strange humans with confusion in their eyes.

Disambiguation

But now I would like to pick up on the topic of this volume without taking the advice to use drug-based »cleaning products« on my own doors to perception. Right at the beginning of my contemplations, I would like to talk about the terms used. For even the artistic literary description of the »doors to perception« is, viewed scientifically, misleading and presents the problem of the question and not its solution. Although it is certainly a metaphor, the failure of the »cleansing attempts« reveal that the approach was wrong. Both the terms »perception« and »doors« are misleading.

Plato's allegory of the cave has always stood for how human perception is restricted and falsified. The prisoners in Plato's cave are chained up and forced to stare at a wall, on which a fire behind them is projecting the shadows of objects being carried by people going past the cave. For Plato, it is clear that only untainted thinking can liberate them from their chains and allow them to recognize the pure, clean truth. The advantage of this well-known metaphor is that it seems to justify every doubt in the objectivity of perception. The world is not given as such but is a product of the conditions in place — in this case, it is something created by shadows. Outside the cave, however, the glistening light would probably blind the viewers' eyes. Thus, what we perceive is always subjective. Scientific objectivity is only attained by penetrating subjectivity hermeneutically and phenomenologically, or by experimentally repeating situations and validating findings.

The German term for perception, *Wahr-Nehmung* — literally, *to take true* — is a misleading term, because perception is neither true nor taken. Something becomes true when the subject processes something that it comes into contact with: what the subject perceives is therefore always what it has already processed. Taking something *as* true would be the more appropriate formulation.

In 1950, Ivo Kohler used his upside-down goggles to show that the human brain is able to turn an image of the outside world that has been rotated by 180 degrees through a pair of glasses the right way up again. We also know that the human eye's physical features would not be able to produce the focused images that we are used to, which is also work done by the brain. Here, it is about the so-called correct depiction of reality. However, even the first sociopsychological conformity experiments carried out by S. Asch (1951) showed that it is very easy to disrupt the way we perceive even very simple objects if we feel the pressure to conform. Subjects described small sticks as bigger and also actually perceived them as bigger because the other (fictitious) subjects in the experiment saw them that way. During later repeats of these experiments, Rumpler (1976) was able to prove that some subjects had in fact not seen the sticks correctly, that their perception was distorted, while others simply gave conforming answers even though they had seen them correctly.[3] The empirical analysis of these experiments showed that there was a correlation between the simplicity of a given social structure and the intensity of the distortion in perception due to the pressure to conform that the subjects perceived. Thus, current and past biographical backgrounds change what is known as perception or sensory knowledge[4] and create faulty perception.

But people can become more skilled at differentiating sensory knowledge due to the necessities of everyday life and the resulting learning processes. One study of steelworkers showed that they perceived the red tones in melted steel so precisely due to the importance of temperature levels in the manufacturing of steel.[5] The Inuit's diverse, differentiated descriptions of snow and its shades of white can be explained by the necessities of their everyday lives because, for the Inuit, being familiar with the quality and temperature of the medium in which their lives take place is a matter of survival. This is a »focusing« of perception that has become necessary due to the conditions in which they live.

Scientific knowledge also needs to have sharp, precise tools at its disposal in order to be successful and to overcome the limitations of its research methods. The recommendation to cleanse the doors to perception that I quoted above is no longer the means of choice; rather, it is about changing the mode of observing the research object and often about selecting new means of study — from the macro to the micro level, from the visible to the invisible.

Today, making something identifiable is not the same as making it visible. The methods of penetrating matter have become so advanced that we have to visualize certain phenomena in order to understand them. In astronomy, the only way to prove the existence of a black hole is to process the data pertaining to electromagnetic radiation into an image that reveals the location and size of the object. Rays of light cannot escape the object due to its enormous mass. Electromagnetic radiation is, so to speak, converted into light and thereby visualized. Thus, ever more powerful computers and technical, visual means of immersion are making many things visible that only few had been able to see before.

If the Enlightenment was a fundamental force behind the liberation of knowledge from all of the religious artifacts that had glued shut the doors of perception — what is it today? In the age of the tin foil hat, fake news, and alternative facts, helping knowledge to break through seems difficult as social media bubbles generate a hermetically sealed, selective »data space« that only members are allowed to enter. The Platonic criteria of the cave allegory seem to have been met for the most part. Objectivity as a general, accepted opinion about a set of facts is becoming increasingly difficult to obtain. It is only by achieving consensus in the scientific community about facts that have been recognized by everybody that we can speak of a provisional agreement — until the next falsification (Popper).

Nobody is able to follow Plato's rejection of sensory knowledge in favor of true knowledge obtained through thought anymore. The metaphysical reality of the Platonic form can still be marveled at in beauty. However, science has to produce repeatable and tangible knowledge and evidence, and to provide data. Studies of neurons have shown that there are, e.g., more internal organismic connections than those to the so-called »doors« of perception (eyes, ears). Perceptions are visualizations and extrapolations of the brain—the results of primarily internal processes. Interpreting the data that has been ascertained is, in turn, the task of hermeneutics.

If this edited volume is about the scientific benefit of visualization, then we must take into account that we are living in a visual age in which media, above all visual media, are playing a dominant role. Visualization thus does not guarantee the truth to which science feels itself beholden. Unfortunately, the darker side of scientific truth seems to be the augmented possibilities of faking—an edited picture can be a lie worth thousands of words. Due to its *apparent* truth, however, it was predestined for all kinds of abuse. Images provide a multitude of possibilities for manipulation. This means that, today, it is almost impossible to differentiate between authentic and faked reality. There are almost no limits to deep fakes in the visual field anymore. Alternative facts will also create alternative realities, and they are already doing so. Because with the accomplishments of today's computers, not to speak of those of future quantum computers, new realities are being tested—people are being prepared for how to behave themselves in future situations. This is science fiction—the merging of ideas and facts—of fantasy and reality, the invention of realities beyond our wildest dreams. Does this make them false too?

Augmented reality in the form of virtual reality is by definition not real, and yet it is reality for its users. The sensory knowledge experienced using a set of AR glasses can be anything from dramatically real to an actual experience of panic. Here, without any use of psychotropic substances at all, there is a short-term change to and expansion of the mind, although »slow thinking« (Kahnemann) knows that this is just an experiment.

Interestingly enough, this is precisely the crossroads at which art and science go their separate ways and demand different things from the manufacturers of those realities. For, in its seduction of the eye, art does not need to limit itself. On the contrary: the more it breaks its banks and the more fantastically delirious and fabulous it becomes, the more spectacular people find it and marvel at it. Visualization in science has a harder time here: at all times, it must be able to provide the evidence that what it is showing and representing does actually reflect something that exists in the world and that it can be repeated at any time.

In the future, when we have all connected our brains to computers, the question of reality will increasingly lose its significance. The question then will be how much of and what parts of the »other reality« we want to integrate into our lives. This question no longer poses itself for a gamer. If a gamer, perhaps wearing AR glasses, enters their game world at level three, they have already decided to enter the other world. Returning to the world they have left is often difficult, especially if they have just won the game as a hero. The hero avatar they select is their alter ego in a better world in which they can use weapons to fight for their self-worth. Some films are already flirting with this better life in the virtual gaming world. *Ready*

Player One by Steven Spielberg successfully transitions between these two worlds. But the major tech corporations do not want to wait for these utopian scenarios to materialize anymore either; they want to help to shape them themselves. For Mark Zuckerberg, science fiction is not enough. He gave Facebook a new name, Meta, pointing toward a new dimension of reality in which people can have and experience the best of both worlds. Thus, one day, and equipped with quantum computers and 5, 6, or 7G technology, we will be able to enter those virtual worlds that we might even have programmed ourselves with avatar heroes we have chosen. Why not visit casinos and night clubs in the meta world with tokens in our pockets? A multiverse is developing, the metaverse of new consumerism and earning opportunities. If we have enough programming skill, we will be able to create our own worlds.

It is difficult for the protagonists of relevant science fiction to say farewell to the all too human needs of these worlds, especially when the conditions they live in in analog reality are miserable (*Ready Player One*, *Snow Crash*[6]). In these fictional narratives, the boundaries between the supposedly real world and the virtual gaming world blur. For in each world, it is possible even today to spend and earn money. Digital currencies are valid in both worlds. This is because the equipment the gaming characters have also plays a crucial role in them improving their performance.

In drug-induced highs and adventures with AR glasses, there are moments when we no longer know whether we are in an artificially induced, altered state of consciousness. The intentionally produced deception is supposed to make us believe that it is an actual reality. But what will happen when we become the ones who decide how long this deception lasts? It will mean that people will be able to choose which world they would like to »continue« living in. This could be possible with the right amount of money. It can be seen in the television series *Upload*. In this series, it already seems quite plausible that, for the transition to the afterlife and life after death, customers are able to move to a virtual world in a paid cloud ahead of schedule before they have said farewell to the real world. At this point of no return, they can make a final decision in favor of existing in the afterlife. But this will make the human miraculously transform into an angelic being. The last production by Georg Lucas, who can make the avatars of ABBA appear on a stage, shows that we will have to get used to Elvis actually being in the building and staying there forever. The evangelicals will then strive to get people to marvel at Jesus as a superstar too. Believing in him will no longer be necessary; following him will be enough of a commitment.

Transformed into digital angels of diverse provenance, more and more bygone but also future creatures will live on in one of many clouds, which might even be recreations of a psychedelically inspired paradise and which we will be able to retrieve into the real world at any time—the last judgment will be a very personal choice. Cloud 7 or Cloud 9 will be the addresses to which we will send emails.

But what is this kind of existence, and what does it mean if the techniques of immersion are so advanced that existing in the afterlife becomes just one of a number of options? How will we then know what is reality? Will we still want to or be able to know what is real if, for very realistic reasons, we prefer the deceptions on offer instead?

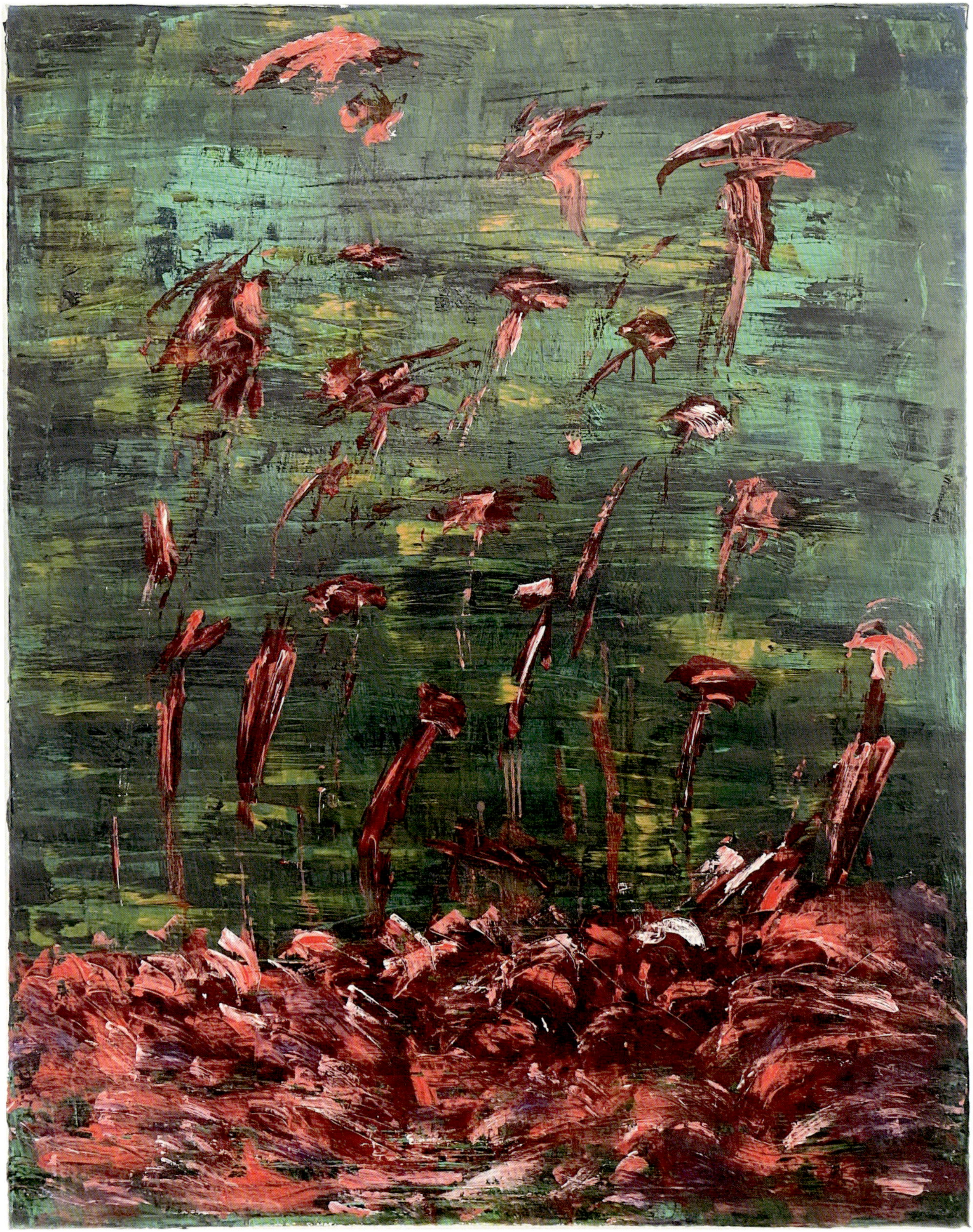

FIG. 6_BLÜTENLANDUNG

© Peter Rumpler, 2021

1 Aristotle, *Metaphysics*, trans. W. D. Ross, accessed January 25, 2023, http://classics.mit.edu/Aristotle/metaphysics.1.i.html.

2 Huxley Aldous, *Brave New World* (London: Chatto & Windus, 1932).

3 Peter Rumpler, *Sozialstruktur, Deprivation und kognitive Abwehr* (dissertation, Vienna, 1977), 119.

4 Klaus Holzkamp, *Sinnliche Erkenntnis* (Frankfurt am Main: Fischer Athenäum, 1989).

5 Rubinstein Sergej, *Sein und Bewusstsein* (Berlin: Thomas, 1972), 92.

6 Cline Ernest, *Ready Player One* (Frankfurt am Main: Fischer Tor, 2017); Neal Ferguson, *Snow Crash* (Frankfurt am Main: Fischer Tor, 2021).

Editors and Authors

ALFRED VENDL is a director and screenwriter with a background in chemistry. Science and research: he studied materials science at Vienna University of Technology, where he obtained his PhD. He then worked as a Research Scientist at Imperial College and the University of London, UK; and at the University of California, San Diego La Jolla, US. He also collaborated with the Max Planck Institute of Metal Research, Stuttgart, Germany. Since 1981, he has been a Full Professor at the University of Applied Arts Vienna, the Head of the Institute of Art and Technology, and an Associated Professor at the Vienna University of Technology. Since 2016, he has been the head of the Science Visualization Lab at the University of Applied Arts Vienna. He is the author of around eighty scientific publications in materials science, archaeometry, and art technology. He has worked as a TV host for scientific discussions, a director of photography, a screenwriter, director, and producer for, among others, the ORF, ZDF, BBC, Discovery, and the Smithsonian.

MARTINA R. FRÖSCHL studied media technology and media design before writing her doctoral thesis *Computer-Animated Scientific Visualizations of Tomographic Scanned Microscopic Organic Entities*. She is a senior scientist at the Science Visualization Lab at the University of Applied Arts Vienna. She has always been inspired by the representation of realities and invisible phenomena. Her most recent work has been based on scientific imaging technologies such as MCT, MRT, SEM, and light microscopic images, in collaborations with imaging experts and biologists. She has exhibited her computer animations in various representations, for instance, in collective media art projects (*Noise Aquarium*, *Virus Dice*, *C_ART*, *Our Earths*, *Butterfly}Pieris{Effect*) as well as in visual effects shots and videos. Her computer animations have been exhibited internationally at venues such as the Deep Space at the Ars Electronica Center Linz (AT), the Barbican Center, London (GB), the SIGGRAPH Art Gallery (US), the EPICenter, Sydney (AU), QUT, Brisbane (AU), Cairotronica (EG), the Paseo Festival Taos (US), the NTU Art Gallery (SG), the Manhattan Pratt Gallery, New York (US), MAK Vienna (AT), Klang-Moor-Schopfe (CH), aMORE Pula (HR), and the ECC Venice (IT). martinafroeschl.com

ROSE ANDERSON is Senior Vice President & Executive Director of the TV & Film Awards and Radio Awards at New York Festivals (NYF). Since joining NYF in 2009, Rose has transformed the competitions by emphasizing the importance of innovative storytelling across all platforms and genres. By instituting Advisory Boards and by expanding the Grand Juries to include more women and more countries, she has brought a greater diversity to the selection process which has translated into an uptick in the level of worldwide participation and the overall growth of the brand's prestige. Rose, an Emmy-Award-winning producer and member of the DGA, is the former Director of the Sports Emmy Awards and Director of Production for the Emmy Awards. Her past production experience includes five Olympic Games for ABC, NBC, and CBS. Her extensive background in TV includes news, sports, arts, and entertainment content for ABC, NBC, CBS, PBS, and ESPN. Rose received a BA from the University of Massachusetts and an MS in Broadcast Journalism from Boston University.

SONJA BÄUMEL is an artist, independent researcher, and art educator based in Amsterdam, the Netherlands, and Vienna, Austria. Her artistic research explores »the living,« the evolving perception of what bodies are made from. She investigates the influence scientific knowledge has had on the way we have perceived and interpreted the human body historically and how this affects current society and the cultural contexts in which we operate. Sonja uses self-experimentation to question skin as a border of the human body, taking risks by engaging directly with the material and by seeking the points where the environment starts and ends. These spaces in between are vivid entanglements full of life. They blur trans-species boundaries, revealing as yet unexplored forms of intelligence and communication. Sonja is currently working as a tutor and head of the Jewellery—Linking Bodies department at the Gerrit Rietveld Academie in Amsterdam, as well as lecturing and holding workshops at different national and international academies and universities. Sonja studied fashion design at the Fashion Institute in Vienna and holds a BA from the University of Arts in Linz, as well as an MA in Conceptual Design in Context from the Design Academy in Eindhoven. Her work has been exhibited internationally at the Ars Electronica Center, the MAK—Museum of Applied Arts, the Museum of Contemporary Art in Taipei, the ZKM Center for Art and Media in Karlsruhe, and the Centre Pompidou in Paris.

MARK CHAVEZ works with contemporary visual development techniques in animation and design. He has used these techniques to explore art and science themes, unifying the intersection between contemporary scientific concepts and a live, anthropomorphic worldview. Mark exhibits his work in interactive, immersive exhibitions and at film festivals, and distributes it digitally on social media. He has also presented his research internationally at conferences, in film screenings, and in art exhibitions. Mark is dedicated to pushing the boundaries of experimental animation and design, and continues to innovate in this exciting field.

INA CONRADI is an award-winning new media artist known for her unique blend of digital painting, experimental S3D animation, installation, and media architecture. She has participated in international events such as the Ars Electronica Festival, the Media Architecture Biennial, and the Art|Sci Gallery UCLA. Her film *Chrysalis* received the prestigious Lumiere Award from the Advanced Imaging Society in Hollywood. Ina has been an Associate Professor at the School of Art, Design, and Media, Nanyang Technological University Singapore, since 2007 and has established a curatorial and educational platform, Media Art Nexus (MAN).

JEREMIAH DIEPHUIS grew up in the great arcades of the American Midwest. After studies in computational linguistics and media, he focused his attention on the utilization and development of games for public spaces. He currently works as a lecturer in the Digital Media department at the Hagenberg Campus of the University of Applied Sciences Upper Austria. He is also cofounder of the nonprofit organization GameStage and a founding member of the research group Playful Interactive Environments.

JAMES K. GIMZEWSKI, PhD, is a distinguished professor in the Department of Chemistry and Biochemistry and joined UCLA in 2001. He founded the Nano/Pico Characterization at the UCLA California NanoSystems Institute (CNSI) and is Scientific Director of the UCLA Art|Sci Center. Gimzewski developed unconventional atomic switch networks for in-materio neuromorphics and unconventional computing using nonlinear network dynamics, contributing to research in brain emergent behavior and self-organized criticality. Before joining UCLA, Gimzewski worked in IBM's corporate research laboratory in Switzerland for nineteen years. He pioneered research on single atoms and molecules using scanning tunneling microscopy (STM). His current research involves energy extraction from vacuum fluctuations and collaborations on neuromorphic systems. Gimzewski has received honorary doctorates from the University of the Mediterranean Aix-Marseille II and the University of Strathclyde. A Fellow of the Royal Academy of Engineering and the Royal Society, he has also received numerous prestigious awards. With a philosophy rooted in curiosity, creativity, and productivity, Prof. Gimzewski believes that there is hope for the future. To achieve this, collaboration between science and the arts, in particular in hybrid forms and humanist collaborations, are crucial in order to deal with our future challenges: the growing emergence of complexity/criticality, dehumanization through AI, and the passing of climate tipping points.

GEORG GLAESER is an Austrian mathematician and Professor of Geometry at the University of Applied Arts Vienna. He studied mathematics and geometry at the Vienna University of Technology, where he received his doctorate and his *Habilitation* in computational geometry. He has written a number of books on computer graphics and biology in relation to mathematics and geometry, e.g., *Geometry and its Applications in Arts, Nature and Technology* (Springer Nature, 2020).

JÜRGEN HAGLER is an academic researcher and curator working at the interface of animation, games, and media art. He is a Professor of Computer Animation and Media Studies, and the head of studies for the bachelor's and master's programs in Digital Arts at the University of Applied Sciences Upper Austria, Hagenberg Campus. He has been involved in the activities of the Ars Electronica since 1997 in a series of different roles. Since 2017, the has been the Director of the Ars Electronica Animation Festival and the initiator and organizer of the Expanded Animation Symposium.

STEPHAN HANDSCHUH is a biologist with a technical focus on microscopic imaging and 3D visualization. He has a broad biological background in diverse fields such as evolutionary biology, comparative and functional morphology, microscopic anatomy, developmental biology, and theoretical biology. In 2012, he became a staff scientist in the imaging core facility at the University of Veterinary Medicine Vienna, where he has a strong technical focus on quantitative microscopic X-ray imaging, and 3D data visualization and analysis. Since 2010, he has also been a member of the Science Visualization Lab at the University of Applied Arts Vienna, where he works on creating scientifically meaningful 3D models of microscopic animal samples.

KARL HEIMBERGER is a specialist in neurology and radiology. In the 1980s, he worked in the Neuro-Oncology Outpatient Clinic and in the Neuro-Angiography and Neuro-CT units of the Department of Neurology at the University of Vienna. In 1991, the then newly established Division of Neuroradiology within the Department of Radiology at the University of Vienna was looking for neurologists who had had additional training in radiology. He met those requirements and became one of the division's first members. In the early 1990s, he initiated the development of digital subtraction CT angiography (DSCTA) with the Siemens Company. This diagnostic method became a preferred tool for neurosurgeons and interventional neuroradiologists for the quick planning of therapy for acutely bleeding intracranial aneurysms. He was invited to work as a Visiting Assistant Professor and Interventional Neuroradiologist in brain tumor treatment with the Blood-Brain-Barrier Unit of the Oregon Health & Science University, Portland (US), from 1993 to 1994. From 2003 to 2007 he was the Acting Chief of the Division of Neuroradiology at the Medical University of Vienna. Between 2004 and his retirement in 2011, he was re-elected as an employee representative in the Works Council of that institution.

PETER HINTERDORFER is currently head of the Institute of Biophysics at Johannes Kepler University (JKU) in Linz. His research focuses on nanoscopic technologies in the life sciences and covers molecular recognition and transport in several molecular and cellular systems, including molecular interactions between viruses, antibodies, transmembrane transporters, and cellular receptors at the single molecule level. His group has also found evidence of the dynamic movement of antibodies on bacterial and viral surfaces, and has localized the distribution of receptor binding sites on cells in nanometer resolution. He has published more than 300 papers, and his publications have been cited more than 10,000 times. His peer review activities for international journals include *Nature*, *Science*, *Nature Nanotechnology*, *Nature Communications*, *Nature Methods*, *Nature Protocols*, *Angewandte Chemie*, and the *Journal of the American Chemical Society*.

CHRISTIAN KÖBERL is Full Professor of Geosciences (Impact Research and Planetary Geology) at the University of Vienna (AT) and a full member of the Austrian Academy of Sciences. From 2010 to 2020, he was Director General of the Natural History Museum Vienna. He studied chemistry, physics, and astronomy, before obtaining his PhD in 1983 and his *Habilitation* in geosciences in 1990. He has had visiting scientist appointments at, among others, the Lunar and Planetary Institute and NASA Johnson Space Center in Houston, and the Carnegie Institution in Washington, and was a visiting professor at, e.g., Dartmouth College (New Hampshire) and the University of the Witwatersrand (SA). His research generally deals with impact craters, especially geochemical, mineralogical, and petrographic studies, as well as the early Earth, meteorites, and planetary geology. His publication record includes over 500 peer-reviewed scientific papers, many of which deal with impact craters; asteroid 15963 was named »Köberl« in his honor. Among other honors, he received the Barringer Medal of the Meteoritical Society for his impact research, and in Austria he was awarded the Cross of Honor First Class for Science and Culture.

WALTER KÖHLER, born in 1962, obtained his PhD in journalism, political science, and history at the University of Vienna. Following his lifelong love for nature, he began his career in the 1980s in the field of natural history filmmaking as a producer, scriptwriter, and director. In 1987, he started working with the renowned ORF strand *Universum*. From 1994 onward, he transformed the strand into the ORF Natural History Unit and led it to international success. In 2011, he left the ORF together with the entire *Universum* team and founded Terra Mater Studios to give natural history film production a new home. Koehler's productions have won a multitude of international awards, from Emmys to Romys. His acclaimed environmental thriller documentaries have won the company awards at the Sundance Festival and many more, and have been shortlisted for the Oscars.

KATHARINA KRÖSL is a senior XR researcher and Area Coordinator for the research area Immersive Analytics at the Viennese research center VRVis Zentrum für Virtual Reality und Visualisierung. She is also a lecturer at the Vienna University of Technology, where she received her PhD in 2020. Her current research interests include virtual and augmented reality, and their applications in interdisciplinary projects on perception, psychology, and medical simulations. During her PhD, she started working on simulating vision impairments in virtual and augmented reality, driven by the goal of having a positive social impact on society by increasing understanding of people with vision impairments and by providing architects and designers with tools to improve the accessibility of their designs. Her work has received awards from the scientific community and has featured in the media several times. In 2022, she received the Futurezone Women in Tech Award, the Liese Prokop Award in the category of Science & Technology, and an Honorable Mention at the IEEE VR Best Dissertation Award. For about three years, Katharina worked in an honorary capacity as an expert in a standardization committee at Austrian Standards, where she contributed to a new standard on visual guidance systems that also takes people with low vision into account, making signage more accessible for everyone. Katharina is a founding member of IEEE Women in Engineering Austria. She mentors young female visual computing students, organizes networking events for women in computer science, and does voluntary work for renowned scientific conferences, such as ISMAR, IEEE VR, and SIGGRAPH.

ALI MAHER, PhD, F. ASCE, is the Director of the Center for Advanced Infrastructure and Transportation (CAIT) and a civil and environmental engineering professor at Rutgers University. His expertise includes geotechnical engineering, civil infrastructure engineering, and infrastructure resilience. He is a recognized leader in promoting innovation in the infrastructure industry and creating collaboration between academia, and the private and public sectors to address pressing transportation challenges. He graduated with a PhD from the University of Michigan in 1988.

THOMAS MATZEK, born in 1964 in Vienna, is a journalist and documentary filmmaker. He has worked for the ORF for over thirty years. His record includes all fields of factual TV programming, such as magazines, reportage series, and documentaries. He has authored and directed more than a dozen films on Austrian contemporary history. After ending his career as a filmmaker, he spent nearly a decade working as a commissioning editor for the Natural History Unit as well as a series editor for ORF's *Universum History* strand. His current post is Head of Specialist Factual, and his department covers science and educational programs in all TV formats—from news to magazines and documentaries.

STEFAN MEREITER is a postdoctoral researcher at the Institute of Molecular Biotechnology (IMBA) and works in the laboratory of Prof. Josef Penninger. He is a specialist in the study of glycan alteration in health and disease, and has studied and worked in numerous countries during his scientific career, including Austria, Japan, Portugal, Sweden, and Germany. Although he is still at the beginning of his scientific career, Stefan Mereiter has already published forty publications with more than 900 citations in total, and has been awarded prestigious prizes and grants such as the Megazyme Prize, the Marie Curie IF, and ESPRIT. With his research, he aims to find new therapeutic avenues to treat incurable deadly diseases using innovative approaches that target glycan structures on the surface of viruses and cancer cells.

PETER MINDEK is a cofounder and CTO of nanographics, a scientific visualization studio. He received his doctorate from the Vienna University of Technology in 2015. In his dissertation, he proposed ways of visually exploring large parameter spaces for visualizations of scientific data. As a postdoctoral researcher, he developed techniques for illustrative molecular visualization and storytelling. Alongside scientific visualization, his current research interests include procedural animation and automated music generation for the enhancement of visual stories.

MARKUS MÜLLER is president (*Rector*) of the Medical University of Vienna (2015–2027). He is also the President of the Supreme Medical Council Austria (2018–2022) and a Member of the Board of Vienna General Hospital (AKH Wien). He was born in 1967 in Klagenfurt (AT) and graduated in 1993 with the mark *summa cum laude*. From 1993 to 2001, he worked as a resident (internal medicine) in Austria, Sweden, and the US, and gained an emergency physician diploma. He was an Associate Professor and Specialist in Clinical Pharmacology (1998) and Internal Medicine (2001), before becoming a Full Professor (2004) and Head of the Department of Clinical Pharmacology, Vienna General Hospital (2002–2015). He is an Honorary Fellow of the American College of Clinical Pharmacology, a Governor of Tel Aviv University (Israel), a Foreign Member of the Russian Academy of Sciences, and an Adjunct Faculty member at the University of Florida (US).

AMIRALI NAJAFI, PhD, is a postdoctoral research associate at the Center for Advanced Infrastructure and Transportation (CAIT) at Rutgers University, New Jersey. His research focuses on developing cyber-physical approaches to testing and simulation. Some of the tools he uses in his research include control theory, digital signal processing, robotics, computer vision, machine learning, and structural engineering and mechanics. He graduated with a PhD from the University of Illinois at Urbana-Champaign in 2021. Before academia, Amirali worked as an engineer in design and construction.

STEVE NICHOLLS is a multi-award-winning producer and director of science and natural history films. He has been making documentaries for forty years and began in the BBC's Natural History Unit, where he worked for ten years before setting up his own company to make science and nature films for broadcasters worldwide, including National Geographic, Discovery, PBS, Apple+, Curiosity Stream, and Terra Mater Factual Studios. Before becoming a filmmaker, Steve worked at the University of Bristol, where he also obtained his PhD in entomology. Later projects involved collaboration between the Zoology and Physics departments and extensive use of scanning electron microscopy. Steve is also an award-winning stills photographer and author. In 2009, *Paradise Found*, charting the history of North America's natural history and the loss of bioabundance, was published by the University of Chicago Press. *Flowers of the Field*, on the biology of British and European wildflowers, was published by Head of Zeus in 2019. In March 2023, *Alien Worlds*, exploring the reasons behind the extraordinary success of insects, was published by Head of Zeus and the University of Princeton Press. Steve has always enjoyed telling stories from right across the world of science in new and imaginative ways, using the latest technology and graphics techniques to make science accessible to a wide audience, whether through film and stills photography or in books.

YOO JIN OH is currently the FWF Elise Richter fellow at the Institute of Biophysics, Johannes Kepler University Linz, with a technical focus on microscopic imaging on the nanoscale. She studied physics by developing scanning probe microscopic techniques and obtained her doctorate in Korea. In 2009, she moved to Europe as a postdoctoral researcher and applied microscopic techniques to living organisms and pathogenic systems. She has been awarded several prestigious EU and Austrian grants (Marie Curie IIF, APART, Elise Richter) and has published <40 peer-reviewed papers. With her broad scientific activities, she is also interested in creating interdisciplinary artworks by collaborating with artists. A recent artistic performance entitled *Coexistence of SARS-CoV-2 Virus* took place at the internationally renowned ARS Electronica Festival.

PETER RUMPLER is a clinical psychologist and has worked as a psychotherapist for over thirty years. He spent five years working in the psychiatric field before becoming a supervisor, organizational consultant, and later teaching therapist, teaching supervisor, and coach. He has training in breath therapy, neo-Reichian body psychotherapy, and trauma therapy. Alongside his work as a psychotherapist, he is also an author and painter. The influence of his decades-long practice of the Japanese martial art Aikido can be found in his painting.

RUTH SCHNELL is a media artist, and Professor and Head of the Digital Arts Department at the University of Applied Arts Vienna. Her work, which includes interactive video and dynamic projection environments, as well as light installations, deals with perception, especially in relation to the moving body. Another central aspect is the examination of political and social issues, and of the constitution of the world and society. One recent focus of her artistic research has been the immersive aspects of fulldome and mixed reality environments. Her works have received multiple awards and have been exhibited internationally. In 1995, she represented Austria at the 46th Venice Biennale.

THOMAS SCHWAHA is a zoologist with a special focus on animal morphology and evolution as well as imaging techniques. He studied biology, specializing in zoology, in particular animal morphology, at the University of Vienna and completed his education with a doctorate in the soft-body morphological aspects of bryozoans, little-known colonial invertebrates. Since 2011, he has held a postdoctoral position at the Department of Integrative Zoology at the same university, where he is responsible for all of the department's microscopes and imaging facilities and also teaches undergraduate and graduate students while continuing to conduct research, primarily on bryozoan morphology. Due to his strong imaging background, he also focuses on new developments in imaging and actively aids other researchers with new approaches toward their research questions. Together with his broad teaching experience, this has allowed him to gain detailed knowledge of the morphology of various animal phyla such as arthropods, mollusks, and annelids.

PETER SICHROVSKY is an Austrian writer and journalist with experience in journalism, politics, and management. He studied pharmacy and chemistry at the University of Vienna and group dynamics at the University of Klagenfurt, and has published a number of books, theater plays, and scripts, some of which have been translated into foreign languages. Several of his books have also been produced as plays and TV documentaries. He has been a foreign correspondent in the US and Asia and a Member of the European Parliament, and manages a listed company in Singapore and Manila.

ANNA STEINHÄUSLER was born in 1990 in Gmunden. She is a Vienna-based visual artist working in the fields of painting, ceramics, and restoration. In 2015, she made a painting and animation film at the University of Applied Arts Vienna. From 2012 to 2013, she studied fine arts at Newcastle University, Newcastle upon Tyne (UK). Examples of symposia, residencies, art fairs, exhibitions and collections: 2022 Khnum 1st International Forum For Sculpture and Ceramics, Cairo (EG); the 2021 Ceramic Association of La Borne, Anagama firing, La Borne (FR); 2019 Artist in Residency, Studios of the Egon Schiele Art Centrum, Cesky Krumlov (CZ); *Fair For Art*, Aula der Wissenschaften, Vienna (AT); *Art Austria*, Palais Liechtenstein, Vienna (AT); 2018 Apple House Gallery, Studio Collection, International Ceramic Research Center, Guldagergaard (DK); *MIND and BODY—I and YOU.*, Estonian Museum of Applied Art and Design, Tallinn (EE).
annasteinhaeusler.com

VICTORIA VESNA, PhD, is a renowned artist and professor at UCLA's Department of Design Media Arts. She is known for her interdisciplinary approach to art, which blends cutting-edge technologies with traditional artistic techniques, exploring the boundaries between art and science. Vesna is also recognized for founding and directing the UCLA Art|Sci Center, an interdisciplinary research center that brings together artists, scientists, and scholars to explore new ways of understanding the world through the convergence of art and science. Vesna's work has been exhibited at major international venues such as the Venice Biennale and LACMA, and she has received numerous commissions and much support from foundations and museums. Her exploration of new technologies, interdisciplinary collaboration, and the potential of sound and vibration in art continue to push the boundaries of what is considered »art.« Vesna believes in the power of art to communicate complex ideas and to create new perspectives on the world we inhabit, inspiring students and audiences alike. In addition to her artistic practice, Vesna is an active researcher and educator. She has published numerous articles, book chapters, and books on the topics of art and technology, including *Database Aesthetics: Art in the Age of Information Overflow* and *Art and Science Now*. Her work and leadership at the UCLA Art|Sci Center have established her as a leading figure in the field of art and science, inspiring new generations of artists and scholars.

MANFRED WAKOLBINGER was born in 1952 in Mitterkirchen, Upper Austria. After an apprenticeship in metalworking and toolmaking, he turned to art. He started by designing jewelry with his wife Anna Heindl, before attending seminars given by Bazon Brock at the University of Applied Arts Vienna, and spending six months in London. Since 1980, he has worked in sculpture and photography. He has had numerous individual and group exhibitions in Europe and the US, including at the Documenta 8 and the Venice Biennale. He has made sculptures for the public space. In 1997, he had a personal exhibition at the Museum of the 20th Century, Vienna. He has had numerous catalogues and book publications. Since 2000, he has been taking photographs under water. In 2003, he published *Bottomtime*, in 2012 *Unter der Oberfläche*, and in 2007, with Christoph Ransmayr, *Damen und Herren unter Wasser—eine Bildergeschichte nach 7 Farbtafeln von Manfred Wakolbinger*. He designed the covers for Ransmayr's *Der fliegende Berg* and *Die letzte Welt*. In 2012, *Up From the Skies* was hosted by the Dominican Church in Krems; in 2013, Vito Zuraj produced the mini opera *Insideout* in Salzburg, with a libretto by Alexander Stockinger, based on the sculpture *Connection*. It was performed at the Salzburg Festival and the Museum of Modern Art, New York. In 2019, the exhibition *Inhale-Exhale* took place at the Museum of Fine Arts, Leipzig; a book was published with texts by Connie Offergeld, Christoph Ransmayr, Elisabeth von Samsonow, and an interview between Jasper Sharp and Manfred Wakolbinger. In 2020, he gave a lecture »Visite—eine Reise« at the Lesehaus at Galerie Krinzinger. The Coronet Theater London has been showing a video piece by Manfred Wakolbinger, *Inside Out*, since 2020; the book *Space is the Place* was published to accompany the *Open Air* exhibition held in the park of the Schloßmuseum Linz in 2020–2021.

Permalinks

Doors to Reality_ALFRED VENDL

P. 13_FIG. 1	https://phaidra.bibliothek.uni-ak.ac.at/o:70480
P. 18_FIG. 6	https://phaidra.bibliothek.uni-ak.ac.at/o:70413
P. 19_FIG. 7	https://phaidra.bibliothek.uni-ak.ac.at/o:70418
P. 19_FIG. 7	https://phaidra.bibliothek.uni-ak.ac.at/o:70416
P. 20_FIG. 8	https://phaidra.bibliothek.uni-ak.ac.at/o:70419
P. 28_FIG. 15	https://phaidra.bibliothek.uni-ak.ac.at/o:70422
P. 29_FIG. 16	https://phaidra.bibliothek.uni-ak.ac.at/o:70425
P. 29_FIG. 17	https://phaidra.bibliothek.uni-ak.ac.at/o:70426
P. 30_FIG. 18	https://phaidra.bibliothek.uni-ak.ac.at/o:70427
P. 30_FIG. 19	https://phaidra.bibliothek.uni-ak.ac.at/o:70428
P. 31_FIG. 20	https://phaidra.bibliothek.uni-ak.ac.at/o:71140
P. 31_FIG. 21	https://phaidra.bibliothek.uni-ak.ac.at/o:70430
P. 32_FIG. 22	https://phaidra.bibliothek.uni-ak.ac.at/o:70459
P. 33_FIG. 23	https://phaidra.bibliothek.uni-ak.ac.at/o:70457
P. 33_FIG. 24	https://phaidra.bibliothek.uni-ak.ac.at/o:70458
P. 36_FIG. 26	https://phaidra.bibliothek.uni-ak.ac.at/o:70440

The Door of Science Visualization_MARTINA R. FRÖSCHL

P. 45_FIG. 2	https://phaidra.bibliothek.uni-ak.ac.at/o:70446
P. 46_FIG. 3	https://phaidra.bibliothek.uni-ak.ac.at/o:70447
P. 48_FIG. 4	https://phaidra.bibliothek.uni-ak.ac.at/o:70463
P. 50_FIG. 5	https://phaidra.bibliothek.uni-ak.ac.at/o:70449
P. 52_FIG. 6	https://phaidra.bibliothek.uni-ak.ac.at/o:71156
P. 56_FIG. 8	https://phaidra.bibliothek.uni-ak.ac.at/o:71346
P. 59_FIG. 9	https://phaidra.bibliothek.uni-ak.ac.at/o:70452
P. 61_FIG. 10	https://phaidra.bibliothek.uni-ak.ac.at/o:70453

Galaxies from the Depths_MANFRED WAKOLBINGER

P. 65	https://phaidra.bibliothek.uni-ak.ac.at/o:70320
P. 65	https://phaidra.bibliothek.uni-ak.ac.at/o:70321
P. 65	https://phaidra.bibliothek.uni-ak.ac.at/o:70322

Expanded Selves: Searching for Encounters _SONJA BÄUMEL

P. 90_FIG. 6	https://phaidra.bibliothek.uni-ak.ac.at/o:71141

Visualization Technologies in Nature Documentaries_WALTER KÖHLER

P. 99_FIG. 1	https://phaidra.bibliothek.uni-ak.ac.at/o:70277
P. 100_FIG. 2	https://phaidra.bibliothek.uni-ak.ac.at/o:71142
P. 101_FIG. 4	https://phaidra.bibliothek.uni-ak.ac.at/o:70310
P. 103_FIG. 3	https://phaidra.bibliothek.uni-ak.ac.at/o:70309
P. 104_FIG. 5	https://phaidra.bibliothek.uni-ak.ac.at/o:70311
P. 107_FIG. 6	https://phaidra.bibliothek.uni-ak.ac.at/o:70312
P. 109_FIG. 7	https://phaidra.bibliothek.uni-ak.ac.at/o:70313
P. 110_FIG. 8	https://phaidra.bibliothek.uni-ak.ac.at/o:70314
P. 111_FIG. 9	https://phaidra.bibliothek.uni-ak.ac.at/o:70315
P. 113_FIG. 10	https://phaidra.bibliothek.uni-ak.ac.at/o:70316
P. 114_FIG. 11	https://phaidra.bibliothek.uni-ak.ac.at/o:70317

Digital Twins for Civil Infrastructure Applications_AMIRALI NAJAFI, ALI MAHER

P. 120_FIG. 2	https://phaidra.bibliothek.uni-ak.ac.at/o:70519

Beyond the Limits of Our Perceptions_STEVE NICHOLLS

P. 132_FIG. 1	https://phaidra.bibliothek.uni-ak.ac.at/o:70409
P. 133_FIG. 2	https://phaidra.bibliothek.uni-ak.ac.at/o:70410

The Serpent and the Dragonfly: Into the Unknown_INA CONRADI, MARK CHAVEZ

P. 164_FIG. 3	https://phaidra.bibliothek.uni-ak.ac.at/o:71150
P. 165_FIG. 4	https://phaidra.bibliothek.uni-ak.ac.at/o:70603
P. 167_FIG. 5	https://phaidra.bibliothek.uni-ak.ac.at/o:70604

Impact Cratering: The Impact of Visualization on Science and Outreach_CHRISTIAN KÖBERL

P. 180_FIG. 6	https://phaidra.bibliothek.uni-ak.ac.at/o:70617

Awards in the Time of COVID_ROSE ANDERSON

P. 217_FIG. 5	https://phaidra.bibliothek.uni-ak.ac.at/o:70402

The Dance of the Spike with Sugar Binders _YOO JIN OH, PETER HINTERDORFER, STEFAN MEREITER

P. 225_FIG. 4	https://phaidra.bibliothek.uni-ak.ac.at/o:70511

*La Borne Worrier*_ANNA STEINHÄUSLER

P. 228_FIG. 1 https://phaidra.bibliothek.uni-ak.ac.at/o:70535

Insights: Stroke, a Medical History in Neurology and Neuroradiology, and the Spirit of Empathy

_KARL HEIMBERGER WITH CHRISTIAN NAŠEL

P. 235_FIG. 1 https://phaidra.bibliothek.uni-ak.ac.at/o:70535
P. 238_FIG. 8 https://phaidra.bibliothek.uni-ak.ac.at/o:70538
P. 240_FIG. 13 https://phaidra.bibliothek.uni-ak.ac.at/o:70540

NANO: Bottom up and in Between

_VICTORIA VESNA, JAMES K. GIMZEWSKI

P. 253_FIG. 1 https://phaidra.bibliothek.uni-ak.ac.at/o:70702
P. 256_FIG. 4 https://phaidra.bibliothek.uni-ak.ac.at/o:70703
P. 261_FIGS. 11–14 https://phaidra.bibliothek.uni-ak.ac.at/o:70704
P. 262_FIGS. 15–16 https://phaidra.bibliothek.uni-ak.ac.at/o:70705

Journeys into the Hidden Microscopic World

_STEPHAN HANDSCHUH, THOMAS SCHWAHA

P. 278_FIG. 3A https://phaidra.bibliothek.uni-ak.ac.at/o:70493
P. 278_FIG. 3C https://phaidra.bibliothek.uni-ak.ac.at/o:70500
P. 279_FIG. 4A https://phaidra.bibliothek.uni-ak.ac.at/o:70494
P. 279_FIG. 4A https://phaidra.bibliothek.uni-ak.ac.at/o:70495
P. 279_FIG. 4C https://phaidra.bibliothek.uni-ak.ac.at/o:70496
P. 280 https://phaidra.bibliothek.uni-ak.ac.at/o:70499
P. 283_FIG. 5 https://phaidra.bibliothek.uni-ak.ac.at/o:70498
P. 284_FIG. 6 https://phaidra.bibliothek.uni-ak.ac.at/o:70501
P. 285_FIG. 7 https://phaidra.bibliothek.uni-ak.ac.at/o:70506
P. 285_FIG. 7 https://phaidra.bibliothek.uni-ak.ac.at/o:70507
P. 285_FIG. 7 https://phaidra.bibliothek.uni-ak.ac.at/o:70508
P. 285_FIG. 7 https://phaidra.bibliothek.uni-ak.ac.at/o:70509
P. 286 https://phaidra.bibliothek.uni-ak.ac.at/o:70502
P. 286 https://phaidra.bibliothek.uni-ak.ac.at/o:70503
P. 286 https://phaidra.bibliothek.uni-ak.ac.at/o:70505

Traversing Invisible Walls: Facilitating Collaborative Experiences in Mixed Reality Environments_JÜRGEN HAGLER, JEREMIAH DIEPHUIS

P. 295_FIG. 1 https://phaidra.bibliothek.uni-ak.ac.at/o:70552

Realisms and Realities: Constructions of Reality in Digital Art_RUTH SCHNELL WITH PATRICIA KÖSTRING

P. 307_FIG. 3A https://phaidra.bibliothek.uni-ak.ac.at/o:71153
P. 308_FIG. 4 https://phaidra.bibliothek.uni-ak.ac.at/o:70660
P. 309_FIG. 5 https://phaidra.bibliothek.uni-ak.ac.at/o:70662
P. 312_FIG. 8 https://phaidra.bibliothek.uni-ak.ac.at/o:70664
P. 313_FIG. 10 https://phaidra.bibliothek.uni-ak.ac.at/o:70663

Vision Impairments in Extended Reality

_KATHARINA KRÖSL

P. 326_FIG. 1 https://phaidra.bibliothek.uni-ak.ac.at/o:70564
P. 331_FIG. 2 https://phaidra.bibliothek.uni-ak.ac.at/o:70565
P. 332_FIG. 3 https://phaidra.bibliothek.uni-ak.ac.at/o:70566
P. 334_FIG. 4 https://phaidra.bibliothek.uni-ak.ac.at/o:70567

Life and Biology_PETER MINDEK

P. 349_FIG. 1 https://phaidra.bibliothek.uni-ak.ac.at/o:70569
P. 351_FIG. 2 https://phaidra.bibliothek.uni-ak.ac.at/o:70573
P. 352_FIG. 3 https://phaidra.bibliothek.uni-ak.ac.at/o:70570
P. 355_FIG. 4 https://phaidra.bibliothek.uni-ak.ac.at/o:71154

Geometry as a Key to Hidden Doors?_GEORG GLAESER

P. 360_FIG. 2 https://phaidra.bibliothek.uni-ak.ac.at/o:70522
P. 360_FIG. 3 https://phaidra.bibliothek.uni-ak.ac.at/o:70523
P. 361_FIG. 5 https://phaidra.bibliothek.uni-ak.ac.at/o:70528
P. 369_FIG. 20 https://phaidra.bibliothek.uni-ak.ac.at/o:70527
P. 374_FIG. 30 https://phaidra.bibliothek.uni-ak.ac.at/o:70524
P. 377_FIG. 33 https://phaidra.bibliothek.uni-ak.ac.at/o:70526
P. 379_FIG. 37 https://phaidra.bibliothek.uni-ak.ac.at/o:70525

»Break on through to the other side«_PETER RUMPLER

P. 397_FIG. 2 https://phaidra.bibliothek.uni-ak.ac.at/o:70442
P. 398_FIG. 3 https://phaidra.bibliothek.uni-ak.ac.at/o:70443
P. 399_FIG. 4 https://phaidra.bibliothek.uni-ak.ac.at/o:70444
P. 399_FIG. 5 https://phaidra.bibliothek.uni-ak.ac.at/o:70445

ALFRED VENDL, MARTINA R. FRÖSCHL (EDS.)

Project Management »Edition Angewandte« on behalf of the University of Applied Arts Vienna: **ANJA SEIPENBUSCH-HUFSCHMIED/ BARBARA WIMMER**, A-Vienna

Content and Production Editor on behalf of the Publisher: **KATHARINA HOLAS**, A-Vienna

Translation from German into English and proofreading/copy editing: **LYDIA J. WHITE**, Wellington, New Zealand
Layout, cover design, and typography: **ANDREA NEUWIRTH**, A-Vienna
Layout assistant: **GABRIEL FISCHER**, A-Vienna
Image editing: **PIXELSTORM LITHO & DIGITAL IMAGING**, A-Vienna
Printing: **HOLZHAUSEN**, the book-printing brand of Gerin Druck GmbH, A-Wolkersdorf
Paper: NSB Fühlkarton | Magno Volume 135 gsm
Type fonts: **SAGONA** by René Bieder | **SOLEIL** by Wolfgang Homola/ TypeTogether

Library of Congress Control Number: 2023935602

Bibliographic information published by the German National Library
The German National Library lists this publication in the Deutsche Nationalbibliografie; detailed bibliographic data are available on the Internet at http://dnb.dnb.de.

ISSN 1866-248X
ISBN 978-3-11-125000-7
e-ISBN (PDF) 978-3-11-125012-0

www.degruyter.com